T0136915

METEOROLOGICAL MONOGRAPHS

VOLUME 56   2017

Multiscale Convection-Coupled Systems in the Tropics: A Tribute to Dr. Michio Yanai

Edited by

Robert G. Fovell
Wen-wen Tung

American Meteorological Society
45 Beacon Street, Boston, Massachusetts 02108

ISBN 978-1-944970-04-8
ISSN 0065-9401

Published by the American Meteorological Society
45 Beacon St., Boston, MA 02108

As of 2015, AMS Meteorological Monographs are available as open access online, with a print-on-demand option. Volumes issued before 2015, numbers 1–55, are available electronically through Springer Nature, of which a selection of casebound volumes are available for purchase at the AMS online bookstore (www.ametsoc.org/bookstore).

Printed in the United States of America
by Allen Press, Inc., Lawrence, KS

TABLE OF CONTENTS

Introduction

ROBERT G. FOVELL

Atmospheric and Oceanic Sciences, University of California, Los Angeles, Los Angeles, California

WEN-WEN TUNG

Earth, Atmospheric, and Planetary Sciences, Purdue University, West Lafayette, Indiana

On 27 January 2011, a special symposium honoring the late Professor Michio Yanai was held, taking place during the 91st American Meteorological Society Annual Meeting in Seattle. The symposium consisted of invited talks (conceived as "lessons") organized around the principal research areas that attracted Professor Yanai's attention and passion. As a tribute not only to his many important contributions but also to his dedicated mentorship of students and junior researchers, many of the lessons represented collaborations between more senior and junior scientists. This monograph was crafted from a subset of oral and poster presentations from the symposium, in the hope of creating a lasting tribute to Professor Yanai and his legacy, and a useful reference for scientists in all stages of their careers. Also included are personal reminisces from Professor Taroh Matsuno (Forward) and a biographical sketch of Professor Yanai's life (Epilogue).

Professor Yanai started his academic career as an assistant professor at the University of Tokyo in 1965, followed by a full professor appointment at the University of California, Los Angeles, in 1970. He began his career studying tropical cyclones (TCs) and his 1964 review paper on TC formation served as the most comprehensive reference on the topic for more than a decade. Much of his groundbreaking work continues to guide research even today, including his observations of the mixed Rossby–gravity wave, his systematic approach of estimating apparent heat sources (Q_1) and moisture sinks (Q_2) and associating them with the bulk properties of convective systems, and his diagnostic studies of the Asian monsoon, in particular

his pioneering works on the impacts of the Tibetan Plateau on the Asian monsoon. In 1986, the American Meteorological Society honored him with the Charney Award. In 1993, he received the Fujiwara Award from the Meteorological Society of Japan. Before the advent of online archives and powerful search engines, his *UCLA Tropical Meteorology and Climate Newsletter* functioned as an invaluable resource to the community since its founding in 1996.

This monograph pays tribute to these distinguished aspects of Professor Yanai's work and life. The first part (chapters 1–8) deals with phenomena, starting with observations from ground-based or spaceborne platforms (chapters 1 and 2) as examples of multiscale convection coupled tropical systems and utilizing, at the start, nothing other than a few elementary conservation laws. As their significance cannot be understated, Q_1 and Q_2 are introduced right away (in chapter 1). Students learn in introductory dynamical meteorology that due to the smallness of the Coriolis parameter, the geostrophic approximation breaks down in the tropics. In lieu of geostrophy, tropical synoptic meteorologists make use of the thermodynamic equation, in which the vertical advection of potential temperature is approximately balanced by the apparent heating (i.e., Q_1) that includes radiative heating, latent heat release by net condensation, and the vertical convergence of the vertical eddy transport of sensible heat. To be a practical reference, the discretization of the thermodynamic equation from which Q_1 can be obtained over a fixed grid in the pressure coordinate on a sphere is shown in the appendix of chapter 8.

In the convectively active tropical troposphere, there is a strong coupling between the thermodynamic equation and the moisture conservation equation, and hence between Q_1 and Q_2. Other than reflecting the consistency between the net latent heat release present in both equations, the beauty of examining this coupling is to

Corresponding author address: Robert Fovell, Atmospheric & Environmental Sciences, University at Albany, SUNY, 1400 Washington Ave., Albany, NY 12222.
E-mail: rfovell@albany.edu

DOI: 10.1175/AMSMONOGRAPHS-D-16-0003.1

reveal the vertical eddy transport of moist static energy due to convection. Much of the complexity in tropical convection coupled motions results from the convective eddy transport interacting with the baroclinic atmospheric states associated, for example, with tropical waves and the Madden–Julian oscillation (chapters 3–5) and monsoon circulations over Asia and the Maritime Continent (chapters 6 and 7), owing to the inherent nonlinearity and the apparent randomness in moist convective processes. The outcomes of the system are manifested in multiple spatiotemporal scales.

In the second part, chapters 9 to 16, a hierarchy of models is presented, ranging from the theoretical ones purposefully constructed to elucidate essential mechanisms in the multiscale convection coupled tropical waves and the Madden–Julian oscillation (chapters 9 and 10) to full-physics cloud-system-resolving or multiscale ("superparameterization") global climate models aimed to simulate the atmosphere with cloud processes in high fidelity (chapters 14 and 15), and those in between. There are multitudes of physical processes operating in the modeled system, most notably here the cloud microphysics–radiation feedback from weather to climate scales (chapters 11–13). Note that cloud microphysics and radiation are processes contributing to the true heat sources as opposed to the apparent ones, which include the heat transports due to unresolved eddies. Chapter 16 articulates the fundamental requirement and the ongoing efforts for the physics to converge from the apparent to the true heat sources as the models with cloud processes are refined.

The Yanai Symposium was not only well received but also very well attended, with one very sorrowful absence: Professor Yanai himself. Humbled yet thrilled by the event, Professor Yanai enthusiastically participated in the planning. However, he was also beset with health problems, and passed away on 13 October 2010, a few short months prior to the symposium. He would also have been honored and humbled by this volume, which is the result of the dedicated efforts of the authors, editors, reviewers, and advisors listed below:

Prof. Akio Arakawa, Dr. Jian-Wen Bao, Dr. James Benedict, Dr. Amit Bhardwaj, Prof. Lance Bosart, Dr. Mark Branson, Dr. Scott Braun, Dr. Yizhe Peggy Bu, Dr. Yang Cao, Prof. Chih-Pei Chang, Dr. Baode Chen, Prof. Tsing-Chang Chen, Dr. Jiun-Dar Chern, Prof. Jan-hwa Chu, Dr. Paul Ciesielski, Prof. Kristen L. Corbosiero, Dr. Mark DeMaria, Prof. Yihui Ding, Prof. Leo Donner, Dr. Timothy Dunkerton, Prof. Russ Elsberry, Prof. Steve Esbensen, Dr. Wojciech Grabowski, Dr. Mircea Grecu, Dr. Yu Gu, Dr. Samson Hagos, Dr. Arthur Hou (deceased), Prof. Robert Houze Jr., Dr. Li-Huan Hsu, Dr. Xian-an Jiang, Prof. Richard Johnson, Dr. Joon-Hee Jung, Dr. Ramesh Kakar, Prof. Marat Khairoutdinov, Prof. Boualem Khouider, Dr. George Kiladis, Prof. Ruby Krishnamurti, Prof. T. N. Krishnamurti, Dr. Vinay Kumar, Prof. Hung-Chi Kuo, Mr. Stephen E. Lang, Dr. William K.-M. Lau, Dr. Seungwon Lee, Dr. Frank J.-L. Li, Prof. Hock Lim, Dr. Changhai Liu, Prof. Fei Liu, Prof. Xiaodong Liu, Prof. Yimin Liu, Dr. Mong-Ming Lu, Prof. Victor Magaña, Prof. Andrew J. Majda, Prof. Eric Maloney, Prof. Brian Mapes, Prof. Taroh Matsuno, Dr. Rachel McCrary, Dr. Mitchell W. Moncrieff, Prof. Kenji Nakamura, Dr. Riko Oki, Dr. William Olson, Dr. Kazuyoshi Oouchi, Prof. David Randall, Prof. Thomas Rickenbach, Dr. Andrew Robertson, Prof. Masaki Satoh, Prof. Courtney Schumacher, Prof. Shoichi Shige, Dr. Anu Simon, Dr. Xiaoliang Song, Dr. Cristiana Stan, Prof. Sam Stechmann, Dr. Graeme Stephens, Dr. Hui Su, Prof. Yukari N. Takayabu, Dr. Wei-Kuo Tao, Dr. Katherine Thayer-Calder, Dr. Aype Thomas, Dr. Duane Waliser, Prof. Chien-Ming Wu, Prof. Guoxiong Wu, Dr. Shaochen Xie, Dr. Kuan-Man Xu, Prof. Yongkang Xue, Prof. Ming-Jen Yang, Prof. Chidong Zhang, Dr. Guang Zhang, Prof. Minghua Zhang, and two anonymous reviewers.

We also thank the AMS publishing and editorial staff led by Mr. Ken Heideman, Mr. Mike Friedman, Ms. Sarah J. Shangraw, and Ms. Melissa Fernau; the Monograph series editors, the late Prof. Peter Lamb and Prof. Robert Rauber; and AMS executive director Dr. Keith Seitter. At UCLA, the Michio Yanai Memorial fund established by Michio's wife, Mrs. Yoko Yanai, provided important financial support for the open access version of this volume. The Yanai fund also supports the annual Yanai Memorial Lecture. Special thanks to Mr. Geoff Girard and the faculty, staff, students, and alumni of the UCLA Department of Atmospheric and Oceanic Sciences for starting and supporting the Yanai Lecture.

For information on supporting the Michio Yanai Memorial Fund, please contact UCLA Physical Sciences Development by phone at +1 (310) 794-9045 or e-mail to physicalsciences@support.ucla.edu. To learn more about the Yanai Lecture, please contact the Department of Atmospheric and Oceanic Sciences at +1 (310) 825-1217 or visit http://www.atmos.ucla.edu/yanai.

Prologue: Tropical Meteorology 1960–2010—Personal Recollections

TAROH MATSUNO

Japan Agency for Marine-Earth Science and Technology, Yokohama, Japan

This volume consists of some papers presented at the AMS Symposium held to honor the memory of the late Professor Michio Yanai as well as additional works inspired by his research. By the nature of this volume, many of the contributed papers describe the development of tropical meteorology over the past half-century or so in connection with Professor Yanai's influence on it. While most of the chapters address specific areas and discuss timely issues, in this prologue I will describe some of Professor Yanai's contributions during the early period of his career from my own point of view. As this is a personal reminiscence, I would like to emphasize how Professor Yanai influenced me.

Both Professor Yanai and I became graduate students at the University of Tokyo to begin our career as meteorologists in 1956 and 1957, respectively. Since we studied and worked together so closely for a long time, in this article I will call him Yanai-san as I have done in our personal interactions.

1. Emergence of CISK and convection parameterization

Operational numerical weather prediction (NWP) began in the late 1950s, within a decade from the first successful computer-generated 500-mb (1 mb = 1 hPa) height prediction (Charney et al. 1950). The Japan Meteorological Agency (JMA) started routine NWP in 1959, following in the footsteps of Sweden (1954) and the United States (1955). In November 1960, the International Symposium on NWP was held at Tokyo. It was very timely, and the world's prominent figures in the field attended. Looking back at the proceedings (Syono et al. 1962), we can see that the symposium represents an important landmark concerning the dynamics and numerical modeling of tropical cyclones and convection. Since the successful realization of NWP implies that the baroclinic instability theory of extratropical cyclones had been verified through practical applications, the tropical cyclone became the next major target of dynamic meteorology and numerical modeling.

Professor Shigekata Syono of the University of Tokyo, our supervisor and the symposium organizer, had published a paper in *Tellus* early in 1953 discussing the formation mechanism of tropical cyclones using linear theory (Syono 1953). He tried to explain the formation in terms of the static instability, which must be correct considering its energy source. When a computer became available, he directly applied this theory of typhoon formation via numerical simulation, with the help of a young graduate student named Takao Takeda, who subsequently published a paper on the mechanism of the long-lasting convection cell or squall line (Takeda 1971).

At the symposium, Professor Syono reported the result, showing that the initially specified, slow upward motion over a 300-km region did not intensify uniformly; instead, extremely strong vertical motion was concentrated in a 15-km central region. Just prior to this

Corresponding author address: Taroh Matsuno, Principal Scientist, Japan Agency for Marine-Earth Science and Technology, 3173-25, Showamachi, Kanazawaku, Yokohama 236-0001, Japan.
E-mail: matsuno@jamstec.go.jp

DOI: 10.1175/AMSMONOGRAPHS-D-15-0012.1

presentation, Dr. Akira Kasahara from the University of Chicago talked about results of his numerical experiments that were almost identical with Professor Syono's, except that he had used a linear stability analysis to show that the maximum growth rate was at the smallest possible scale. Questions and intensive discussions followed, which included contributions by Professor Jule Charney.

Charney commented that the dominance of the smallest scale was nothing but a manifestation of Bénard-like convection owing to conditional instability, which corresponds to cumulus clouds in the real atmosphere. Then he talked about his ongoing study with Dr. Yoshimitsu Ogura, suggesting that they had encountered a similar difficulty that they were currently working to overcome. Charney stated that "a typhoon and the cumulus clouds do not compete but they cooperate ... actually cooperate to maintain the energy of the large-scale system." Then he identified the crux of the problem: "But ... cumulus clouds are very small-scale phenomena. How do you handle that in the numerical prediction scheme? Isn't it very difficult to deal with both small and large scales?" For more about this and the origin of cumulus parameterization, see Kasahara (2000).

Following this, Yanai-san made a presentation on a detailed analysis of the formation process of Typhoon Doris, which was a part of his doctoral thesis. An interesting discussion followed, but more important was conversation and discussion that took place when Professor Charney visited Professor Syono at the University of Tokyo prior to the symposium. Later, Yanai-san talked often about this visit, recalling that he was surprised and deeply impressed that Charney, the field's top theoretician, had showed a keen interest in his detailed analysis of this complicated, real-world phenomenon. Considering that Professor Charney was in the midst of a struggle concerning how to handle cumulus clouds as part of larger-scale dynamics, Yanai-san's work must have represented what he most wanted to know. In any event, Yanai-san was thankful that, as a result of Charney's interest, Professor Syono had changed his evaluation of Yanai-san's work, which he had thought did not include enough dynamics.

As to Yanai-san's thoughts on this problem, I do not have clear memory. However, as another part of his thesis work, he wrote a paper titled "Dynamical aspects of typhoon formation" (Yanai 1961), in which he proposed inertial instability, found at upper levels near the center of a typhoon, as integral to the large-scale dynamics of tropical cyclone development. In the conclusion of this paper, he wrote that the "actual condensation process takes place in individual cumulus clouds, and their horizontal sizes are utterly different from large scale of the pre-existing vertical motion. Therefore ... in the dynamics of typhoons, some artificial modification in treatment seems to be necessary" (Yanai 1961, p. 304). In this way, Yanai-san was involved in the birth of both cumulus parameterization and conditional instability of second kind (CISK).

It was a few years after the symposium that these key concepts of the new tropical meteorology emerged in the papers by Ooyama (1964) and Charney and Eliassen (1964). The work of Dr. Ooyama was presented in early 1963 at an international symposium, and its content had a great impact on Japanese meteorologists, including myself. Collaborating with a new graduate student, Masanori Yamasaki, we began a numerical experiment of typhoon formation, based on the primitive equations and a cumulus parameterization scheme which we discovered later to be similar to H. L. Kuo's scheme. By conducting two (or three) integrations with different sea surface temperature (SST) conditions, we concluded that typhoons can develop under the initial conditions represented by the Jordan (1958) tropical atmosphere when the SST exceeded 27°C. We reported our results to Professor Syono, asking to present the result at the Berkeley International Union of Geodesy and Geophysics (IUGG) meeting, but he declined because he was still sticking with the original experiment, questioning the use of parameterization.

As noted previously, there were many important papers concerning tropical meteorology and convection presented at the Tokyo NWP symposium. In the "Mesoscale Phenomena" session, Professors Ogura and Chaney presented their early results, entitled "A numerical model of thermal convection in the atmosphere," which used the nonhydrostatic equations to examine convection caused by a moving head of a cold front. It was soon after this that the paper by Ogura and Phillips (1962) on the anelastic approximation appeared. Professor

Kuo's presentation was titled "On cellular convection and heat transfer," which was a finite amplitude theory of Bénard convection, a cutting-edge problem in fluid mechanics at that time. Although he could not come to the symposium, Professor Yoshi Sasaki contributed the paper "Effects of condensation, evaporation and rainfall on development of mesoscale disturbances: A numerical experiment." As this title suggests, this was perhaps the first study to discuss the mechanisms underlying the typical features of mesoscale systems known from observation by performing a numerical experiment. Thus, the early 1960s was the time that a new stream in the study of tropical meteorology and convection emerged following the success of NWP.

2. Equatorial waves in the stratosphere

While he was affiliated with the Meteorological Research Institute of the JMA, Yanai-san spent two years at Colorado State University at the invitation of Professor Herbert Riehl. Soon after returning to Japan, in April 1965, he was appointed as an associate professor of meteorology at the University of Tokyo, where I was working as a research associate. As a new university professor, he looked for something new to study with his graduate students. In those days, the stratosphere was a research frontier in meteorology. New and unique phenomena were being discovered owing to expansion of the routine radiosonde observations up to the 100-mb level. The opening of the "space era" by the launch of Sputnik on 4 October 1957 also drew the attention of meteorologists to the upper atmosphere.

In the upper atmosphere, two mysterious phenomena stood out: sudden stratospheric warmings (SSWs) and the quasi-biennial oscillation (QBO) of zonal winds at equatorial stratosphere. Working in tropical meteorology, Yanai-san decided to address the mechanism of the QBO. The core of the mystery concerns how and why the zonal winds alternate directions over the equator. Among a limited number of possibilities, the eddy transport of momentum was supposed to be most likely, and for that there must be "eddies." Thus, Yanai-san started to search for eddies from observational data, with his graduate student Taketo Maruyama.

As luck would have it, a dataset very well suited for this study already existed, consisting of special observations conducted by the U.S. Army to monitor winds up to high levels over the Pacific for nuclear weapon tests. In his thesis work on the analysis of typhoon formation, Yanai-san used those data printed and bound in books that were found on a bookshelf. Thus, in a very short time, the westward movement of disturbances having north–south wind fluctuations was discovered in the raw data plots. The first result was submitted for publication in July 1966, about a year after Yanai-san moved to the University of Tokyo (Yanai and Maruyama 1966).

It is interesting that almost at the same time, across the Pacific at the University of Washington in Seattle, similarly young, new faculty members Professors James Holton and John M. (Mike) Wallace began to investigate the mechanism of the QBO along with Professor Richard Lindzen of the University of Chicago, and their diagnostic study indicated the need for momentum forcing. Thus, a search for disturbances in the equatorial stratosphere as the source of momentum began, involving a graduate student named Kousky. They discovered another kind of disturbance with only zonal wind fluctuations and identified it as an equatorial Kelvin wave (Wallace and Kousky 1968). The research activities of these two groups and their search for tropical waves are described briefly in a story (Madden and Julian 2005) about the discovery of the Madden–Julian oscillation (MJO). In the biographical memoir of the late Professor Holton published quite recently, an early history of investigation of the QBO mechanism is vividly described, including those analysis studies to search for waves (Wallace 2014).

During the period 1965–66, I was working on the theoretical analysis of wave motions at equatorial latitudes where the Coriolis force becomes extremely small and changes sign as my thesis work for my doctorate degree. The motivation for this study was a theoretical interest in the dynamics under this particular condition, which had not been discussed while midlatitude geostrophic dynamics was the major player in the early NWP era. By treating a simple shallow water model on the equatorial β plane, some normal mode solutions were found to have characteristics unique to the equatorial latitudes (Matsuno 1966). They are now known as the (equatorial) Kelvin wave and mixed Rossby–gravity (MRG) wave.

The MRG wave is westward propagating and characterized by meridional wind fluctuations, so I considered that the wave found by Yanai and Maruyama must correspond to this mode. [Maruyama also quickly identified them as MRGs; see Maruyama (1967).] However, in those days I was anxious about the discovery of the equatorial Kelvin wave in the real atmosphere or ocean, as that would become decisive proof of the correctness of the theory. The Wallace and Kousky paper that reported the discovery came out in 1968, after I moved to Kyushu University. Although Yanai-san was occupied with his QBO research, he was very kind to read the draft of my paper and gave me valuable advice to help me to finish the thesis and publish it in time to meet a requirement for my getting a new job at Kyushu University. Thus, at the time of the first report (Yanai and Maruyama 1966), identification of the newly found wave with a mode in the equatorial wave theory was not a top priority.

3. GARP

The greatest innovations in modern atmospheric science and weather prediction are NWP and satellite observation. In the late 1960s, considering the emergence of these two powerful new tools, the world's leading scientists led by Professor Charney created a vision for the future of global-scale weather prediction and discussed how to realize it. This led to the establishment of the Global Atmospheric Research Programme (GARP), to be implemented by international cooperation. As part of its participation, Japan (JMA) decided to launch one of five Geostationary Meteorological Satellites designed to survey the whole globe starting from the First GARP Global Experiment (FGGE) period, 1977–78.

From an early stage (1967), Yanai-san worked actively to contribute to the planning of GARP as a tropical meteorology expert. He proposed a special field observation program to be conducted in the Marshall Islands area in the Pacific, tentatively called the Tropical Meteorology Experiment (TROMEX). Implementation proved difficult owing to the international political situation, and it was replaced by the GARP Atlantic Tropical Experiment (GATE), to be conducted in the tropical Atlantic. Owing to the relocation of the tropical field project, and also because of difficulties at the university because of students' rioting, Yanai-san decided to leave Japan to join the University of California, Los Angeles (UCLA), as a professor.

Soon after arriving at UCLA, Yanai-san actively worked on GATE and, at the same time through cooperation with UCLA Professor Akio Arakawa, he made an analysis of the heat and water vapor budgets in regions of cloud clusters and easterly waves. Collaborating with new graduate students, he first applied basically the same analysis method but to diagnose the "cloud mass flux" following Professor Arakawa's new formulation (Yanai et al. 1973). It was very timely that Dr. Tsuyoshi Nitta, Yanai-san's eldest former student, visited UCLA and devised a novel calculation scheme to quantify spectral mass fluxes with different cloud-top heights from observed data. He applied it to the GATE data to obtain reasonable results, suggesting a need for including cloud downdrafts (Nitta 1977). These series of works by Yanai-san's colleagues and students established the method to diagnose the role of convective clouds in various types of tropical weather systems as a standard analysis tool. It has been applied in many studies, and the notations Q_1 and Q_2 originally used in Yanai-san's doctoral thesis work became a common language among tropical meteorologists, as reviewed in Johnson et al. (2016, chapter 1).

As part of his GARP activity, in 1968 Yanai-san worked as a member of an ad hoc Study Group on Tropical Disturbances, along with Professors P. R. Pisharoty and Tetsuya Fujita. They examined the then very new satellite images of tropical clouds and attempted a classification of them. According to the report, they identified three types: "cloud clusters," "monsoon clusters," and "popcorn cumulonimbi." At this time, Yanai-san was still in Japan and I listened to his explanations, which stimulated my special interest in cloud clusters.

4. Twinkling cloud clusters—What are you?

In the mid-1960s, when cloud imagery by satellite observation became available to meteorologists for the first time, some cloud images appeared as anticipated but some others were previously unimaginable. To me, clouds associated with extratropical cyclones and fronts looked very much like those drawn in many textbooks based on the cyclone model of the Bergen school. In contrast to this, the appearance of cloud clusters in the tropics was completely beyond imagination, at least for me. Their sizes are 100–500 km or even larger, which are comparable to tropical cyclones or smallest extratropical cyclones, so that they must represent some kind of as-yet-unknown weather system. I was shocked and puzzled by their existence. After a while, I came to think that they must represent a kind of mesoscale convective system and confirmed that by asking specialists in this field.

However, another question arose. In midlatitudes, mesoscale convective systems develop in small areas under limited conditions where the vertical stratification becomes a state of (large) latent instability, perhaps because of particular situations of the larger-scale flow fields. In contrast to this, cloud clusters are ubiquitous in the tropics. They develop everywhere and every day are scattered over tropical areas around the world, just as extratropical cyclones are seen always occupying mid- and high-latitude areas.

Noting this comparison, we come to consider that cloud clusters are nothing but the manifestation of convective activity transporting energy upward to compensate for radiative energy loss aloft. Usually in textbooks this function is attributed to "convective clouds" or "cumulus and cumulonimbus clouds," but actually, these clouds are not distributed individually and uniformly as Bénard cells, nor are all of them organized by larger-scale, well-defined disturbances like tropical cyclones. Most convective clouds always aggregate by themselves to create bigger and very unique cloud systems having their own mechanism and life cycle. Needless to say, the special mechanism for producing mesoscale convective systems originates from the coupling of hydrodynamical and cloud physical processes unique to Earth's atmosphere.

Bearing this point in mind, and recalling that in most general circulation models midlatitude baroclinic waves are treated explicitly by the dynamical equations, I came to think that the generation and dynamics of tropical cloud clusters should also be treated explicitly.

Such thoughts went back to the mid- to late 1960s, when the ubiquitous nature of cloud clusters became apparent. The first question was whether and why tropical cloud clusters must be formed naturally and inevitably when the atmosphere undergoes differential heating. To answer this question, a long-term integration over a large area would be required and, because of computing limitations, I postponed trying to do that. After about 20 years, in the 1980s when computing resources at the University of Tokyo became available for conducting numerical experiments for this purpose, a graduate student Kensuke Nakajima (currently at Kyushu University) developed a 2D cloud dynamics model and performed several experiments with and without some cloud physical processes. From the results, we concluded that if all of a minimum set of cloud physical processes operating within mesoscale convective systems are included, cloud-cluster-like convective patterns appear, but they do not develop when some of the processes are missing. Instead, in those cases, convection of different types appears, including Bénard convection (Nakajima and Matsuno 1988).

I recall that I heard from someone who visited the United States sometime in the 1960s that Dr. Douglas Lilly was also wondering the same thing and said that the best way to answer the question was to perform a numerical experiment. I agreed with this view because it seemed impossible to make a laboratory experiment corresponding to cloud cluster formation, and it also seemed impossible to prove theoretically that the cloud cluster type of convection should occur under the given conditions, in the manner of the theory of Bénard convection or the baroclinic instability theory of extratropical cyclones.

In the course of this research, in 1985 I visited the NASA Goddard Space Flight Center (GSFC) and talked about this. At this time I first met Dr. W.-K. Tao, who, like us, was engaged in development of cloud dynamics models to be used for a wide area of long-term integrations. Since then, he continued his research as one of the leaders in this field (e.g., Tao and Moncrieff 2009). The director of Dr. Tao's branch was Dr. Joanne Simpson, who was truly the founder of atmospheric convection modeling. She published the first paper on the numerical simulation of the motion of a buoyant bubble in the Rossby memorial volume (Malkus and Witt 1959), which got me to recognize that numerical experimentation would become a powerful tool in various fields of atmospheric sciences.

5. TRMM, Earth Simulator, and NICAM

Soon after we met at GSFC, Dr. Simpson and I worked together toward the realization of a U.S.–Japan joint space program, the Tropical Rainfall Measuring Mission (TRMM), as representatives of each side's group of scientists. After a few years, I passed that role to Professor Tsuyoshi Nitta, a former student of Yanai-san and then the most active tropical meteorologist representing Japan. At an early stage, we meteorologist colleagues—Nitta, Sumi, and I—worried that the original plan for the inclination of the satellite orbit, 28°, would be too low to cover the baiu precipitation zone and requested the space agency people to change it to 35°. Since we were so naïve concerning the space program, we were afraid that our request regarding a most basic element of the satellite might be difficult to accept. Regardless of our anxiety, they readily consented to our request. The TRMM satellite was successfully launched on 27 November 1997 and is only now reaching the end of its mission at the time I write these words, after a 17-yr-long operation surpassing many times the originally planned lifespan of 3 years. Tao et al. (2016, chapter 2) describe the latest achievements utilizing the TRMM data by the collaborative team. Currently, the leader of Japan's science team is Professor Yukari Takayabu [coauthor of Takayabu et al. (2016), chapter 3], who was a student of the late Professor Nitta, and so is a grand-student of Yanai-san, as it were.

In the late 1990s, I participated in committees for research projects; one was to develop the world's fastest supercomputer, which was named the Earth Simulator (ES), completed in 2002, and another was a project to promote modeling of the global environment and prediction of global environmental changes. Until this time, meteorologists in Japan had been suffering from a shortage of computer resources for a long time. By my own very crude estimate, we were 10 years or more behind the United States with regard to our computing environment. Suddenly we were going to be given the best computer in the world! To respond to this situation, we would have to challenge ourselves with a new problem of great scientific value, something that was made possible only by the world's most capable computer.

Consulting with Professor Akimasa Sumi, my closest colleague who was working together in leading these projects, I proposed a global atmospheric model with a horizontal mesh size of less than 5 km that could resolve tropical cloud clusters as one of the targets for model development. The 5-km mesh size is marginal for resolving the 10-km-wide upward motion areas typically found in convective systems. However, it was twice the resolution of the formal target (10 km) employed as the reference for the computer architecture design, based on which the maximum effective speed was determined to be 5 teraFLOPS (floating-point operations per second). Doubling the resolution meant increasing the computing time by a factor of 10. Thus, it seemed daunting to run the models at the kind of resolutions that could produce really new, significant results. However, lucky for us that the estimate was wrong! When completed, the ES ran 3 times faster than the officially declared target.

A new model, the Nonhydrostatic Icosahedral Atmosphere Model (NICAM; Satoh et al. 2008), was developed from scratch starting in the year 2000, and the first scientifically significant result was an aquaplanet experiment that came out in early 2005 (Tomita et al.

2005). More lucky and essential for this achievement than the faster computer speed was that young, talented scientists led by Masaki Satoh came to join this ambitious project, dedicating themselves to a work without knowing whether it would produce results accepted by the community or not.

To evaluate the effects of increasing the model's resolution aimed at a fundamental improvement in climate modeling, the Athena project was conducted in 2009–10 under the leadership of Professor Jagadish Shukla. NICAM was adopted as a global atmospheric model with explicit convection as well as the European Center for Medium-Range Weather Forecast's highest-resolution model with parameterized convection. Results of the project are reported in Dirmeyer et al. (2012) and Kinter et al. (2013). Some recent results using NICAM are described in Oouchi and Satoh (2016, chapter 14).

6. Midlatitude versus tropical meteorology

It may be fair to say that the aquaplanet experiment corresponds to a tropical version of Norman Phillips' numerical experiment of the midlatitude general circulation published in 1956 (Phillips 1956). Prior to the NICAM experiment, there was another aquaplanet experiment by Hayashi and Sumi (1986), which was based on a (hydrostatic) primitive equation model including parameterized convection. Results of these two aquaplanet experiments are more or less similar, except that in the NICAM experiment individual cloud clusters of $O(100)$-km size were reproduced correctly. That is, they have mesoscale convective system structure. Thus, the NICAM experiment established that cloud cluster in the tropics is a fundamental structure of the convective system that naturally emerges under realistic conditions, just as baroclinic waves emerged naturally in Phillips' experiment.

In addition to this, both aquaplanet experiments produced active convection areas of $O(1000)$-km size and, at the same time, a wavenumber-1 modulation of them encircling the equator. Both of these large-scale structures moved eastward at about $15 \, \text{m s}^{-1}$ along the equator (taking 30–35 days for one cycle). The $O(1000)$-km size convection area was named the "super (cloud) cluster," whose observed counterpart was identified in satellite cloud imageries by Nakazawa (1988).

More systematic analysis of the space–time variability of convective clouds in the tropical belt by Takayabu (1994) led to a finding that a major part of the variability follows the dispersion relation of equatorial waves. By use of a novel technique, Wheeler and Kiladis (1999) have succeeded in showing the existence of this variability clearly, which is now referred to as convectively coupled equatorial waves (CCEWs; Kiladis et al. 2009). Through comparison with observational studies of the tropics, we recognize that the aquaplanet experiment can successfully reproduce (Kelvin type) CCEWs in addition to cloud clusters.

Originally, the Hayashi–Sumi aquaplanet experiment was conducted with an anticipation that the MJO might appear naturally as a free oscillation of the tropical atmosphere coupled with convection, and the result was once taken to support this idea. However, owing to the discovery of CCEWs, the MJO is now considered to be a different entity both from the dispersion diagram and structures. As to the background stationary fields, trade winds and an ITCZ located just over the equator are also seen in the model results.

This situation is quite different from the case of Phillips' experiment. In his case, almost all of the major phenomena of meteorologists' interest—the generation of extratropical cyclones, the three-cell structure of the zonal mean meridional circulation, and the strong westerly flow (the jet stream)—were reproduced, solving many long-standing open questions with just one experiment. In contrast to this, in the tropics case, there are many more important phenomena, such as tropical cyclones, easterly waves, and the MJO, that are not found in the aquaplanet experiments. The lack of tropical cyclones and easterly waves is considered to be due to the lack of background vorticity fields. Namely, in the aquaplanet experiment, the trade wind field has complete symmetry about the equator, so that the ITCZ that appears just over the equator is not associated with vorticity, whereas in the real situation the ITCZ is found at off-equatorial latitudes having vorticity. Though it is not yet clear, perhaps the existence of the warm water mass extending from the Indian Ocean to the Maritime Continent area may be a necessary condition to the generation of the MJO. Although they appear in nature, the monsoon circulation and the Walker circulation, which are important in tropical meteorology, are also (understandably) missing. From this comparison, we realize that understanding and simulation of the tropical weather and circulation are more difficult and complicated.

Perhaps the ultimate origin of this dissimilarity lies in the difference of the scales of energy sources. In midlatitude dynamics, the energy source for circulations is in the baroclinic instability that produces waves whose scales are several thousand kilometers. Finer structures like jet streams and fronts are generated from the largest-scale energy-containing eddies. Further, because of the stable stratification, all dynamical evolution of flow and temperature fields is constrained by potential vorticity conservation, which give us a feeling that their evolution is "stiff."

In contrast, in the tropics the ultimate energy source is vertical instability that first produces convective clouds with a 10-km horizontal size. Then, via coupling with other processes, larger-scale structure and flows are generated. The generation of the cloud cluster is the first level in the size hierarchy above the individual cloud scale. If there is a background vorticity field with the proper configuration, cloud clusters may couple with it to strengthen the rotational motion. The surface inhomogeneity is much more influential in the generation of tropical convection and associated circulations compared with its effect on large-scale baroclinic waves in the midlatitude.

Working as a meteorology professor for a long time, I have experienced that giving lectures in an organized style is easy for midlatitude dynamics but difficult for the tropics. Thus, even how to discuss tropical dynamics is a challenge, which is now presented as "multiscale convection-coupled systems" in this volume.

Acknowledgments. I would express my sincere gratitude to the coeditors of this volume, Professors Wen-wen Tung and Robert Fovell, for generously allowing me to write this prologue in a very personal style, and for further taking time to read the original draft carefully and for giving me valuable comments for improving it.

REFERENCES

Charney, J. G., and A. Eliassen, 1964: On the growth of the hurricane depression. *J. Atmos. Sci.*, **21**, 68–75, doi:10.1175/1520-0469(1964)021<0068:OTGOTH>2.0.CO;2.

——, R. Fjörtoft, and J. von Neumann, 1950: Numerical integration of the barotropic vorticity equation. *Tellus*, **2**, 237–254, doi:10.1111/j.2153-3490.1950.tb00336.x.

Dirmeyer, P. A., and Coauthors, 2012: Simulating diurnal cycle of rainfall in climate models: Resolution versus parameterizaion. *Climate Dyn.*, **39**, 399–418, doi:10.1007/s00382-011-1127-9.

Hayashi, Y.-Y., and A. Sumi, 1986: The 30–40 day oscillations simulated in an "aqua-planet" model. *J. Meteor. Soc. Japan*, **64**, 451–467.

Johnson, R. H., P. E. Ciesielski, and T. M. Rickenbach, 2016: A further look at Q_1 and Q_2 from TOGA COARE. *Multiscale Convection-Coupled Systems in the Tropics: A Tribute to Dr. Michio Yanai, Meteor. Monogr.*, No. 56, Amer. Meteor. Soc., doi:10.1175/AMSMONOGRAPHS-D-15-0002.1.

Jordan, C. L., 1958: Mean soundings for the West Indies area. *J. Meteor.*, **15**, 91–97, doi:10.1175/1520-0469(1958)015<0091:MSFTWI>2.0.CO;2.

Kasahara, A., 2000: On the origin of cumulus parameterization for numerical prediction models. *General Circulation Model Development: Past, Present and Future*, D. A. Randall, Ed., Academic Press, 199–224.

Kiladis, G. N., M. C. Wheeler, P. T. Haertel, K. H. Straub, and P. E. Roundy, 2009: Convectively coupled equatorial waves. *Rev. Geophys.*, **47**, RG2003, doi:10.1029/2008RG000266.

Kinter, J. L., III, and Coauthors, 2013: Revolutionalizing climate modeling with Project Athena. *Bull. Amer. Meteor. Soc.*, **94**, 231–245, doi:10.1175/BAMS-D-11-00043.1.

Madden, R. A., and P. R. Julian, 2005: Historical perspective. *Intraseasonal Variability in the Atmosphere–Ocean Climate System*, W. K. M. Lau and D. E. Waliser, Eds., Springer, 1–18.

Malkus, J. S., and G. Witt, 1959: The evolution of a convective element: A numerical calculation. *The Atmosphere and Sea in Motion*, Rockefeller Institute Press, 425–439.

Maruyama, T., 1967: Large-scale disturbances in the equatorial lower stratosphere. *J. Meteor. Soc. Japan*, **45**, 391–408.

Matsuno, T., 1966: Quasi-geostrophic motions in the equatorial area. *J. Meteor. Soc. Japan*, **44**, 25–43.

Nakajima, K., and T. Matsuno, 1988: Numerical experiment concerning the origin of cloud clusters in the tropical atmosphere. *J. Meteor. Soc. Japan*, **66**, 309–329.

Nakazawa, T., 1988: Tropical super-clusters within intraseasonal variations over the western Pacific. *J. Meteor. Soc. Japan*, **66**, 823–839.

Nitta, T., 1977: Response of cumulus updraft and downdraft to GATE A/B-scale motion systems. *J. Atmos. Sci.*, **34**, 1163–1186, doi:10.1175/1520-0469(1977)034<1163:ROCUAD>2.0.CO;2.

Ogura, Y., and N. A. Phillips, 1962: Scale analysis of deep and shallow convection in the atmosphere. *J. Atmos. Sci.*, **19**, 173–179, doi:10.1175/1520-0469(1962)019<0173:SAODAS>2.0.CO;2.

Oouchi, K., and M. Satoh, 2016: A synoptic-scale cold-reservoir hypothesis on the origin of the mature-stage super cloud cluster: A case study with a global nonhydrostatic model. *Multiscale Convection-Coupled Systems in the Tropics: A Tribute to Dr. Michio Yanai, Meteor. Monogr.*, No. 56, Amer. Meteor. Soc., doi:10.1175/AMSMONOGRAPHS-D-15-0008.1.

Ooyama, K., 1964: A dynamical model for the study of tropical cyclone development. *Geofis. Int.*, **4**, 187–198.

Phillips, N. A., 1956: The general circulation of the atmosphere: A numerical experiment. *Quart. J. Roy. Meteor. Soc.*, **82**, 123–164, doi:10.1002/qj.49708235202.

Satoh, M., T. Matsuno, H. Tomita, H. Miura, T. Nasuno, and S. Iga, 2008: Nonhydrostatic icosahedral atmospheric model (NICAM) for global cloud resolving simulations. *J. Comput. Phys.*, **227**, 3486–3514, doi:10.1016/j.jcp.2007.02.006.

Syono, S., 1953: On the formation of tropical cyclones. *Tellus*, **5**, 179–195, doi:10.1111/j.2153-3490.1953.tb01047.x.

——, and Coauthors, Eds., 1962: *Proceedings of International Symposium on Numerical Weather Prediction in Tokyo, 1960*. Meteorological Society of Japan, 656 pp.

Takayabu, Y., 1994: Large-scale wave disturbances associated with equatorial waves. Part 1: Spectral features of cloud disturbances. *J. Meteor. Soc. Japan*, **72**, 433–449.

——, G. N. Kiladis, and V. Magaña, 2016: Michio Yanai and tropical waves. *Multiscale Convection-Coupled Systems in the Tropics: A Tribute to Dr. Michio Yanai, Meteor. Monogr.*, No. 56, Amer. Meteor. Soc., doi:10.1175/AMSMONOGRAPHS-D-15-0019.1.

Takeda, T., 1971: Numerical simulation of a precipitating convective cloud: The formation of a "long-lasting" cloud. *J. Atmos. Sci.*, **28**, 350–375, doi:10.1175/1520-0469(1971)028<0350:NSOAPC>2.0.CO;2.

Tao, W.-K., and M. W. Moncrieff, 2009: Multi-scale cloud system modeling. *Rev. Geophys.*, **47**, RG4002, doi:10.1029/2008RG000276.

——, and Coauthors, 2016: TRMM latent heating retrieval: Applications and comparisons with field campaigns and large-scale analyses. *Multiscale Convection-Coupled Systems in the Tropics: A Tribute to Dr. Michio Yanai, Meteor. Monogr.*, No. 56, Amer. Meteor. Soc., doi:10.1175/AMSMONOGRAPHS-D-15-0013.1.

Tomita, H., H. Miura, S. Iga, T. Nasuno, and M. Satoh, 2005: A global cloud resolving simulation: Preliminary results from an aqua planet experiment. *Geophys. Res. Lett.*, **32**, L08805, doi:10.1029/2005GL022459.

Wallace, J. M., 2014: James R. Holton, 1938–2004, U.S. National Academy of Sciences Biographical Memoirs, 27 pp. [Available online at http://nasonline.org/publications/biographical-memoirs/.]

——, and V. E. Kousky, 1968: Observational evidence of Kelvin waves in the tropical stratosphere. *J. Atmos. Sci.*, **25**, 900–907, doi:10.1175/1520-0469(1968)025<0900:OEOKWI>2.0.CO;2.

Wheeler, M. C., and G. N. Kiladis, 1999: Convectively coupled equatorial waves: Analysis of cloud and temperature in the wavenumber-frequency domain. *J. Atmos. Sci.*, **56**, 374–399, doi:10.1175/1520-0469(1999)056<0374:CCEWAO>2.0.CO;2.

Yanai, M., 1961: Dynamical aspects of typhoon formation. *J. Meteor. Soc. Japan*, **39**, 282–309.

——, and T. Maruyama, 1966: Stratospheric wave disturbances propagating over the equatorial Pacific. *J. Meteor. Soc. Japan*, **44**, 291–294.

——, S. Esbensen, and J.-H. Chu, 1973: Determination of bulk properties of tropical cloud clusters from large-scale heat and moisture budgets. *J. Atmos. Sci.*, **30**, 611–627, doi:10.1175/1520-0469(1973)030<0611:DOBPOT>2.0.CO;2.

Chapter 1

A Further Look at Q_1 and Q_2 from TOGA COARE

RICHARD H. JOHNSON AND PAUL E. CIESIELSKI

Colorado State University, Fort Collins, Colorado

THOMAS M. RICKENBACH

Department of Geography, Planning, and Environment, East Carolina University, Greenville, North Carolina

ABSTRACT

Two features of Yanai et al.'s profiles of Q_1 and Q_2—the commonly observed double-peak structure to Q_2 and an inflection in the Q_1 profile below the melting level—are explored using estimates of convective and stratiform rainfall partitioning based on Massachusetts Institute of Technology (MIT) radar reflectivity data collected during TOGA COARE. The MIT radar data allow the Q_1 and Q_2 profiles to be classified according to stratiform rain fraction within the radar domain and, within the limitations of the datasets, allow interpretations to be made about the relative contributions of convective and stratiform precipitation to the mean profiles. The sorting of Q_2 by stratiform rain fraction leads to the confirmation of previous findings that the double-peak structure in the mean profile is a result of a combination of separate contributions of convective and stratiform precipitation. The convective contribution, which has a drying peak in the lower troposphere, combines with a stratiform drying peak aloft and low-level moistening peak to yield a double-peak structure. With respect to the inflection in the Q_1 profile below the 0°C level, this feature appears to be a manifestation of melting. It is the significant horizontal dimension of the stratiform components of tropical convective systems that yields a small but measurable imprint on the large-scale temperature and moisture stratification upon which the computations of Q_1 and Q_2 are based. The authors conclude, then, that the rather subtle features in the Q_1/Q_2 profiles of Yanai et al. are directly linked to the prominence of stratiform precipitation within tropical precipitation systems.

1. Introduction

A conceptual breakthrough in understanding how tropical cloud systems interact with their environment was achieved through the landmark paper of Yanai et al. (1973), wherein the now-familiar Q_1 (apparent heat source) and Q_2 (apparent moisture sink) were defined. However, well prior to the 1973 study, Yanai (1961) introduced Q_1 and Q_2 in a paper titled "A detailed analysis of typhoon formation," which investigated the dynamic and thermodynamic properties of the formation of Typhoon Doris (1958) in the western Pacific. Vertical profiles of Q_1 and Q_2—the first ever to be presented—were shown in Yanai (1961) for the period of the transformation of Doris from a cold-core to warm-core tropical cyclone.

Following Yanai et al. (1973), but including ice processes, we write the equations for the apparent heat source and moisture sink as follows:

$$Q_1 \equiv \frac{\partial \overline{s}}{\partial t} + \overline{\mathbf{v}} \cdot \nabla \overline{s} + \overline{\omega} \frac{\partial \overline{s}}{\partial p}$$

$$= L_v(\overline{c} - \overline{e}) + (L_v + L_f)(\overline{d} - \overline{s}_*)$$

$$+ L_f(\overline{f} - \overline{m}) + Q_R - \frac{\partial}{\partial p}(\overline{s'\omega'}), \quad \text{and} \quad (1\text{-}1)$$

$$Q_2 \equiv -L_v \left(\frac{\partial \overline{q}}{\partial t} + \overline{\mathbf{v}} \cdot \nabla \overline{q} + \overline{\omega} \frac{\partial \overline{q}}{\partial p} \right)$$

$$= L_v(\overline{c} - \overline{e}) + L_v(\overline{d} - \overline{s}_*) + L_v \frac{\partial}{\partial p}(\overline{q'\omega'}), \quad (1\text{-}2)$$

where c, e, d, s_*, f, and m are condensation, evaporation, deposition, sublimation, freezing, and melting rates,

Corresponding author address: Richard H. Johnson, Department of Atmospheric Science, Colorado State University, 3915 W. Laport Ave., Fort Collins, CO 80523.
E-mail: johnson@atmos.colostate.edu

DOI: 10.1175/AMSMONOGRAPHS-D-15-0002.1

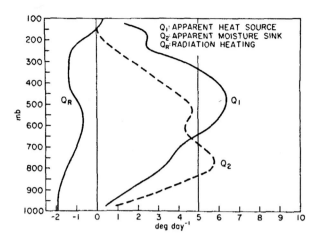

FIG. 1-1. Apparent heat source Q_1, apparent moisture sink Q_2, and net radiative heating rate Q_R for the Marshall Islands. From Yanai et al. (1973).

respectively; q is the water vapor mixing ratio; $s \equiv c_p T + gz$ is the dry static energy; Q_1 is the apparent heat source; Q_2 is the apparent moisture sink; Q_R is the net radiative heating rate; L_v and L_f are the latent heats of vaporization and fusion, respectively; the overbar refers to a horizontal average; and the primes denote a deviation from this average.

Yanai et al. (1973) used 1956 Marshall Islands sounding data to compute Q_1 and Q_2 and then applied a simplified cloud model to diagnose the bulk properties of cumulus ensembles, which provided substantial insight into the properties of tropical convective systems and how they impact their environment. The profiles of Q_1 and Q_2 obtained by Yanai et al. (1973) for the Marshall Islands for the period 15 April–22 July 1956 are shown in Fig. 1-1, along with radiative heating Q_R estimates from Dopplick (1972, 1979). They found a primary peak in Q_1 in the upper troposphere, whereas the principal Q_2 peak was in the lower troposphere. Yanai et al. (1973) pointed out that the separation of the Q_1 and Q_2 peaks is indicative of vigorous deep convection [i.e., strong vertical eddy fluxes of moist static energy $h(\equiv s + L_v q)$]. Since the profiles are averages over a 3-month period, they represent contributions from both convective and stratiform components of precipitation systems, which have distinctly different heating and moistening profiles (Houze 1982, 1989; Johnson 1984). Just how the structure of the heating profile varies in the vertical is of considerable interest, since that determines the dynamical response of the environment to convection (Hartmann et al. 1984; Raymond and Jiang 1990; Nicholls et al. 1991; Schumacher et al. 2004).

A curious feature of the Q_2 profile is its double-peaked structure, which has also been found for the GARP Atlantic Tropical Experiment (GATE)

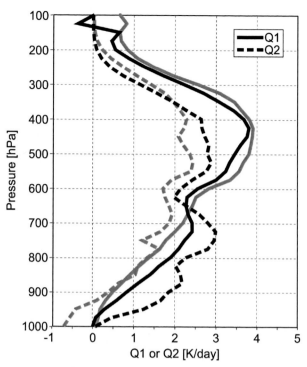

FIG. 1-2. Apparent heat source Q_1 and apparent moisture sink Q_2 for MISMO in the Indian Ocean (dark curve) and TOGA COARE Intensive Flux Array (light curve). From Katsumata et al. (2011).

(Esbensen et al. 1988), as well as for other experiments such as the more recent R/V *Mirai* Indian Ocean Cruise for the Study of the Madden–Julian oscillation (MJO) Onset (MISMO) (Katsumata et al. 2011). This feature has been attributed by Johnson (1984) and Esbensen et al. (1988) to the separate contributions of

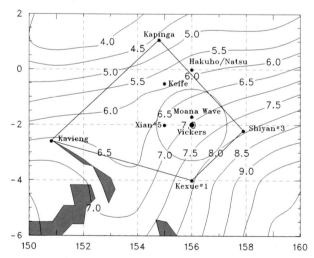

FIG. 1-3. TOGA COARE Intensive Flux Array quadrilateral, atmospheric sounding sites, and 4-month mean TOGA COARE rainfall (mm day^{-1}) from the Global Precipitation Climatology Project (GCPC). The circle indicates the 120-km radius radar study area. MIT radar was on board R/V *Vickers*.

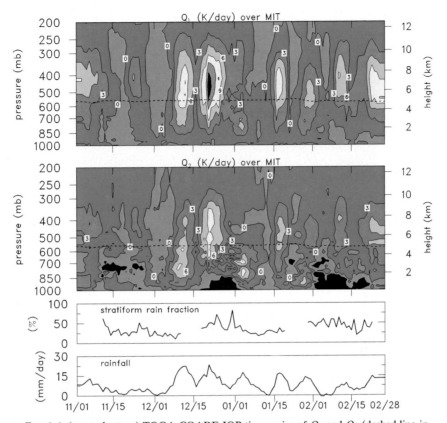

FIG. 1-4. (top to bottom) TOGA COARE IOP time series of Q_1 and Q_2 (dashed line indicates the 0° level), time series of stratiform rain fraction, and moisture-budget-derived rainfall. Budget fields and rainfall have been smoothed with a 5-day running mean filter. All fields are for MIT radar area.

convective and stratiform precipitation systems. Namely, condensation and deposition aloft associated with stratiform precipitation produces the upper peak while condensation in convective precipitation produces the lower peak. Placed in a time perspective, the vertical motion and hence heating and drying typically undergo an evolution (on time scales ranging from diurnal to the life cycle of tropical cloud clusters to the passage of easterly waves) from a low-level peak in the vertical motion (and heating) in the early stages to an upper-level peak in the later stages (Frank 1978; Nitta 1978; Houze 1982; Johnson and Young 1983). This evolution has since been documented on longer time scales ranging up to the 30–60-day MJO, as reviewed by Kiladis et al. (2009).

It is the purpose of this article to delve further into the relationship between the Q_1 and Q_2 profiles and convective/stratiform rain partitioning using radar and sounding data from an experiment, which is now celebrating its twenty-third anniversary—the 1992–93 Tropical Ocean and Global Atmosphere Coupled Ocean–Atmosphere Response Experiment (TOGA COARE)—conducted in the western Pacific (Webster and Lukas 1992). However, we first note that the experiment-mean TOGA COARE

Q_1 and Q_2 profiles (Johnson and Ciesielski 2000) are different from those of the Marshall Islands, GATE, and MISMO. To illustrate this difference, a comparison of the mean profiles for MISMO and TOGA COARE is shown in Fig. 1-2 (from Katsumata et al. 2011). Notably absent from the TOGA COARE results are low-level peaks in heating and drying. This behavior was also found over the northern South China Sea during the May–June 1998 South China Sea Monsoon Experiment (SCSMEX; Johnson and Ciesielski 2002). Johnson and Lin (1997) explained the unique feature of the TOGA COARE profiles as a consequence of frequent trade wind–like, nonprecipitating cumulus clouds during the quiescent and westerly wind burst phases of the MJO. These clouds were prevalent during dry midtropospheric conditions and served to moisten the lower troposphere, thereby producing prolonged periods of negative Q_2, as also observed during the Barbados Oceanographic and Meteorological Experiment (BOMEX; Nitta and Esbensen 1974). In contrast, during the active phase of the MJO, Q_1 and Q_2 distributions much like those of Yanai et al. (1973) were observed [Fig. 3 of Johnson and Lin (1997)].

In this chapter we will use radar data from the Massachusetts Institute of Technology (MIT) radar on board the R/V *Vickers* during TOGA COARE (Rickenbach and Rutledge 1998) to relate the heating and moistening profiles to the stratiform/convective rain fractions derived by that radar to further elucidate the factors contributing to the observed mean Q_1 and Q_2 profiles for that experiment.

2. Data and analysis procedures

a. Radar data

Radar reflectivity data from the MIT C-band Doppler radar, deployed on board the R/V *Vickers* and processed as described in Rickenbach and Rutledge (1998), are used in this study. The position of the *Vickers* within the TOGA COARE Intensive Flux Array (IFA) is shown in Fig. 1-3 along with other sounding sites in the region used in the budget analyses. Rickenbach and Rutledge (1998, see their Fig. 2) indicate the time periods of the radar and sounding operations during the 4-month TOGA COARE intensive observing period (IOP; November 1992–February 1993). Convective/stratiform rain partitioning was carried out using a slight modification of the method of Steiner et al. (1995) applied to 10-min radar reflectivity volumes. The mean stratiform rain fraction for the three cruises of the *Vickers* was determined to be 28%, which is less than the ~40% estimate for this region by Schumacher and Houze (2003) obtained from the Tropical Rainfall Measuring Mission (TRMM) Precipitation Radar (PR). However, as we will see later, the MIT-based estimate is closer to the TRMM-based value if we remove the periods of trade wind cumulus from the average. We would expect this to be the case since the 4–5-km footprint of the PR tends to smooth out and lower the reflectivity of the precipitating cumulus (the so-called "beamfilling" problem) compared to the higher-resolution MIT shipborne radar.

b. Sounding data and averaging methodology

The 6-hourly sounding data were objectively analyzed onto a 1° × 1° grid at 25-hPa intervals from 1000 to 25 hPa over the large-scale TOGA COARE domain using procedures described in Ciesielski et al. (1997), and incorporating humidity sensor corrections as outlined in Ciesielski et al. (2003). The resulting gridded basic and derived fields were then averaged over the circular MIT radar coverage area shown in Fig. 1-3. The distribution of sounding sites in proximity to the *Vickers* is reasonably good, so analyses over this region are deemed fairly reliable, apart from other issues such as temporal data gaps and random sampling errors (Mapes et al. 2003).

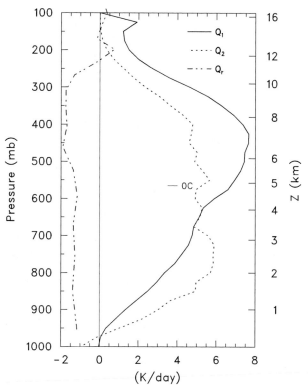

FIG. 1-5. Average Q_1 and Q_2 profiles for MIT radar coverage area for times when precipitation from the atmospheric moisture budget over this area exceeded 3.5 mm day^{-1}. The Q_R profile is from L'Ecuyer and Stephens (2003) based on 13-yr (1998–2011) mean over this region. The 0°C level is indicated.

3. Results for the 4-month TOGA COARE intensive observing period

A time series of Q_1 and Q_2 for the TOGA COARE IOP is shown in Fig. 1-4, along with the stratiform rainfall fraction based on MIT radar data for the three cruises of the *Vickers* and average rainfall for the MIT radar coverage area derived from the moisture budget, where the Woods Hole Oceanographic Institution Improved Meteorology (IMET) buoy is used for surface evaporation. A prominent MJO event occurred in December, characterized by a transition from shallow convection (low-level peak in Q_2) early in the month followed by deep convection midmonth and then stratiform precipitation (drying above moistening) at the end of the month (Lin and Johnson 1996). Consistent with this evolution is an increasing trend in the stratiform rainfall fraction through December as determined by the MIT radar.

Conspicuous in Fig. 1-4 are several ~10-day to 2-week periods of negative Q_2 in the lower troposphere, namely, near 750 hPa in mid-November and near 900 hPa in late December and early February. These features were described by Johnson and Lin (1997) as episodes of trade wind–like shallow cumulus existing during very

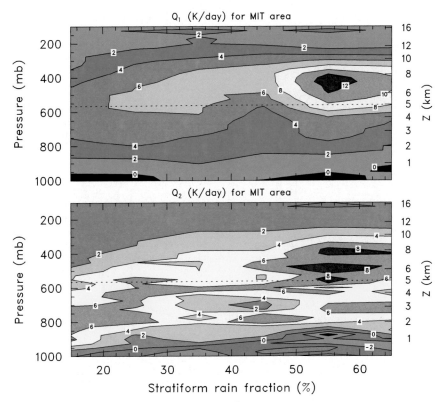

FIG. 1-6. The (top) Q_1 and (bottom) Q_2 (K day^{-1}) as a function of stratiform rain fraction for the MIT radar coverage area during TOGA COARE for times when moisture-budget-derived rainfall over that area exceeded 3.5 mm day^{-1} (dashed line indicates the 0°C level).

dry low- to midtroposphere conditions. The moistening in November occurred during light surface wind conditions, whereas the episodes in late December and February were during strong winds. They noted that when winds were strong during these shallow cumulus periods, there was greater moistening (negative Q_2) near the cloud base in association with subinversion-type (Esbensen 1978) or forced-type (Stull 1985) cumuli that were, in part, a manifestation of overshooting boundary layer eddies. When winds were light, the cumuli were slightly deeper ["inversion penetrating" after Esbensen (1978) or "active" after Stull (1985)] and the moistening peaked in the upper part of the cloud layer [similar to the Atlantic trades; Nitta and Esbensen (1974)]. Higher sea surface temperatures (by 1 K) and weaker shear observed during the November period likely contributed to the more active cumuli at that time.

4. Relating Q_1 and Q_2 to stratiform rain fraction

The frequent occurrence of trade wind regimes over the warm pool during TOGA COARE can be attributed to "dry intrusions" that were observed in association with the MJO (Parsons et al. 1994; Numaguti et al. 1995;

Yoneyama and Fujitani 1995; Mapes and Zuidema 1996), which helped to inhibit the growth of convection (e.g., Redelsperger et al. 2002). Johnson and Lin (1997) focused on Q_1 and Q_2 during these trade wind–like periods by averaging the profiles for times when precipitation was less than 3.5 mm day^{-1}, the average surface evaporation rate during TOGA COARE. Alternatively, to explore the characteristics of deep convection during TOGA COARE, we remove the effects of these periods by including only those times when the precipitation exceeded 3.5 mm day^{-1}. The resulting Q_1/Q_2 profiles are shown in Fig. 1-5. They now look much like those of Yanai et al. (1973) (see Fig. 1-1), namely, possessing a double-peak structure to Q_2 (though with several additional peaks) and an inflection in the Q_1 profile near 650 hPa.[1] Therefore, to interpret the Yanai

---

[1] Both the Marshall Islands and the western Pacific Q_1 profiles also show peaks near 150 hPa. These peaks may not represent any physical processes, rather they may be artifacts of applying the boundary condition $\omega = 0$ at the tropopause or 100 hPa such that some upward motion acting on weak stability above the lapse-rate minimum (Fueglistaler et al. 2009) produces a local heating maximum.

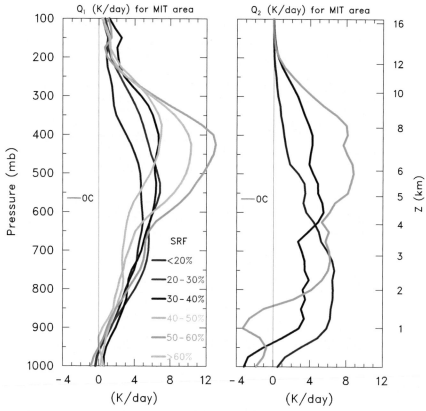

FIG. 1-7. The Q_1 and Q_2 profiles as a function of stratiform rain fraction (SRF) for the MIT radar coverage area during TOGA COARE for times when moisture-budget-derived rainfall over that area exceeded 3.5 mm day^{-1}. SRF values plotted for Q_1 are averages for the following SF bins: <20%, 20%–30%, 30%–40%, 40%–50%, 50%–60%, and >60%. Values for Q_2 are plotted only for <20%, 30%–40%, and 50%–60% for the sake of clarity. Curves have been smoothed in the vertical with a 1–2–1 filter.

et al. (1973) profiles in light of the TOGA COARE results, we subsequently restrict analyses to the dates and times that make up Fig. 1-5 (184 total 6-h periods). In support of this procedure, we note that an inflection in the Q_1 profile was also found by Schumacher et al. (2007) for the heavy-rain threshold cases of SCSMEX and the 1999 Kwajalein Experiment (KWAJEX).

The Q_1 and Q_2 profiles as a function of stratiform rain fraction (SRF) are shown in Fig. 1-6. As SRF increases, there is a transition from vertically separated, single peaks in heating and drying in the low- to midtroposphere to nearly coincident peaks at higher levels, consistent with an evolution in the precipitation from mostly convective to mainly stratiform in character (Luo and Yanai 1984). The evolution for Q_2 is more complex than that for Q_1, with multiple drying peaks aloft above a moistening peak in the lower troposphere. The low-level drying peaks between 700 and 800 hPa for SRF > 40% likely reflect the effects of convection coexisting with stratiform precipitation (Leary and Houze 1979).

Another way of viewing this transition is by comparing profiles for different values of SRF (Fig. 1-7). In this plot SRF values shown are averages for the following SRF bins: <20%, 20%–30%, 30%–40%, 40%–50%, 50%–60%, and >60%. The number of cases in each bin is 41, 53, 29, 17, 15, and 29, respectively. Although there is noise in the profiles (especially for Q_2, where results for only three bins are shown), the transition from convective to stratiform structure with increasing SRF can clearly be seen. Also evident for large values of SRF is an inflection in the Q_1 profile below 0°C near 4 km. This feature has been attributed to cooling by melting, which has been seen in observations (e.g., Johnson et al. 1996) and cloud-resolving models (e.g., Shie et al. 2003). The inflection in Q_1 is also observed in the Yanai et al. (1973) Marshall Islands Q_1 profile (Fig. 1-1).

Because of the abundant precipitation over the warm pool, melting leaves a measurable and persistent effect on the static stability. Melting takes place over a relatively shallow layer [~300–400 m deep; Willis and Heymsfield (1989)] below the 0°C level. Freezing of

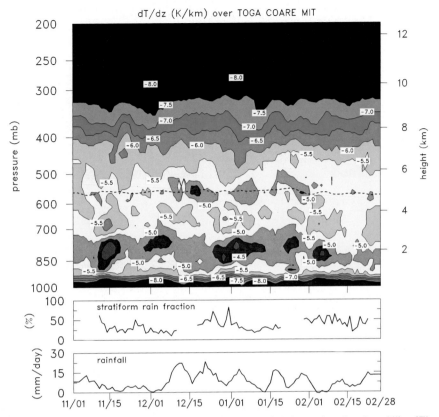

FIG. 1-8. (top) TOGA COARE intensive observing period time series of static stability dT/dz (K km^{-1}) (dashed line indicates the 0° level), (middle) time series of stratiform rain fraction, and (bottom) moisture-budget-derived rainfall. All fields are for MIT radar area.

hydrometeors occurs at levels above 0°C, but over a deeper layer; nevertheless, the combined effects of freezing above melting serves to increase the static stability at the 0°C level and also from $L_v(\overline{f} - \overline{m})$ in (1-1), the vertical gradient in Q_1 there. However, when computing Q_1 from sounding data, it is the lhs of (1-1) that is calculated and, since the vertical advection term $\overline{\omega}\partial\overline{s}/\partial p$

is usually the dominant term in that calculation, the impact of freezing and melting perturbs $\partial\overline{s}/\partial p$ and hence the Q_1 profile. To illustrate this effect, a time series of static stability, represented as dT/dz in Fig. 1-8, shows a persistent stable layer in proximity to 0°C. This stable layer is strongest during periods of heaviest rainfall, so its linkage to melting is perceptible. Also seen in Fig. 1-8

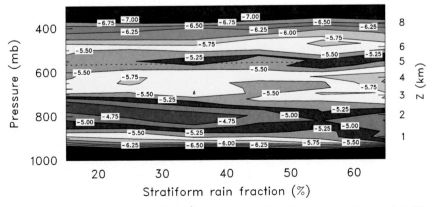

FIG. 1-9. Static stability dT/dz (K km^{-1}) as a function of stratiform rain fraction for the MIT radar coverage area during TOGA COARE for times when moisture-budget-derived rainfall over that area exceeded 3.5 mm day^{-1}.

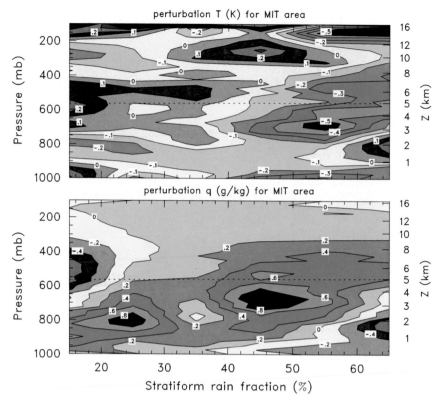

FIG. 1-10. (top) Perturbation temperature (K; deviation from IOP mean) and (bottom) perturbation water vapor mixing ratio (g kg^{-1}) as a function of stratiform rain fraction for the MIT radar coverage area during TOGA COARE for times when moisture-budget-derived rainfall over that area exceeded 3.5 mm day^{-1} (dashed line indicates the 0°C level).

is a persistent, though fluctuating in intensity, trade wind stable layer, which tends to be strongest during periods of light rainfall and negative low-level Q_2 (Fig. 1-4).

To further elucidate the relationship between the melting stable layer and stratiform precipitation, dT/dz is plotted as a function of SRF in Fig. 1-9. As the SRF increases, the strength of the melting stable layer increases, as one would expect considering the associated larger horizontal areas covered by precipitation when the SRF is high. On the other hand, the trade wind stable layer weakens as SRF increases, consistent with the idea that this layer occurs episodically over the warm pool (Johnson and Lin 1997) and is present primarily outside the active phase of the MJO when the SRF is low. In addition, there is a noticeable descent of the height of the trade stable layer as SRF increases. This behavior may be related to the finding of Johnson and Lin (1997) that during light-rain (and light wind) periods of the MJO with small SRF, there is a greater proportion of active trade cumulus (Esbensen 1978; Stull 1985), which rise to higher elevations than during the westerly wind burst phase with high SRF, when "forced" cumuli that are extensions of boundary layer turbulence predominate.

Because the stratiform precipitation systems cover large areas, their impact on the temperature field is significant. This effect is seen in Fig. 1-10, where for SRF > 50%, a shallow cool anomaly is present in a layer below the melting level, accompanied by a deeper moist layer aloft. For the largest values of SRF, there are both warm and dry anomalies centered just below 800 hPa. These features are manifestations of "onion soundings," often observed in association with the stratiform precipitation regions of tropical squall lines or cloud clusters (Zipser 1977). This structure can also be seen in Fig. 1-11, a plot of the mean relative humidity for SRF > 60% and < 20%. For large values of SRF, moist conditions exist above 0°C, indicative of stratiform anvil clouds aloft, with dry conditions below. In contrast, the mostly convective situations (SRF < 20%) are characterized by a relatively dry mid- to upper troposphere but slightly moister conditions below 3 km. The two contrasting profiles in Fig. 1-11 are consistent with the bimodal relative humidity behavior for TOGA COARE reported by Brown and Zhang (1997). For the lowest values of SRF (Fig. 1-10), dry conditions are observed in the midtroposphere, which act to suppress cloud development.

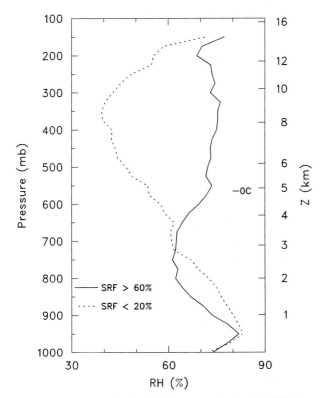

FIG. 1-11. Relative humidity (with respect to ice for $T < 0°C$) for stratiform rain fraction >60% (solid curve) and <20% (dashed curve).

Also observed for large SRF values (Fig. 1-10) is a layer of cooling near the tropopause (~150 hPa), a feature commonly observed atop mesoscale convective systems in the tropics (Johnson and Kreite 1982; Holloway and Neelin 2007). In addition, a cool anomaly is observed near the surface for increasing values of SRF, indicative of extensive modification of the boundary layer by downdrafts as the stratiform rain amounts and associated area coverage increases (Zipser 1977; Fitzjarrald and Garstang 1981).

In the computation of Q_2 from large-scale fields, the dominant term is $\overline{\omega}\partial\overline{q}/\partial p$. The individual contributions to this term are plotted as a function of SRF in Fig. 1-12. There is a clear upward shift in the vertical motion peak (Fig. 1-12, top) as SRF increases. The average SRF for the cases considered here (those for rainfall exceeding $3.5\,\mathrm{mm\,day}^{-1}$) is 36%, somewhat higher than the 28% for the entire MIT radar period. Viewing the $\overline{\omega}$ variation and noting that half the rain is above SRF = 36% and half below, it is readily apparent that the double-peak structure to Q_2 arises from the individual convective and stratiform contributions. The time series of dq/dp (Fig. 1-12, middle) shows a trend toward a local maximum near the melting layer as SRF increases, reflecting the increasing presence of stratiform clouds with bases near that level (Johnson et al. 1996). To confirm that the

term involving $\overline{\omega}\partial\overline{q}/\partial p$ is the principal contributor to Q_2, the SRF-sorted behavior of this term is shown in Fig. 1-12 (bottom). Overall, the broad pattern and evolution of this term is similar to that of Q_2 (Fig. 1-6, bottom), showing distinctly different vertical structures for small and large values of SRF. The greater vertical structure in Q_2 is likely a result of horizontal advection of dry air in thin layers, often occurring near the melting level, which Mapes and Zuidema (1996) and Johnson et al. (1996) found to be prevalent over the western Pacific warm pool during TOGA COARE.

5. Summary and conclusions

It has been over four decades since the landmark paper of Yanai et al. (1973) diagnosing apparent heat sources Q_1 and moisture sinks Q_2, as well as convective transports based on a simplified cloud model, using Marshall Islands data. In subsequent years, numerous field campaigns in the tropics and midlatitudes have been conducted computing Q_1 and Q_2, with results similar to those of Yanai et al. In this chapter we explore aspects of those heating and moistening profiles using estimates of convective and stratiform rainfall partitioning based on MIT radar reflectivity data from the R/V *Vickers* during TOGA COARE (Rickenbach and Rutledge 1998). Particular attention is given to (i) the commonly observed double-peak structure to Q_2 and (ii) an inflection in the Q_1 profile below the melting level, and the relationship of these features to the convective cloud populations.

The MIT radar data allow the Q_1 and Q_2 profiles to be classified according to stratiform rain fraction and, within the limitations of the datasets, interpretations to be made about the relative contributions of convective and stratiform precipitation to the mean profiles. Radar data from the three cruises of the *Vickers* are used along with gridded fields constructed from the TOGA COARE sounding array and averaged over the MIT radar coverage area. To remove periods when trade wind–like cumulus were present during TOGA COARE, times were selected for analysis when the average rainfall rate over the MIT area exceeded $3.5\,\mathrm{mm\,day}^{-1}$. This criterion resulted in 184 total 6-hourly periods, for which the average stratiform rain fraction was 36%.

The sorting of Q_2 by stratiform rain fraction leads to the conclusion that its double-peak structure is a result of the separate contributions of convective and stratiform precipitation, as suggested by Johnson (1984), Esbensen et al. (1988), and others. The convective contribution gives a lower drying peak, which when combined with the stratiform contribution of a drying peak aloft (along with moistening at low levels), yields the double-peak structure. With respect to the inflection

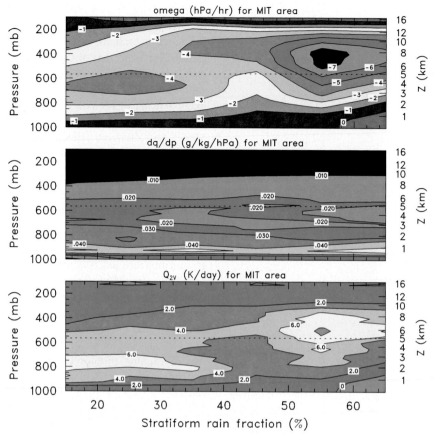

FIG. 1-12. (top) $\overline{\omega}$ (hPa h^{-1}), (middle) $d\overline{q}/dp$ (g kg^{-1} hPa^{-1}), and (bottom) $-L_v\overline{\omega}\partial\overline{q}/\partial p$ (K day^{-1}) as a function of stratiform rain fraction for the MIT radar coverage area during TOGA COARE for times when moisture-budget-derived rainfall over that area exceeded 3.5 mm day^{-1} (dashed line indicates the 0°C level).

in the Q_1 profile below 600 hPa or 4 km, this feature appears to be a manifestation of cooling by melting just below the 0°C level, which has a prominent impact due to the broad areal coverage of stratiform precipitation.

We conclude, then, that the rather subtle features in the Q_1/Q_2 profiles of Yanai et al. (1973) are directly linked to effects of stratiform precipitation systems, years ago identified by Zipser (1969, 1977) and Houze (1977) as integral components of tropical convection. These systems cover large areas and leave behind perturbations to the temperature and moisture stratification in the tropics, which are readily detectable using sounding data. With the recent completion of the Dynamics of the Madden–Julian Oscillation (DYNAMO) field campaign in the Indian Ocean (October 2011–March 2012), the opportunity exists to further extend this analysis to another tropical ocean basin.

Acknowledgments. This article is prepared as a tribute to Michio Yanai in honor of his enormous contributions to the field of atmospheric science. We thank Courtney Schumacher and two anonymous reviewers for their helpful comments on the manuscript. The research has been supported by the National Science Foundation under Grants AGS-0966758, AGS-1059899, and AGS-1118141, and the National Aeronautics and Space Administration under Grant NNX10AG81G.

REFERENCES

Brown, R. G., and C. Zhang, 1997: Variability of midtropospheric moisture and its effect on cloud-top height distribution during TOGA COARE. *J. Atmos. Sci.*, **54**, 2760–2774, doi:10.1175/1520-0469(1997)054<2760:VOMMAI>2.0.CO;2.

Ciesielski, P. E., L. Hartten, and R. H. Johnson, 1997: Impacts of merging profiler and rawinsonde winds on TOGA COARE analyses. *J. Atmos. Oceanic Technol.*, **14**, 1264–1279, doi:10.1175/1520-0426(1997)014<1264:IOMPAR>2.0.CO;2.

——, R. H. Johnson, P. T. Haertel, and J. Wang, 2003: Corrected TOGA COARE sounding humidity data: Impact on diagnosed properties of convection and climate over the warm pool. *J. Climate*, **16**, 2370–2384, doi:10.1175/2790.1.

Dopplick, T. G., 1972: Radiative heating of the global atmosphere. *J. Atmos. Sci.*, **29**, 1278–1294, doi:10.1175/1520-0469(1972)029<1278:RHOTGA>2.0.CO;2.

——, 1979: Radiative heating of the global atmosphere: Corrigendum. *J. Atmos. Sci.*, **36**, 1812–1817, doi:10.1175/1520-0469(1979)036<1812:RHOTGA>2.0.CO;2.

Esbensen, S., 1978: Bulk thermodynamic effects and properties of small tropical cumuli. *J. Atmos. Sci.*, **35**, 826–837, doi:10.1175/1520-0469(1978)035<0826:BTEAPO>2.0.CO;2.

——, J.-T. Wang, and E. I. Tollerud, 1988: A composite life cycle of nonsquall mesoscale convective systems over the tropical ocean. Part II: Heat and moisture budgets. *J. Atmos. Sci.*, **45**, 537–548, doi:10.1175/1520-0469(1988)045<0537:ACLCON>2.0.CO;2.

Fitzjarrald, D. R., and M. Garstang, 1981: Vertical structure of the tropical boundary layer. *Mon. Wea. Rev.*, **109**, 1512–1526, doi:10.1175/1520-0493(1981)109<1512:VSOTTB>2.0.CO;2.

Frank, W. M., 1978: The life cycles of GATE convective systems. *J. Atmos. Sci.*, **35**, 1256–1264, doi:10.1175/1520-0469(1978)035<1256:TLCOGC>2.0.CO;2.

Fueglistaler, S., A. E. Dessler, T. J. Dunkerton, I. Folkins, Q. Fu, and P. W. Mote, 2009: Tropical tropopause layer. *Rev. Geophys.*, **47**, 1004, doi:10.1029/2008RG000267.

Hartmann, D. L., H. H. Hendon, and R. A. Houze Jr., 1984: Some implications of the mesoscale circulations in tropical cloud clusters for large-scale dynamics and climate. *J. Atmos. Sci.*, **41**, 113–121, doi:10.1175/1520-0469(1984)041<0113:SIOTMC>2.0.CO;2.

Holloway, C. E., and J. D. Neelin, 2007: The convective cold top and quasi-equilibrium. *J. Atmos. Sci.*, **64**, 1467–1487, doi:10.1175/JAS3907.1.

Houze, R. A., Jr., 1977: Structure and dynamics of a tropical squall-line system. *Mon. Wea. Rev.*, **105**, 1540–1567, doi:10.1175/1520-0493(1977)105<1540:SADOAT>2.0.CO;2.

——, 1982: Cloud clusters and large-scale vertical motion in the tropics. *J. Meteor. Soc. Japan*, **60**, 396–410.

——, 1989: Observed structure of mesoscale convective systems and implications for large-scale heating. *Quart. J. Roy. Meteor. Soc.*, **115**, 425–461, doi:10.1002/qj.49711548702.

Johnson, R. H., 1984: Partitioning tropical heat and moisture budgets into cumulus and mesoscale components: Implications for cumulus parameterization. *Mon. Wea. Rev.*, **112**, 1590–1601, doi:10.1175/1520-0493(1984)112<1590:PTHAMB>2.0.CO;2.

——, and D. C. Kreite, 1982: Thermodynamic and circulation characteristics of winter monsoon tropical mesoscale convection. *Mon. Wea. Rev.*, **110**, 1898–1911, doi:10.1175/1520-0493(1982)110<1898:TACCOW>2.0.CO;2.

——, and G. S. Young, 1983: Heat and moisture budgets of tropical mesoscale anvil clouds. *J. Atmos. Sci.*, **40**, 2138–2147, doi:10.1175/1520-0469(1983)040<2138:HAMBOT>2.0.CO;2.

——, and X. Lin, 1997: Episodic trade wind regimes over the western Pacific warm pool. *J. Atmos. Sci.*, **54**, 2020–2034, doi:10.1175/1520-0469(1997)054<2020:ETWROT>2.0.CO;2.

——, and P. E. Ciesielski, 2000: Rainfall and radiative heating rate estimates from TOGA COARE atmospheric budgets. *J. Atmos. Sci.*, **57**, 1497–1514, doi:10.1175/1520-0469(2000)057<1497:RARHRF>2.0.CO;2.

——, and ——, 2002: Characteristics of the 1998 summer monsoon onset over the northern South China Sea. *J. Meteor. Soc. Japan*, **80**, 561–578, doi:10.2151/jmsj.80.561.

——, ——, and K. A. Hart, 1996: Tropical inversions near the 0°C level. *J. Atmos. Sci.*, **53**, 1838–1855, doi:10.1175/1520-0469(1996)053<1838:TINTL>2.0.CO;2.

Katsumata, M., P. E. Ciesielski, and R. H. Johnson, 2011: Evaluation of budget analysis during MISMO. *J. Appl. Meteor. Climatol.*, **50**, 241–254, doi:10.1175/2010JAMC2515.1.

Kiladis, G. N., M. C. Wheeler, P. T. Haertel, K. H. Straub, and P. E. Roundy, 2009: Convectively coupled equatorial waves. *Rev. Geophys.*, **47**, 2003, doi:10.1029/2008RG000266.

L'Ecuyer, T. S., and G. L. Stephens, 2003: The tropical atmosphere energy budget from the TRMM perspective. Part I: Algorithm and uncertainties. *J. Climate*, **16**, 1967–1985, doi:10.1175/1520-0442(2003)016<1967:TTOEBF>2.0.CO;2.

Leary, C. A., and R. A. Houze Jr., 1979: The structure and evolution of convection in a tropical cloud cluster. *J. Atmos. Sci.*, **36**, 437–457, doi:10.1175/1520-0469(1979)036<0437:TSAEOC>2.0.CO;2.

Lin, X., and R. H. Johnson, 1996: Kinematic and thermodynamic characteristics of the flow over the western Pacific warm pool during TOGA COARE. *J. Atmos. Sci.*, **53**, 695–715, doi:10.1175/1520-0469(1996)053<0695:KATCOT>2.0.CO;2.

Luo, H., and M. Yanai, 1984: The large-scale circulation and heat sources over the Tibetan Plateau and surrounding areas during the early summer of 1979. Part II: Heat and moisture budgets. *Mon. Wea. Rev.*, **112**, 966–989, doi:10.1175/1520-0493(1984)112<0966:TLSCAH>2.0.CO;2.

Mapes, B. E., and P. Zuidema, 1996: Radiative-dynamical consequences of dry tongues in the tropical troposphere. *J. Atmos. Sci.*, **53**, 620–638, doi:10.1175/1520-0469(1996)053<0620:RDCODT>2.0.CO;2.

——, P. E. Ciesielski, and R. H. Johnson, 2003: Sampling errors in rawinsonde-array budgets. *J. Atmos. Sci.*, **60**, 2697–2714, doi:10.1175/1520-0469(2003)060<2697:SEIRB>2.0.CO;2.

Nicholls, M. E., R. A. Pielke, and W. R. Cotton, 1991: Thermally forced gravity waves in an atmosphere at rest. *J. Atmos. Sci.*, **48**, 1869–1884, doi:10.1175/1520-0469(1991)048<1869:TFGWIA>2.0.CO;2.

Nitta, T., 1978: A diagnostic study of the interaction of cumulus updrafts and downdrafts with large-scale motions in GATE. *J. Meteor. Soc. Japan*, **56**, 232–242.

——, and S. Esbensen, 1974: Heat and moisture budget analyses using BOMEX data. *Mon. Wea. Rev.*, **102**, 17–28, doi:10.1175/1520-0493(1974)102<0017:HAMBAU>2.0.CO;2.

Numaguti, A., R. Oki, K. Nakamura, K. Tsuboki, T. Asai, and Y.-M. Kodama, 1995: 4-5 day-period variations and low-level dry air observed in the equatorial western Pacific during the TOGA-COARE IOP. *J. Meteor. Soc. Japan*, **73**, 267–290.

Parsons, D., and Coauthors, 1994: The integrated sounding system: Description and preliminary observations from TOGA COARE. *Bull. Amer. Meteor. Soc.*, **75**, 553–567, doi:10.1175/1520-0477(1994)075<0553:TISSDA>2.0.CO;2.

Raymond, D. J., and H. Jiang, 1990: A theory for long-lived mesoscale convective systems. *J. Atmos. Sci.*, **47**, 3067–3077, doi:10.1175/1520-0469(1990)047<3067:ATFLLM>2.0.CO;2.

Redelsperger, J.-L., D. Parsons, and F. Guichard, 2002: Recovery processes and factors limiting cloud-top height following the arrival of a dry intrusion observed during TOGA COARE. *J. Atmos. Sci.*, **59**, 2438–2457, doi:10.1175/1520-0469(2002)059<2438:RPAFLC>2.0.CO;2.

Rickenbach, T. M., and S. A. Rutledge, 1998: Convection in TOGA COARE: Horizontal scale, morphology, and rainfall production. *J. Atmos. Sci.*, **55**, 2715–2729, doi:10.1175/1520-0469(1998)055<2715:CITCHS>2.0.CO;2.

Schumacher, C., and R. A. Houze Jr., 2003: Stratiform rain in the tropics as seen by the TRMM precipitation radar. *J. Climate*, **16**, 1739–1756, doi:10.1175/1520-0442(2003)016<1739:SRITTA>2.0.CO;2.

——, ——, and I. Kraucunas, 2004: The tropical dynamical response to latent heating estimates derived from the TRMM precipitation radar. *J. Atmos. Sci.*, **61**, 1341–1358, doi:10.1175/1520-0469(2004)061<1341:TTDRTL>2.0.CO;2.

——, M. H. Zhang, and P. E. Ciesielski, 2007: Heating structures of the TRMM field campaigns. *J. Atmos. Sci.*, **64**, 2593–2610, doi:10.1175/JAS3938.1.

Shie, C.-L., W.-K. Tao, J. Simpson, and C.-H. Sui, 2003: Quasi-equilibrium states in the tropics simulated by a cloud-resolving model. Part I: Specific features and budget analysis. *J. Climate*, **16**, 817–833, doi:10.1175/1520-0442(2003)016<0817:QESITT>2.0.CO;2.

Steiner, M., R. A. Houze, and S. E. Yuter, 1995: Climatological characterization of three-dimensional storm structure from operational radar and rain gauge data. *J. Appl. Meteor.*, **34**, 1978–2007, doi:10.1175/1520-0450(1995)034<1978:CCOTDS>2.0.CO;2.

Stull, R. B., 1985: A fair-weather cumulus cloud classification scheme for mixed-layer studies. *J. Climate Appl. Meteor.*, **24**, 49–56, doi:10.1175/1520-0450(1985)024<0049:AFWCCC>2.0.CO;2.

Webster, P. J., and R. Lukas, 1992: TOGA COARE: The Coupled Ocean–Atmosphere Response Experiment. *Bull. Amer. Meteor. Soc.*, **73**, 1377–1416, doi:10.1175/1520-0477(1992)073<1377:TCTCOR>2.0.CO;2.

Willis, P. T., and A. J. Heymsfield, 1989: Structure of the melting layer in mesoscale convective system stratiform precipitation. *J. Atmos. Sci.*, **46**, 2008–2025, doi:10.1175/1520-0469(1989)046<2008:SOTMLI>2.0.CO;2.

Yanai, M., 1961: A detailed analysis of typhoon formation. *J. Meteor. Soc. Japan*, **39**, 187–214.

——, S. Esbensen, and J.-H. Chu, 1973: Determination of bulk properties of tropical cloud clusters from large-scale heat and moisture budgets. *J. Atmos. Sci.*, **30**, 611–627, doi:10.1175/1520-0469(1973)030<0611:DOBPOT>2.0.CO;2.

Yoneyama, K., and T. Fujitani, 1995: The behavior of dry westerly air associated with convection observed during TOGA-COARE R/V *Natsushima* cruise. *J. Meteor. Soc. Japan*, **73**, 291–304.

Zipser, E. J., 1969: The role of organized unsaturated convective downdrafts in the structure and rapid decay of an equatorial disturbance. *J. Appl. Meteor.*, **8**, 799–814, doi:10.1175/1520-0450(1969)008<0799:TROOUC>2.0.CO;2.

——, 1977: Mesoscale and convective-scale downdrafts as distinct components of squall-line circulation. *Mon. Wea. Rev.*, **105**, 1568–1589, doi:10.1175/1520-0493(1977)105<1568:MACDAD>2.0.CO;2.

Chapter 2

TRMM Latent Heating Retrieval: Applications and Comparisons with Field Campaigns and Large-Scale Analyses

W.-K. Tao,[a] Y. N. Takayabu,[b] S. Lang,[a,c] S. Shige,[d] W. Olson,[a,e] A. Hou,[a]
G. Skofronick-Jackson,[a] X. Jiang,[f,g] C. Zhang,[h] W. Lau,[i] T. Krishnamurti,[j]
D. Waliser,[f] M. Grecu,[a,k] P. E. Ciesielski,[l] R. H. Johnson,[l] R. Houze,[m] R. Kakar,[n]
K. Nakamura,[o] S. Braun,[l] S. Hagos,[h] R. Oki,[p] and A. Bhardwaj[j]

[a] Mesoscale Atmospheric Processes Laboratory, NASA Goddard Space Flight Center, Greenbelt, Maryland
[b] Atmosphere and Ocean Research Institute, The University of Tokyo, Kashiwa, Chiba, Japan
[c] Science Systems and Applications Inc., Lanham, Maryland
[d] Division of Earth and Planetary Sciences, Kyoto University, Kyoto, Japan
[e] UMBC Joint Center for Earth Systems Technology, University of Maryland at Baltimore, Baltimore, Maryland
[f] Jet Propulsion Laboratory, California Institute of Technology, Pasadena, California
[g] Joint Institute for Regional Earth System Science and Engineering, University of California, California
[h] Rosenstiel School of Marine and Atmospheric Science, University of Miami, Miami, Florida
[i] Earth System Science Interdisciplinary Center, Joint Global Change Research Institute, University of Maryland,
College Park, College Park, Maryland
[j] Department of Meteorology, Florida State University, Tallahassee, Florida
[k] Goddard Earth Sciences Technology and Research, Morgan State University, Baltimore, Maryland
[l] Department of Atmospheric Science, Colorado State University, Fort Collins, Colorado
[m] Department of Atmospheric Science, University of Washington, Seattle, Washington
[n] NASA Headquarters, Science Mission Directorate, Washington, D.C.
[o] Hydrospheric Atmospheric Research Center, Nagoya University, Nagoya, Japan
[p] Earth Observation Research Center, Japan Aerospace Exploration Agency, Tsukuba, Japan

ABSTRACT

Yanai and coauthors utilized the meteorological data collected from a sounding network to present a pioneering work in 1973 on thermodynamic budgets, which are referred to as the apparent heat source (Q_1) and apparent moisture sink (Q_2). Latent heating (LH) is one of the most dominant terms in Q_1. Yanai's paper motivated the development of satellite-based LH algorithms and provided a theoretical background for imposing large-scale advective forcing into cloud-resolving models (CRMs). These CRM-simulated LH and Q_1 data have been used to generate the look-up tables in Tropical Rainfall Measuring Mission (TRMM) LH algorithms. A set of algorithms developed for retrieving LH profiles from TRMM-based rainfall profiles is described and evaluated, including details concerning their intrinsic space–time resolutions. Included in the paper are results from a variety of validation analyses that define the uncertainty of the LH profile estimates. Also, examples of how TRMM-retrieved LH profiles have been used to understand the life cycle of the MJO and improve the predictions of global weather and climate models as well as comparisons with large-scale analyses are provided. Areas for further improvement of the TRMM products are discussed.

Corresponding author address: Dr. Wei-Kuo Tao, NASA/Goddard Space Flight Center, Mesoscale Atmospheric Processes Laboratory (Code 612), Greenbelt, MD 20771.
E-mail: wei-kuo.tao-1@nasa.gov

DOI: 10.1175/AMSMONOGRAPHS-D-15-0013.1

1. Introduction

Release of latent heat during precipitation formation is of immense consequence to the nature of large- and small-scale atmospheric circulations, particularly in the tropics where various large-scale tropical modes dominated by latent heating (LH) persist and vary on a global scale. Latent heat release and its variation are without doubt some of the most important physical processes within the atmosphere and thus play a central role in Earth's water and energy cycle. The launch of the Tropical Rainfall Measuring Mission satellite (TRMM), a joint U.S.–Japan project, in November 1997 made it possible for quantitative measurements of tropical rainfall to be obtained on a continuous basis over the global tropics. TRMM has provided much-needed accurate measurements of rainfall as well as estimates of the four-dimensional structure of LH over the global tropics. Over the last few years, standard LH products from TRMM measurements have become established as a valuable resource for scientific research and applications [see a review by Tao et al. (2006) and the papers published in the *Journal of Climate* special collection on TRMM diabatic heating]. Such products enable new insights and investigations into the complexities of convective system life cycles, diabatic heating controls and feedbacks related to mesoscale and synoptic-scale circulations and their prediction, the relationship of tropical patterns of LH to the global circulation and climate, and strategies for improving cloud parameterizations in environmental prediction models. TRMM's success provided the impetus for another major international satellite mission known as the Global Precipitation Measurement (GPM), launched by NASA and JAXA in February 2014 (http://gpm.nasa.gov). As the centerpiece of NASA's Weather and Global Water/Energy Cycle research programs, GPM consists of a constellation of satellites provided by a consortium of international partners to provide the next generation of spaceborne precipitation measurements with better sampling (3-hourly over a specific location), higher accuracy (with a Ku-Ka band radar), finer spatial resolution (up to $0.1° \times 0.1°$), and greater coverage (from the tropics to high latitudes) relative to TRMM.

LH is dominated by phase changes between water vapor and small liquid or frozen cloud-sized particles. It consists of the condensation of cloud droplets; evaporation of cloud droplets and raindrops; freezing of cloud droplets and raindrops; melting of ice, snow, and graupel/hail; and the deposition and sublimation of ice particles. It is important to keep in mind that eddy heat flux convergence from cloud motions can also redistribute the heating or cooling associated with LH vertically and horizontally. LH cannot be measured directly with current techniques, including current remote sensing or in situ instruments, which explains why nearly all satellite retrieval schemes depend heavily on some type of cloud-resolving model (CRM) (Tao et al. 2006).

However, the apparent heat source or Q_1, of which LH is an important component, can be derived indirectly by measuring vertical profiles of temperature and the associated wind fields from extensive rawinsonde networks through a residual method (Yanai et al. 1973). This residual approach was first described in seminal papers by Professor Yanai (Yanai 1961; Yanai et al. 1973) and expressed by

$$Q_1 = \overline{\pi} \left[\frac{\partial \overline{\theta}}{\partial t} + \overline{\mathbf{V}} \cdot \nabla \overline{\theta} + \overline{w} \frac{\partial \overline{\theta}}{\partial z} \right], \qquad (2\text{-}1)$$

where $\overline{\pi}$ is the nondimensional pressure, $\mathbf{V}$ is the horizontal wind vector, and the overbars denote large-scale horizontal averages. The right-hand side (RHS) is the total derivative of θ, the potential temperature, (times the non-dimensional pressure) measurable from radiosonde data. Here the large-scale vertical motion w is diagnosed from the horizontal winds via the kinematic method with appropriate boundary conditions on w at the surface and the tropopause. There is an accompanying equation for the apparent moisture sink or drying (Q_2), which is similar to Eq. (2-1) except that $\overline{\theta}$ is replaced by water vapor specific humidity ($\overline{q}$) and Q_1 is replaced by negative Q_2. To derive Eq. (2-1), Yanai et al. (1973) stated that "we consider an ensemble of cumulus clouds, which is embedded in a tropical large-scale motion system, then we imagine a horizontal area that is large enough to contain the ensemble of clouds, but small enough to be regarded as a fraction of the large-scale system."

Both the vertical velocity in the third term on the RHS and the horizontal and vertical advection terms on the RHS of Eq. (2-1) have been used to force CRMs (or cumulus ensemble models) to study the response of convective systems to large and mesoscale processes (Soong and Tao 1980). This CRM approach to studying cloud and precipitation processes is called *cloud ensemble modeling* [Soong and Tao 1980; Tao and Soong 1986; Tao et al. 1987; Krueger 1988; Moncrieff et al. 1997; also see review papers by Tao (2003, 2007) and Tao and Moncrieff (2009)]. It allows many clouds of various sizes and stages to exist at any given time. The advantage is that modeled rainfall Q_1 and Q_2 usually agree well with observations (Tao 2003; Randall et al. 2003; and others). The model results also include cloud statistics representing different types of cloud systems over their life cycle. Large-scale forcing derived from many field programs (e.g., GATE, TOGA COARE, SCSMEX, TWP-ICE, and others; see the appendix for expansions used in this chapter) have been used to drive CRMs. These CRM-simulated datasets are especially valuable for LH algorithm developers (see Tao et al. 1990, 1993, 2000, 2006, 2010; Shige et al. 2004, 2007, 2008, 2009; Grecu and Olson 2006).

The Madden–Julian oscillation (MJO) is one of the most prominent climate variability modes and exerts pronounced influences on global climate and weather systems (e.g., Zhang 2005; Lau and Waliser 2011). Current general circulation models (GCMs), however, exhibit rather limited capability in representing the MJO (e.g., Slingo et al. 2005; Lin et al. 2006; Kim et al. 2009). Meanwhile, the fundamental physics of the MJO are still elusive. The essential roles of various diabatic heating components for the MJO have been suggested based on GCM studies, including shallow convective heating (e.g., Zhang and Mu 2005; Benedict and Randall 2009; Li et al. 2009; Zhang and Song 2009), stratiform heating (e.g., Fu and Wang 2009; Seo and Wang 2010), and radiative heating (e.g., Lee et al. 2001; Raymond 2001; Sobel and Gildor 2003). A transition in the vertical heating structure during MJO evolution—namely, from shallow to deep, and then to stratiform—has been reported based on TOGA COARE observations (Lin et al. 2004; Kiladis et al. 2005). However, this vertical tilting structure in the MJO heating field was not clearly evident in sounding observations during the Mirai Indian Ocean Cruise for the Study of the MJO Convection Onset (MISMO; Katsumata et al. 2009), as well as in a composite study over both the Indian and western Pacific Oceans (Morita et al. 2006) and in a case study over the Indian Ocean during the 1998/99 winter (Jiang et al. 2009) based on earlier versions of TRMM heating estimates. A comprehensive characterization of the vertical heating structure of the MJO would be of considerable value in elucidating its essential physics.

This paper describes the second major goal of TRMM of obtaining credible LH estimates as well as their applications within TRMM's zone of coverage, the standard TRMM LH products, and areas for further improvement. Section 2 describes CRM-simulated heating structure estimates in comparison with diagnostic Q_1 estimates based on observed radiosonde profiles of wind, pressure, and temperature. Section 3 gives an overview of the five LH retrieval algorithms developed for TRMM applications while section 4 details the relevant field campaigns used in their development as well as efforts by algorithm developers to validate their LH algorithms. Section 5 highlights applications of the LH products including comparisons with large-scale reanalyses regarding the life cycle of the MJO and improving monsoon forecasts and the physics of global models. Finally, the conclusions as well as final remarks intended to stimulate further research are given in section 6.

2. Relating quantitative cloud heating estimates to CRMs

CRMs are one of the most important tools used to establish quantitative relationships between diabatic heating and rainfall. This is because LH is dominated by phase changes between water vapor and small, cloud-sized particles; these particles as well as their changes are difficult to detect directly using remote sensing techniques (although some passive microwave frequencies respond to path-integrated cloud water and *CloudSat* can detect such particles in the tops of clouds). CRMs, however, employing sophisticated microphysical schemes (that are by no means yet perfect) can explicitly simulate the conversion of cloud condensate into raindrops and various forms of precipitating ice. It is these different forms of precipitation that are most readily detected from space, and which ultimately reach the surface in the form of rain in the tropics. CRMs have been used for TRMM for both rainfall and heating retrieval algorithm development.

Under the Boussinesq approximation, the heat (temperature) budget can be explicitly calculated by a CRM (e.g., Tao and Simpson 1989):

$$Q_1 - Q_R = \overline{\pi}[-(1/\overline{\rho})(\partial\overline{\rho}\overline{w'\theta'}/\partial z) - \overline{\nabla \cdot V'\theta'}]$$
$$+ (1/c_p)[L_v(c - e) + L_f(f - m) + L_s(d - s)], \quad (2\text{-}2)$$

where the primes indicate deviations from the large-scale environment due to smaller-scale cloud processes. The variable θ is potential temperature, $\overline{\rho}$ is density, $\overline{\pi} = (p/p_{00})^{R/c_p}$ is nondimensional pressure (where p and p_{00} are dimensional and reference pressures, respectively, with p_{00} taken as 1000 hPa), and c_p and R are the specific heat of dry air at constant pressure and the gas constant of dry air, respectively. The variables L_v, L_f, and L_s are the latent heats of condensation, freezing, and sublimation, respectively, while the variables c, e, f, m, d, and s are the condensation of cloud droplets; evaporation of cloud droplets and rain drops; freezing of water droplets and rain drops; melting of ice crystals, snowflakes, graupel, and hail; deposition of ice crystals; and sublimation of all ice hydrometeors, respectively. The quantity $(1/c_p)[L_v(c - e) + L_f(f - m) + L_s(d - s)]$ is defined as the LH due to microphysical phase changes while the first two terms on the RHS of Eq. (2-2) are the vertical and horizontal eddy heat flux divergence, respectively. The horizontal divergence term is neglected when Eq. (2-2) is spatially averaged over an area suitable for diagnostic analysis.

Figure 2-1 shows CRM-simulated time-domain mean profiles of heating/cooling due to the individual microphysical processes (i.e., condensation, evaporation, deposition, sublimation, melting, and freezing) in the convective and stratiform regions of a tropical MCS using the Goddard Cumulus Ensemble model (GCE;

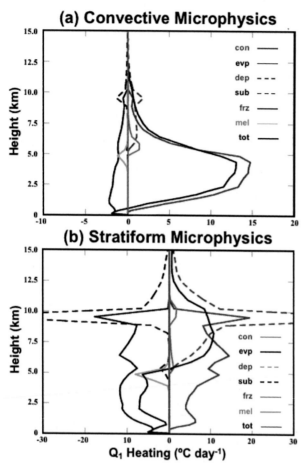

FIG. 2-1. Goddard Cumulus Ensemble model (GCE)-simulated time-mean profiles of LH components averaged over the (a) convective and (b) stratiform region. The components consist of condensation (solid red), evaporation (solid blue), deposition (dashed red), sublimation (dashed blue), freezing (solid orange), melting (solid turquoise), and total (solid black).

Tao and Simpson 1993). Condensation and evaporation have the largest magnitudes in the convective region with evaporation and sublimation about one-third the values of the condensation and deposition rates, respectively. Melting and freezing are small compared to condensation, evaporation, deposition, and sublimation; however, melting is responsible for converting precipitating ice to rain, which can then fall to the surface. Figure 2-2 shows vertical profiles of LH, vertical eddy heat flux divergence, radiation, and Q_1 averaged over a 9-day period during SCSMEX over the northern enhanced sounding array (NESA). LH is the largest term in the Q_1 budget via the heat released by condensation and deposition (as shown in Fig. 2-1). Its peak is around 6.5 km. The radiative term (Q_R) accounts for about 1°–3° of cooling per day. The eddy transport is the smallest term, but it does redistribute heat through cloud updrafts and downdrafts. The CRM-simulated Q_1 profile is

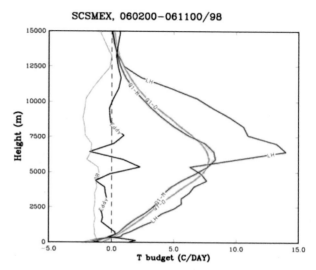

FIG. 2-2. GCE-simulated time-domain-mean profiles of net condensation or LH ($c - e + d - s + f - m$, red), eddy heat flux divergence (blue), Q_R (yellow), and Q_1 (purple). The observed Q_1 (green) estimated from a sounding network is also shown for comparison. Adapted from Tao (2007).

in very good agreement with the observed [i.e., Fig. 3 in Tao (2007)].

3. Overview of the TRMM LH retrieval algorithms

The primary TRMM instruments used to measure rainfall are the TRMM Microwave Imager (TMI), Precipitation Radar (PR), and the Visible and Infrared Scanner (VIRS; Kummerow et al. 1998; for additional details see http://trmm.gsfc.nasa.gov). Five different TRMM LH algorithms designed for applications with satellite-estimated surface rain rate and precipitation profile inputs have been developed, compared, validated, and applied in the past decade [see a review by Tao et al. (2006)]. They are the 1) Goddard convective–stratiform heating (CSH) algorithm, 2) spectral latent heating (SLH) algorithm, 3) Goddard TRAIN (Trained Radiometer) algorithm, 4) hydrometeor heating (HH) algorithm, and 5) precipitation radar heating (PRH) algorithm. CRM-simulated vertical heating profiles are required in the form of look-up tables (LUTs) for the CSH, SLH, and TRAIN heating algorithms. CRM-simulated rainfall and vertical heating structures are also used for these heating algorithms for validation via consistency checks. Neither the HH nor the PRH algorithm uses precalculated LH profiles in LUTs. Instead, both schemes estimate the net flux of water mass into (out of) layers and assume that under steady-state conditions net fluxes are compensated for by a local decrease (increase) of hydrometeors by microphysical processes. Thus, a decrease in mass is associated with

TABLE 2-1. Summary of the five LH algorithms [see Tao et al. (2006) for further details and salient references]. Data inputs, retrieved products, and salient references are included. The conventional relationship between Q_1 (apparent heat source), LH, and Q_R (radiative heating) is expressed by $Q_1 - Q_R = LH + EHT$, where the final term represents eddy heat transport by clouds (vertically integrated EHT is zero; i.e., it provides no explicit influence on surface rainfall). TMI is the TRMM Microwave Imager and PR the TRMM Precipitation Radar.

	Required TRMM data	Algorithm products	Key references in algorithm description	Algorithm developers
CSH (convective-stratiform heating)	PR, TMI, PR-TMI	Q_1, LH, Q_2	Tao et al. (1990, 1993, 2000, 2001, 2010)	W.-K. Tao and S. E. Lang
SLH (spectral latent heating)	PR	LH, $Q_1 - Q_R$ Q_2	Shige et al. (2004, 2007, 2008, 2009)	S. Shige and Y. N. Takayabu
TRAIN (trained radiometer algorithm)	TMI (PR training)	$Q_1 - Q_R$, LH	Grecu and Olson (2006), Olson et al. (2006) Grecu et al. (2009)	M. Grecu and W. Olson
HH (hydrometeor heating)	PR-TMI	LH	Yang and Smith (1999b), Yang et al. (2006)	E. A. Smith and Y. Song
PRH (precipitation radar heating)	PR	LH	Satoh and Noda (2001)	S. Satoh and A. Noda

evaporation, melting, or sublimation cooling, whereas an increase is associated with condensation, freezing, or deposition heating. The overall strengths and weaknesses of these five different heating algorithms are shown in Table 2 of Tao et al. (2006).

Table 2-1 gives a summary of the five algorithms, including the type(s) of TRMM input data used to generate their associated heating product(s), the type of heating product(s) produced, and the salient reference(s) describing their design. Additional improvements made to the SLH, TRAIN, and CSH algorithms as well as brief descriptions of the HH and PRH algorithms are presented next.

a. The CSH algorithm

Diagnostic budget studies (e.g., Houze 1982; Johnson 1984) and cloud modeling studies [see review by Tao (2003)] have shown that characteristic LH profiles in the stratiform regions of tropical MCSs are considerably different than the characteristic LH profiles in the convective regions. In general, for both observed and simulated convective systems, evaporative cooling in the lower troposphere below a bow-shaped positive heating profile in the middle and upper cloud layers (peaking in the middle to upper troposphere) is the dominant feature within stratiform precipitation regions (i.e., the archetypical reverse S-shaped stratiform LH profile), while a combination of vertically continuous condensation and deposition heating (peaking in the middle troposphere) is the dominant signature for convective rain areas (i.e., the archetypical, deep, all-positive, bow-shaped convective LH profile). Based on these findings, the CSH algorithm was developed and described by Tao et al. (1993).

Recently, the CSH algorithm was redesigned and improved (Tao et al. 2010). The key difference between the new and old versions (Tao et al. 1993, 2000, 2001) involves the new LUTs[1] and how they are accessed. First, there are many more heating profiles (approximately 700 total compared to 20 in the previous version[2]) in the new LUTs due to their being separated into detailed intensity and stratiform bins. And second, the profiles are distributed and thus accessed according to conditional rain rates. Together these lead to several potential advantages regarding heating structure. Obviously, having many more profiles in the LUTs allows for the possibility of having many more heating structures. For example, rather than just having shallow (i.e., <5 km) or deep heating profiles, the new LUTs allow the depth of heating to vary considerably.[3] Using conditional rain rates is what allows those structures to be better differentiated. For example, given a stratiform fraction and an average rain rate over a region (i.e., a $0.5° \times 0.5°$ area), knowing that average rain rate is due to a small area of intense rain (e.g., a single intense convective cell) rather than a larger area of weak rain (e.g., a broader field of weaker convective cells) allows the algorithm to select a more representative heating

---

[1] To date, field program data that have been examined in conjunction with the CSH algorithm include 1) GATE, 2) EMEX, 3) PRE-STORM, 4) TOGA COARE, 5) SCSMEX, 6) TRMM-LBA, 7) KWAJEX, and 8) DOE-ARM.

[2] These profiles were obtained by distributing heating/cooling profiles from model subdomains (64 km or the approximate grid size of the TRMM rain retrievals) into the same conditional rain intensity and stratiform percentage bins used to differentiate the surface rainfall distributions. Separate LUTs were constructed for each of the three main components: latent, eddy (horizontal and vertical combined), and radiative.

[3] Mean echo-top heights from the PR and from the model correlate nicely over almost the entire range of LUT bins (not shown).

structure. In the older version, these two rain areas would have been treated the same. The newer LUTs include CRM-generated LH, eddy heating, and radiative heating/cooling at common levels on a common grid. They can thus easily provide the eddy and radiative terms to other LH algorithm groups[4] (at the same rainfall intensity and stratiform percentage).

b. The SLH algorithm

Spectral representation of precipitation profiles obtained from the PR algorithm by use of a small set of distinct profile properties, as reported by Takayabu (2002), provide the basis for the SLH algorithm, which was introduced and modified by Shige et al. (2004, 2007, 2008, 2009). This algorithm is currently intended for use with PR-retrieved rain rate profiles only and estimates LH, $Q_1 - Q_R$, and Q_2. Akin to the CSH algorithm, a set of three LUTs is produced using the GCE associated with three types of rainfall: 1) convective, 2) shallow-stratiform, and 3) anvil. Specifically, however, the LUTs are indexed according to vertical rain profile information: precipitation top height (PTH) for convective and shallow stratiform rain and melting-level rain intensity for anvil (deep stratiform with a PTH higher than the melting level) rain. The nomenclature "spectral" stems from the spectrally indexed table, designed to reduce the dependency on GCE/CRM simulations from specific field campaigns.

In the latest version of the SLH algorithm, deep stratiform rain is further divided into two new categories: deep stratiform with decreasing precipitation from the melting level toward the surface and deep stratiform with increasing precipitation from the melting level toward the surface (Shige et al. 2013). It computes deep stratiform cooling magnitudes as a function of Pm (melting level) − Ps (surface rain rate), assuming the evaporative cooling rate below the melting level in deep stratiform regions is proportional to the reduction in the precipitation profile toward the surface from the melting level (based on 1D water substance conservation). However, increasing precipitation profiles are found in some portions of stratiform regions, especially in regions adjacent to convective regions where 1D water substance conservation may be invalid. An LUT[5] for deep stratiform

with increasing precipitation toward the surface from the melting level is produced with the amplitude determined by Ps.

c. The TRAIN algorithm

The TRAIN heating algorithm is designed specifically for application with TMI passive microwave (PMW) radiance observations. First, precipitation and heating profiles are derived from PR reflectivity profiles, using a method similar to that of Shige et al. (2004), over a one-month span of PR observations. In this method, month-long CRM (i.e., GCE) simulations of precipitation/heating during SCSMEX (18 May–17 June 1998), TOGA COARE (19 December 1992–18 January 1993), and KWAJEX (6 August–5 September 1999) are used to relate vertical reflectivity structure and surface rain rate to vertical heating structure. Since TMI-observed microwave brightness temperatures (Tb) are collocated with PR observations over the PR swath, TMI Tb are assigned to each precipitation/heating profile in the large PR-derived database. The database then serves as a kind of LUT to be used in a Bayesian method to estimate precipitation and LH from the TMI. Given a set of TMI-observed Tb, an estimated precipitation/heating profile is constructed by compositing database precipitation/heating profiles associated with Tb values that are consistent with the TMI-observed Tb values and their uncertainties.

Originally developed for application with SSM/I data, the Bayesian method was adapted for application with TMI radiances and integrated within the GPROF TMI precipitation retrieval algorithm (see Olson et al. 1999, 2006). Versions of the GPROF heating algorithm were used by Rodgers et al. (1998, 2000) to diagnose the relationship between LH distributions and storm intensification within Hurricane Opal and Supertyphoon Paka. More recently, Grecu and Olson (2006) and Grecu et al. (2009) demonstrated that Q_1 profiles from TRAIN were consistent with independent estimates derived from SCSMEX and MISMO rawinsonde analyses. Note that Q_1 was estimated by combining TRAIN estimates of $Q_1 - Q_R$ with Q_R estimates from the Hydrologic Cycle and Earth Radiation Budget (HERB) algorithm of L'Ecuyer and Stephens (2003, 2007).

d. The HH algorithm

The HH algorithm, including its verification and global application, is described in Yang and Smith (1999a,b, 2000). These studies describe how cloud-scale vertical velocity can be estimated using multiple-linear regression based on hydrometeor profile densities as independent input variables. For applications with TRMM level-2 retrievals, the current scheme uses truncated

---

[4] Since the various algorithms produce different heating, it was recommended by the TRMM Latent Heating Working Group at the Fifth TRMM LH workshop (Annapolis, MD, 27–28 Aug 2007) that CSH should be used to provide the eddy and radiative terms to the other algorithms.

[5] It is based on four 9-day (10–18 Dec 1992, 27 Dec 1992–4 Jan 1993, 9–17 Feb 1993, and 18–26 Feb 1993) and one 8-day (19–26 Dec 1992) TOGA COARE CRM simulation(s).

TABLE 2-2. Location, duration, and references of field campaigns. One of the major objectives of SCSMEX, KWAJEX, and LBA was to provide forcing for CRMs and validation for TRMM LH profiles.

Field experiment	Location	Period	Reference
GATE	Tropical Atlantic	26 Jun–19 Sep 1974	Houze and Betts (1981)
TOGA COARE	Equatorial west Pacific	1 Nov 1992–28 Feb 1993	Webster and Lukas (1992)
SCSMEX (N and S)	South China Sea	2–25 May and 5–22 Jun 1998	Lau et al. (2000)
LBA	Amazonia	1 Nov 1998–28 Feb 1999	Silva Dias et al. (2002)
KWAJEX	Marshall Islands	24 Jul–15 Sep 1999	Yuter et al. (2005)
TWP-ICE	Darwin	21 Jan–12 Feb 2006	May et al. (2008)
MISMO	Equatorial Indian Ocean	24 Oct–25 Nov 2006	Yoneyama et al. (2008)
ARM-SGP-97	Central United States	18 Jun–17 Jul 1997	Ackerman and Stokes (2003)
ARM-SGP-02	Central United States	25 May–15 Jun 2002	Ackerman and Stokes (2003)

Legendre polynomial representations of precipitation mass fluxes from the surface to precipitation top height before taking vertical derivatives, thus preventing retrieval noise from producing unrealistic heating rates. For applications with PR data, no account is made for LH by deposition/sublimation and freezing/melting above and below the melting level since the sensitivity of the PR is only 17 dBZ, which is insufficient for detection of most frozen precipitation, particularly for small and/or less dense graupel particles. For applications with TMI data, terminal velocities of precipitating rain and graupel are calculated assuming that both size spectra are distributed according to a Marshall–Palmer distribution.

e. The PRH algorithm

The PRH algorithm uses PR-based retrievals (precipitation profiles and convective/stratiform rain classification) to estimate the vertical LH structure (Satoh and Noda 2001). It requires an initial-guess vertical velocity profile that is used to evaluate a hydrometeor conservation equation under steady-state conditions. In stratiform regions, the LH profile is derived directly from the hydrometeor conservation equation (similar to the HH algorithm). In convective regions, if a net increase of hydrometeors due to microphysics is inferred from the conservation equation, then the associated LH profile is calculated based upon the vertical motion profile, assuming saturated adiabatic ascent. An iterative method is then used to adjust the original vertical motion profile to ensure that the vertically integrated net heating and surface rain rate are consistent.

4. Field campaigns and validation

As discussed in section 1, advective forcing in temperature and water vapor have been used as forcing for CRMs to simulate cloud and precipitation properties including LH, Q_1, and Q_R for TRMM LH algorithm developers. These simulated LH profiles including their convective and stratiform components and their

relationship to precipitation have been used to generate LUTs for LH algorithms. In addition, these simulated data and their associated observed Q_1 have been used for validation. This section briefly describes the GCE simulations, field data used, and validation of LH algorithms.

a. Relevant campaigns and their environment

Table 2-2 shows the location, duration, and references for the various field campaigns used in the study and development TRMM LH algorithms. SCSMEX was conducted in May–June 1998. Two major convective events, prior to and during monsoon onset (18–26 May 1998) and post monsoon onset (2–11 June 1998), were observed. The SCSMEX forcing data were obtained from a variational analysis approach (Zhang and Lin 1997; Zhang et al. 2001) and used to drive the GCE for 44 days starting at 0600 UTC 6 May 1998. TOGA COARE was conducted from November 1992 through February 1993 over the central Pacific. The most intense convection during TOGA COARE occurred in mid and late December 1992, prior to the peak in a westerly wind burst around 1 January 1993. Several major convective events occurred around 11–16 and 20–25 December 1992, mainly due to the low-level large-scale convergence of easterlies and westerlies (Lin and Johnson 1996). For TOGA COARE, the large-scale forcing used in the GCE was derived from the intensive flux array (IFA) sounding network (Ciesielski et al. 2003). GATE was conducted in 1974 over the east Atlantic. Cloud systems (nonsquall clusters, a squall line, and scattered convection) for the period 1–8 September 1974 during phase III of GATE have also been simulated using the GCE (Li et al. 1999; Tao 2003). Large-scale GATE forcings from Sui and Yanai (1986) were used to drive the GCE. The environmental conditions for SCSMEX, TOGA COARE, and GATE can be found in Tao et al. (2004). The TOGA COARE surface flux algorithm (Fairall et al. 1996) was used to calculate sea surface fluxes for these oceanic cases.

TABLE 2-3. CRM-simulated rainfall amount and stratiform percentage (%) for SCSMEX (1998), ARM (1997, 2002), TOGA COARE (1992), and GATE (1974). Adapted from Tao et al. (2010).

Field campaign	Simulated rainfall amount (mm day^{-1})	Stratiform rain percentage (%)	Estimated rainfall amount (mm day^{-1})
SCSMEX (NESA)	12.31	42.6	11.35
ARM (1997)	4.31	41.3	4.32
ARM (2002)	4.85	36.0	4.77
TOGA COARE (1992–93)	7.72	47.6	9.32
GATE (1974)	10.56	41.4	11.38

KWAJEX was sponsored by NASA in cooperation with the U.S. Army Kwajalein Atoll/Kwajalein Missile Range and NOAA and was conducted from 23 July to 15 September 1999. It was designed to obtain an empirical physical characterization of precipitating convective clouds over the tropical ocean and to improve physical assumptions made within the TRMM satellite algorithms. TRMM LBA took place in Amazonia in Brazil and focused on the dynamical, microphysical, electrical, and diabatic heating characteristics of tropical convection in the region. Diagnostic analyses from sounding data for KWAJEX and TRMM LBA are reported in Schumacher et al. (2007). TWP-ICE was a comprehensive observing campaign around Darwin, Australia, to study weather and climate change through improved understanding and modeling of cloud and aerosol processes in tropical cloud systems (May et al. 2008); the large-scale forcing derived from its sounding array was described in Xie et al. (2010). The GCE has been used to study convective systems from LBA (Lang et al. 2007, 2011), TWP-ICE (Zeng et al. 2011, 2013), and KWAJEX (Zeng et al. 2008, 2009a,b, 2011).

The ARM program established the SGP site to observe clouds and precipitation for climate research. The site is centered at 36.6°N, 96.5°W. Two summer field campaigns were conducted at the site in 1997 and 2002 and are referred to here as ARM-SGP-97 and -02. The ARM forcing data were also obtained from the variational analysis approach of Zhang and Lin (1997) and Zhang et al. (2001). Surface fluxes taken from site-wide averages of observed fluxes from the ARM energy balance Bowen ratio (EBBR) stations were imposed into the GCE (Zeng et al. 2007, 2011). The ARM-SGP-97 numerical simulation was started at 2330 UTC 18 June 1997 and lasted for 29 days. The ARM-SGP-02 simulation was started at 2030 UTC 25 May 2002 and lasted for 20 days. For the ARM cases, the surface wind did not interact with the boundary layer.

Table 2-3 shows grid-averaged total rainfall and stratiform rain percentage for each of the GCE-simulated cases used to generate the initial CSH version 2 (Tao et al. 2010) LH LUTs. The oceanic cases have more rainfall than the continental ones. This is due primarily to the fact that the oceanic environments have higher precipitable water contents (i.e., more moisture) than the continental [see Table 1 in Tao et al. (2004)]. That is why the SCSMEX simulation has the largest rainfall amount. Although the TOGA COARE environment is generally moister than that for GATE, it has less rainfall because the model simulation starts in November, which did not have many active convective events. In general, the tropical oceanic cases should have a higher stratiform amount (i.e., 40%–50%) than the midlatitude continental cases. However, the ARM cases also have a large stratiform rain fraction (from 36% to 41%) because they include frontal cases. Houze (1997), Zipser et al. (1981), and Gamache and Houze (1983) estimated that widespread stratiform rain accounted for about 32%–49% of the total rainfall during GATE. The fraction of stratiform rainfall from midlatitude squall lines has been estimated at 29%–43% (Rutledge and Houze 1987; Johnson and Hamilton 1988). The GCE-simulated results are in good agreement with these observations. Figure 2-3 shows the geographic

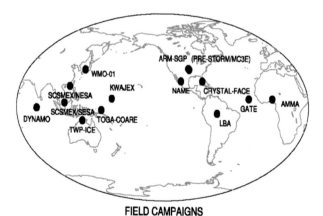

FIELD CAMPAIGNS

FIG. 2-3. Geographic locations of 12 field campaigns used to provide data to drive and evaluate CRM simulations. These include the ARM-SGP (Southern Great Plains) campaigns conducted in the summer of 1997, the spring of 2000, and the summer of 2002; GATE (1974); KWAJEX (1999); TOGA COARE (conducted in 1992 and 1993); TWP-ICE (2006), SCSMEX/NESA and /SESA (1998), AMMA (2006), MC^{3}E (2011), and AMIE/DYNAMO (2011). MISMO has the same location as DYNAMO. (See the appendix for expansions of acronyms.)

locations of field campaigns used to provide data to drive and evaluate CRM simulations.

b. Validation of LH algorithms

Validation of LH profiles retrieved from satellite data is not straightforward because there is no instrument (i.e., no "latent heatometer") or direct means to measure this quantity, and as a result there is no primary calibration standard by which the validation process can be adjudicated. Two methods, consistency checks using CRMs and comparisons with diagnostic budget estimates, have been used for validation.

1) COMPARISON OF CRM HEATING WITH RECONSTRUCTED AND DIAGNOSTIC HEATING

Consistency checks involving CRM-generated heating profiles and both algorithm-reconstructed and diagnostically estimated heating profiles are a useful step in evaluating the performance of a given LH algorithm. In this process, as time-varying CRM-simulated precipitation processes (multiple-day time series) are used to obtain the required input parameters for a given LH algorithm, the algorithm can then be used to reconstruct the actual heating profiles within the CRM simulation using various model quantities (e.g., surface rainfall) as pseudo observations from the model. Finally, both sets of conformal estimates (model and algorithm) can be compared to coincident estimates of diagnostically based heating derived from radiosonde observations. Such observations from various field experiments, as well as simulations of individual precipitation systems, have been used for such consistency checks (Tao et al. 1990, 1993, 2000; Olson et al. 1999, 2006; Shige et al. 2004, 2007, 2008).

It is evident in Fig. 2-4 that the temporal variations of both the CSH- and SLH-reconstructed LH profiles are generally similar to the variations in the GCE simulation profiles.[6] For example, both capture the evolution of a quasi-2-day oscillation, which occurred during the period 1800 UTC 23–1800 UTC 25 December 1992, an oscillation earlier noted by Takayabu et al. (1996). However, as pointed out by Shige et al. (2004), there are noteworthy improvements in the SLH-reconstructed profiles for the shallow-convective stage from 1800 UTC 23 to 0600 UTC 24 December 1992 and in the anvil decay stage from 0600 UTC to 1800 UTC 25 December 1992. Shallow convective heating is more explicitly retrieved by the SLH algorithm because it uses observed

---

[6]Note that the CSH profiles shown here were from an older version that did not use conditional rain rates like the current version as discussed in section 3a.

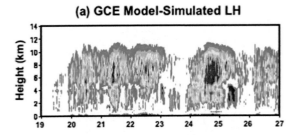

(a) GCE Model-Simulated LH

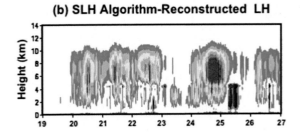

(b) SLH Algorithm-Reconstructed LH

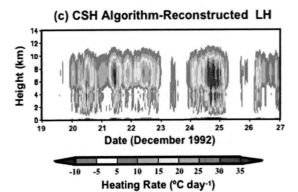

(c) CSH Algorithm-Reconstructed LH

Date (December 1992)

Heating Rate (°C day⁻¹)

FIG. 2-4. Evolution of LH profiles (5-min intervals) over the TOGA COARE IFA for an 8-day period (19–27 December 1992) from the (a) GCE simulation, (b) SLH algorithm reconstruction, and (c) CSH algorithm reconstruction. The contour interval is 5°C day⁻¹. GCE-simulated convective/shallow–stratiform/anvil stratus fractions, surface rain rates (RRs), PTHs, and melting-level RRs are used as inputs to the SLH algorithm with profiles averaged over a 512-km grid mesh. Adapted from Shige et al. (2004).

information on precipitation depth (i.e., the PTH parameter), and heating profiles in the decaying stage without surface rain (e.g., 1200 UTC 25 December) can be retrieved by the SLH algorithm by using the precipitation rate at the melting level. Both the CSH- and SLH-reconstructed results are smoother than the GCE simulations because the associated LUTs contain averaged profiles for each height/rain bin.

2) COMPARISON OF SATELLITE-RETRIEVED HEATING WITH DIAGNOSTICALLY CALCULATED HEATING

One of the TRMM field campaigns, SCSMEX, which included two sounding networks, the NESA and SESA (northern and southern enhanced sounding array), was

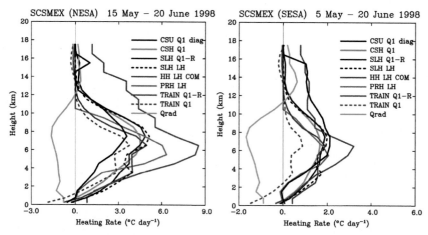

SCSMEX (NESA) 15 May – 20 June 1998   SCSMEX (SESA) 5 May – 20 June 1998

FIG. 2-5. Space–time-averaged heating profiles for the (left) case 1a (SCSMEX-NESA) and (right) case 1b (SCSMEX-SESA) regions. Profiles for different heating terms are obtained from five different satellite algorithms: CSH (solid green), HH (solid violet), TRAIN (solid red) and TRAIN + L'Ecuyer's Q_R (dashed red), SLH (solid and dashed blue), and PRH (solid orange). Q_1 profiles from CSU's diagnostic calculations are the solid black lines (DIAG) from within the NESA/SESA sounding networks. Satellite-derived Q_R profiles from CSU (Qrad, solid turquoise line,) are a gridded product.

conducted in May and June 1998. One of the main underlying scientific objectives of the experiment was to help validate TRMM precipitation and LH algorithms (i.e., vertical profiles of multihydrometeor densities, rain rates, and LH). Diagnostic Q_1 calculations based on the sounding networks were provided by Professor Richard Johnson at Colorado State University (Johnson and Ciesielski 2002; Ciesielski and Johnson 2006).

Two examples of validation results are presented in Fig. 2-5 for the SCSMEX NESA and SESA regions. These diagrams illustrate space–time-averaged vertical profiles of different heating terms obtained from the five different algorithms (i.e., Q_1 from CSH, LH from HH, Q_1 and $Q_1 - Q_R$ from TRAIN, $Q_1 - Q_R$ and LH from SLH, and LH from PRH). In addition, the sounding-diagnosed (DIAG) mean Q_1 profile produced by Colorado State University (CSU) is shown for comparison. For NESA, the results indicate that 1) only SLH exhibits close agreement with the diagnostic (i.e., DIAG) altitude of peak heating; 2) CSH, SLH, and TRAIN show close agreement with each other between low and middle levels; 3) CSH, HH, TRAIN, and PRH exhibit close agreement in the altitude of peak heating among themselves; and 4) all satellite algorithms except HH exhibit relatively close agreement in amplitude of peak heating among themselves, whereas HH exhibits considerably larger amplitude. In the case of SESA, the results indicate that 1) HH and TRAIN exhibit very close agreement with each other in terms of level of peak heating; 2) CSH exhibits close and SLH very close agreement with the DIAG amplitude of peak heating,

although the DIAG peak heating layer is somewhat broader aloft than either those of CSH or SLH (or any algorithm); 3) HH is the only algorithm to exhibit positive upper-level heating similar to DIAG, but it also exhibits the largest amplitude of peak heating relative to the other algorithms; 4) Q_R-augmented TRAIN's Q_1 term exhibits the smallest amplitude of peak heating relative to the other algorithms; 5) PRH's lower-level heating agrees well with DIAG; and 6) all of the other algorithms except TRAIN Q_1 have small low-level heating. It should be noted that in addition to the algorithms themselves, differences between the retrieved and observed profiles could also arise from insufficient satellite sampling of the budget domain both in space and time. The inconsistency of the physical quantities of the results (i.e., having different heating products from different heating algorithms) must also be resolved in future intercomparisons.

Table 2-4 lists the altitude of maximum mean heating for the algorithm retrievals and the diagnostic calculations including other cases from a validation study. For all algorithms except HH, the stratiform percentage is needed as a crucial term in determining the respective altitude of peak heating. Generally, a greater stratiform percentage is associated with a higher altitude of maximum heating. For the HH algorithm, which derives its LH profile from the vertical derivative of total rain mass flux adjusted by any cloud layer lift or descent, its level of peak heating is largely determined by the height at which the rain mass flux begins to decrease upward. For several cases, the altitudes of maximum mean heating

TABLE 2-4. Altitude of maximum mean heating (km). Diagnostic Q_1 is calculated from both within the associated sounding arrays and the gridded rectangular study areas for the two SCSMEX cases but only for the associated sounding arrays for KWAJEX and ARM. The SCSMEX cases have two values: one value is for the entire period over the entire grid and the other only when there is good sounding coverage.

Case	CSH (Q_1)	HH (LH)	TRAIN (Q_1, $Q_1 - Q_R$)	SLH ($Q_1 - Q_R$, LH)	PRH (LH)	Diagnostic (Q_1)
SCSMEX–NESA	6.6	6.6	6.5, 6.6	7.5, 7.5	6.5	7.6, 7.7
SCSMEX–SESA	7.5	6.6	6.7, 6.6	7.6, 7.6	6.0	6.5, 6.7
KWAJEX	6.7	5.5	—, 3.6/6.5 (2 max)	7.5, 7.5	6.6	4.5
ARM–spring 2000	7.0	3.0	—	5.5, 5.6	4.5	6.0
ARM–summer 2002	6.5	5.5	—	5.6, 5.6	5.6	8.1

for the algorithms are within 1 km of the diagnostic peak heating levels. However, greater departures are also found, particularly for the less robust KWAJEX case[7] in which the diagnostic calculation indicates a 4.5-km level of maximum heating. Future work will be required to determine if this seemingly low altitude for maximum heating is actually realistic or a bias in the KWAJEX diagnostic analysis. In addition to mean profiles, CFADs (contoured frequency with altitude diagrams; Yuter and Houze 1995) are another useful way to validate LH profiles by comparing heating PDFs.

5. Applications of LH products

A special collection on TRMM diabatic heating was published in the *Journal of Climate*; it comprises papers that derive, test, and compare different diabatic heating products derived from TRMM data. These papers highlight the challenges in separating contributions from deep convective, stratiform, and shallow convective clouds in using TRMM-derived products to study the distribution of diabatic heating and its impact on atmospheric circulations in the tropics. Table 2-5 lists the authors and titles of the papers published in this special collection. In this section, some of the applications of TRMM heating data are highlighted from these papers.

a. Comparing TRMM algorithm, sounding and reanalysis estimates of latent heating profiles over the tropics

Our knowledge of vertical structures of tropical diabatic heating is limited. Vertical structures of diabatic heating from numerical models, including data assimilation products, are strongly influenced by cumulus parameterization, a significant source of model error and uncertainty. Observational (indirect) estimates of diabatic heating profiles as Q_1 using radiosonde data

(Yanai et al. 1973) or radar data (Mapes and Houze 1995; Mather et al. 2007; Schumacher et al. 2007) from field campaigns are rare and do not provide a global perspective on the long-term means and variability of vertical diabatic heating structures. On the other hand, heating profiles from TRMM retrievals or data assimilation products provide global and long-term coverage. Their reliability must be quantitatively assessed for their proper application. Their similarities and disagreement define an uncertainty envelope of our current knowledge of diabatic heating. Hagos et al. (2010) systematically compared diabatic heating profiles derived from TRMM, sounding observations, and global reanalyses, and their results are summarized here.

1) DATA

Time series of Q_1 estimated from radiosonde observations are available from eight field campaign networks (Table 2-2). All data represent averages over areas of roughly $10^3–10^5 \, km^2$ in different tropical climate regimes (Fig. 2-3), including open ocean with small or no islands (GATE, TOGA-COARE, KWAJEX, and MISMO), coastal and monsoon regions (SCSMEX, TWP-ICE), and continental rain forest (LBA). The time interval of all Q_1 data is 6 h and the vertical levels are from 1000 to 100 hPa with a 25-hPa increment.

Estimates of diabatic heating associated with precipitation and total diabatic heating were made from several global reanalyses, including three recently released high-quality reanalysis datasets [ERA-Interim (hereinafter ERA-I), MERRA, and CFSR] as well as earlier reanalysis datasets (NCEP-2, JRA-25, and ERA-40). All reanalysis products overlap with TRMM from 1 January 1998 to 31 December 2007. For all reanalyses, diabatic heating was estimated as Q_1 from the 3D wind and temperature fields. CFSR and MERRA provide direct output of total diabatic heating (Q_T[8]).

---

[7] KWAJEX had a relatively low ratio of satellite sampling relative to the sounding array.

[8] Q_T is total diabatic heating direct output from reanalyses as a component of temperature tendency. Q_1 is calculated as the residual of the heat budget.

TABLE 2-5. Authors and titles of papers published in the special collection on TRMM diabatic heating in the *Journal of Climate*. Dr. Tony Del Genio was a guest editor for this special collection.

Authors	Topic/title
T. L'Ecuyer and G. McGarragh	A 10-year climatology of tropical radiative heating and its vertical structure from TRMM observations
S. Shige, Y. N. Takayabu, S. Kida, W.-K. Tao, X. Zeng, C. Yokoyama, and T. L'Ecuyer	Spectral retrieval of latent heating profiles from TRMM PR data. Part IV: Comparisons of lookup tables from two- and three-dimensional cloud-resolving model simulations
M. Grecu, W. Olson, C.-L. Shie, T. L'Ecuyer, and W.-K. Tao	Combining satellite microwave radiometer and radar observations to estimate atmospheric heating profiles
W.-K. Tao, S. Lang, X. Zeng, S. Shige, and Y. N. Takayabu	Relating convective and stratiform rain to latent heating
T. Krishnamurti, A. Chakraborty, and A. K. Mishra	Improving multimodel forecasts of the vertical distribution of heating using the TRMM profiles
X. Jiang, D. Waliser, W. Olson, W.-K. Tao, T. L'Ecuyer, J.-L. Li, B. Tian, Y. L. Yung, A. Tompkins, S. Lang, and M. Grecu	Vertical heating structures associated with the MJO as characterized by TRMM estimates, ECMWF reanalyses, and forecasts: A case study during 1998/99 winter
S. Hagos, C. Zhang, W.-K. Tao, S. Lang, B. Olson, Y. Takayabu, S. Shige, M. Katsumata and T. L'Ecuyer	Estimates of tropical diabatic heating profiles: Commonalities and uncertainties
K.-M. Lau and H.-T. Wu	Characteristics of precipitation, cloud, and latent heating associated with the Madden–Julian oscillation
M. Zuluaga, C. Hoyos, and P. Webster	Spatial and temporal distribution of latent heating in the South Asian monsoon region
Y. N. Takayabu, S. Shige, W.-K. Tao and N. Hirota	Shallow and deep latent heating modes over tropical oceans observed with TRMM PR spectral latent heating data
Y.-M. Kodama, M. Katsumata, S. Mori, S. Sato, Y. Hirose, and H. Ueda	Climatology of warm rain and associated latent heating derived from TRMM-PR observations
S. Xie, T. Hume, C. Jakob, S. A. Klein, R. B. McCoy and M. Zhang	Observed large-scale structures and diabatic heating and drying profiles during TWP-ICE
R. H. Johnson, P. E. Ciesielski, T. S. L'Ecuyer, and A. J. Newman	Diurnal cycle of convection during the 2004 North American Monsoon Experiment

The TRMM heating products are available only in regions with precipitation (and hence are predominantly LH). When compared to the TRMM heating retrievals, diabatic heating from the reanalyses is set to zero if there is no precipitation. The focus of the study is the vertical structure of diabatic/LH in the tropics, not its actual magnitude and spatial distribution. For comparison, the TRMM products are regridded onto the $(2.5° \times 2.5°)$ horizontal reanalysis grids and interpolated onto 17 reanalysis pressure levels. In comparing heating profiles from the soundings with those from the TRMM and reanalysis products, one should bear in mind that the estimates from the TRMM and reanalysis products are either purely LH (PRH) or diabatic heating only when there is precipitation (SLH, CSH, TRAIN, NCEP-2, JRA-25, ERA-40, and MERRA), while those from the soundings are purely total heating. For brevity, however, all the variables (Q_T, Q_1,[9] and LH) are referred to as diabatic heating.

2) GENERAL CHARACTERISTICS

Several tropical precipitation regions (Fig. 2-6) were defined to facilitate discussions on regional heating

---

[9] Q_R is relatively small in regions of large precipitation.

characteristics. Global time means of tropical diabatic heating profiles can be perceived as averages over these regions (with LH dominating, suitable for TRMM retrievals) or the entire tropics (with both latent and radiative heating from the reanalyses). Mean profiles averaged over the precipitation regions reveal that the largest disagreement among the TRMM retrievals and reanalyses is low-level heating. Some products (e.g., SLH, TRAIN, NCEP-2, MERRA, and JRA-25) exhibit distinct or even dominant heating peaks below the 700-hPa level, which are very weak or absent in others (Fig. 2-7). Another related disagreement is the number of heating peaks in the vertical. Some products show two or more peaks (e.g., SLH, TRAIN, MERRA, and JRA-25) and others only one. These two major disagreements among the TRMM retrievals and reanalyses can be repeatedly seen in various comparisons with different configurations. Global zonal mean heating profiles from the reanalyses agree with each other well in their contrast between the tropics and extratropics and between the oceans and land (Fig. 2-8). However, large disagreement in their heating peaks, either the level or the number, is obvious.

Another way to characterize the heating profiles is to compare them as functions of the precipitation rate. In the Atlantic region, for example, TRMM and reanalysis

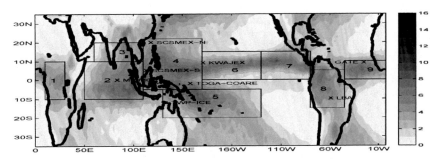

FIG. 2-6. 10-yr mean precipitation from TRMM (3G68, mm day^{-1}). Boxes indicate the analysis domains. Locations of the field campaign sounding sites are marked by an X. Adapted from Hagos et al. (2010).

estimates, except PRH, show their heating peaks becoming elevated as the precipitation rate increases (Fig. 2-9). This relationship between the LH profiles and the precipitation intensity is in agreement with the results of Short and Nakamura (2000), which showed a correlation of 0.71 between PR echo-top height and conditional rain rate over the Atlantic and eastern Pacific Oceans. The increase in heating peak with rain rate is gradual in some products (e.g., SLH, TRAIN, and MERRA) but fast or even abrupt in others (e.g., CSH, ERA-40, JRA-25). Double peaks exist at certain rain rates in PRH, ERA-40, and JRA-25. The only estimates that produce significant stratiform cooling in the lower troposphere at high precipitation rates are PRH and TRAIN. A similar diagnostic was performed over Africa (Hagos et al. 2010; not shown). There is no low-level heating peak in any of the three TRMM estimates (TRAIN has no estimate of heating over land). In the lower tail of precipitation rates, the TRMM estimates have elevated heating and low-level cooling. In the reanalysis estimates, there is an abrupt transition in the diabatic heating profiles with sensible heat fluxes and radiative cooling dominating below about 1 mm day^{-1} and elevated LH at higher precipitation rates, because shallow LH is essentially absent there. Therefore in general, these estimates differ from each other mainly in where their heating peaks are and whether they have just a single peak or double peaks. The differences in the oceanic low-level heating among the TRMM products are, however, in the amount and structure of the shallow LH, which is most abundant in SLH, small in CSH and TRAIN, and essentially absent in PRH. On the other hand, while all the reanalyses have low-level heating peak near the surface, the magnitude and height vary.

3) TEMPORAL VARIABILITY

The temporal characteristics of daily heating profiles can be described in terms of their primary modes of variability. Such primary modes can be extracted using

various forms of empirical orthogonal function (EOF) analysis (Zhang and Hagos 2009; Hagos et al. 2010). Two leading rotated EOF modes—one deep, one shallow—emerge from heating profiles based on sounding observations, TRMM retrievals, and reanalyses (Fig. 2-10). The differences among the mean profiles (Figs. 2-10a,e) are larger than those among the deep modes (Figs. 2-10b,f) as well as the shallow modes (Figs. 2-10c,g). However, there are discrepancies among them. The deep modes of CSH and PRH are outliers in their lack of heating at low levels. The peak of the deep heating of PRH, CSH, and NCEP-1 (at 300 hPa) is higher than that of the sounding average (near 400 hPa). For the shallow modes, PRH has a peak near 600 hPa and JRA-25 at 850 hPa while those of the other estimates as well as the soundings are at 700 hPa.

For the purpose of interpreting the variability characteristics of LH, a diabatic heating profile is designated as due either to radiation, if the vertically integrated diabatic heating is negative, or to latent heat release. In a tropical convective region such as the western Pacific warm pool, the vertical structure of heating is primarily determined by LH. An EOF analysis shows that almost all of the variability in total diabatic heating is due to LH. For both, the first two EOF modes explain about 95% of the variance (Hagos et al. 2010). This is not surprising because, while the vertical structure of LH varies significantly, the profile of clear-sky cooling shows little variability.

An oblique rotated EOF (OREOF) analysis yielded the first mode resembling a stratiform heating profile with low-level cooling and the second OREOF resembles convective heating (Fig. 1 in Schumacher et al. 2007). Almost the entire diabatic heating data are composed of the two profiles. This is not by accident. If indeed mesoscale LH is primarily composed of stratiform and convective heating, they naturally should constitute the large-scale diabatic heating as well. Hagos et al. (2010) demonstrated that the bimodal variability and the structure of the leading EOF modes alone can

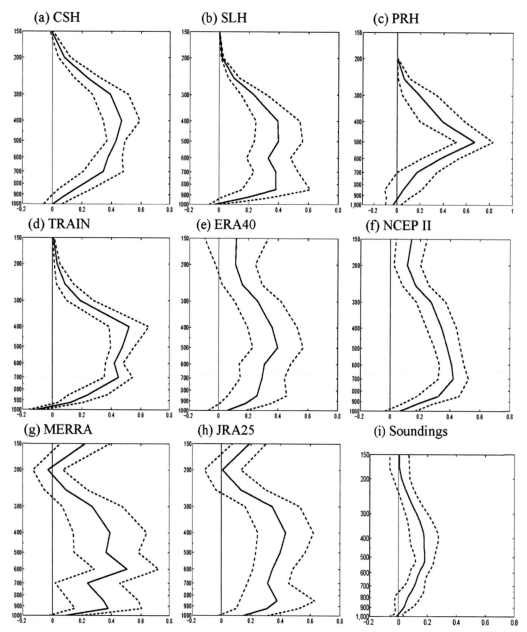

FIG. 2-7. (a)–(h) Normalized mean heating profiles averaged over the tropical precipitation regions (shown by the boxes in Fig. 2-10) and (i) the mean profile of the diabatic heating from all the soundings. The normalization was done by dividing each heating profile by its norm, which is the square root of the sum of the squared heating at all levels. Dashed lines are the standard deviation. Adapted from Hagos et al. (2010).

represent the mean diabatic heating in different climate regimes of the tropics (Fig. 2-11). For all the sounding-based heating time series, the two modes account for almost all their means. It follows that the total heating is primarily composed of these two building blocks.

Comparisons of diabatic/LH derived from in situ soundings, satellite observations, and global reanalyses have revealed that, in general, they agree with each other on their bimodal variability. The common bimodal behavior comes from the composition of large-scale heating by convective and stratiform clouds. This is implicitly built into TRMM LH algorithms that depend on PR reflectivity; hence, the bimodal variability in those products is not surprising. The commonalities among the various products, however, appear to end at the bimodal variability. The structures of the two leading modes, the mean profiles, and the seasonal cycle vary significantly among the products. The large

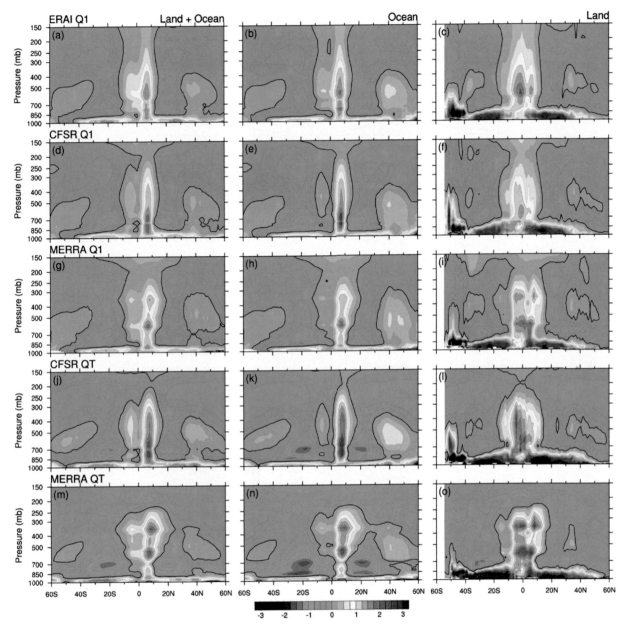

FIG. 2-8. (left) Time and zonal means of land and ocean diabatic heating (K day^{-1}) from (top to bottom) ERAI Q_1, CFSR Q_1, MERRA Q_1, CFSR Q_T, and MERRA Q_T; (center), (right), as at left, but respectively over the oceans only and land only. Adapted from Ling and Zhang (2013).

uncertainties defined by their disagreement inevitably affect their applications. The limited availability of sounding-based heating profiles and the large spatial variability in the vertical structure of diabatic and LH preclude any assumption on the realism of diabatic and LH profiles from TRMM and reanalyses in a region without any observations. Evaluation of diabatic and LH profiles from TRMM and reanalyses must be done in the context of their related large-scale circulation.

b. MJO life cycle

1) VERTICAL DIABATIC HEATING STRUCTURE OF THE MJO THROUGH ITS LIFE CYCLE

Using diabatic heating datasets from three TRMM-based estimates (TRAIN, SLH, CSH) and three recent reanalyses (ERA-I, MERRA, CFS-R), Jiang et al. (2011) conducted a composite analysis of vertical anomalous heating structures associated with the MJO based on

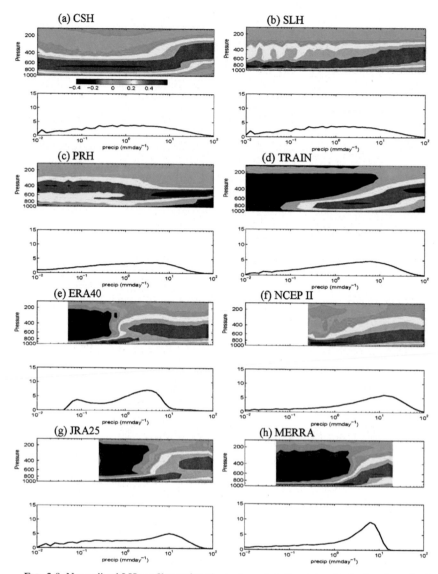

FIG. 2-9. Normalized LH profiles as functions of precipitation intensity (units of standard deviation) and the PDF of precipitation over the Atlantic. The dashed lines indicate a precipitation rate of 1 mm day^{-1}. For the soundings, vertically integrated diabatic heating is used as a proxy for precipitation. Adapted from Hagos et al. (2010).

strong MJO events during boreal winter (November–April) from 1998 to 2007/08. The strong MJO events were selected and their phases (ranging from 1 to 8) determined by the real-time multivariate Wheeler-Hendon index (Wheeler and Hendon 2004; hereinafter the WH index). Figure 2-12 illustrates vertical-temporal anomalous MJO heating profiles (shaded) based on six datasets over the western Pacific (WP; 150°–160°E) and eastern equatorial Indian Ocean (EEIO; 80°–90°E), all averaged over 10°S–10°N. The black curve in each panel denotes the evolution of TRMM 3B42 rainfall anomalies (scales on the right). The results suggest that, over the WP, the heating profiles based on three reanalyses exhibit a

similar vertical tilting structure (Figs. 2-12a–c). The low-level heating below 800 hPa appears around phase 3 and peaks at phase 4 prior to the maximum MJO convection in phase 5. Meanwhile, a maximum heating near 450 hPa after phase 5 is discerned in all three reanalysis datasets. In addition to the upper-level heating maximum, a second peak around 600 hPa is also apparent in MERRA (Fig. 2-12b). In contrast, the vertical tilt in the heating profiles varies among the three TRMM products. While the tilt is evident in CSH, the heating does not extend to the upper troposphere as in other datasets (Fig. 2-12f). Although the emergence of shallow heating prior to maximum convection is also discerned in the SLH

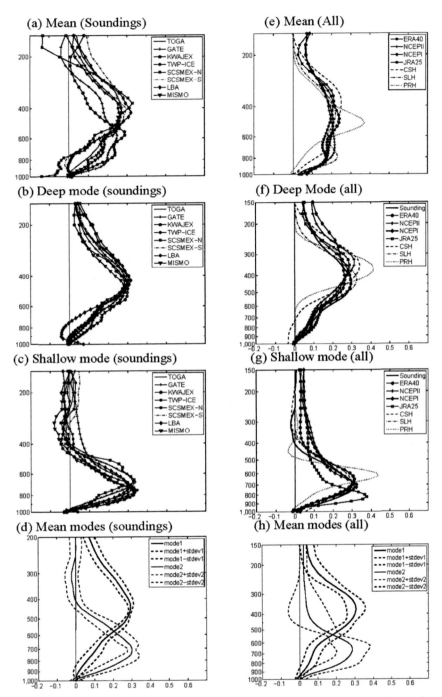

FIG. 2-10. (a) Mean, (b) deep, (c) shallow, and (d) the average of the deep and shallow mode profiles and their standard deviations from the soundings. (e)–(h) As in (a)–(d), but for the reanalyses and TRMM products. Adapted from Hagos et al. (2010).

heating ($Q_1 - Q_R$) profiles, it has much weaker amplitude below 600 hPa (Fig. 2-12e). Meanwhile, a rather weak tilt is seen in the TRAIN profiles (Fig. 2-12d); instead of a slight lag in maximum convection evident in other datasets, the upper-level heating maximum is largely in phase with convection in TRAIN.

Over the EEIO, the transition from a shallow to deep heating structure during MJO evolution is again evident based on the three reanalysis datasets (Figs. 2-12g–2-16i). However, some differences in the upper-level heating profiles are noticed between the EEIO and WP. While the heating maxima around 400 hPa lag the

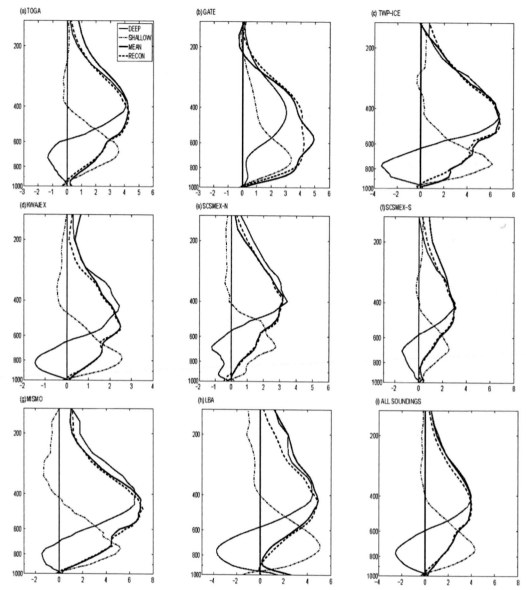

FIG. 2-11. Reconstruction of mean sounding profiles using the first two oblique rotated EOFs. Adapted from Hagos et al. (2010).

rainfall peaks over the WP, they appear with the peaks in MJO convection over the EEIO (cf. Figs. 2-12a–c and 2-12g–i). The vertical transition from shallow to deep heating structures as seen in the reanalyses is not readily apparent in three of the TRMM-based datasets over this region (Figs. 2-12j–l).

Differences in vertical heating structures of the MJO between TRMM estimates and reanalyses are also noted in a similar composite study by Ling and Zhang (2011). By illustrating vertical/temporal MJO heating structures at three longitudes (90°, 120°, and 150°E), but averaged over 15°S–15°N instead of 10°S–10°N as in Fig. 2-12, significant differences in composite vertical

MJO heating structures were found among several reanalysis datasets in addition to differences between reanalyses and TRMM estimates as mentioned above.

2) RAIN AND CLOUD CHARACTERISTICS AND LH PROFILES DURING DIFFERENT PHASES OF THE MJO

Lau and Wu (2010) used TRMM observations to examine the characteristics of clouds, rainfall and LH associated with the eight MJO phases defined by the real-time multivariate WH index. In a 2D cloud–rain probability distribution function (PDF) using brightness temperature (Tb) and echo-top height (ETH) for the

Western Pacific (150–160E; 10S–10N)

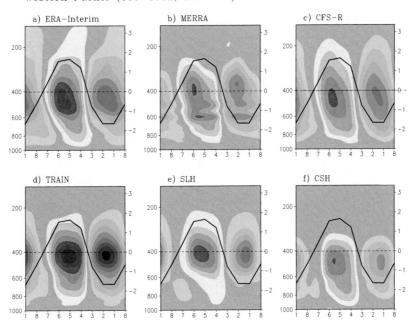

Indian Ocean (80–90E; 10S–10N)

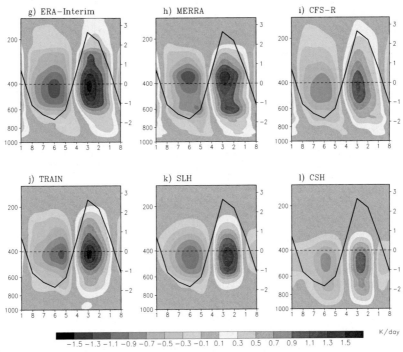

FIG. 2-12. Vertical–temporal (MJO phase) evolution of anomalous heating Q_1 or $Q_1 - Q_R$ for TRMM SLH (shaded, in K day^{-1}) over the (a)–(f) WP (150°–160°E) and (g)–(l) EEIO (80°–90°E) based on three reanalysis datasets and three TRMM estimates. The black curve in each panel represents the evolution of TRMM 3B42 rainfall anomalies (mm day^{-1}; see scales on right y axis). All variables are averaged over 10°S–10°N. The time (MJO phases) in the x axis of each panel runs from right to left so that these plots also mimic longitude–height cross sections for an eastward moving system. Adapted from Jiang et al. (2011).

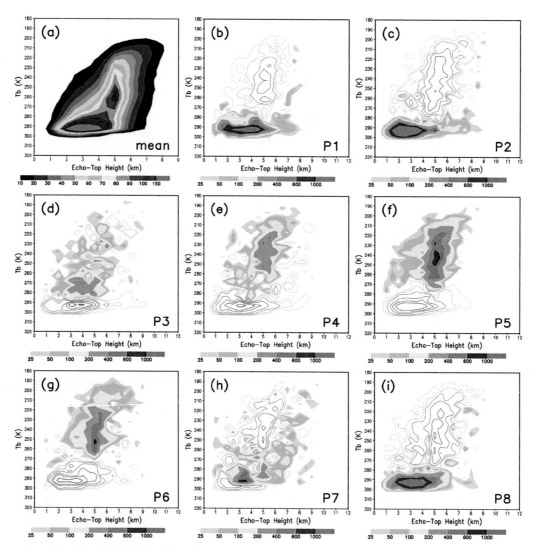

FIG. 2-13. Joint PDF (JPDF) of Tb and ETH over the equatorial western Pacific: (a) mean state of the eight MJO phases and (b)–(i) the difference between the JPDF for each of the eight phases (P1–P8) and the mean state. Positive values are color shaded and negative values are contoured. The units for the mean state are in 0.01% of the total occurrence counts. For P1–P8, the units are number of counts. Adapted from Lau and Wu (2010).

region 10°S–10°N, 120°–150°E (Fig. 2-13), a bimodal distribution with an abundance of warm-low and cold-middle cloud and rain types is evident for all active (amplitude of the WH index >1) phases of the MJO (Fig. 2-13a). The highest population is from the warm-low type, with Tb > 273 K and ETHs below the freezing level (~5 km). The cold-middle type, identified as congestus, has a high population centered rather narrowly near the melting level with a wide range of cloud tops colder than 273 K. Four main regimes—WL (warm-rain low-level cloud), MM (mixed-phase rain, middle-level cloud), CM (cold cloud-top and medium storm height), and CH (cold rain, high-cloud)—defined based on Tb and ETH, are consistent with the four tropical precipitation systems—shallow, cumulus congestus, deep

stratiform, and deep convective—classified in the observed study of Masunaga et al. (2005). There is also a nonnegligible warm rain, middle-level cloud (WM) regime, which counts for about 9% of the total population.

The changes in rain characteristics over the MJO life cycle, shown in anomalous PDFs defined as the deviation of the PDF at a particular phase from that of the mean in (Figs. 2-13b–i), are characterized by the following stages. (a) An abundant occurrence of the WL type (color shaded) and a large deficit in the MM and CM types (black and white contours) during the early build-up stage (i.e., phases 1 and 2) is followed by a transition from a bottom-heavy to a top-heavy distribution with a large increase in MM and CM types representing an increase in mixed-phase precipitation due

TABLE 2-6. The characteristics of MJO phases in terms of brightness temperature (Tb), echo-top height (ETH), and type of cloud systems. WL stands for warm-rain low-level cloud, MM for mixed-phase rain, middle-level cloud, CM for cold cloud-top and medium storm height, and CH for cold rain, high-cloud.

Phase(s)	Life cycle	Tb (K)	ETH (km)	Types of cloud system
1–2	Genesis	290–300	2.5–4.5	Abundant occurrence of WL
3–4	Developing	260–280	4.5	Large increase of MM and CM
5	Mature or peak convection	<275	>5	Large increase of CM and CH
6	Start of decaying	Wide range of Tb	>5	
7	Decaying	<275	>6–7	Increase of WL
8	Similar to phase 1	290–300	2.5–4.5	WL

to developing deep convection (phases 3 and 4). (b) At the maximum large-scale organization (phase 5), the CM and CH types increase considerably, with the coexistence of mixed-phase and ice-phase precipitation and shallow, middle, and deep convective clouds. (c) Deep convection starts to diminish in phase 6, with the convective system dominated by CM and MM types and the presence of both precipitating and nonprecipitating high-level anvil clouds, and continues in phase 7 with a substantial amount of deep convective rain (ETHs above 6–7 km) and appearance of low-level rain. (d) The completion of an MJO cycle occurs with the WL-type rain reestablishing itself in phase 8 and continuing into phase 1 again. Table 2-6 shows the main characteristics of each MJO phase in terms of Tb, ETH, and cloud system type.

Figure 2-14 shows mean TRMM daily LH profiles[10] as well as the anomalous heating (deviation in each phase from the mean) during the eight phases of the MJO cycle. Contributions from heating with ETHs less than and greater than 5 km are shown separately to demonstrate the relative contributions from shallow (liquid-phase and mixed-phase rain) and deep (ice-phase and mixed-phase rain) convection. The magnitude of the mean heating by shallow convection is about 30%–35% of that due to deep convection. In phases 1 and 2, anomalous lower tropospheric heating is due equally to shallow convection (ETH < 5 km) and low-level heating of deep convection (ETH > 5 km). A switch from a bottom-heavy (warm and shallow convective rain) to a top-heavy (mixed convective and stratiform rain) heating profile occurs from phases 2 through 4, consistent with the cloud PDF distributions shown in Fig. 2-13. The anomalous low-level heating from shallow convection in phase 3 is most likely from the abundant mixed-phase rain. During phases 4 and 5, the heating profiles show maximum heating at about 7–8 km and cooling below 2–3 km, typical of that associated with stratiform rain systems (Houze 1989; Tao et al. 2006; Jakob and Schumacher 2008). In the decaying phases

(6 and 7), mid- and upper-tropospheric heating diminishes and low-level heating reverses sign, reflecting the reduction in warm-rain processes with the deep heating profiles changing sign. Phase 8 completes the MJO cycle with a large reduction in deep heating and the beginning of low-level heating processes.

c. Improving the ability of large-scale models to simulate/predict weather and climate

The explicit use of TRMM LH information for initialization and/or assimilation in global models is a relatively new research topic. Further study is needed to quantify how much improvement can be obtained in predictions with LH profile-based data assimilation. However, this is worth considering from a theoretical perspective since it is the vertical distribution of diabatic heating that determines the nature of many low-latitude circulations. Thus, it seems reasonable to expect that improved data assimilation techniques involving accurate LH profile data will ultimately improve NWP forecasts on a consistent basis. Two examples are provided to show how TRMM-retrieved LH profiles are being used as data assimilation variables to bring about prediction improvements in global weather and climate models and to improve the understanding of physical processes in tropical circulations.

1) SHALLOW AND DEEP LATENT HEATING MODES AND THE LARGE-SCALE CIRCULATION

Utilizing $Q_1 - Q_R$ data estimated from the SLH algorithm, Takayabu et al. (2010) showed that nondrizzle precipitation over tropical and subtropical oceans consists of two dominant modes of rain systems: deep systems and congestus. They found that while rain from congestus simply increases with sea surface temperature (SST), deep convective precipitation is inhibited by large-scale atmospheric subsidence even though SSTs are warm enough to support congestus. Figure 2-15 compares $Q_1 - Q_R$ at 7.5 and 2 km: the former represents the effect of deep organized precipitation, while the latter represents the effect of congestus rain. Over the central-to-eastern Pacific, in the Southern

---

[10] Prototype CSH-derived LH was used for this study because the new CSH algorithm was not available at that time.

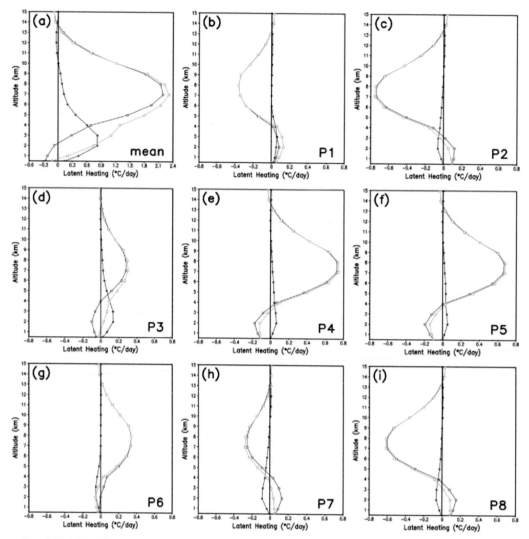

FIG. 2-14. MJO LH profiles based on the CSH algorithm and daily averaged ETH: (a) mean state of the eight MJO phases and (b)–(i) the difference between the heating profile of each phase, P1–P8, and the mean state. The three curves in each panel are red for ETHs < 5 km, blue for ETHs > 5 km, and green for total. Adapted from Lau and Wu (2010).

Hemisphere, along the equator, and also near the Hawaiian Islands, for example, the congestus rain distribution neatly follows the SST distribution, even where deep convective rain is almost completely suppressed.

Figure 2-16 shows 7-yr mean $Q_1 - Q_R$ profiles over 30°N–30°S for September–November stratified against 500-hPa vertical velocity. The results confirm the existence of two dominant modes in tropical nondrizzle precipitation and an effective suppression of the deep mode associated with large-scale subsidence, which is accompanied by middle to lower tropospheric drying. These results are in concert with recent studies indicating significant entrainment of environmental air by cumulus clouds (Sherwood 1999; Zipser 2003; Bretherton et al. 2004; Takemi et al. 2004; Takayabu et al. 2006). A reduction in buoyancy via dry air entrainment prevents

cumulus from penetrating above the freezing level by not allowing them to gain additional latent heat from freezing (Zipser 2003). This is possibly why the development to deep convection is discretized at midlevels.

More recently, Hirota et al. (2011) compared the distributions of tropical precipitation from 19 models as part of CMIP3 and found that the double ITCZ bias is linked to the cumulus convection scheme: the more sensitive the deep convective scheme is to midtropospheric humidity, the less double ITCZ bias the model has. This result is consistent with Del Genio et al. (2012), who attributed successful MJO model simulations to adequate representation of deep convection in relation to the midtropospheric humidity. Hirota et al. (2011) showed that the double ITCZ bias is absent in MIROC5, the current version of the climate model developed by the Atmosphere and

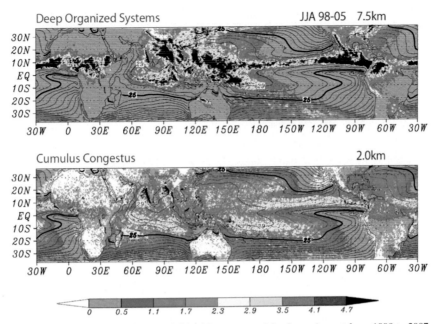

FIG. 2-15. $Q_1 - Q_R$ at (a) 7.5 and (b) 2.0 km averaged for June–August from 1998 to 2007 overlaid on SST (contours). Color scale labels show $Q_1 - Q_R$ values in degrees day^{-1}; SST contour intervals are every 1°C with the 28° and 25°C contours shown in thick lines in the upper and lower panels, respectively. Adapted from Takayabu et al. (2010).

Ocean Research Institute (AORI), National Institute for Environmental Studies (NIES), and the Japan Agency for Marine-Earth Science and Technology (JAMSTEC), with a new entrainment scheme introduced by Chikira and Sugiyama (2010) into the cumulus parameterization. The essential impact of cumulus entrainment on deep precipitation is further examined by Hirota et al. (2014) in a sensitivity study utilizing the atmospheric part of MIROC5. It was confirmed that the entrainment rate controls the double ITCZ even for the same SST distribution.

As shown in Fig. 2-15, congestus heating (Fig. 2-15b) follows the SST, which results in a double ITCZ–like distribution, while the deep heating does not show such a double ITCZ–like distribution (Fig. 2-15a). Deep heating is well suppressed along the equator in the southeastern Pacific, as well as around the Hawaiian Islands, which is consistent with the above interpretation of climate model results. In these two regions, although congestus precipitation is enhanced by relatively warm SSTs, large-scale subsidence with a dry middle troposphere strongly discourages further development of the congestus and suggests that real world cumulus convective systems entrain considerable amounts of environmental air.

2) IMPROVING MONSOON FORECASTS AND MODEL PHYSICS USING THE MULTIMODEL SUPERENSEMBLE APPROACH

The vertical distribution of heating predicted by a suite of global models (Krishnamurti et al. 2007) was improved using a multimodel superensemble technique (Krishnamurti et al. 2000a,b). The same approach but with a suite of mesoscale models in place of global models is being used to construct forecasts of Q_1. A standard version of ARW was used in this study (Krishnamurti et al. 2012). Table 2-7 lists a set of model configurations put together from the available choices of different cumulus parameterizations and microphysics packages within ARW.

The multimodel superensemble has a training and a forecast phase. The training phase covered the period from 1 July to 31 August 2004 and 1 July to 28 August 2005. During the training phase, a superensemble was constructed for the geopotential height z and the temperature T. Note that Q_1 is the substantial derivative of

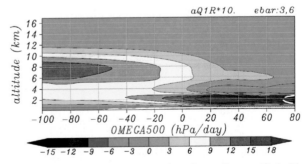

FIG. 2-16. 7-yr conditional mean $Q_1 - Q_R$ profiles stratified with vertical velocity (dp/dt) at 500 hPa averaged for September–November for 30°N–30°S at all longitudes over the ocean in association with all rain. The values for the color scale are scaled by a factor of 10. Adapted from Takayabu et al. (2010).

TABLE 2-7. Numerical experiments conducted by different combinations of cumulus parameterization and microphysics schemes.

Experiment	Cumulus parameterization scheme	Microphysics scheme
Model1	Kain–Fritsch (Kain and Fritsch 2004)	Kessler (Kessler 1969)
Model2	Betts–Miller–Janjic (Janjic 1994, 2000)	Kessler
Model3	Grell–Devenyi ensemble (Grell and Devenyi 2002)	Kessler
Model4	Kain–Fritsch	WSM5 (Hong et al. 2004)
Model5	Betts–Miller–Janjic	WSM5
Model6	Grell–Devenyi ensemble	WSM5

the dry static energy ($gz + c_p T$) where g is gravity and c_p the specific heat of dry air at constant pressure. The computation of Q_1 entails the calculation of the local change and the advective changes (horizontal and vertical advection) of the dry static energy. The observed counterparts of Q_1 for all these forecast time intervals are from the CSH algorithm.

First, the model- and superensemble-based forecasts of precipitation over India were validated. In Figs. 2-17a–d, the vertical bars show skill via the root-mean-square (rms) errors for the six model configurations and the superensemble for each of the forecast days. Figures 2-18a–d show the daily skill based on the area-averaged correlations of the observed and simulated rainfall. The observed rain comes from the rain gauge–based

estimates of Rajeevan et al. (2008). Of interest in Figs. 2-17 and 2-18 is the slow increase in rms errors in the forecasts from day 1 to day 6 and the slow decline of the areal correlations during this forecast period. The multimodel superensemble performs the best in comparison to all of the member models in the forecast suite and exhibits very little decline in the correlations from day 1 to day 6; the rms errors of the multimodel superensemble also do not show much of an increase with forecast time. The model- and multimodel superensemble-based vertical distributions are compared with those from the CSH estimates.

Figures 2-19a–d illustrate the vertical profiles of area-averaged Q_1 (K day^{-1}) over the Indian subdomain (6.85°–25.13°N, 70°–90.17°E). These represent four

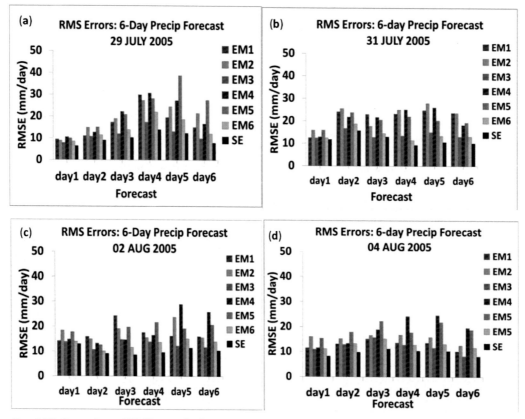

FIG. 2-17. Comparison of RMS errors for forecasts from the super ensemble and six member models over 6 days with initial conditions at (a) 29 July 2005, (b) 31 July 2005, (c) 2 August 2005, and (d) 4 August 2005.

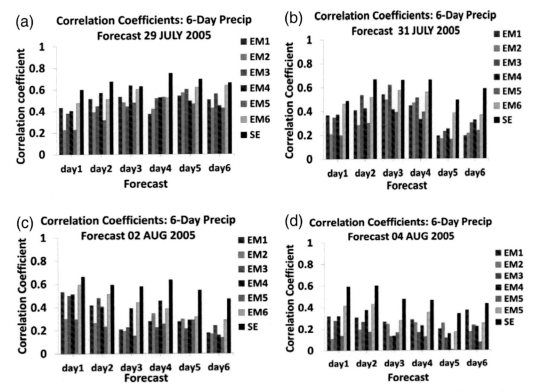

FIG. 2-18. Comparison of spatial correlation coefficients between observed and simulated rain for forecasts from the super ensemble and six member models over 6 days with initial conditions at **(a)** 29 July 2005, **(b)** 31 July 2005, **(c)** 2 August 2005, and **(d)** 4 August 2005.

selected map times during a 4-day forecast phase of the multimodel superensemble. The mesoscale models have higher values for Q_1 than do the CSH values. There are many details in the model-based profiles in the vertical. The satellite-based CSH profiles are smooth and look parabolic along the vertical as was also noted in Krishnamurti et al. (2010). The multimodel super-ensemble recognizes these differences between the CSH and the member model vertical profiles and removes the large differences that are persistent systematic errors. As a result the forecasts through day 4 from the multi-model superensemble come out very close to the CSH profiles. The straight ensemble mean would reside be-tween the forecast profiles of the member models and would contain large errors. In conclusion it is safe to state that given observed measures of heating such as the CSH profiles, it is possible to produce accurate forecasts of Q_1 from the construction of a multimodel superensemble.

6. Summary and future research

a. Summary

This chapter presents some of the recent improve-ments in TRMM LH algorithms and their relationship

with the pioneering works of Yanai et al. (1973). Results from an intercomparison of the LH algorithms are also presented. Differences in the derived heating profiles from the different algorithms, including their associated level of maximum heating, could be due to the physical assumptions as well as the different LUTs (i.e., CRM-simulated heating profiles used to generate the LUTs). This intercomparison will be continued in collaboration with those working on observed heating estimates, which could help to identify the salient physical pro-cesses leading to the similarities and differences produced by the retrieval algorithms. In addition, data from GPM field campaigns and ground validation sites (e.g., MC^{3}E) and others (e.g., TWP-ICE, DYNAMO[11]) that provide extensive and high-quality in situ microphysical observa-tions will be valuable in improving and validating CRM microphysics. This is important because representative

---

[11] In a recent study by R. Johnson and P. Ciesielski, both CSH- and SLH-retrieved heating profiles were found to be in excellent agreement with sounding estimated heating profiles from DYNAMO. However, CSH- and SLH-retrieved LH profiles were quite different for another location over land.

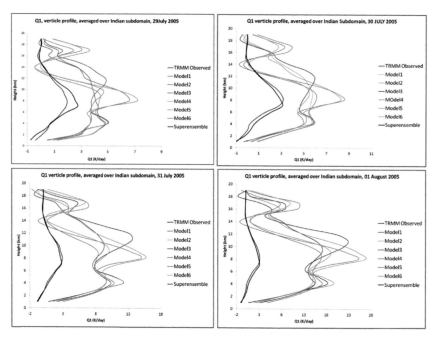

FIG. 2-19. Forecasts of the vertical distribution of heating Q_1 (K day^{-1}) from six mesoscale models, the multimodel superensemble, and the satellite-based CSH algorithm over the Indian subdomain (6.85°–25.13°N, 70°–90.17°E). The respective panels show forecasts in sequence for days 1 through 4.

microphysics is essential in reproducing, within a modeling framework, the key four-dimensional features of LH.

This paper also presented highlights published in the *Journal of Climate* special collection on TRMM diabatic heating. In particular, the comparison of heating profiles derived from TRMM LH algorithms, sounding networks, and reanalyses over the tropics were discussed. One key finding was that the major differences between the heating structures from the various estimates are related to low-level heating and the level of maximum heating. Low-level heating is important to the MJO cycle. Given the uncertainties in TRMM-based diabatic heating estimates, the central role of diabatic heating in the MJO, and the demands for reducing model deficiencies in simulating and forecasting the MJO, there is a great interest and urgent need to examine the MJO vertical heating structure and related processes in current GCMs and to explore how their

structure and fidelity relate to the models' MJO representation and forecast skill. To help address these objectives, a model and observation comparison project on vertical heating structures and diabatic processes associated with the MJO is being organized through a joint effort by the WCRP–WWRP YOTC MJO Task Force and GEWEX Atmospheric System Study Project (Petch et al. 2011; www.ucar.edu/yotc/mjodiab.html).

TRMM-based LH products are also beginning to provide data assimilation guidance in numerical predictions as well as improvements in the underlying cumulus parameterization schemes. For example, these products have been used to identify the importance of the treatment of shallow convection in cumulus parameterization schemes in simulating a more realistic ITCZ. In addition, LH products have been used to improve 4-day monsoon forecasts and model physics with the construction of a multimodel superensemble.

TABLE 2-8. Summary of PMM cloud heating products from the CSH and SLH algorithms. (Orbital heating is not a standard PMM product.)

	Spatial scale	Temporal scale	Algorithm	Products
Gridded	0.5° × 0.5°, 19 vertical layers	Monthly	SLH-PR CSH-Combined	LH, $Q_1 - Q_R$, Q_2
				LH, Q_1, Q_R, Q_2
Orbital	Pixel, 19 vertical layers	Instantaneous	SLH-PR CSH-Combined	LH, $Q_1 - Q_R$, Q_2
				LH, Q_1, Q_R, Q_2
Gridded orbital	0.5° × 0.5°, 19 vertical layers	Instantaneous with time stamps on each grid	SLH-PR CSH-Combined	LH, $Q_1 - Q_R$, Q_2
				LH, Q_1, Q_R, Q_2

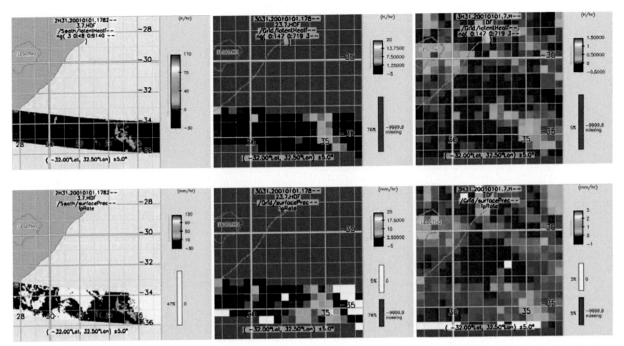

FIG. 2-20. LH products from the version 2 CSH algorithm based on rainfall data from the TRMM combined algorithm: (left) instantaneous pixel scale LH off the southeast coast of Africa 1 Jan 2001 at a height near 2.5 km from the orbital product, (center) same but for the 3G31 gridded (0.5° × 0.5°) orbital product, and (right) same but for monthly mean LH from the 3H31 gridded monthly product. The new CSH algorithm uses conditional rain rates and LUTs based on GCE results divided into fine intensity and stratiform bins (Tao et al. 2010). The corresponding surface rainfall is shown below each of the LH products.

b. Standard LH products

The PMM joint science team has decided to have two standard LH algorithms: the Goddard CSH algorithm and the SLH algorithm. Table 2-8 lists the required data and type of heating products for these two algorithms. Note that one of the major inputs for these standard products is the improved rainfall estimate. Figure 2-20 shows an example of the LH products generated from the new version of the CSH algorithm.

Standard LH products from TRMM will represent a valuable new source of data for the research community, products that, a decade ago, were considered beyond reach. These data products will enable compelling new investigations into the complexities of storm life cycles, diabatic heating controls, and feedbacks related to mesoscale and synoptic-scale circulations and the influence of diabatic heating on Earth's general circulation and climate. In particular, the LH estimates will be of great help as a benchmark for a model intercomparison study on vertical MJO heating structures as shown in section 5b and for the model intercomparison experiment mentioned above. The standard LH products could help to determine how well the model-simulated heating structures agree with observations and determine how different they are relative to the spread of observational (reanalysis and TRMM) values, which would also address the question of how useful the observations are at this point.

c. Future directions

Since temperature (heating) and water vapor (moistening/drying) are closely related (Yanai et al. 1973), it is proposed to produce both heating and moistening profiles using GPM rainfall products. Both LH and moistening profiles are also needed for improving large-scale model simulations and forecasts (Rajendran et al. 2004). GPM will produce higher temporal (3-hourly) and spatial resolution (up to 0.05°) rainfall products. Several issues therefore need to be addressed. CRM results have shown that the horizontal eddy term is quite small if averaged over a large area. In addition, the CRM results have indicated that the horizontal and vertical eddy transport terms usually counteract each other (mass continuity) over small spatial scales (cloud scales). The contribution by horizontal and vertical eddy heat and moisture transport to the heat and moisture budgets must be examined at various horizontal resolutions (e.g., TRMM and GPM satellite footprint sizes). It may be necessary to produce heating and moistening profiles including all of the eddy transport and microphysics terms. Also, the accuracy of the heating retrievals could be affected by differences in the convective–stratiform separation. As

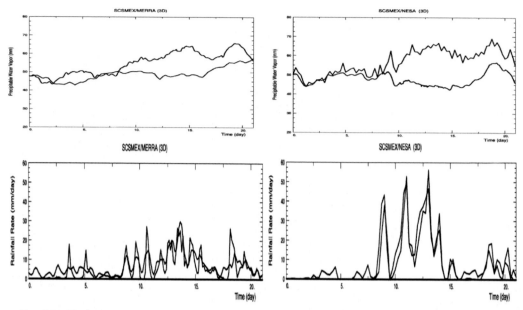

FIG. 2-21. Time series of simulated and observed (top) precipitable water and (bottom) precipitation rate using forcing derived from MERRA. The black lines are model-simulated, and the blue lines are observed. The left panels depict MERRA-forced simulations, and the right panels depict simulations using forcing derived from soundings during SCSMEX (NESA).

such, the separation algorithms used in the LH algorithms should be as close as possible to those applied to the satellite data and at a comparable scale but this is likely only a secondary issue with regard to the current standard algorithms, as their separation schemes are fairly close to the satellite's.

Only a limited number of CRM-simulated cases were used to build the SLH and CSH algorithm (see section 3) LUTs. Observations from additional field experiments (e.g., TWP-ICE, MC³E, DYNAMO) and future GV site(s) will be needed to provide new types of initial conditions to CRMs to expand the number of cases and environments represented in the LUTs. In addition, large-scale reanalysis products such as MERRA can be used both to improve cloud-resolving simulations by placing them in a larger-scale dynamical context and to expand the range of environmental conditions explored by CRM simulations beyond field experiments. Figure 2-21 depicts preliminary results comparing simulations forced by the SCSMEX sounding network versus simulations forced by MERRA on GCE grid boundaries in the same location. The rainfall and precipitable water amounts obtained from simulations forced by MERRA agree with the MERRA values themselves at least as well as (and even better with regard to precipitable water) those obtained from simulations forced by the sounding network do with those from the sounding budget itself. This suggests that the GCE-MERRA approach has the potential to provide

reasonably good-quality simulations to the heating algorithms for a variety of locations and conditions, including those regions with large surface rainfall, such as the Indian Ocean, SPCZ, South America, Africa, and snow events, which are not well represented in the current LUTs. In addition to expanding the number and type of environments, further improvements to the CSH LUTs will be made by using the improved and validated physics in the GCE and running cases at higher resolution. The GCE+MERRA simulated cloud datasets in the Cloud Data Library (http://cloud.gsfc.nasa.gov) can be used to improve the performance of satellite-based rainfall and LH retrievals through more representative LUTs and to improve moist process parameterizations for GCMs.

Acknowledgments. The authors extend their heartfelt appreciation to the late Professor Michio Yanai of the University of California at Los Angeles. The enthusiasm of Dr. Yanai over the years and his willingness to discuss ideas concerning the subject of atmospheric latent heating and its implications with the various authors has had a profound influence on ideas central to this paper. We also appreciate the leadership of the late Dr. Joanne Simpson, a close colleague of Dr. Yanai and the first TRMM project scientist in the United States, and the late Prof. Tsuyoshi Nitta, who was a student of Dr. Yanai and the TRMM project scientist in Japan. Drs. Simpson, Robert Adler, Robert Houze, and Nitta promoted LH as being one of the two key products for TRMM. The authors also thank two anonymous reviewers for their constructive comments that improved this paper considerably.

Acknowledgment is also made to the NASA Center for Climate Simulation (NCCS) at the NASA Goddard Space Flight Center for computer time used in this research. Finally, the authors appreciate the implementation of LH retrieval algorithms into the Precipitation Processing System also at the NASA Goddard Space Flight Center and the TRMM Mission Operation System at the Japan Aerospace Exploration Agency (JAXA) for the production of standardized LH products. This research has been supported by an assortment of PMM Science Team grants under the auspices of both the National Aeronautics and Space Administration (NASA) and the Japan Aerospace Exploration Agency (JAXA). D. Waliser's contribution was carried out on behalf of the Jet Propulsion Laboratory, California Institute of Technology, under a contract with NASA. X. Jiang was also supported by the NSF Climate and Large-Scale Dynamics Program under Awards ATM-0934285 and AGS-1228302.

APPENDIX

Definitions of Acronyms

AMIE	ARM MJO Investigation Experiment
AMMA	African Monsoon Multidisciplinary Analysis
AORI	Atmosphere and Ocean Research Institute
ARM	Atmospheric Radiation Measurement
ARW	Advanced Research WRF
CFADs	Contoured frequency with altitude diagrams
CFSR	NCEP Climate Forecast System Reanalyses
CISK	Conditional instability of the second kind
CMIP3	Coupled Model Intercomparison Project phase 3
CRM	Cloud-resolving model
CSH	Convective–stratiform heating
DIAG	Sounding-diagnosed
DOE	Department of Energy
DYNAMO	Dynamics of the MJO
EBBR	Energy balance Bowen ratio
ECMWF	European Centre for Medium-Range Weather Forecasts
EEIO	Eastern equatorial Indian Ocean
EMEX	Equatorial Monsoon Experiment
EOF	Empirical orthogonal function
ERA-40	40-Year ECMWF Re-Analysis
ERA-I	ECMWF interim reanalysis
ETH	Echo-top height
GATE	Global Atmospheric Research Program (GARP) Atlantic Tropical Experiment
GCE	Goddard Cumulus Ensemble model
GCM	General circulation model
GPM	Global precipitation measurement
GPROF	Goddard profiling algorithm
GV	Ground validation
HERB	Hydrologic cycle and Earth Radiation Budget
HH	Hydrometeor heating
IFA	Intensive flux array
ITCZ	Intertropical convergence zone
JAMSTEC	Japan Agency for Marine-Earth Science and Technology
JAXA	Japan Aerospace Exploration Agency
JRA-25	Japanese 25-Year Reanalysis
KWAJEX	Kwajalein Experiment
LBA	Large-Scale Biosphere–Atmosphere Experiment
LH	Latent heating
LUT	Lookup table
MC³E	Midlatitude Continental Convective Clouds Experiment
MIROC5	Model for Interdisciplinary Research on Climate, version 5
MCS	Mesoscale convective system
MERRA	Modern-Era Retrospective Analysis for Research and Applications
MISMO	Mirai Indian Ocean Cruise for the Study of the MJO Convection Onset
MJO	Madden–Julian oscillation
NASA	National Aeronautics and Space Administration
NCEP-FNL	NCEP Global Forecast System (GFS) final
NCEP	National Centers for Environmental Prediction
NCEP-2	NCEP–Department of Energy (DOE) reanalysis
NESA	Northern enhanced sounding array
NIES	National Institute for Environmental Studies
NOAA	National Oceanographic and Atmospheric Administration
NOAH	NCEP–Oregon State University (Dept. of Atmospheric Sciences)/Air Force/Hydrologic Research Laboratory (Office of Hydrologic Development)
OREOF	Oblique rotated EOF
PBL	Planetary boundary layer
PDF	Probability distribution function
PRH	Precipitation radar heating
PRE-STORM	Preliminary Regional Experiment for STORM-Central
PMW	Passive microwave

PR	Precipitation Radar
PTH	Precipitation top height
Q_1	Apparent heat source
Q_2	Apparent moisture sink
Q_R	Radiation
rms	root-mean-square
RRTM	Rapid Radiative Transfer Model
SCSMEX	South China Sea Monsoon Experiment
SESA	Southern enhanced sounding array
SGP	Southern Great Plains
SLH	Spectral latent heating
SPCZ	South Pacific convergence zone
SST	Sea surface temperature
TMI	TRMM Microwave Imager
TOGA COARE	Tropical Ocean Global Atmosphere–Coupled Ocean–Atmosphere Response Experiment
TRAIN	Trained radiometer
TRMM	Tropical Rainfall Measuring Mission
TWP-ICE	Tropical Warm Pool–International Cloud Experiment
WCRP	World Climate Research Program
WH index	Wheeler–Hendon index
WP	Western Pacific
WRF	Weather Research and Forecasting Model
WWRP	World Weather Research Program
VIRS	Visible and Infrared Scanner
YOTC	Year of Tropical Convection

REFERENCES

Ackerman, T. P., and G. M. Stokes, 2003: The Atmospheric Radiation Measurement Program. *Phys. Today*, **56**, 38–44, doi:10.1063/1.1554135.

Benedict, J. J., and D. A. Randall, 2009: Structure of the Madden–Julian oscillation in the superparameterized CAM. *J. Atmos. Sci.*, **66**, 3277–3296, doi:10.1175/2009JAS3030.1.

Bretherton, C., M. E. Peters, and L. E. Back, 2004: Relationships between water vapor path and precipitation over the tropical oceans. *J. Climate*, **17**, 1517–1528, doi:10.1175/1520-0442(2004)017<1517:RBWVPA>2.0.CO;2.

Chikira, M., and M. Sugiyama, 2010: A cumulus parameterization with state-dependent entrainment rate. Part I: Description and sensitivity to temperature and humidity profiles. *J. Atmos. Sci.*, **67**, 2171–2193, doi:10.1175/2010JAS3316.1.

Ciesielski, P. E., and R. H. Johnson, 2006: Contrasting characteristics of convection over the northern and southern South China Sea during SCSMEX. *Mon. Wea. Rev.*, **134**, 1041–1062, doi:10.1175/MWR3113.1.

——, ——, P. T. Haertel, and J. Wang, 2003: Corrected TOGA COARE sounding humidity data: Impact on diagnosed properties of convection and climate over the warm pool. *J. Climate*, **16**, 2370–2384, doi:10.1175/2790.1.

Del Genio, A. D., Y.-H. Chen, D.-H. Kim, and M.-S. Yao, 2012: The MJO transition from shallow to deep convection in

CloudSat/CALIPSO data and GISS GCM simulations. *J. Climate*, **25**, 3755–3770, doi:10.1175/JCLI-D-11-00384.1.

Fairall, C., E. F. Bradley, D. P. Rogers, J. B. Edson, and G. S. Young, 1996: Bulk parameterization of air–sea fluxes for Tropical Ocean–Global Atmosphere Coupled–Ocean Atmosphere Response Experiment. *J. Geophys. Res.*, **101**, 3747–3764, doi:10.1029/95JC03205.

Fu, X. H., and B. Wang, 2009: Critical roles of the stratiform rainfall in sustaining the Madden–Julian oscillation: GCM experiments. *J. Climate*, **22**, 3939–3959, doi:10.1175/2009JCLI2610.1.

Gamache, J. F., and R. A. Houze Jr., 1983: Water budget of a mesoscale convective system in the tropics. *J. Atmos. Sci.*, **40**, 1835–1850, doi:10.1175/1520-0469(1983)040<1835:WBOAMC>2.0.CO;2.

Grecu, M., and W. S. Olson, 2006: Bayesian estimation of precipitation from satellite passive microwave observations using combined radar–radiometer retrievals. *J. Appl. Meteor. Climatol.*, **45**, 416–433, doi:10.1175/JAM2360.1.

——, ——, C.-L. Shie, T. L'Ecuyer, and W.-K. Tao, 2009: Combining satellite microwave radiometer and radar observations to estimate atmospheric latent heating profiles. *J. Climate*, **22**, 6356–6376, doi:10.1175/2009JCLI3020.1.

Grell, G. A., and D. Devenyi, 2002: A generalized approach to parameterizing convection combining ensemble and data assimilation techniques. *Geophys. Res. Lett.*, **29**, 1693, doi:10.1029/2002GL015311.

Hagos, S., and Coauthors, 2010: Estimates of tropical diabatic heating profiles: Commonalities and uncertainties. *J. Climate*, **23**, 542–558, doi:10.1175/2009JCLI3025.1.

Hirota, N., Y. N. Takayabu, M. Watanabe, and M. Kimoto, 2011: Precipitation reproducibility over tropical oceans and its relationship to the double ITCZ problem in CMIP3 and MIROC5 climate models. *J. Climate*, **24**, 4859–4873, doi:10.1175/2011JCLI4156.1.

——, ——, ——, ——, and M. Chikira, 2014: Role of convective entrainment in spatial distributions of and temporal variations in precipitation over tropical oceans. *J. Climate*, **27**, 8707–8723, doi:10.1175/JCLI-D-13-00701.1.

Hong, S.-Y., J. Dudhia, and S.-H. Chen, 2004: A revised approach to ice microphysical processes for the bulk parameterization of clouds and precipitation. *Mon. Wea. Rev.*, **132**, 103–120, doi:10.1175/1520-0493(2004)132<0103:ARATIM>2.0.CO;2.

Houze, R. A., Jr., 1982: Cloud clusters and large-scale vertical motions in the tropics. *J. Meteor. Soc. Japan*, **60**, 396–409.

——, 1989: Observed structure of mesoscale convective systems and implications for large-scale heating. *Quart. J. Roy. Meteor. Soc.*, **115**, 425–461, doi:10.1002/qj.49711548702.

——, 1997: Stratiform precipitation in regions of convection: A meteorological paradox? *Bull. Amer. Meteor. Soc.*, **78**, 2179–2196, doi:10.1175/1520-0477(1997)078<2179:SPIROC>2.0.CO;2.

——, and A. K. Betts, 1981: Convection in GATE. *Rev. Geophys.*, **19**, 541–576, doi:10.1029/RG019i004p00541.

Jakob, C., and C. Schumacher, 2008: Precipitation and latent heating characteristics of the major tropical western Pacific cloud regimes. *J. Climate*, **21**, 4348–4364, doi:10.1175/2008JCLI2122.1.

Janjic, Z. I., 1994: The step-mountain eta coordinate model: Further developments of the convection, viscous sublayer and turbulence closure schemes. *Mon. Wea. Rev.*, **122**, 927–945, doi:10.1175/1520-0493(1994)122<0927:TSMECM>2.0.CO;2.

——, 2000: Comments on "Development and evaluation of a convection scheme for use in climate models." *J. Atmos.*

Sci., **57**, 3686, doi:10.1175/1520-0469(2000)057<3686: CODAEO>2.0.CO;2.

Jiang, X., and Coauthors, 2009: Vertical heating structures associated with the MJO as characterized by TRMM estimates, ECMWF reanalyses, and forecasts: A case study during 1998/99 winter. *J. Climate*, **22**, 6001–6020, doi:10.1175/2009JCLI3048.1.

——, and Coauthors, 2011: Vertical diabatic heating structure of the MJO: Intercomparison between recent reanalyses and TRMM estimates. *Mon. Wea. Rev.*, **139**, 3208–3223, doi:10.1175/2011MWR3636.1.

Johnson, R. H., 1984: Partitioning tropical heat and moisture budgets into cumulus and mesoscale components: Implications for cumulus parameterization. *Mon. Wea. Rev.*, **112**, 1590–1601, doi:10.1175/1520-0493(1984)112<1590:PTHAMB>2.0.CO;2.

——, and P. J. Hamilton, 1988: The relationship of surface pressure features to the precipitation and airflow structure of an intense mid-latitude squall line. *Mon. Wea. Rev.*, **116**, 1444–1473, doi:10.1175/1520-0493(1988)116<1444:TROSPF>2.0.CO;2.

——, and P. E. Ciesielski, 2002: Characteristics of the 1998 summer monsoon onset over the northern South China Sea. *J. Meteor. Soc. Japan*, **80**, 561–578, doi:10.2151/jmsj.80.561.

Kain, J. S., and M. Fritsch, 2004: The Kain–Fritsch convective parameterization: An update. *J. Appl. Meteor.*, **43**, 170–181, doi:10.1175/1520-0450(2004)043<0170:TKCPAU>2.0.CO;2.

Katsumata, M., R. H. Johnson, and P. E. Ciesielski, 2009: Observed synoptic-scale variability during the developing phase of an ISO over the Indian Ocean during MISMO. *J. Atmos. Sci.*, **66**, 3434–3448, doi:10.1175/2009JAS3003.1.

Kessler, E., 1969: *On the Distribution and Continuity of Water Substance in Atmospheric Circulation, Meteor. Monogr.*, No. 32, Amer. Meteor. Soc., 84 pp.

Kiladis, G. N., K. H. Straub, and P. T. Haertel, 2005: Zonal and vertical structure of the Madden–Julian oscillation. *J. Atmos. Sci.*, **62**, 2790–2809, doi:10.1175/JAS3520.1.

Kim, D., and Coauthors, 2009: Application of MJO simulation diagnostics to climate models. *J. Climate*, **22**, 6413–6436, doi:10.1175/2009JCLI3063.1.

Krishnamurti, T. N., C. M. Kishtawal, W. D. Shin, and C. E. Williford, 2000a: Improving tropical precipitation forecasts from a multianalysis superensemble. *J. Climate*, **13**, 4217–4227, doi:10.1175/1520-0442(2000)013<4217:ITPFFA>2.0.CO;2.

——, ——, Z. Zhang, T. Larow, D. Bachiochi, E. Williford, S. Gadgil, and S. Surendran, 2000b: Multimodel ensemble forecasts for weather and seasonal climate. *J. Climate*, **13**, 4196–4216, doi:10.1175/1520-0442(2000)013<4196:MEFFWA>2.0.CO;2.

——, C. Gnanaseelan, and A. Chakraborty, 2007: Prediction of the diurnal cycle using a multimodel superensemble. Part I: Precipitation. *Mon. Wea. Rev.*, **135**, 3613–3632, doi:10.1175/MWR3446.1.

——, A. Chakraborty, and A. K. Mishra, 2010: Improving multimodel forecasts of the vertical distribution of heating using the TRMM profiles. *J. Climate*, **23**, 1079–1094, doi:10.1175/2009JCLI2878.1.

——, A. Simon, A. Thomas, A. Mishra, D. Sikka, D. Niyogi, A. Chakraborty, and L. Li, 2012: Modeling of forecast sensitivity on the march of monsoon isochrones from Kerala to New Delhi: The first 25 days. *J. Atmos. Sci.*, **69**, 2465–2487, doi:10.1175/JAS-D-11-0170.1.

Krueger, S. K., 1988: Numerical simulation of tropical cumulus clouds and their interaction with the subcloud layer. *J. Atmos. Sci.*, **45**, 2221–2250, doi:10.1175/1520-0469(1988)045<2221: NSOTCC>2.0.CO;2.

Kummerow, C., W. Barnes, T. Kozu, J. Shiue, and J. Simpson, 1998: The Tropical Rainfall Measurement Mission (TRMM) sensor package. *J. Atmos. Oceanic Technol.*, **15**, 809–817, doi:10.1175/1520-0426(1998)015<0809:TTRMMT>2.0.CO;2.

Lang, S., W.-K. Tao, R. Cifelli, W. Olson, J. Halverson, S. Rutledge, and J. Simpson, 2007: Improving simulations of convective system from TRMM LBA: Easterly and westerly regimes. *J. Atmos. Sci.*, **64**, 1141–1164, doi:10.1175/JAS3879.1.

——, ——, and X. Zeng, 2011: Reducing the biases in simulated radar reflectivities from a bulk microphysics scheme: Tropical convective systems. *J. Atmos. Sci.*, **68**, 2306–2320, doi:10.1175/JAS-D-10-05000.1.

Lau, K.-M., and Coauthors, 2000: A report of the field operations and early results of the South China Sea Monsoon Experiment (SCSMEX). *Bull. Amer. Meteor. Soc.*, **81**, 1261–1270, doi:10.1175/1520-0477(2000)081<1261:AROTFO>2.3.CO;2.

Lau, W. K.-M., and H.-T. Wu, 2010: Characteristics of precipitation, cloud, and latent heating associated with the Madden–Julian oscillation. *J. Climate*, **23**, 504–518, doi:10.1175/2009JCLI2920.1.

——, and D. E. Waliser, 2011: *Intraseasonal Variability in the Atmosphere–Ocean Climate System.* 2nd ed. Springer, 613 pp.

L'Ecuyer, T. S., and G. L. Stephens, 2003: The tropical oceanic energy budget from the TRMM perspective. Part I: Algorithm and uncertainties. *J. Climate*, **16**, 1967–1985, doi:10.1175/1520-0442(2003)016<1967:TTOEBF>2.0.CO;2.

——, and ——, 2007: The tropical atmospheric energy budget from the TRMM perspective. Part II: Evaluating GCM representations of the sensitivity of regional energy and water cycles to the 1998–99 ENSO cycle. *J. Climate*, **20**, 4548–4571, doi:10.1175/JCLI4207.1.

Lee, M.-I., I.-S. Kang, J.-K. Kim, and B. E. Mapes, 2001: Influence of cloud–radiation interaction on simulating tropical intraseasonal oscillation with an atmospheric general circulation model. *J. Geophys. Res.*, **106**, 14 219–14 233, doi:10.1029/2001JD900143.

Li, C., X. Jia, J. Ling, W. Zhou, and C. Zhang, 2009: Sensitivity of MJO simulations to diabatic heating profiles. *Climate Dyn.*, **32**, 167–187, doi:10.1007/s00382-008-0455-x.

Li, X., C.-H. Sui, K.-M. Lau, and M.-D. Chou, 1999: Large-scale forcing and cloud–radiation interaction in the tropical deep convective regime. *J. Atmos. Sci.*, **56**, 3028–3042, doi:10.1175/1520-0469(1999)056<3028:LSFACR>2.0.CO;2.

Lin, J., B. E. Mapes, M. Zhang, and M. Newman, 2004: Stratiform precipitation, vertical heating profiles, and the Madden–Julian oscillation. *J. Atmos. Sci.*, **61**, 296–309, doi:10.1175/1520-0469(2004)061<0296:SPVHPA>2.0.CO;2.

——, and Coauthors, 2006: Tropical intraseasonal variability in 14 IPCC AR4 climate models. Part I: Convective signals. *J. Climate*, **19**, 2665–2690, doi:10.1175/JCLI3735.1.

Lin, X., and R. H. Johnson, 1996: Heating, moistening, and rainfall over the western Pacific during TOGA COARE. *J. Atmos. Sci.*, **53**, 3367–3383, doi:10.1175/1520-0469(1996)053<3367: HMAROT>2.0.CO;2.

Ling, J., and C. Zhang, 2011: Structural evolution in heating profiles of the MJO in global reanalyses and TRMM retrievals. *J. Climate*, **24**, 825–842, doi:10.1175/2010JCLI3826.1.

——, and ——, 2013: Diabatic heating profiles in recent global reanalyses. *J. Climate*, **26**, 3307–3325, doi:10.1175/JCLI-D-12-00384.1.

Mapes, B. E., and R. A. Houze Jr., 1995: Diabatic divergence profiles in western Pacific mesoscale convective systems. *J. Atmos. Sci.*, **52**, 1807–1828, doi:10.1175/1520-0469(1995)052<1807: DDPIWP>2.0.CO;2.

Masunaga, H., T. S. L'Ecuyer, and C. D. Kummerow, 2005: Variability in the characteristics of precipitation systems in the tropical Pacific. Part I: Spatial structure. *J. Climate*, **18**, 823–840, doi:10.1175/JCLI-3304.1.

Mather J. H., S. A. McFarlane, M. Miller, and K. L. Johnson, 2007: Cloud properties and associated radiative heating rates in the tropical western Pacific. *J. Geophys. Res.*, **112**, D05201, doi:10.1029/2006JD007555.

May, P. T., J. H. Mather, G. Vaughan, C. Jakob, G. M. McFarquar, and G. G. Mace, 2008: The Tropical Warm Pool International Cloud Experiment. *Bull. Amer. Meteor. Soc.*, **89**, 629–645, doi:10.1175/BAMS-89-5-629.

Moncrieff, M. W., S. K. Krueger, D. Gregory, J.-L. Redelsperger, and W.-K. Tao, 1997: GEWEX Cloud System Study (GCSS) Working Group 4: Precipitating convective cloud systems. *Bull. Amer. Meteor. Soc.*, **78**, 831–845, doi:10.1175/1520-0477(1997)078<0831:GCSSGW>2.0.CO;2.

Morita, J., Y. N. Takayabu, S. Shige, and Y. Kodama, 2006: Analysis of rainfall characteristics of the Madden–Julian oscillation using TRMM satellite data. *Dyn. Atmos. Oceans*, **42**, 107–126, doi:10.1016/j.dynatmoce.2006.02.002.

Olson, W. S., C. D. Kummerow, Y. Hong, and W.-K. Tao, 1999: Atmospheric latent heating distributions in the tropics derived from satellite passive microwave radiometer measurements. *J. Appl. Meteor.*, **38**, 633–664, doi:10.1175/1520-0450(1999)038<0633:ALHDIT>2.0.CO;2.

——, and Coauthors, 2006: Precipitation and latent heating distributions from satellite passive microwave radiometry. Part I: Improved method and uncertainties. *J. Appl. Meteor.*, **45**, 702–720, doi:10.1175/JAM2369.1.

Petch, J., D. Waliser, X. Jiang, P. Xavier, and S. Woolnough, 2011: A global model inter-comparison of the physical processes associated with the MJO. *GEWEX News*, No. 21, International GEWEX Project Office, Silver Spring, MD, 3–5.

Rajeevan, M., J. Bhate, and A. K. Jaswal, 2008: Analysis of variability and trends of extreme rainfall events over India using 104 years of gridded daily rainfall data. *Geophys. Res. Lett.*, **35**, L18707, doi:10.1029/2008GL035143.

Rajendran, R., T. N. Krishnamurti, V. Misra, and W.-K. Tao, 2004: An empirical cumulus parameterization scheme based on TRMM latent heating profiles. *J. Meteor. Soc. Japan*, **82**, 989–1006.

Randall, D., and Coauthors, 2003: Confronting models with data: The GEWEX Cloud Systems Study. *Bull. Amer. Meteor. Soc.*, **84**, 455–469, doi:10.1175/BAMS-84-4-455.

Raymond, D. J., 2001: A new model of the Madden–Julian oscillation. *J. Atmos. Sci.*, **58**, 2807–2819, doi:10.1175/1520-0469(2001)058<2807:ANMOTM>2.0.CO;2.

Rodgers, E. B., W. S. Olson, V. M. Karyampudi, and H. F. Pierce, 1998: Satellite-derived latent heating distribution and environmental influences in Hurricane Opal (1995). *Mon. Wea. Rev.*, **126**, 1229–1247, doi:10.1175/1520-0493(1998)126<1229:SDLHDA>2.0.CO;2.

——, ——, J. Halverson, J. Simpson, and H. Pierce, 2000: Environmental forcing of Supertyphoon Paka's (1997) latent heat structure. *J. Appl. Meteor.*, **39**, 1983–2006, doi:10.1175/1520-0450(2001)040<1983:EFOSPS>2.0.CO;2.

Rutledge, S. A., and R. A. Houze Jr., 1987: A diagnostic modeling study of the trailing stratiform rain of a midlatitude squall line. *J. Atmos. Sci.*, **44**, 2640–2656, doi:10.1175/1520-0469(1987)044<2640:ADMSOT>2.0.CO;2.

Satoh, S., and A. Noda, 2001: Retrieval of latent heating profiles from TRMM radar data. *Proc. 30th Int. Conf. on Radar Meteorology*, Munich, Germany, Amer. Meteor. Soc., 340–342.

Schumacher, C., M. H. Zhang, and P. E. Ciesielski, 2007: Heating structures of the TRMM field campaigns. *J. Atmos. Sci.*, **64**, 2593–2610, doi:10.1175/JAS3938.1.

Seo, K.-H., and W. Wang, 2010: The Madden–Julian oscillation simulated in the NCEP Climate Forecast System Model: The importance of stratiform heating. *J. Climate*, **23**, 4770–4793, doi:10.1175/2010JCLI2983.1.

Sherwood, S. C., 1999: Convective precursors and predictability in the tropical western Pacific. *Mon. Wea. Rev.*, **127**, 2977–2991, doi:10.1175/1520-0493(1999)127<2977:CPAPIT>2.0.CO;2.

Shige, S., Y. N. Takayabu, W.-K. Tao, and D. E. Johnson, 2004: Spectral retrieval of latent heating profiles from TRMM PR data. Part I: Development of a model-based algorithm. *J. Appl. Meteor.*, **43**, 1095–1113, doi:10.1175/1520-0450(2004)043<1095:SROLHP>2.0.CO;2.

——, ——, ——, and C.-L. Shie, 2007: Spectral retrieval of latent heating profiles from TRMM PR data. Part II: Algorithm improvement and heating estimates over tropical ocean regions. *J. Appl. Meteor. Climatol.*, **46**, 1098–1124, doi:10.1175/JAM2510.1.

——, ——, and ——, 2008: Spectral retrieval of latent heating profiles from TRMM PR data. Part III: Moistening estimates over the tropical ocean regions. *J. Appl. Meteor. Climatol.*, **47**, 620–640, doi:10.1175/2007JAMC1738.1.

——, ——, S. Kida, W.-K. Tao, X. Zeng, C. Yokoyama, and T. L'Ecuyer, 2009: Spectral retrieved of latent heating profiles from TRMM PR data. Part VI: Comparisons of lookup tables from two- and three-dimensional simulations. *J. Climate*, **22**, 5577–5594, doi:10.1175/2009JCLI2919.1.

——, S. Kida, H. Ashiwake, T. Kubota, and K. Aonashi, 2013: Improvement of TMI rain retrievals in mountainous areas. *J. Appl. Meteor. Climatol.*, **52**, 242–254, doi:10.1175/JAMC-D-12-074.1.

Short, D. A., and K. Nakamura, 2000: TRMM radar observations of shallow precipitation over the tropical oceans. *J. Climate*, **13**, 4107–4124, doi:10.1175/1520-0442(2000)013<4107:TROOSP>2.0.CO;2.

Silva Dias, M., and Coauthors, 2002: Cloud and rain processes in a biosphere–atmosphere interaction context in the Amazon region. *J. Geophys. Res.*, **107**, 8072, doi:10.1029/2001JD000335.

Slingo, J. M., P. M. Inness, and K. R. Sperber, 2005: Modeling. *Intraseasonal Variability in the Atmosphere–Ocean Climate System*, W. K. M. Lau and D. E. Waliser, Eds., Springer, 361–388.

Sobel, A. H., and H. Gildor, 2003: A simple time-dependent model of SST hot spots. *J. Climate*, **16**, 3978–3992, doi:10.1175/1520-0442(2003)016<3978:ASTMOS>2.0.CO;2.

Soong, S.-T., and W.-K. Tao, 1980: Response of deep tropical clouds to mesoscale processes. *J. Atmos. Sci.*, **37**, 2016–2034, doi:10.1175/1520-0469(1980)037<2016:RODTCC>2.0.CO;2.

Sui, C.-H., and M. Yanai, 1986: Cumulus ensemble effects on the large-scale vorticity and momentum fields of GATE. Part I: Observational evidence. *J. Atmos. Sci.*, **43**, 1618–1642, doi:10.1175/1520-0469(1986)043<1618:CEEOTL>2.0.CO;2.

Takayabu, Y. N., 2002: Spectral representation of rain features and diurnal variations observed with TRMM PR over the equatorial areas. *Geophys. Res. Lett.*, **29**, 301–304, doi:10.1029/2001GL014113.

——, K.-M. Lau, and C.-H. Sui, 1996: Observation of a quasi-2-day wave during TOGA COARE. *Mon. Wea. Rev.*, **124**, 1892–1913, doi:10.1175/1520-0493(1996)124<1892:OOAQDW>2.0.CO;2.

——, J. Yokomori, and K. Yoneyama, 2006: A diagnostic study on interactions between atmospheric thermodynamic structure and cumulus convection over the tropical western Pacific Ocean and over the Indochina Peninsula. *J. Meteor. Soc. Japan*, **84A**, 151–169, doi:10.2151/jmsj.84A.151.

——, S. Shige, W.-K. Tao, and N. Hirota, 2010: Shallow and deep latent heating modes over tropical oceans observed with TRMM PR spectral latent heating data. *J. Climate*, **23**, 2030–2046, doi:10.1175/2009JCLI3110.1.

Takemi, T., O. Hirayama, and C. Liu, 2004: Factors responsible for the vertical development of tropical oceanic cumulus convection. *Geophys. Res. Lett.*, **31**, L11109, doi:10.1029/2004GL020225.

Tao, W.-K., 2003: Goddard Cumulus Ensemble (GCE) model: Application for understanding precipitation processes. *Cloud Systems, Hurricanes, and the Tropical Rainfall Measuring Mission (TRMM)*, Meteor. Monogr., No. 51, Amer. Meteor. Soc., 107–138.

——, 2007: Cloud resolving modeling. *J. Meteor. Soc. Japan*, **85B**, 305–330, doi:10.2151/jmsj.85B.305.

——, and S.-T. Soong, 1986: A study of the response of deep tropical clouds to mesoscale processes: Three-dimensional numerical experiments. *J. Atmos. Sci.*, **43**, 2653–2676, doi:10.1175/1520-0469(1986)043<2653:ASOTRO>2.0.CO;2.

——, and J. Simpson, 1989: Modeling study of a tropical squall-type convective line. *J. Atmos. Sci.*, **46**, 177–202, doi:10.1175/1520-0469(1989)046<0177:MSOATS>2.0.CO;2.

——, and ——, 1993: The Goddard Cumulus Ensemble Model. Part I: Model description. *Terr. Atmos. Ocean. Sci.*, **4**, 35–72.

——, and M. Moncrieff, 2009: Multi-scale cloud-system modeling. *Rev. Geophys.*, **47**, RG4002, doi:10.1029/2008RG000276.

——, J. Simpson, and S.-T. Soong, 1987: Statistical properties of a cloud ensemble: A numerical study. *J. Atmos. Sci.*, **44**, 3175–3187, doi:10.1175/1520-0469(1987)044<3175:SPOACE>2.0.CO;2.

——, ——, S. Lang, M. McCumber, R. Adler, and R. Penc, 1990: An algorithm to estimate the heating budget from vertical hydrometeor profiles. *J. Appl. Meteor.*, **29**, 1232–1244, doi:10.1175/1520-0450(1990)029<1232:AATETH>2.0.CO;2.

——, S. Lang, J. Simpson, and R. Adler, 1993: Retrieval algorithms for estimating the vertical profiles of latent heat release: Their applications for TRMM. *J. Meteor. Soc. Japan*, **71**, 685–700.

——, ——, ——, W. S. Olson, D. Johnson, B. Ferrier, C. Kummerow, and R. Adler, 2000: Vertical profiles of latent heat release and their retrieval for TOGA COARE convective systems using a cloud resolving model, SSM/I and shipborne radar data. *J. Meteor. Soc. Japan*, **78**, 333–355.

——, and Coauthors, 2001: Retrieved vertical profiles of latent heating release using TRMM rainfall products for February 1998. *J. Appl. Meteor.*, **40**, 957–982, doi:10.1175/1520-0450(2001)040<0957:RVPOLH>2.0.CO;2.

——, D. Johnson, C.-L. Shie, and J. Simpson, 2004: Atmospheric energy budget and large-scale precipitation efficiency of convective systems during TOGA COARE, GATE, SCSMEX, and ARM: Cloud-resolving model simulations. *J. Atmos. Sci.*, **61**, 2405–2423, doi:10.1175/1520-0469(2004)061<2405:TAEBAL>2.0.CO;2.

——, and Coauthors, 2006: Retrieval of latent heating from TRMM measurements. *Bull. Amer. Meteor. Soc.*, **87**, 1555–1572, doi:10.1175/BAMS-87-11-1555.

——, S. Lang, X. Zeng, S. Shige, and Y. Takayabu, 2010: Relating convective and stratiform rain to latent heating. *J. Climate*, **23**, 1874–1893, doi:10.1175/2009JCLI3278.1.

Webster, P. J., and R. Lukas, 1992: TOGA-COARE: The Coupled Ocean–Atmosphere Response Experiment. *Bull. Amer. Meteor. Soc.*, **73**, 1377–1416, doi:10.1175/1520-0477(1992)073<1377:TCTCOR>2.0.CO;2.

Wheeler, M. C., and H. H. Hendon, 2004: An all-season real-time multivariate MJO index: Development of an index for monitoring and prediction. *Mon. Wea. Rev.*, **132**, 1917–1932, doi:10.1175/1520-0493(2004)132<1917:AARMMI>2.0.CO;2.

Xie, S., T. Hume, C. Jakob, S. A. Klein, R. B. McCoy, and M. Zhang, 2010: Observed large-scale structures and diabatic heating and drying profiles during TWP-ICE. *J. Climate*, **23**, 57–79, doi:10.1175/2009JCLI3071.1.

Yanai, M., 1961: Dynamical aspects of typhoon formation. *J. Meteor. Soc. Japan*, **39**, 282–309.

——, S. Esbensen, and J.-H. Chu, 1973: Determination of bulk properties of tropical cloud clusters from large-scale heat and moisture budgets. *J. Atmos. Sci.*, **30**, 611–627, doi:10.1175/1520-0469(1973)030<0611:DOBPOT>2.0.CO;2.

Yang, S., and E. A. Smith, 1999a: Moisture budget analysis of TOGA COARE area using SSM/I retrieved latent heating and large-scale Q_2 estimates. *J. Atmos. Oceanic Technol.*, **16**, 633–655, doi:10.1175/1520-0426(1999)016<0633:MBAOTC>2.0.CO;2.

——, and ——, 1999b: Four-dimensional structure of monthly latent heating derived from SSM/I satellite measurements. *J. Climate*, **12**, 1016–1037, doi:10.1175/1520-0442(1999)012<1016:FDSOML>2.0.CO;2.

——, and ——, 2000: Vertical structure and transient behavior of convective–stratiform heating in TOGA COARE from combined satellite–sounding analysis. *J. Appl. Meteor.*, **39**, 1491–1513, doi:10.1175/1520-0450(2000)039<1491:VSATBO>2.0.CO;2.

——, W. S. Olson, J.-J. Wang, T. L. Bell, E. A. Smith, and C. D. Kummerow, 2006: Precipitation and latent heating distributions from satellite passive microwave radiometry. Part II: Evaluation of estimates using independent data. *J. Appl. Meteor.*, **45**, 721–739, doi:10.1175/JAM2370.1.

Yoneyama, K., and Coauthors, 2008: MISMO field experiment in the equatorial Indian Ocean. *Bull. Amer. Meteor. Soc.*, **89**, 1889–1903, doi:10.1175/2008BAMS2519.1.

Yuter, S. E., and R. A. Houze Jr., 1995: Three-dimensional kinematic and microphysical evolution of Florida cumulonimbus. Part II: Frequency distributions of vertical velocity, reflectivity, and differential reflectivity. *Mon. Wea. Rev.*, **123**, 1941–1963, doi:10.1175/1520-0493(1995)123<1941:TDKAME>2.0.CO;2.

——, ——, E. A. Smith, T. T. Wilheit, and E. Zipser, 2005: Physical characterization of tropical oceanic convection observed in KWAJEX. *J. Appl. Meteor.*, **44**, 385–415, doi:10.1175/JAM2206.1.

Zeng, X., and Coauthors, 2007: Evaluating clouds in long-term cloud-resolving model simulations with observational data. *J. Atmos. Sci.*, **64**, 4153–4177, doi:10.1175/2007JAS2170.1.

——, W.-K. Tao, S. Lang, A. Y. Hou, M. Zhang, and J. Simpson, 2008: On the sensitivity of atmospheric ensembles to cloud microphysics in long-term cloud-resolving model simulations. *J. Meteor. Soc. Japan*, **86A**, 45–65, doi:10.2151/jmsj.86A.45.

——, and Coauthors, 2009a: A contribution by ice nuclei to global warming. *Quart. J. Roy. Meteor. Soc.*, **135**, 1614–1629, doi:10.1002/qj.449.

——, and Coauthors, 2009b: An indirect effect of ice nuclei on atmospheric radiation. *J. Atmos. Sci.*, **66**, 41–61, doi:10.1175/2008JAS2778.1.

——, W.-K. Tao, T. Matsui, S. Xie, S. Lang, M. Zhang, D. Starr, and X. Li, 2011: Estimating the ice crystal enhancement factor in the tropics. *J. Atmos. Sci.*, **68**, 1424–1434, doi:10.1175/2011JAS3550.1.

——, ——, S. Powell, R. A. Houze Jr., P. Ciesielski, N. Guy, H. Pierce, and T. Matsui, 2013: A comparison of the water budgets between clouds from AMMA and TWP-ICE. *J. Atmos. Sci.*, **70**, 487–503, doi:10.1175/JAS-D-12-050.1.

Zhang, C., 2005: Madden–Julian oscillation. *Rev. Geophys.*, **43**, RG2003, doi:10.1029/2004RG000158.

——, and S. M. Hagos, 2009: Bi-modal structure and variability of large-scale diabatic heating in the tropics. *J. Atmos. Sci.*, **66**, 3621–3640, doi:10.1175/2009JAS3089.1.

Zhang, G. J., and M. Mu, 2005: Effects of modifications to the Zhang–McFarlane convection parameterization on the simulation of the tropical precipitation in the National Center for Atmospheric Research Community Climate Model, version 3. *J. Geophys. Res.*, **110**, D09109, doi:10.1029/2004JD005617.

——, and X. Song, 2009: Interaction of deep and shallow convection is key to Madden–Julian oscillation simulation. *Geophys. Res. Lett.*, **36**, L09708, doi:10.1029/2009GL037340.

Zhang, M. H., and J. L. Lin, 1997: Constrained variational analysis of sounding data based on column-integrated budgets of mass, moisture, and momentum: Approach and application to ARM measurements. *J. Atmos. Sci.*, **54**, 1503–1524, doi:10.1175/1520-0469(1997)054<1503:CVAOSD>2.0.CO;2.

——, ——, R. T. Cederwall, J. J. Yio, and S. C. Xie, 2001: Objective analysis of ARM IOP data: Method and sensitivity. *Mon. Wea. Rev.*, **129**, 295–311, doi:10.1175/1520-0493(2001)129<0295:OAOAID>2.0.CO;2.

Zipser, E. J., 2003: Some views on "hot towers" after 50 years of tropical field programs and two years of TRMM data. *Cloud Systems, Hurricanes and the Tropical Rainfall Measuring Mission (TRMM)—A Tribute to Dr. Joanne Simpson, Meteor. Monogr.*, No. 51, Amer. Meteor. Soc., 49–58.

——, R. J. Meitin, and M. A. LeMone, 1981: Mesoscale motion fields associated with a slowly moving GATE convective band. *J. Atmos. Sci.*, **38**, 1725–1750, doi:10.1175/1520-0469(1981)038<1725:MMFAWA>2.0.CO;2.

Chapter 3

Michio Yanai and Tropical Waves

YUKARI N. TAKAYABU

Atmosphere and Ocean Research Institute, University of Tokyo, Kashiwa, Chiba, Japan

GEORGE N. KILADIS

NOAA/Earth System Research Laboratory/Physical Sciences Division, Boulder, Colorado

VICTOR MAGAÑA

Instituto de Geografía, Universidad Nacional Autónoma de México, Ciudad de México, Mexico

ABSTRACT

Insights by Professor Michio Yanai on tropical waves, which have been vital ingredients for progress in tropical meteorology over the last half-century, are recollected. This study revisits various aspects of research on tropical waves over the last five decades to examine, in Yanai's words, "the nature of 'A-scale' tropical wave disturbances and the interaction of the waves and the 'B-scale' phenomena (cloud clusters)," the fundamental problem posed by Yanai at the design phase of the GARP Atlantic Tropical Experiment (GATE) in 1971. The various contributions of Michio Yanai to the current understanding of the dynamics of the tropical atmosphere are briefly reviewed to show how his work has led to several current theories in this field.

1. Introduction

Professor Michio Yanai, who referred to himself as a "meteorological freak" since he was a junior high school student, started his career as a scientist with a study of the genesis of Typhoon Doris from an easterly wave. He recalled that Professor Jule Charney was in attendance during his first presentation about typhoon formation during the First International Symposium on Numerical Weather Prediction in Tokyo, held in 1960. Motivated by the fact that numerical simulations of tropical cyclones were having great difficulty in achieving storm genesis, including those efforts by Charney's Massachusetts Institute of Technology (MIT) group, Michio was determined to pursue the observational and theoretical bases for the formation of storms from preexisting equatorial disturbances. This particular problem had a huge impact on his development as a tropical meteorologist and stimulated his interest in tropical waves, which became a long-lasting focus of his scientific career. After obtaining his Ph.D. on the study of typhoon formation at the University of Tokyo in 1960, Michio was invited to be a research scientist in the Typhoon Research Division in the Meteorological Research Institute at the Japan Meteorological Agency (JMA). His stay at JMA was short, however, since in the following year, Professor Herbert Riehl at Colorado State University invited Michio to become a post-doctoral researcher for 2 years in 1962–64. Of course, Professor Riehl was the leading expert on easterly waves at the time, so this experience undoubtedly further motivated Michio's interests in the study of tropical waves. In 1965, Michio was invited back to the University of Tokyo as associate professor in the Faculty of Science, where he and his student, Taketo Maruyama,

Corresponding author address: Yukari N. Takayabu, Atmosphere and Ocean Research Institute, University of Tokyo, 5-1-5 Kashiwanoha, Kashiwa, Chiba 277-8568, Japan.
E-mail: yukari@aori.u-tokyo.ac.jp

DOI: 10.1175/AMSMONOGRAPHS-D-15-0019.1

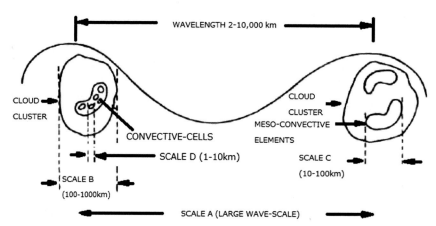

FIG. 3-1. The scales of tropospheric motions in the tropics, taken from the February 1970 GARP technical report. Adopted from Sikdar and Suomi (1971).

documented the existence of mixed Rossby–gravity waves, also called Yanai–Maruyama (YM) waves (see section 3). At that time, the University of Tokyo was strongly affected by student riots, which prompted Michio to accept an invitation by Professor Yale Mintz to join the Department of Meteorology faculty at the University of California, Los Angeles (UCLA), in 1969, where he remained for the rest of his career.

Just 2 years after he moved to UCLA, Michio contributed "A review of recent studies of tropical meteorology relevant to the planning of GATE"(Yanai 1971a) to the International Council for Science (ICSU)–World Meteorological Organization (WMO) Joint Organizing Committee (JOC) for Global Atmospheric Research Program (GARP) where he was appointed as a member of a study group on tropical disturbances. He stated that his primary interests on this subject were "on the A-scale tropical wave disturbances and the interaction of the waves and the B-scale phenomena (cloud clusters)." (Yanai 1971a) During seminars and as part of his tropical meteorology class, Michio often used a schematic such as the one in Fig. 3-1 for representing the multiscale interaction between tropical waves and cumulus ensembles. The nature of the interaction between the cloud scale and the large scale remained a primary focus of Michio's interest throughout the rest of his career.

It was remarkably prescient that at the design stage of the GARP Atlantic Tropical Experiment (GATE), Michio selected the word "cloud cluster" to refer to a grouping of cumulonimbus, as a key ingredient of tropical waves for scale interactions. One of the major advances of GATE was its field observations that contributed to the understanding of the structure and aggregation of mesoscale systems over the eastern tropical Atlantic (Gamache and Houze 1985; Zipser 1977; Houze and Betts 1981). Another major contribution from Michio's work during GATE was the analysis of

the mass flux of convection associated with African easterly waves (AEWs). Utilizing the apparent heat source Q_1 and moisture sink Q_2 and a spectral cloud model, Michio and his students (Yanai 1961a, 1971b; Yanai et al. 1973, 1976; Nitta 1977) diagnosed convective mass flux from the soundings (Tao et al. 2016, chapter 2). It was shown that the vertical profiles of convective mass flux in the Atlantic differed from that over the Marshall Islands of the west-central Pacific, with a layered structure of divergence in the middle troposphere in GATE not found in the Marshall Island data. While maximum upward motion is found near 700 hPa in GATE and a secondary maximum is found in the upper troposphere, the only peak found in the western Pacific was at around 400 hPa (Nitta 1977; Yanai and Johnson 1993). Michio's former graduate student Nitta (1978) further studied the GATE Phase III data to establish the relationships between the structure of the large-scale wave disturbances and cloud height, a key relationship affecting the interaction between the cloud scale and the synoptic scale.

The scale interaction between clouds and waves occupied Michio's attention for his entire career. In this article, we will revisit Professor Yanai's contributions to the study of tropical waves and their relationship with tropical convection.

2. Easterly waves and tropical storm genesis

As remarked upon, during the 1960s Michio was heavily involved in both observational and theoretical work related to tropical waves. As his early interests were in typhoons, he was well aware that easterly waves were known to be precursors to tropical storm formation (Riehl 1948; Palmer 1952). As part of his doctoral thesis, Michio studied in detail the 1958 transition of an easterly wave into Typhoon Doris by analyzing Marshall Island special observation data (Yanai 1961a), describing three stages of tropical storm development. He then proceeded

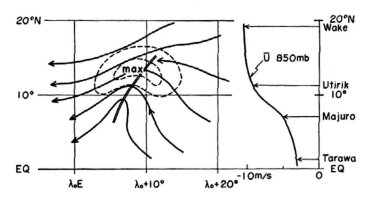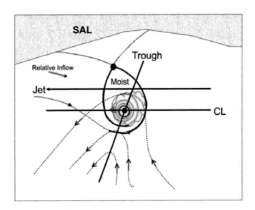

FIG. 3-2. (left) The schematic horizontal view of an easterly wave. Dotted lines are isovels. The latitudinal distribution of mean zonal wind at the 850-hPa level is shown in the right-hand side of the figure. Adopted from Yanai (1961b). (right) A schematic for formation of a tropical storm within a wave pouch. The dashed contours represent the streamlines in the ground-based frame of reference, which usually have an inverted-V pattern. The solid streamlines delineate the wave pouch as viewed in the frame of reference moving at the same speed with the wave. The pouch can protect the mesoscale vortices inside from the hostile environment, such as dry air associated with the Saharan air layer. Deep convection (gray shading) is sustained within the pouch. Owing to convergent flow, the pouch may have an opening that allows the influx of environmental air and vorticity. The easterly jet (Jet), the critical latitude (CL), and the wave trough axis (Trough) are also indicated in the schematic. The intersection of the critical latitude and the trough axis pinpoints the pouch center as the preferred location for tropical cyclogenesis. Adopted from Wang et al. (2010).

to lay out the dynamical basis for such development (Yanai 1961a,b). While this problem is still being worked on today (see Fovell et al. 2016), Michio's ideas on the subject were truly visionary since they clearly spelled out the conditions for the instability of a wave within a moist easterly current and also laid the foundation for the study of the coupling of convection with the large-scale wave field.

In Yanai (1961b, p. 285), Michio first considered the concept, which he attributed to Priestley (1959), of free versus forced convection, with free convection due to buoyancy and forced convection "where the motion is otherwise imposed" by large-scale dynamics, even in a stably stratified atmosphere. The expressions for Q_1 and Q_2 were also first presented in Yanai (1961a). Michio pointed out that the scale of condensational heating initially favored within an easterly wave was much smaller than that of the wave and incipient typhoon themselves but that this heating ultimately must be responsible for the generation of a warm-core protovortex within the initially cold-core easterly wave. This then enabled the rapid deepening of the system (Yanai 1963, 1964). He also remarked in Yanai (1961b, p. 303) that "the banana-shaped storm region associated with an easterly wave appears to be cut off from the main easterly jet" (see Fig. 3-2), perhaps the first hint of the easterly wave "pouch" concept where developing mesoscale vortices are protected from the external hostile environment, allowing development (Dunkerton et al. 2009; Wang et al. 2010).

Throughout the 1960s, Michio was thinking not only about the genesis of tropical storms but also about the genesis of the other equatorial disturbances that were

becoming more evident in observational data. Nitta and Yanai (1969) suggested that the disturbances observed in rainfall in the Marshall Islands (9.00°N, 168.00°E) region were due to the barotropic instability of the easterly zonal current found in the vicinity of the ITCZ. This conclusion found support in later work (e.g., Ferreira and Schubert 1997; Wang and Magnusdottir 2005), while other hypotheses include the roles of baroclinicity aided by convectively generated potential vorticity and/or inertial instability either of the pure kind or along isentropes, concepts that have still recently been pursued as explanations for easterly wave generation (e.g., Tomas and Webster 1997; Toma and Webster 2010). Michio's earliest work thus provided a solid foundation for the idea of large-scale hydrodynamic instability within the ITCZ and provided inspiration for much of the research on equatorial waves and tropical storm development over the following decades to the present.

3. Yanai–Maruyama waves

In the mid-1960s, Michio and his graduate student Taketo Maruyama set out on a search for observational evidence of equatorial disturbances in the central equatorial Pacific using the Marshall Islands data. In this study, they "indeed found interesting facts" corresponding to eddy disturbances in the lower stratospheric winds (Yanai and Maruyama 1966, p. 291). What they found were roughly 5-day-period meridional wind oscillations in the lower stratosphere corresponding to waves traveling westward at around $23 \, \text{m s}^{-1}$ with a wavelength of 10 000 km or so (Fig. 3-3).

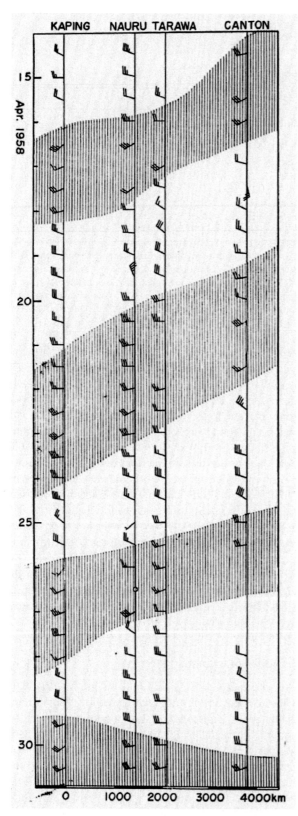

FIG. 3-3. Time series of 70 000-ft (or 21 km) winds at Kapinga-marangi, Nauru, Tarawa, and Canton Island for 15–30 Apr 1958. Winds with southerly components are shaded. Adopted from Yanai and Maruyama (1966).

Over the next several years, Michio along with his students Maruyama, Tsuyoshi Nitta, Masato Murakami, and Yoshikazu Hayashi published a classic series of papers documenting these YM waves in observations and theoretically.

It is interesting to note that Yanai and Maruyama's (Yanai and Maruyama 1966; Maruyama and Yanai 1967; Maruyama 1967, 1968) initial work was partly motivated by the search for an explanation for the stratospheric quasi-biennial oscillation (QBO), which had been recently discovered by Ebdon (1960) and Reed et al. (1961). It was theorized that the regular downward-propagating signal of alternating westerly and easterly winds of the QBO should be due to momentum flux convergence produced by equatorial waves (Reed 1962; Tucker 1968). Yanai and Maruyama (1966, p. 291) were also motivated by the fact that "one still lacks observational evidence of large-scale eddies in the tropical stratosphere." While YM waves were presumed to provide part of the necessary acceleration, it is now known that a broad spectrum of forcing must be involved ranging from equatorial waves (Lindzen and Matsuno 1968; Lindzen and Holton 1968; Holton and Lindzen 1972), including intermediate inertia–gravity waves (Dunkerton 1997; Kawatani et al. 2009) and high-frequency gravity waves [see Baldwin et al. (2001) for a review].

At the same time, during the early 1960s, theoretical studies had suggested the existence of a broad class of trapped waves along the equator in addition to off-equatorial easterly waves (Yoshida 1959; Rosenthal 1960, 1965; Bretherton 1964). It was very fortunate that Michio was working "elbow to elbow" with Taroh Matsuno at the University of Tokyo, who was developing the linear theory of the behavior of shallow-water waves on an equatorial beta plane for his Ph.D. work. Matsuno's (1966) theory predicted the existence of several types of equatorially trapped waves, including those now called Kelvin and equatorial Rossby (ER) waves, along with a family of inertio-gravity (IG) and mixed Rossby–gravity (MRG) waves.

Although Michio was acknowledged by Matsuno in his 1966 paper, they apparently did not initially realize that the YM waves provided the first observational confirmation of Matsuno's theory. Quite soon afterward, though, Maruyama (1967, p. 405) examined the observations further and declared that the waves they were observing were indeed "of the mixed characteristics of the Rossby-type wave and the westward moving inertio-gravity wave" isolated by Rosenthal and Matsuno theoretically. This discovery was certainly a beautiful example of where theory and observations worked hand in hand to make rapid advances in a field.

FIG. 3-4. Latitude–time sections of winds at 6, 16, and 18 km in the vicinity of the Line Islands during the period 19 to 27 Jun. The odd and even numerals denote the centers of anticlockwise and clockwise eddies, respectively. Adopted from Yanai and Hayashi (1969).

Further work in the 1960s concentrated on isolating the horizontal structure of YM waves, and their vertical structure, through spectral analysis case studies of data from a wider array of radiosonde sites (Yanai et al. 1968; Yanai and Hayashi 1969; Wallace and Chang 1969; Yanai and Murakami 1970a,b). These studies confirmed that YM waves consisted of eddies with geopotential perturbations centered along the equator with signals being strongest in the upper troposphere (Fig. 3-4). However, coherent signals of the YM waves also appeared to be present near the surface (Fig. 3-5; see section 5). It is notable that in all of these papers there is ongoing speculation that while one energy source for the waves is likely to be latent heating, "there is another possibility of searching for the energy source of the equatorial disturbances in the interaction with higher-latitude disturbances. The low coherence between the equatorial disturbances and those in the subtropics does not necessarily reject a possible impulsive excitation of the equatorial atmosphere by disturbances extending from the subtropics" (Yanai et al. 1968, p. 321). This idea was given further justification

by the two-level model study of Mak (1969), who found a potentially large excitation of YM and ER waves through a realistic lateral forcing at 30°N and 30°S. As we shall see below, the idea of lateral forcing does seem to be quite relevant for a variety of equatorial disturbances.

4. Theoretical developments on equatorial waves

Along with YM waves, observational evidence for other modes predicted by Matsuno's theory was accumulating quickly by 1970. Wallace and Kousky (1968) documented zonal wind fluctuations in the lower stratosphere using radiosonde data that they identified as Kelvin waves. Yanai et al. (1968), Yanai and Hayashi (1969), and Yanai and Murakami (1970a,b) expanded the spectral analysis approach to a wider range of frequencies and included zonal wind fluctuations and made use of the theoretical expectation that, in a resting basic state, Matsuno modes should have either symmetric or antisymmetric structures about the equator, a common practice today.

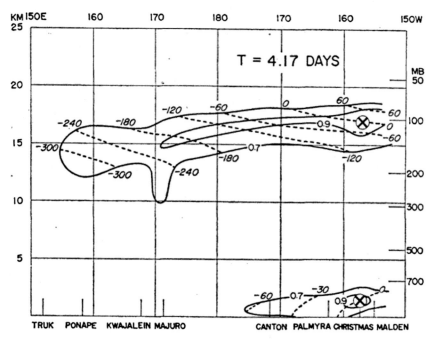

FIG. 3-5. Longitude–height cross section of the coherence (solid lines) and the phase difference (dashed lines) of the v component measured from the 17-km level at Christmas and those measured from the 1.5-km level for the period of 4.17 days. Adopted from Yanai et al. (1968).

As summarized by Wallace (1971), there had been much work done up to this point utilizing spectral approaches developed by Yanai and collaborators. These papers verified many of the predictions of the extensions and applications to Matsuno's theory that were being rapidly developed at that exciting time. For example, Yanai and Murakami (1970b) attempted to account for the scale of the waves using dry, linear theory established by Lindzen (1967) for an isothermal atmosphere, with good results for the YM waves but rather mixed results for Kelvin and ER waves because of the small sample size and low horizontal resolution. Their estimate of an equivalent depth of 41 m, based on the vertical wavelengths for YM waves, was very close to later estimates using satellite cloudiness data (Takayabu 1994a; Wheeler and Kiladis 1999). This is a relatively small value given the observed stratification of the tropical atmosphere (Lindzen 1967; Holton 1970); however, this discrepancy was not commented upon immediately. At around the same time, Hayashi (1971) had incorporated for the effect of heating within YM waves using a conditional instability of the second kind (CISK) formulation for the effect of convective heating, which gave reasonable values for an equivalent depth and for the horizontal scaling of the waves. This pioneering work of Michio's student, along with that of Yamasaki (1969), ultimately led to the formulation of so-called wave-CISK (see Lindzen 1974). It turns out

that the mechanisms that result in such scaling of convectively coupled waves, which undoubtedly arise through the interaction between the convective scale and the large scale, have remained a topic of active research since the 1960s (Neelin and Held 1987) [see discussions in Takayabu (1994a), Wheeler and Kiladis (1999), Kiladis et al. (2009), and Raymond et al. (2009)].

Another result that came out of the pioneering studies of Michio and his collaborators was the evidence that observed equatorial waves exhibited strong tilts in the vertical. A westward tilt with height in the upper troposphere and lower stratosphere was well established for YM waves, and a careful analysis by Yanai et al. (1968) and Nitta (1970a) indicated an eastward tilt within the lower troposphere. Michio pointed out that these tilts suggest that the waves "are excited at the upper tropospheric levels and their wave energy propagates both upward and downward from the excitation levels" (Yanai and Murakami 1970b, p. 345). This was further evidence that, once excited, latent heating within tropical convection was an important energy source for the growth and maintenance of the waves. Michio was particularly impressed with the results of Nitta (1972), who calculated cross spectra in the vertical between Q_1 and the generation of eddy available potential energy (EAPE), finding strong coherence between them at periods near 5 (from YM waves) and 12.5 days (Kelvin waves). Michio commented on the implications for the role of convection: "It is very interesting that the two diagrams are nearly identical, that is, at all frequency ranges the EAPE is generated by diabatic heating and is immediately converted into the EKE [eddy kinetic energy]" (Yanai 1971a, p. 10).

In another prescient statement, Yanai and Murakami (1970a) saw the need for more detailed space–time spectral analysis techniques, which were then in fact being developed by Michio's student Yoshikazu Hayashi (Hayashi 1971). They realized that when enough data became available "to establish a physical interpretation of the time series data, we have to analyze the spectral estimates into different wave modes in a manner similar to the analysis of atmospheric tides" (Yanai and Murakami 1970a, p. 196) as was done, for example, by Longuet-Higgins (1968). This statement foresaw the matching of theoretical dispersion properties of equatorial waves to space–time characteristics of observed fields.

5. Convection and equatorial waves

At around 1970, the motivation to study equatorial waves was further reinforced by some initial work using satellite data. Michio was struck by the work of Chang (1970), who showed an association between tropical wave disturbances and cloud cluster activity (Fig. 3-6). Shortly thereafter, Tanaka and Ryuguji (1971, 1973), Murakami and Ho (1972a,b), and Wallace and Chang (1972), among others, found strong evidence for equatorial wave activity in satellite cloudiness, using some of the first available observations from the *TIROS-1* satellite. The importance of easterly waves for determining the organization of precipitation within the tropics was already long recognized, and Michio and others saw the potential for other equatorial modes to also impact rainfall. The important advance that satellite data offered made it even more relevant to establish the nature of the convective coupling to wave dynamics at low latitudes.

Michio's efforts to understand the relationships between wave and convective activity within the theoretical framework of Matsuno's theory were aided by the arrival of his first Ph.D. student at UCLA, Abraham Zangvil. Zangvil and Yanai (1980, 1981) analyzed the space–time spectra of dynamical and brightness temperature fields using one of the first gridded analysis fields of the tropics produced by Krishnamurti (1971) and a gridded satellite brightness temperature product obtained from NCAR. The use of a grid instead of spotty radiosonde data made it possible to decompose the fields into wavenumber–frequency space, and they

found clear evidence of cloudiness variations coupled to YM, Kelvin, and ER waves and strong coherence between cloudiness and their dynamical fields. Another outgrowth of the work with Zangvil was further evidence confirming the previous findings of Nitta and others, for the equatorward flux of wave energy of zonal wavenumbers 3–6, matching the scale of YM waves (Fig. 3-7).

The early evidence gathered by Yanai and collaborators on equatorial waves provided inspiration for much of the tropical research in the late 1980s and 1990s. An eastward-propagating mode responsible for the large-scale organization of cloud clusters, called super cloud clusters (SCC) by Nakazawa (1988), was later determined to be coupled with tropospheric Kelvin wave disturbances by Takayabu and Murakami (1991) and Takayabu et al. (1999). Liebmann and Hendon (1990) and Hendon and Liebmann (1991) also demonstrated the coupling of tropospheric YM waves and cloudiness utilizing gridded ECMWF wind analyses and outgoing longwave radiation (OLR) data and confirmed the tilted vertical structures and scales of the waves that were isolated in the more limited early observations of Yanai.

Further study revealed a rich variety in YM waves ranging from those coupled to convection to the free waves of the upper troposphere and lower stratosphere. Dunkerton (1991) analyzed radiosonde data from a large set of equatorial stations and found that lower-stratospheric YM waves at 70 hPa tended to appear episodically as localized wave packets and were characterized by a broad range of phase speeds. Similar results were obtained by Randel (1992) but for longer-period (6–10 day) upper-tropospheric YM waves over the eastern Pacific/South America sectors, which were uncoupled to convection and determined to be quite distinct from those over the west Pacific warm pool. Randel (1992) also found strong evidence for extratropical forcing of the wave activity, with wave activity coincident with meridional momentum fluxes from the Southern Hemisphere subtropics. Dunkerton (1993) and Dunkerton and Baldwin (1995) documented the detailed vertical structure of YM and easterly waves in station as well as gridded OLR and ECMWF analyses. Apart from the lower-stratospheric waves, these studies again inferred a broad range of scales associated with tropospheric YM waves, with at least two types of waves of identified. One of these appeared to be uncoupled to convection and more prevalent in the upper troposphere, as were the YM waves studied by Randel (1992), while the other appeared to be more convectively coupled and had maximum amplitude in the lower troposphere.

FIG. 3-6. Time–longitude section of satellite photographs of the period 1 Jul–14 Aug 1967 for the 10°–5°N latitude band in Pacific. The following data are missing: 4 Jul (150°E–155°W), 8 Jul (150°E–160°W), 7 Jul (150°E–150°W, 130°–100°W), 29 Jul (130°–100°W), and 11 Aug (150°E–150°W). Adopted from Chang (1970).

As more complete gridded satellite and dynamical analyses were quickly becoming available in the 1980s, the work of Zangvil and Yanai (1980, 1981) helped to lay the groundwork for the application of Hayashi's (1971) space–time spectral techniques to gridded satellite and meteorological analyses based on satellite and rawinsonde data. Yanai and Lu (1983) documented the existence of Kelvin, YM, and ER waves in the equatorial

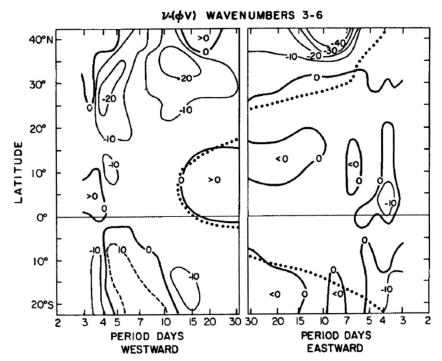

FIG. 3-7. Latitude–frequency section of the meridional flux of wave energy multiplied by frequency (positive northward) for $s = 3$–6 ($m^3 s^{-3}$). Adopted from Zangvil and Yanai (1980).

upper troposphere and related their variability to variations in the equatorward flux of wave energy through the subtropics (see section 6). Takayabu (1994a) plotted the space–time spectral power of satellite brightness temperature in a wavenumber–frequency diagram averaged over latitude, in addition to the more commonly used wavenumber–latitude diagram for a given wavenumber range. These spectra were then directly comparable to dispersion relationships derived by Matsuno, and a wide variety of equatorial waves were revealed including Kelvin, YM, ER, easterly waves [tropical depression (TD) type], and various species of inertio-gravity waves (Fig. 3-8). It was also confirmed that the equivalent depth of these waves indeed corresponded to relatively shallow values of around 15–30 m. A similar application was taken to look specifically at westward inertio-gravity (WIG) waves (Takayabu 1994b). These studies created the impetus for further application of space–time spectral approaches and enabled the development of objective techniques to filter for convectively coupled equatorial waves (CCEWs) (Fig. 3-9; Wheeler and Kiladis 1999). The motivation for much of the more recent work on convectively coupled equatorial waves (Yang et al. 2007; Kiladis et al. 2009; Chen and Tam 2012) and the MJO (see Maloney and Zhang 2016) can indeed be traced back to Michio's long standing interest in the relationship between convection

and the large-scale flow field and to the analysis techniques developed over the years as a result of his huge influence.

6. Extratropical forcing of equatorial waves

Based on a scale analysis, Charney (1963) was among the first to hypothesize that a large portion of the equatorial circulation could be forced by extratropical circulations. Charney (1969) then went on to reason that, in the presence of a realistic zonal-mean basic state, extratropical waves might be prevented from penetrating into the tropics through the existence of a critical line, where the zonal phase speed of an equatorward-propagating disturbance matched the background zonal wind speed (see also Charney and Drazin 1961). It was further argued that, in the absence of convection, such disturbances would be difficult to couple in the vertical once they reached lower latitudes. Charney cited Michio's spectral analysis of Pacific meridional wind (Yanai et al. 1968), which found distinctly different time scales between the lower and upper equatorial troposphere, as evidence of the relatively weak coupling in the vertical of dry tropical motions.

However, Nitta (1970a,b, 1972) demonstrated a lateral flux of wave energy into the tropics from the extratropics in the upper troposphere, and later theoretical

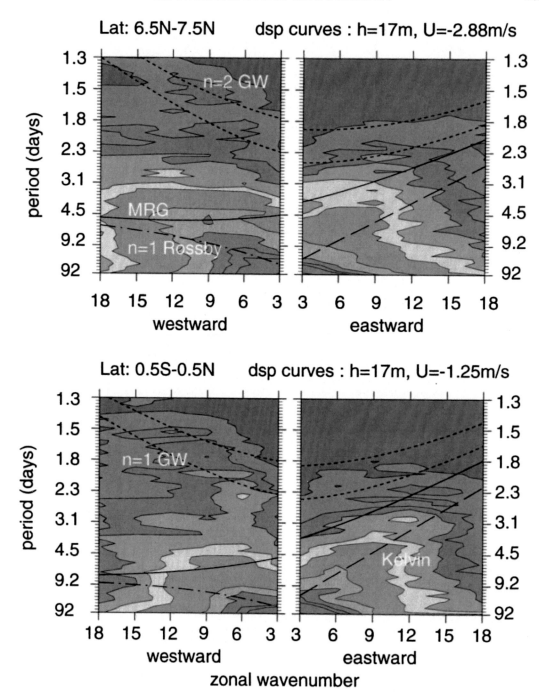

FIG. 3-8. Wavenumber–frequency distribution of space–time power spectral density along the equatorial latitude of (a) 0.5°N–0.5°S and (b) 6.5°–7.5°N. Spectral values are averaged for the entire period from 1981 to 1989, except for 1984. Contours are plotted for 5.3, 7.9, 10.5, 15.7, 21.0, 42.0, 63.0, 84.0, and $105.0 \times 10^{12} \mathrm{K^2 s\,m}$. Doppler shifted dispersion curves of equatorial waves for $n = -1, 0,$ and 1 modes and $n = 2$ inertio-gravity wave mode are superimposed. (Long dashed line indicates Kelvin wave, solid line is MRG, short dashed line is $n = 1, 2$ IGW, and dashed–dotted line is $n = 1$ Rossby wave.) Adopted from Takayabu (1994a).

and modeling work provided evidence that lateral forcing from the extratropics should be possible, and indeed prevalent, in regions of sufficiently broad-scale upper-level westerly flow at low latitudes (e.g., Webster and Holton 1982; Itoh and Ghil 1988; Zhang and Webster 1992). These regions, termed westerly ducts, present no critical line to the equatorward propagation of quasi-stationary Rossby wave energy (Webster and

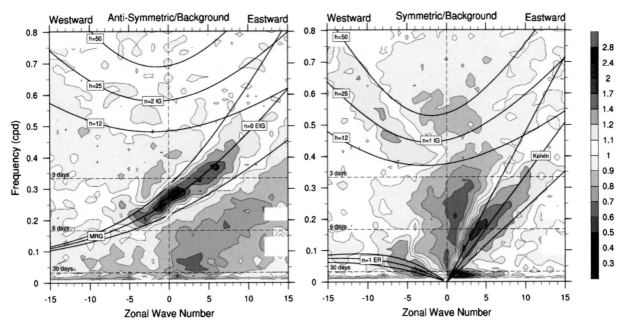

FIG. 3-9. (a) The antisymmetric OLR power divided by the background power. Superimposed are the dispersion curves of the even meridional mode–numbered equatorial waves for the three equivalent depths of $h = 12, 25,$ and 50 m. (b) As in (a), but for the symmetric component of OLR and the corresponding odd meridional mode–numbered equatorial waves. Frequency spectral bandwidth is $1/96$ cpd. Adopted from Wheeler and Kiladis (1999).

Holton 1982). For example, Fig. 3-10 shows the perturbation zonal wind generated in a shallow-water model by a forcing situated at 20°N poleward of a westerly duct, with a clear cross-equatorial response.

During the 1980s and 1990s, Michio's group further documented the importance of lateral forcing as a triggering mechanism for equatorially trapped waves. Zangvil and Yanai (1980) and Yanai and Lu (1983) showed the general patterns of tropical–extratropical interactions that result in Rossby, YM, and Kelvin waves forced by wave energy flux at equatorial latitudes (Fig. 3-7). In various analyses (Zangvil and Yanai 1980; Yanai and Lu 1983; Magaña and Yanai 1995), the wave energy flux of long waves ($s = 3–6$) at 200 mb exhibited a dominance of equatorward concentration in the 4–6-day-period range. Numerous theoretical and modeling studies followed up on the idea that this wave energy could trigger equatorial waves (e.g., Webster and Chang 1988; Itoh and Ghil 1988; Zhang and Webster 1992; Horinouchi et al. 2000).

The importance of lateral forcing led Magaña and Yanai (1991) to explore the possible role of the MJO on the meridional convergence of wave energy flux into the tropics (see Krishnamurti et al. 2016). They showed how the 30–60-day oscillations in tropical convection over the western Pacific may influence the subtropical circulation over a range of longitudes, which in turn could lead to the formation of the westerly duct even during

the Northern Hemisphere summer months. Utilizing ECMWF analysis data, Magaña and Yanai (1995) documented how a midlatitude Rossby wave train can penetrate deeply into the tropics and project onto YM waves, especially in the western Hemisphere within a sufficiently wide equatorial westerly duct or even when the mean easterlies are weak (~ 2 m s^{-1}) (Fig. 3-11).

Other evidence for lateral forcing of equatorial motions has become quite abundant over the past three

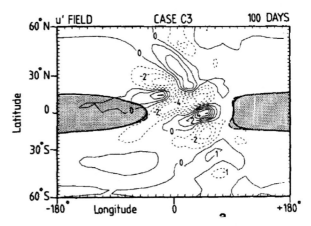

FIG. 3-10. Distribution of the perturbation zonal wind component u' at 100 days for case C3 that considers an $s = 3$ forcing with $\phi = 0$ located at 20°N with basic state C. Adopted from Webster and Holton (1982).

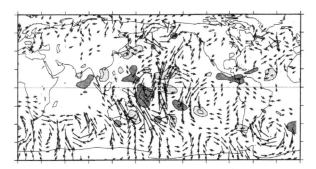

FIG. 3-11. Linear regression between the bandpass-filtered anomalies of v at the reference point x and bandpass-filtered u, v, and OLR anomalies at all points. Dark (light) shaded regions correspond to negative (positive) OLR anomalies. Contour intervals for OLR anomalies: $4\,W\,m^{-2}$. Only wind vectors with magnitudes larger than $0.8\,m\,s^{-1}$ are plotted. For periods of active MRGs. Adopted from Magaña and Yanai (1995).

decades in synoptic-scale studies. For example, Kiladis and Weickmann (1992) demonstrated that extratropical Rossby wave energy propagating equatorward within the westerly duct of the eastern Pacific was responsible for convective activity within the ITCZ. Tomas and Webster (1994) showed that such disturbances could efficiently cross the equator at upper levels but are hindered at low levels by the critical line of the so-called easterly dome due to the trade winds. Similar equatorward wave propagation is also seen in the Atlantic sector, where equatorial westerlies are also present during northern winter. The interaction of this wave activity with the ITCZ is now well established [see the review by Knippertz (2007)].

There is also observational evidence that ER waves can be triggered by extratropical waves penetrating into low latitudes (Yanai and Lu 1983; Kiladis and Wheeler 1995; Kiladis 1998). More recently, Straub and Kiladis (2003) and Liebmann et al. (2009) present evidence of laterally forced equatorial Kelvin waves. For example, in Fig. 3-12, a convectively coupled Kelvin wave signal originating over the west Pacific warm pool is preceded by an extratropical Rossby wave train originating many days earlier over South America and propagating through the southern Indian Ocean storm track. It is especially notable that this wave train is very similar to the one associated with YM waves in Fig. 3-11. These events frequently occur even in the presence of a critical line, and it has been hypothesized that the high-latitude forcing can project directly onto the equatorial wave, yielding an equatorial response in either easterly or westerly low-latitude flow (e.g., Hoskins and Yang 2000). Thus, the early insightful observations of Michio and his collaborators, such as Nitta, that indicated excitation

of equatorial modes by lateral forcing have since been verified since by many studies, highlighting the important role of this source of equatorial wave variability, as was suspected early on by Michio.

7. Revisiting the classical easterly and equatorial waves

In preceding sections, we have reviewed the work on tropical waves where Michio had tremendous contributions. Various studies over the years have revealed that there are different mechanisms responsible for generating or maintaining tropical waves. Off-equatorial 3–5-day gyres traveling along the Atlantic ITCZ may be considered to be classical easterly waves, as these were the first to be described in detail using the Caribbean rawinsonde and surface network (Riehl 1945). Later, atmospheric wave disturbances very similar to Caribbean easterly waves were found in the central to western equatorial Pacific, utilizing the radiosonde network around the Marshall Islands (Palmer 1952). In Palmer (1952), these disturbances were called equatorial waves, since those waves were found closer to the equator, although their characteristics were described as very similar to easterly waves. In the 1960s, the distinction between equatorial waves and easterly waves was generally not very clear (e.g., Nitta and Yanai 1969), mainly because of the similarity of their periodicity and their coexistence as will be described later. However, Michio was careful about the distinction between them in papers (e.g., Yanai et al. 1968). In this section, we revisit these classical easterly waves and their relationship to equatorial waves, and YM waves in particular, through accumulated studies on tropical wave disturbances.

One of the first ideas for the generation mechanism was due to Palmer (1951), who speculated that easterly waves could be maintained through the horizontal shear within an easterly current. Later, Michio pursued the possibility of the generation of easterly waves in the Marshall Island region from barotropic instability (Nitta and Yanai 1969), based on a careful application of the finite differencing approximation (Yanai and Nitta 1968), and showed that the meridional structure of easterly flow in this region could become barotropically unstable, providing one potential source for easterly wave generation with observed spatial scales and propagation speeds.

At the time of GATE, similar arguments were put forth to account for the origin and maintenance of AEWs. Burpee (1972) argued that the African easterly jet satisfied the Charney–Stern condition for barotropic instability (Charney and Stern 1962). Based on the GATE analyses, a consensus was established that

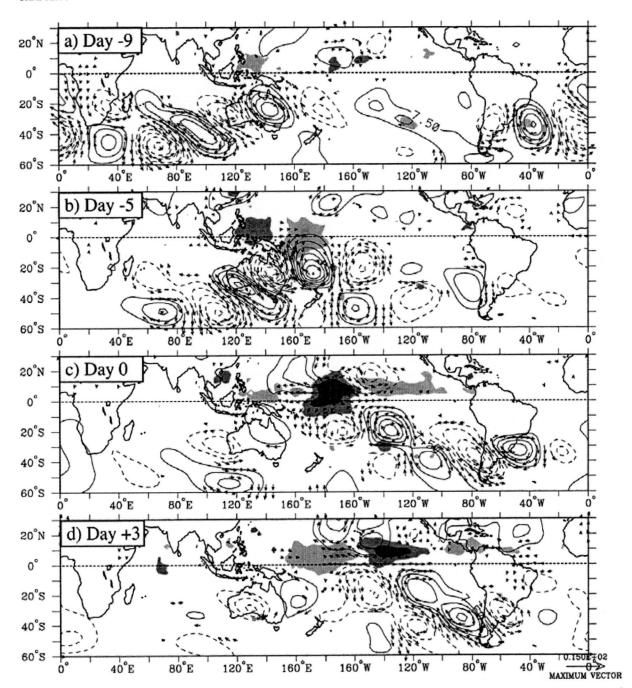

FIG. 3-12. Regressed values of OLR (shading) and 200-hPa streamfunction (contours) and winds (vectors), based on a $-40\,\mathrm{W\,m}^{-2}$ anomaly in Kelvin wave filter OLR at the base point on day 0, for (a) day -9, (b) day -5, (c) day 0, and (d) day $+3$. OLR is shaded at ±6 and $15\,\mathrm{W\,m}^{-2}$; dark shading represents negative OLR anomalies. Stream function contour interval is $7.5 \times 10^{5}\,\mathrm{m}^{2}\,\mathrm{s}^{-1}$; the zero contour has been omitted. The longest wind vectors correspond to a $10\,\mathrm{m\,s}^{-1}$ wind, and are plotted only where either the u or v component is significant at the 95% level or greater. Adopted from Straub and Kiladis (2003).

AEWs were dynamically maintained with barotropic and baroclinic conversions from the lower-tropospheric easterly jet (e.g., Norquist et al. 1977; Reed et al. 1977). However, it appears that linear instability is not necessarily supported by realistic observed basic states over

Africa once reasonable damping is assumed (see Hall et al. 2006), pointing to a role by moist processes in the triggering of AEWs (Thorncroft et al. 2008) as well as their maintenance (e.g., Hsieh and Cook 2008; Cornforth et al. 2009; Berry and Thorncroft 2012).

On the other hand, tropical waves observed over the Marshall Islands, around 150°–175°E, were shown to be maintained primarily by deep convection (Nitta 1972), consistent with the view that there is less horizontal and vertical shear over the western to central Pacific than over Africa. Similarly for the convectively coupled equatorial waves, convective heating in organized convective systems, which consists of deep convection and deep stratiform clouds, is considered the major player in the maintenance of these disturbances.[1]

Differences in the structure of easterly waves from the eastern Pacific to the western Pacific had long been recognized starting with Reed and Recker (1971), who showed the structure change of easterly waves from the central Pacific at Majuro to the western Pacific at Koror and Truk. In the central Pacific, the disturbances have more eastward-tilted structure than in the western Pacific, where the disturbances are more erect, as noted by Dunkerton (1993; see also Serra et al. 2008).

In addition to the vertical structure, early studies with station data (e.g., Nitta 1970a,b; Yanai and Murakami 1970a,b; Wallace 1971) suggested the existence of two kinds of westward-propagating waves. Nitta (1970a) showed that while longer (~8000 km) waves dominate over the central Pacific, shorter (~4000 km) waves dominate over the western Pacific. Liebmann and Hendon's (1990) work visually indicated that former wave disturbance had a YM (MRG) wave structure and was coupled with convection. Takayabu and Nitta (1993) identified the TD-type disturbances that corresponded more with easterly waves and dominated in the western Pacific, distinct from the MRG wave in the central Pacific, although they indicated some cases that the former disturbance was transformed from the latter. One of the conclusions that emerged from Dunkerton and Baldwin (1995) was that some MRG waves in the central to western Pacific appear to transition

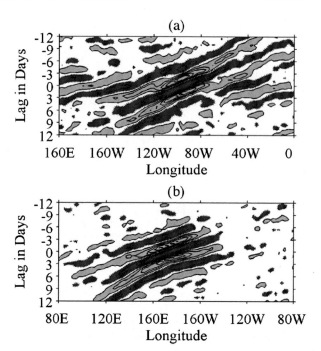

FIG. 3-13. Hovmoeller diagram of TD-filtered OLR regressions averaged from 7.5° to 12.5°N for the base points at (a) 10°N, 95°W and (b) 7.5°N, 172.5°E. Dark gray contours are negative anomalies; lighter gray contours are positive anomalies. Contour lines are drawn at intervals of $2\,\mathrm{W\,m^{-2}}$ starting at $\pm1\,\mathrm{W\,m^{-2}}$. Adopted from Serra et al. (2008).

continuously into the TD-type disturbances. This result corroborated earlier work by Dunkerton (1993) from rawinsonde data alone, that showed evidence of a continuous transition from a tilted to a first baroclinic mode vertical structure of disturbances.

The relative occurrence of MRG waves and TD-type disturbances depends on the background zonal wind structure over the tropical Pacific with MRG waves favored in low-level easterlies and TDs favored in westerlies. Takayabu and Nitta (1993) found such an interannual variation associated with ENSO. Recalling that there are at least some inconsistencies in earlier studies for the naming of the 3–5-day disturbances over the Marshall Island region (e.g., Palmer 1952 versus Nitta and Yanai 1969), some of these discrepancies may have their origin due to this interannual variability.

Studies of easterly waves in the eastern Pacific ITCZ are more prevalent in more recent years, which was partly due to the lack of a rawinsonde station network in this region. In agreement with classical studies dating at least to 1970, Serra et al. (2008, 2010) showed that some easterly waves originating in the Atlantic propagate across Costa Rica and Panama to the eastern Pacific (Fig. 3-13), although they are not the primary origin of Pacific disturbances. In the eastern Pacific at around

---

[1] More recently, the role of a third type of cloud, shallow/congestus, has been emphasized (Johnson et al. 1999). Usually, these clouds are assumed to play a primary role in moistening the lower troposphere above the boundary layer ahead of convectively coupled disturbances, although there is some dispute on this point (Hohenegger and Stevens 2013). Kiladis et al. (2005) and Mapes et al. (2006) remarked on the fact that a wide variety of cloud systems with different scales have similar life cycles, consisting of a progression of shallow/congestus, deep, and stratiform clouds as they pass by (e.g., Zipser 1969; Takayabu et al. 1996; Straub and Kiladis 2003; Lin and Johnson 1996). Frenkel et al. (2012) showed that realistic structures of various cloud systems are well simulated by stochastic cloud models, including the effects of the three cloud types, as demonstrated by Majda and Khouider (2002) and Khouider and Majda (2006) (see Fovell et al. 2016; Zhang and Song 2016).

80°–95°W, they show that barotropic conversion has similar magnitudes in the lower troposphere as the contribution from convection.

The eastern Pacific ITCZ region is characterized by more prevalent contributions by shallow convection to the heating profile when compared to the western Pacific warm pool region, as illuminated by recent studies. Zhang et al. (2004) indicated the existence of a shallow meridional circulation in the eastern Pacific at around 2–4-km altitude, utilizing dropsonde soundings obtained in the Eastern Pacific Investigation of Climate Processes in the Coupled Ocean–Atmosphere System (EPIC) program. Kubar et al. (2007) showed a larger ratio of midlevel-topped clouds in the eastern Pacific compared to the western or central Pacific. Back and Bretherton (2009a,b) emphasized the significance of a shallow mode of convection in the eastern Pacific ITCZ region, accompanying convergence in the boundary layer associated with strong SST gradients associated with the so-called cold tongue in the equatorial eastern Pacific. Shallow heating may also be enhancing the low-level convergence through convection–circulation coupling (Wu 2003).

An analysis of TRMM precipitation features (PFs), which are contiguous rain areas defined by the TRMM PR data and are provided by the University of Utah group (e.g., Nesbitt et al. 2006), described significant differences of convective systems between the western Pacific (WPAC) warm pool and over the eastern Pacific (EPAC) ITCZ (Yokoyama and Takayabu 2012a,b). Over the WPAC warm pool, tall and deep structures are prominent. On the other hand, over the EPAC ITCZ, small and shallow features rich in cumulus congestus and highly organized features with moderate heights are dominant, collocated with regions of shallow convergence associated with large meridional SST gradients.

The significant contributions of shallow convection observed in temporally averaged states over the eastern Pacific ITCZ are related to transient atmospheric disturbances. Serra and Houze (2002) analyzed the observational data obtained during the Tropical East Pacific Process Studies (TEPPS). Their time–height plots of sounding data indicate that premoistening below the 0°C level precedes the precipitation events associated with 3–6-day wave disturbances. Roundy and Frank (2004) showed an interesting snapshot of easterly wave disturbances over the central to eastern equatorial Pacific, with a map of TD-type band-filtered precipitable water (PW), OLR, and 850-hPa winds on 7 December 1992. In the map, disturbances are clearly detected in precipitable water signals and associated with an MRG wavelike circulation over the equator.

Utilizing NOAA OLR data, total precipitable water data derived from SSM/I, surface winds from the

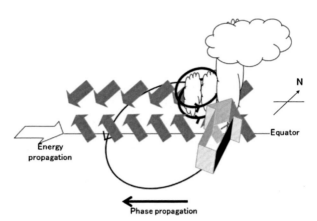

FIG. 3-14. Schematic for the relationships among cumulus convection, the synoptic-scale coupled disturbances, and the large-scale environment over the eastern Pacific in the boreal autumn. Small and large circles indicate a vortex disturbance and a MRG wave–type disturbance, respectively. The thick light gray arrow represents deep convergence associated with the MRG wave–type disturbance. Dark gray arrows indicate shallow convergence, which is largely driven by the strong SST gradient. Black solid and open arrows indicate phase and energy propagation, respectively. Small and large cloud-shaped figures are depicted as congestus and well-organized systems, respectively. Adopted from Yokoyama and Takayabu (2012b).

QuikSCAT, 25-yr Japanese reanalysis, and TRMM latent heating data during the boreal autumn, Yokoyama and Takayabu (2012b) statistically obtained the coupled structure of the ITCZ disturbances over the eastern Pacific centered around 130°W, as summarized in Fig. 3-14. It consists of vortices tilted from SW to NE along the ITCZ latitude, and larger-scale MRG wave disturbances, over the equator. At both these latitudes, westward-propagating 3–6-day disturbances were observed, but the phase speeds and wavelengths of the vertical and MRG wavelike disturbances suggest they have separate origins. Two-tiered convergence–divergence profiles are observed with these coupled disturbances. In the western part of the disturbances, deep convection is suppressed and shallow convection is observed. Much deeper convergence up to 700 hPa is observed in the eastern part, associated with southerly inflow of the MRG waves. This helps to allow deep convection in the rear of the vortex disturbance at the ITCZ and enhances the coupled wave disturbances. When these two components become coupled, they tend to propagate westward together over a period of several days.

Recent studies have revealed that tropical deep convection is more correlated with free-tropospheric humidity rather than buoyancy instability of the shallower boundary layer parcels represented by CAPE (e.g., Sherwood 1999; Bretherton et al. 2004; Takemi et al. 2004; Neggers et al. 2007; Takayabu et al. 2010). The above-mentioned coupled structure of the ITCZ disturbances show that shallow convection associated with one disturbance is aided by the deep convergence of a large-scale disturbance, which helps to prepare the environment of the moist free troposphere for deep convection to develop.

To summarize, it is suggested that dynamic instability and barotropic conversion in the shallow boundary layer generate and maintain the shallow vortex disturbances in the eastern Pacific ITCZ, which are coupled to congestus heating. When conditions allow, associated with the arrival of energy from larger-scale MRG waves, deeper convergence in the lower troposphere allows deep convection in a certain phase of the shallow vortex.

The coupling of shallower and deeper disturbances in the eastern Pacific ITCZ is consistent with previous descriptions of these disturbance structures. While disturbances in the west Pacific have erect structures with a single maximum of vertical velocity in the upper troposphere, AEWs (Thompson et al. 1979; Kiladis et al. 2006) and wave disturbances in the eastern Pacific (Tai and Ogura 1987; Serra et al. 2008; Yokoyama and Takayabu 2012b) are associated with double maxima of vertical velocity in the lower and upper troposphere. With both AEW and classical eastern Pacific easterly waves, there are indications of lower-level barotropic instability triggering the disturbance, which is coupled with shallow convection, especially in the eastern Pacific cases. Then further coupling with deep convection enhances the disturbances with deep convective heating.

From the very early stage of the YM wave studies, Michio was interested in the relationship between upper-level YM waves and the lower-level easterly waves. As was already shown in Fig. 3-5, Yanai et al. (1968) showed a coexistence of long distance propagation of large-scale YM waves from the upper troposphere to lower stratosphere and short distance correlation of wave disturbances in the lower troposphere at around 1.5 km, at a very similar period range around 4 days. Although Michio did not explicitly express that the latter lower-tropospheric coherent disturbances was a part of YM or MRG waves, he suggested that these spectral peaks may correspond to the passage of equatorial waves of the type discussed by Palmer (1952), considering that the latitudes of stations showing these peaks are close to the equator. From Fig. 3-5, the zonal scale of these lower-tropospheric wave disturbances is estimated at wavenumber 4, which is distinct from the smaller-scale easterly waves. Utilizing results from recent studies (Roundy and Frank 2004; Yokoyama and Takayabu 2012b), we can now revisit the equatorial waves described by Palmer (1952), which should correspond to the lower-tropospheric MRG

waves, and distinguish these from the easterly waves over the ITCZ. They can be consistently understood as two disturbances with separate origins, which are coupled and amplified through an enhancement of deep convection over a limited longitudinal range of the eastern Pacific ITCZ. Given the limited data availability available to them, we can certainly appreciate the marvelous insights of early tropical meteorologists in assigning two separate names, easterly waves and equatorial waves, to these disturbances in the early days of the 1950s and 1960s.

Professor Yanai, in his review for the design of GATE, emphasized the importance of understanding multiscale interaction in tropical waves and coupled tropical convection: "Finally, I emphasize that the B-scale network must be meshed within a good A-scale network. Only by doing so, will we be able to associate a particular type of convective heating with a particular type of wave disturbance" (Yanai 1971a, p. 29). Successive studies have shown abundant variations in tropical waves, and the selection of waves has been related to the large-scale basic state. One of the facts initially emphasized by Michio, that the scale of tropical convection was much smaller compared to that of tropical waves, has ultimately led to recent field projects such as the Pre-Depression Investigation of Cloud-Systems in the Tropics (PREDICT) experiment (Montgomery et al. 2012). PREDICT was designed to understand which waves are apt to develop into the tropical cyclones and which are not, in terms of convective effects on the pregenesis disturbances, the very problem studied by Michio in the early 1960s in his papers on the genesis of Typhoon Doris. Another recent field project, Dynamics of the MJO (DYNAMO)/ Cooperative Indian Ocean Experiment on Intraseasonal Variability in Year 2011 (CINDY) (e.g., Zhang et al. 2013; Yoneyama et al. 2013), is designed to examine the role of mesoscale to synoptic-scale interactions in the genesis and maintenance of the planetary-scale MJO (see Krishnamurti et al. 2016). Thus, Michio's insightful views have ultimately played a pivotal role in motivating much recent observational, theoretical, and modeling research aimed at unraveling the workings of multiscale interactions within the tropical atmosphere.

Acknowledgments. We thank Tim Dunkerton and another anonymous reviewer for their thorough reading of the manuscript and their many excellent suggestions that led to significant improvements in this chapter.

REFERENCES

Back, L. E., and C. S. Bretherton, 2009a: On the relationship between SST gradients, boundary layer winds, and convergence over the tropical oceans. *J. Climate*, **22**, 4182–4196, doi:10.1175/2009JCLI2392.1.

——, and ——, 2009b: A simple model of climatological rainfall and vertical motion patterns over the tropical oceans. *J. Climate*, **22**, 6477–6497, doi:10.1175/2009JCLI2393.1.

Baldwin, M. P., and Coauthors, 2001: The quasi-biennial oscillation. *Rev. Geophys.*, **39**, 179–229, doi:10.1029/1999RG000073.

Berry, G. J., and C. D. Thorncroft, 2012: African easterly wave dynamics in a mesoscale numerical model: The upscale role of convection. *J. Atmos. Sci.*, **69**, 1267–1283, doi:10.1175/JAS-D-11-099.1.

Bretherton, C. S., M. E. Peters, and L. E. Back, 2004: Relationships between water vapor path and precipitation over the tropical oceans. *J. Climate*, **17**, 1517–1528, doi:10.1175/1520-0442(2004)017<1517:RBWVPA>2.0.CO;2.

Bretherton, F. P., 1964: Low frequency oscillations trapped near the equator. *Tellus*, **16**, 181–185, doi:10.1111/j.2153-3490.1964.tb00159.x.

Burpee, R. W., 1972: The origin and structure of easterly waves in the lower troposphere of North Africa. *J. Atmos. Sci.*, **29**, 77–89, doi:10.1175/1520-0469(1972)029<0077:TOASOE>2.0.CO;2.

Chang, C.-P., 1970: Westward propagating cloud patterns in the tropical Pacific as seen from time-composite satellite photographs. *J. Atmos. Sci.*, **27**, 133–138, doi:10.1175/1520-0469(1970)027<0133:WPCPIT>2.0.CO;2.

Charney, J. G., 1963: A note on large-scale motions in the tropics. *J. Atmos. Sci.*, **20**, 607–609, doi:10.1175/1520-0469(1963)020<0607:ANOLSM>2.0.CO;2.

——, 1969: A further note on large-scale motions in the tropics. *J. Atmos. Sci.*, **26**, 182–185, doi:10.1175/1520-0469(1969)026<0182:AFNOLS>2.0.CO;2.

——, and P. G. Drazin, 1961: Propagation of planetary-scale disturbances from the lower into the upper atmosphere. *J. Geophys. Res.*, **66**, 83–109, doi:10.1029/JZ066i001p00083.

——, and M. E. Stern, 1962: On the stability of internal baroclinic jets in a rotating atmosphere. *J. Atmos. Sci.*, **19**, 159–172, doi:10.1175/1520-0469(1962)019<0159:OTSOIB>2.0.CO;2.

Chen, G., and C.-Y. Tam, 2012: A new perspective on the excitation of low-tropospheric mixed Rossby–gravity waves in association with energy dispersion. *J. Atmos. Sci.*, **69**, 1397–1403, doi:10.1175/JAS-D-11-0331.1.

Cornforth, R. J., B. J. Hoskins, and C. D. Thorncroft, 2009: The impact of moist processes on the African easterly jet and African easterly wave system. *Quart. J. Roy. Meteor. Soc.*, **135**, 894–913, doi:10.1002/qj.414.

Dunkerton, T. J., 1991: Intensity variation and coherence of 3–6 day equatorial waves. *Geophys. Res. Lett.*, **18**, 1469–1472, doi:10.1029/91GL01780.

——, 1993: Observation of 3–6-day meridional wind oscillations over the tropical Pacific, 1973–1992: Vertical structure and interannual variability. *J. Atmos. Sci.*, **50**, 3292–3291, doi:10.1175/1520-0469(1993)050<3292:OODMWO>2.0.CO;2.

——, 1997: The role of gravity waves in the quasi-biennial oscillation. *J. Geophys. Res.*, **102**, 26 053–26 076, doi:10.1029/96JD02999.

——, and M. P. Baldwin, 1995: Observation of 3–6-day meridional wind oscillations over the tropical Pacific, 1973–1992: Horizontal structure and propagation. *J. Atmos. Sci.*, **52**, 1585–1601, doi:10.1175/1520-0469(1995)052<1585:OODMWO>2.0.CO;2.

——, M. T. Montgomery, and Z. Wang, 2009: Tropical cyclogenesis in a tropical wave critical layer: Easterly waves. *Atmos. Chem. Phys.*, **9**, 5587–5646, doi:10.5194/acp-9-5587-2009.

Ebdon, R. A., 1960: Notes on the wind flow at 50 mb in tropical and sub-tropical regions in January 1957 and January 1958. *Quart. J. Roy. Meteor. Soc.*, **86**, 540–542, doi:10.1002/qj.49708637011.

Ferreira, R. N., and W. H. Schubert, 1997: Barotropic aspects of ITCZ breakdown. *J. Atmos. Sci.*, **54**, 261–285, doi:10.1175/1520-0469(1997)054<0261:BAOIB>2.0.CO;2.

Fovell, R. G., Y. P. Bu, K. L. Corbosiero, W.-W. Tung, Y. Cao, H.-C. Kuo, L.-H. Hsu, and H. Su, 2016: Influence of cloud microphysics and radiation on tropical cyclone structure and motion. *Multiscale Convection-Coupled Systems in the Tropics: A Tribute to Dr. Michio Yanai, Meteor. Monogr.*, No. 56, Amer. Meteor. Soc., doi:10.1175/AMSMONOGRAPHS-D-15-0006.1.

Frenkel, Y., A. Majda, and B. Khouider, 2012: Using the stochastic multicloud model to improve tropical convective parameterization: A paradigm example. *J. Atmos. Sci.*, **69**, 1080–1104, doi:10.1175/JAS-D-11-0148.1.

Gamache, J. F., and R. A. Houze, 1985: Further analysis of the composite wind and thermodynamic structure of the 12 September GATE squall line. *Mon. Wea. Rev.*, **113**, 1241–1259, doi:10.1175/1520-0493(1985)113<1241:FAOTCW>2.0.CO;2.

Hall, N. M. J., G. N. Kiladis, and C. D. Thorncroft, 2006: Three-dimensional structure and dynamics of African easterly waves. Part II: Dynamical modes. *J. Atmos. Sci.*, **63**, 2231–2245, doi:10.1175/JAS3742.1.

Hayashi, Y., 1971: A generalized method of resolving disturbances into progressive and retrogressive waves by space Fourier and time cross-spectral analyses. *J. Meteor. Soc. Japan*, **49**, 125–128.

Hendon, H.-H., and B. Liebmann, 1991: The structure and annual variation of antisymmetric fluctuations of tropical convection and their association with Rossby–gravity waves. *J. Atmos. Sci.*, **48**, 2127–2140, doi:10.1175/1520-0469(1991)048<2127:TSAAVO>2.0.CO;2.

Hohenegger, C., and B. Stevens, 2013: Preconditioning deep convection with cumulus congestus. *J. Atmos. Sci.*, **70**, 448–464, doi:10.1175/JAS-D-12-089.1.

Holton, J. R., 1970: A note on forced equatorial waves. *Mon. Wea. Rev.*, **98**, 614–615, doi:10.1175/1520-0493(1970)098<0614:ANOFEW>2.3.CO;2.

——, and R. S. Lindzen, 1972: An updated theory for the quasi-biennal cycle of the tropical stratosphere. *J. Atmos. Sci.*, **29**, 1076–1080, doi:10.1175/1520-0469(1972)029<1076:AUTFTQ>2.0.CO;2.

Horinouchi, T., F. Sassi, and B. A. Boville, 2000: Synoptic-scale Rossby waves and the geographic distribution of lateral transport routes between the tropics and the extratropics in the lower stratosphere. *J. Geophys. Res.*, **105**, 26 579–26 592, doi:10.1029/2000JD900281.

Hoskins, B. J., and G. Y. Yang, 2000: The equatorial response to higher latitude forcing. *J. Atmos. Sci.*, **57**, 1197–1213, doi:10.1175/1520-0469(2000)057<1197:TERTHL>2.0.CO;2.

Houze, R. A., Jr., and A. K. Betts, 1981: Convection in GATE. *Rev. Geophys.*, **19**, 541–576, doi:10.1029/RG019i004p00541.

Hsieh, J.-S., and K. H. Cook, 2008: On the instability of the African easterly jet and the generation of African waves: Reversals of the potential vorticity gradient. *J. Atmos. Sci.*, **65**, 2130–2151, doi.org/10.1175/2007JAS2552.1.

Itoh, H., and M. Ghil, 1988: The generation mechanism of mixed Rossby–gravity waves in the equatorial troposphere. *J. Atmos. Sci.*, **45**, 585–604, doi:10.1175/1520-0469(1988)045<0585:TGMOMR>2.0.CO;2.

Johnson, R. H., T. Rickenbach, S. A. Rutledge, P. Ciesielski, and W. Schubert, 1999: Trimodal characteristics of tropical convection. *J. Climate*, **12**, 2397–2418, doi:10.1175/1520-0442(1999)012<2397:TCOTC>2.0.CO;2.

Kawatani, Y., M. Takahashi, K. Sato, S. P. Alexander, and T. Tsuda, 2009: Global distribution of atmospheric waves in the equatorial upper troposphere and lower stratosphere:

AGCM simulation of sources and propagation. *J. Geophys. Res.*, **114**, D01102, doi:10.1029/2008JD010374.

Khouider, B., and A. Majda, 2006: A simple multicloud parameterization for convectively coupled tropical waves. Part I: Linear analysis. *J. Atmos. Sci.*, **63**, 1308–1323, doi:10.1175/JAS3677.1.

Kiladis, G. N., 1998: Observations of Rossby waves linked to convection over the eastern tropical Pacific. *J. Atmos. Sci.*, **55**, 321–339, doi:10.1175/1520-0469(1998)055<0321:OORWLT>2.0.CO;2.

——, and K. M. Weickmann, 1992: Extratropical forcing of tropical Pacific convection during northern winter. *Mon. Wea. Rev.*, **120**, 1924–1939, doi:10.1175/1520-0493(1992)120<1924:EFOTPC>2.0.CO;2.

——, and M. Wheeler, 1995: Horizontal and vertical structure of observed tropospheric equatorial Rossby waves. *J. Geophys. Res.*, **100**, 22 981–22 997, doi:10.1029/95JD02415.

——, K. H. Straub, and P. T. Haertel, 2005: Zonal and vertical structure of the Madden–Julian oscillation. *J. Atmos. Sci.*, **62**, 2790–2809, doi:10.1175/JAS3520.1.

——, C. D. Thorncroft, and N. M. J. Hall, 2006: Three-dimensional structure and dynamics of African easterly waves. Part I: Observations. *J. Atmos. Sci.*, **63**, 2212–2230, doi:10.1175/JAS3741.1.

——, M. C. Wheeler, P. T. Haertel, K. H. Straub, and P. E. Roundy, 2009: Convectively coupled equatorial waves. *Rev. Geophys.*, **47**, RG2003, doi:10.1029/2008RG000266.

Knippertz, P., 2007: Tropical–extratropical interactions related to upper-level troughs at low latitudes. *Dyn. Atmos. Oceans*, **43**, 36–62, doi:10.1016/j.dynatmoce.2006.06.003.

Krishnamurti, T. N., 1971: Observational study of the tropical upper tropospheric motion field during the Northern Hemisphere summer. *J. Appl. Meteor.*, **10**, 1066–1096, doi:10.1175/1520-0450(1971)010<1066:OSOTTU>2.0.CO;2.

——, R. Krishnamurti, A. Simon, A. Thomas, and V. Kumar, 2016: A mechanism of the MJO invoking scale interactions. *Multiscale Convection-Coupled Systems in the Tropics: A Tribute to Dr. Michio Yanai*, Meteor. Monogr., No. 56, Amer. Meteor. Soc., doi:10.1175/AMSMONOGRAPHS-D-15-0009.1.

Kubar, T. L., D. A. Hartman, and R. Wood, 2007: Radiative and convective driving of tropical high clouds. *J. Climate*, **20**, 5510–5526, doi:10.1175/2007JCLI1628.1.

Liebmann, B., and H.-H. Hendon, 1990: Synoptic-scale disturbances near the equator. *J. Atmos. Sci.*, **47**, 1463–1479, doi:10.1175/1520-0469(1990)047<1463:SSDNTE>2.0.CO;2.

——, G. N. Kiladis, L. M. V. Carvalho, C. Jones, C. S. Vera, I. Bladé, and D. Allured, 2009: Origin of convectively coupled Kelvin waves over South America. *J. Climate*, **22**, 300–315, doi:10.1175/2008JCLI2340.1.

Lin, X., and R. H. Johnson, 1996: Kinematic and thermodynamic characteristics of the flow over the western Pacific warm pool during TOGA COARE. *J. Atmos. Sci.*, **53**, 695–715, doi:10.1175/1520-0469(1996)053<0695:KATCOT>2.0.CO;2.

Lindzen, R. S., 1967: Planetary waves on beta planes. *Mon. Wea. Rev.*, **95**, 441–451, doi:10.1175/1520-0493(1967)095<0441:PWOBP>2.3.CO;2.

——, 1974: Wave-CISK and tropical spectra. *J. Atmos. Sci.*, **31**, 1447–1449, doi:10.1175/1520-0469(1974)031<1447:WCATS>2.0.CO;2.

——, and J. R. Holton, 1968: A theory of the quasi-biennial oscillation. *J. Atmos. Sci.*, **25**, 1095–1107, doi:10.1175/1520-0469(1968)025<1095:ATOTQB>2.0.CO;2.

——, and T. Matsuno, 1968: On the nature of large scale wave disturbances in the equatorial lower stratosphere. *J. Meteor. Soc. Japan*, **46**, 215–220.

Longuet-Higgins, M. S., 1968: The eigenfunctions of Laplace's tidal equations over a sphere. *Philos. Trans. Roy. Soc. London*,

A262, 511–607. [Available online at http://www.jstor.org/stable/73582.]

Magaña, V., and M. Yanai, 1991: Tropical–midlatitude interaction on the time scale of 30 to 60 days during the northern summer of 1979. *J. Climate*, **4**, 180–201, doi:10.1175/1520-0442(1991)004<0180:TMIOTT>2.0.CO;2.

——, and ——, 1995: Mixed Rossby–gravity waves triggered by lateral forcing. *J. Atmos. Sci.*, **52**, 1473–1486, doi:10.1175/1520-0469(1995)052<1473:MRWTBL>2.0.CO;2.

Majda, A., and B. Khouider, 2002: Stochastic and mesoscopic models or tropical convection. *Proc. Natl. Acad. Sci. USA*, **99**, 1123–1128, doi:10.1073/pnas.032663199.

Mak, M.-K., 1969: Laterally driven stochastic motions in the tropics. *J. Atmos. Sci.*, **26**, 41–64, doi:10.1175/1520-0469(1969)026<0041:LDSMIT>2.0.CO;2.

Maloney, E. D., and C. Zhang, 2016: Dr. Yanai's contributions to the discovery and science of the MJO. *Multiscale Convection-Coupled Systems in the Tropics: A Tribute to Dr. Michio Yanai*, Meteor. Monogr., No. 56, Amer. Meteor. Soc., doi:10.1175/AMSMONOGRAPHS-D-15-0003.1.

Mapes, B., S. Tulich, J.-L. Lin, and P. Zuidema, 2006: The mesoscale convection life cycle: Building block or prototype for large-scale tropical waves? *Dyn. Atmos. Oceans*, **42**, 3–29, doi:10.1016/j.dynatmoce.2006.03.003.

Maruyama, T., 1967: Large-scale disturbances in the equatorial lower stratosphere. *J. Meteor. Soc. Japan*, **45**, 391–408.

——, 1968: Time sequence of power spectra of disturbances in the equatorial lower stratosphere in relation to the quasi-biennial oscillation. *J. Meteor. Soc. Japan*, **46**, 327–341.

——, and M. Yanai, 1967: Evidence of large-scale wave disturbances in the equatorial lower stratosphere. *J. Meteor. Soc. Japan*, **45**, 196–199.

Matsuno, T., 1966: Quasi-geostrophic motions in the equatorial area. *J. Meteor. Soc. Japan*, **44**, 25–43.

Montgomery, M. T., and Coauthors, 2012: The Pre-Depression Investigation of Cloud-Systems in the Tropics (PREDICT) experiment: Scientific basis, new analysis tools, and some first results. *Bull. Amer. Meteor. Sci.*, **93**, 153–172, doi:10.1175/BAMS-D-11-00046.1.

Murakami, T., and F. P. Ho, 1972a: Spectrum analysis of cloudiness over the northern Pacific. *J. Meteor. Soc. Japan*, **50**, 285–300.

——, and ——, 1972b: Spectrum analysis of cloudiness over the Pacific. *J. Meteor. Soc. Japan*, **50**, 301–311.

Nakazawa, T., 1988: Tropical super clusters within intraseasonal variations over the western Pacific. *J. Meteor. Soc. Japan*, **66**, 823–839.

Neelin, J. D., and I. M. Held, 1987: Modeling tropical convergence based on the moist static energy budget. *Mon. Wea. Rev.*, **115**, 3–12, doi:10.1175/1520-0493(1987)115<0003:MTCBOT>2.0.CO;2.

Neggers, R. A. J., J. D. Neelin, and B. Stevens, 2007: Impact mechanisms of shallow cumulus convection on tropical climate dynamics. *J. Climate*, **20**, 2623–2642, doi:10.1175/JCLI4079.1.

Nesbitt, S. W., R. Cifelli, and S. A. Rutledge, 2006: Storm morphology and rainfall characteristics of TRMM precipitation features. *Mon. Wea. Rev.*, **134**, 2702–2721, doi:10.1175/MWR3200.1.

Nitta, T., 1970a: Statistical study of tropospheric wave disturbances in the tropical Pacific region. *J. Meteor. Soc. Japan*, **48**, 47–60.

——, 1970b: A study of generation and conversion of eddy available potential energy in the tropics. *J. Meteor. Soc. Japan*, **48**, 524–528.

——, 1972: Energy budget of wave disturbances over the Marshall Islands during the years of 1956 and 1958. *J. Meteor. Soc. Japan*, **50**, 71–84.

——, 1977: Response of cumulus updraft and downdraft to GATE A/B-scale motion systems. *J. Atmos. Sci.*, **34**, 1163–1186, doi:10.1175/1520-0469(1977)034<1163:ROCUAD>2.0.CO;2.

——, 1978: A diagnostic study of interaction of cumulus updrafts and downdrafts with large-scale motions in GATE. *J. Meteor. Soc. Japan*, **56**, 232–242.

——, and M. Yanai, 1969: A note on the barotropic instability of the tropical easterly current. *J. Meteor. Soc. Japan*, **47**, 127–130.

Norquist, D. C., E. E. Recker, and R. J. Reed, 1977: The energetics of African wave disturbances as observed during Phase III of GATE. *Mon. Wea. Rev.*, **105**, 334–342, doi:10.1175/1520-0493(1977)105<0334:TEOAWD>2.0.CO;2.

Palmer, C. E., 1951: Tropical meteorology. *Compendium of Meteorology*, Amer. Meteor. Soc., 859–880.

——, 1952: Tropical meteorology. *Quart. J. Roy. Meteor. Soc.*, **78**, 126–163, doi:10.1002/qj.49707833603.

Priestley, C. H. B., 1959: *Turbulent Transfer in the Lower Atmosphere.* University of Chicago Press, 130 pp.

Randel, W. J., 1992: Upper tropospheric equatorial waves in ECMWF analyses. *Quart. J. Roy. Meteor. Soc.*, **118**, 365–394, doi:10.1002/qj.49711850409.

Raymond, D. J., S. L. Sessions, A. H. Sobel, and Z. Fuchs, 2009: The mechanics of gross moist stability. *J. Adv. Model. Earth Syst.*, **1** (9), doi:10.3894/JAMES.2009.1.9.

Reed, R. J., 1962: Evidence of geostrophic motion in the equatorial stratosphere. *Quart. J. Roy. Meteor. Soc.*, **88**, 324–327, doi:10.1002/qj.49708837711.

——, and E. E. Recker, 1971: Structure and properties of synoptic-scale wave disturbances in the equatorial western Pacific. *J. Atmos. Sci.*, **28**, 1117–1133, doi:10.1175/1520-0469(1971)028<1117:SAPOSS>2.0.CO;2.

——, W. J. Campbell, L. A. Rasmussen, and D. G. Rogers, 1961: Evidence of downward-propagating annual wind reversal in the equatorial stratosphere. *J. Geophys. Res.*, **66**, 813–818, doi:10.1029/JZ066i003p00813.

——, D. C. Norquist, and E. E. Recker, 1977: The structure and properties of African wave disturbances as observed during Phase III of GATE. *Mon. Wea. Rev.*, **105**, 317–333, doi:10.1175/1520-0493(1977)105<0317:TSAPOA>2.0.CO;2.

Riehl, H., 1945: Waves in the easterlies and the polar front in the tropics. Department of Meteorology, University of Chicago Misc. Rep. 17, 79 pp.

——, 1948: On the formation of typhoons. *J. Meteor.*, **5**, 247–265, doi:10.1175/1520-0469(1948)005<0247:OTFOT>2.0.CO;2.

Rosenthal, S. L., 1960: Some estimates of the power spectra of large-scale disturbances in low latitudes. *J. Meteor.*, **17**, 259–263, doi:10.1175/1520-0469(1960)017<0259:SEOTPS>2.0.CO;2.

——, 1965: Some preliminary theoretical considerations of tropospheric wave motions in equatorial latitudes. *Mon. Wea. Rev.*, **93**, 605–612, doi:10.1175/1520-0493(1965)093<0605:SPTCOT>2.3.CO;2.

Roundy, P. E., and W. M. Frank, 2004: A climatology of waves in the equatorial region. *J. Atmos. Sci.*, **61**, 2105–2132, doi:10.1175/1520-0469(2004)061<2105:ACOWIT>2.0.CO;2.

Serra, Y. L., and R. A. Houze, 2002: Observations of variability on synoptic timescales in the east Pacific ITCZ. *J. Atmos. Sci.*, **59**, 1723–1743, doi:10.1175/1520-0469(2002)059<1723:OOVOST>2.0.CO;2.

——, G. N. Kiladis, and M. F. Cronin, 2008: Horizontal and vertical structure of easterly waves in the Pacific ITCZ. *J. Atmos. Sci.*, **65**, 1266–1284, doi:10.1175/2007JAS2341.1.

——, ——, and K. L. Hodges, 2010: Tracking and mean structure of easterly waves over the Intra-America Sea. *J. Climate*, **23**, 4823–4840, doi:10.1175/2010JCLI3223.1.

Sherwood, S. C., 1999: Convective precursors and predictability in the tropical western Pacific. *Mon. Wea. Rev.*, **127**, 2977–2991, doi:10.1175/1520-0493(1999)127<2977:CPAPIT>2.0.CO;2.

Sikdar, D. N., and V. E. Suomi, 1971: Time variation of tropical energetics as viewed from a geostationary satellite. *J. Atmos. Sci.*, **28**, 170–180, doi:10.1175/1520-0469(1971)028<0170:TVOTEA>2.0.CO;2.

Straub, K. H., and G. N. Kiladis, 2003: Interaction between the boreal summer intraseasonal oscillation and higher-frequency tropical wave activity. *Mon. Wea. Rev.*, **131**, 945–960, doi:10.1175/1520-0493(2003)131<0945:IBTBSI>2.0.CO;2.

Tai, K.-S., and Y. Ogura, 1987: An observational study of easterly waves over the eastern Pacific in the northern summer using FGGE data. *J. Atmos. Sci.*, **44**, 339–361, doi:10.1175/1520-0469(1987)044<0339:AOSOEW>2.0.CO;2.

Takayabu, Y. N., 1994a: Large-scale cloud disturbances associated with equatorial waves. Part I: Spectral features of the cloud disturbances. *J. Meteor. Soc. Japan*, **72**, 433–449.

——, 1994b: Large-scale cloud disturbances associated with equatorial waves. Part II: Westward-propagating inertio-gravity waves. *J. Meteor. Soc. Japan*, **72**, 451–465.

——, and M. Murakami, 1991: The composite structure of super cloud clusters observed over the Pacific Ocean in June 1-20, 1986 and their relationship with easterly waves. *J. Meteor. Soc. Japan*, **69**, 105–125.

——, and T. Nitta, 1993: 3-5 day-period disturbances coupled with convection over the tropical Pacific Ocean. *J. Meteor. Soc. Japan*, **71**, 221–246.

——, K.-M. Lau, and C.-H. Sui, 1996: Observation of a quasi-2-day wave during TOGA COARE. *Mon. Wea. Rev.*, **124**, 1892–1913, doi:10.1175/1520-0493(1996)124<1892:OOAQDW>2.0.CO;2.

——, T. Iguchi, M. Kachi, A. Shibata, and H. Kanzawa, 1999: Abrupt termination of the 1997–98 El Niño in response to a Madden–Julian oscillation. *Nature*, **402**, 279–282, doi:10.1038/46254.

——, S. Shige, W.-K. Tao, and N. Hirota, 2010: Shallow and deep latent heating modes over tropical oceans observed with TRMM PR spectral latent heating data. *J. Climate*, **23**, 2030–2046, doi:10.1175/2009JCLI3110.1.

Takemi, T., O. Hirayama, and C. Liu, 2004: Factors responsible for the vertical development of tropical oceanic cumulus convection. *Geophys. Res. Lett.*, **31**, L11109, doi:10.1029/2004GL020225.

Tanaka, H., and O. Ryuguji, 1971: Spectrum analysis of tropical cloudiness (I). *J. Meteor. Soc. Japan*, **49**, 13–19.

——, and ——, 1973: Spectrum analysis of tropical cloudiness (II). *J. Meteor. Soc. Japan*, **51**, 93–100.

Tao, W.-K., and Coauthors, 2016: TRMM latent heating retrieval: Applications and comparisons with field campaigns and large-scale analyses. *Multiscale Convection-Coupled Systems in the Tropics: A Tribute to Dr. Michio Yanai*, Meteor. Monogr., No. 56, Amer. Meteor. Soc., doi:10.1175/AMSMONOGRAPHS-D-15-0013.1.

Thompson, R. M., S. W. Payne, E. E. Recker, and R. J. Reed, 1979: Structure and properties of synoptic-scale wave disturbances in the intertropical convergence zone of the eastern Atlantic. *J. Atmos. Sci.*, **36**, 53–72, doi:10.1175/1520-0469(1979)036<0053:SAPOSS>2.0.CO;2.

Thorncroft, C. D., N. M. J. Hall, and G. N. Kiladis, 2008: Three-dimensional structure and dynamics of African easterly waves. Part III: Genesis. *J. Atmos. Sci.*, **65**, 3596–3607, doi:10.1175/2008JAS2575.1.

Toma, V. E., and P. J. Webster, 2010: Oscillations of the intertropical convergence zone and the genesis of easterly waves.

Part I: Diagnostics and theory. *Climate Dyn.*, **34**, 587–604, doi:10.1007/s00382-009-0584-x.

Tomas, R. A., and P. J. Webster, 1994: Horizontal and vertical structure of cross-equatorial wave propagation. *J. Atmos. Sci.*, **51**, 1417–1430, doi:10.1175/1520-0469(1994)051<1417:HAVSOC>2.0.CO;2.

——, and ——, 1997: The role of inertial instability in determining the location and strength of near-equatorial convection. *Quart. J. Roy. Meteor. Soc.*, **123**, 1445–1482, doi:10.1002/qj.49712354202.

Tucker, G. B., 1968: The 26-month zonal wind oscillation in the lower stratosphere of the Southern Hemisphere. *J. Atmos. Sci.*, **25**, 293–298, doi:10.1175/1520-0469(1968)025<0293:TMZWOI>2.0.CO;2.

Wallace, J. M., 1971: Spectral studies of tropospheric wave disturbances in the tropical western Pacific. *Rev. Geophys.*, **9**, 557–612, doi:10.1029/RG009i003p00557.

——, and V. E. Kousky, 1968: Observational evidence of Kelvin waves in the tropical stratosphere. *J. Atmos. Sci.*, **25**, 900–907, doi:10.1175/1520-0469(1968)025<0900:OEOKWI>2.0.CO;2.

——, and C.-P. Chang, 1969: Spectrum analysis of large-scale wave disturbances in the tropical lower troposphere. *J. Atmos. Sci.*, **26**, 1010–1025, doi:10.1175/1520-0469(1969)026<1010:SAOLSW>2.0.CO;2.

——, and L. A. Chang, 1972: On the application of satellite data on cloud brightness to the study of tropical wave disturbances. *J. Atmos. Sci.*, **29**, 1400–1403, doi:10.1175/1520-0469(1972)029<1400:OTAOSD>2.0.CO;2.

Wang, C.-C., and G. Magnusdottir, 2005: ITCZ breakdowns in three-dimensional flows. *J. Atmos. Sci.*, **62**, 1497–1512, doi:10.1175/JAS3409.1.

Wang, Z., M. T. Montgomery, and T. J. Dunkerton, 2010: Genesis of pre-hurricane Felix (2007). Part I: The role of the easterly wave critical layer. *J. Atmos. Sci.*, **67**, 1711–1729, doi:10.1175/2009JAS3420.1.

Webster, P. J., and J. R. Holton, 1982: Cross-equatorial response to middle-latitude forcing in a zonally varying basic state. *J. Atmos. Sci.*, **39**, 722–733, doi:10.1175/1520-0469(1982)039<0722:CERTML>2.0.CO;2.

——, and H.-R. Chang, 1988: Equatorial energy accumulation and emanation regions: Impacts of a zonally varying basic state. *J. Atmos. Sci.*, **45**, 803–829, doi:10.1175/1520-0469(1988)045<0803:EEAAER>2.0.CO;2.

Wheeler, M., and G. N. Kiladis, 1999: Convectively coupled equatorial waves: Analysis of clouds and temperature in the wavenumber–frequency domain. *J. Atmos. Sci.*, **56**, 374–399, doi:10.1175/1520-0469(1999)056<0374:CCEWAO>2.0.CO;2.

Wu, Z., 2003: A shallow CISK, deep equilibrium mechanisms for the interaction between large-scale convection and large-scale circulations in the tropics. *J. Atmos. Sci.*, **60**, 377–392, doi:10.1175/1520-0469(2003)060<0377:ASCDEM>2.0.CO;2.

Yamasaki, M., 1969: Large-scale disturbances in a conditionally unstable atmosphere in low latitudes. *Pap. Meteor. Geophys.*, **20**, 289–336, doi:10.2467/mripapers1950.20.4_289.

Yanai, M., 1961a: A detailed analysis of typhoon formation. *J. Meteor. Soc. Japan*, **39**, 187–214.

——, 1961b: Dynamical aspects of typhoon formation. *J. Meteor. Soc. Japan*, **39**, 282–309.

——, 1963: A comment on the creation of warm core in incipient tropical cyclone. *J. Meteor. Soc. Japan*, **41**, 183–187.

——, 1964: Formation of tropical cyclones. *Rev. Geophys.*, **2**, 367–414, doi:10.1029/RG002i002p00367.

——, 1971a: A review of recent studies of tropical meteorology relevant to the planning of GATE. Experimental Design Proposed by the Interim Scientific and Management Group (ISMG), Vol. 2, Annex I, World Meteorological Organization, 1–43.

——, 1971b: The mass, heat and moisture budgets and the convective heating within tropical cloud clusters. Preprints, *Seventh Tech. Conf. on Hurricanes and Tropical Meteorology*, St. Michael, Barbados, Amer. Meteor. Soc.

——, and T. Maruyama, 1966: Stratospheric wave disturbances propagating over the equatorial Pacific. *J. Meteor. Soc. Japan*, **44**, 291–294.

——, and T. Nitta, 1968: Finite difference approximation for the barotropic instability problem. *J. Meteor. Soc. Japan*, **46**, 389–403.

——, and Y. Hayashi, 1969: Large-scale equatorial waves penetrating from the upper troposphere into the lower stratosphere. *J. Meteor. Soc. Japan*, **47**, 167–182.

——, and M. Murakami, 1970a: A further study of tropical wave disturbances by the use of spectrum analysis. *J. Meteor. Soc. Japan*, **48**, 185–197.

——, and ——, 1970b: Spectrum analysis of symmetric and antisymmetric equatorial waves. *J. Meteor. Soc. Japan*, **48**, 331–347.

——, and M.-M. Lu, 1983: Equatorially trapped waves at the 200 mb level and their association with meridional convergence of wave energy flux. *J. Atmos. Sci.*, **40**, 2785–2803, doi:10.1175/1520-0469(1983)040<2785:ETWATM>2.0.CO;2.

——, and R. H. Johnson, 1993: Impacts of cumulus convection on thermodynamic fields. *The Representation of Cumulus Convection in Numerical Models, Meteor. Monogr.*, No. 46, Amer. Meteor. Soc., 39–62.

——, T. Maruyama, T. Nitta, and Y. Hayashi, 1968: Power spectra of large-scale disturbances over the tropical Pacific. *J. Meteor. Soc. Japan*, **46**, 308–323.

——, S. Esbensen, and J. Chu, 1973: Determination of average bulk properties of tropical cloud clusters from large-scale heat and moisture budgets. *J. Atmos. Sci.*, **30**, 611–627, doi:10.1175/1520-0469(1973)030<0611:DOBPOT>2.0.CO;2.

——, C.-H. Chu, T. E. Stark, and T. Nitta, 1976: Response of deep and shallow tropical maritime cumuli to large-scale processes. *J. Atmos. Sci.*, **33**, 976–991, doi:10.1175/1520-0469(1976)033<0976:RODAST>2.0.CO;2.

Yang, G.-Y., B. Hoskins, and J. Slingo, 2007: Convectively coupled equatorial waves. Part II: Propagation characteristics. *J. Atmos. Sci.*, **64**, 3424–3437, doi:10.1175/JAS4018.1.

Yokoyama, C., and Y. N. Takayabu, 2012a: Relationships between rain characteristics and environment. Part I: TRMM precipitation feature and the large-scale environment over the tropical Pacific. *Mon. Wea. Rev.*, **140**, 2831–2840, doi:10.1175/MWR-D-11-00252.1.

——, and ——, 2012b: Relationships between rain characteristics and environment. Part II: Atmospheric disturbances associated with shallow convection over the eastern tropical Pacific. *Mon. Wea. Rev.*, **140**, 2841–2859, doi:10.1175/MWR-D-11-00251.1.

Yoneyama, K., C. Zhang, and C. N. Long, 2013: Tracking pulses of the Madden–Julian oscillation. *Bull. Amer. Meteor. Soc.*, **94**, 1871–1891, doi:10.1175/BAMS-D-12-00157.1.

Yoshida, K., 1959: A theory of the Cromwell Current (the equatorial undercurrent) and of the equatorial upwelling—An interpretation in a similarity to a coastal circulation. *J. Oceanogr. Soc. Japan*, **15**, 159–170.

Zangvil, A., and M. Yanai, 1980: Upper tropospheric waves in the tropics. Part I: Dynamical analysis in the wavenumber-frequency domain. *J. Atmos. Sci.*, **37**, 283–298, doi:10.1175/1520-0469(1980)037<0283:UTWITT>2.0.CO;2.

——, and ——, 1981: Upper tropospheric waves in the tropics. Part II: Association with clouds in the wavenumber-frequency domain. *J. Atmos. Sci.*, **38**, 939–953, doi:10.1175/1520-0469 (1981)038<0939:UTWITT>2.0.CO;2.

Zhang, C., and P. J. Webster, 1992: Laterally forced equatorial perturbations in a linear model. Part I: Stationary transient flow. *J. Atmos. Sci.*, **49**, 585–607, doi:10.1175/1520-0469(1992)049<0585:LFEPIA>2.0.CO;2.

——, M. McGauley, and N. A. Bond, 2004: Shallow meridional circulation in the tropical eastern Pacific. *J. Climate*, **17**, 133–139, doi:10.1175/1520-0442(2004)017<0133:SMCITT>2.0.CO;2.

——, J. Gottschalck, E. D. Maloney, M. W. Moncrieff, F. Vitart, D. E. Waliser, B. Wang, and M. C. Wheeler, 2013: Cracking the MJO nut. *Geophys. Res. Lett.*, **40**, 1223–1230, doi:10.1002/grl.50244.

Zhang, G., and X. Song, 2016: Parameterization of microphysical processes in convective clouds in global climate models. *Multiscale Convection-Coupled Systems in the Tropics: A Tribute to Dr. Michio Yanai, Meteor. Monogr.*, No. 56, Amer. Meteor. Soc., doi:10.1175/AMSMONOGRAPHS-D-15-0015.1.

Zipser, E. J., 1969: The role of organized unsaturated convective downdrafts in the structure and rapid decay of an equatorial disturbance. *J. Appl. Meteor.*, **8**, 799–814, doi:10.1175/1520-0450(1969)008<0799:TROOUC>2.0.CO;2.

——, 1977: Mesoscale and convective-scale downdrafts as distinct components of squall-line circulation. *Mon. Wea. Rev.*, **105**, 1568–1589, doi:10.1175/1520-0493(1977)105<1568:MACDAD>2.0.CO;2.

Chapter 4

Dr. Yanai's Contributions to the Discovery and Science of the MJO

ERIC D. MALONEY

Department of Atmospheric Science, Colorado State University, Fort Collins, Colorado

CHIDONG ZHANG

Rosenstiel School of Atmospheric and Marine Science, University of Miami, Miami, Florida

ABSTRACT

This chapter reviews Professor Michio Yanai's contributions to the discovery and science of the Madden–Julian oscillation (MJO). Professor Yanai's work on equatorial waves played an inspirational role in the MJO discovery by Roland Madden and Paul Julian. Professor Yanai also made direct and important contributions to MJO research. These research contributions include work on the vertically integrated moist static energy budget, cumulus momentum transport, eddy available potential energy and eddy kinetic energy budgets, and tropical–extratropical interactions. Finally, Professor Yanai left a legacy through his students, who continue to push the bounds of MJO research.

1. Introduction

Among his many achievements in atmospheric science, Professor Yanai made substantial contributions, both directly and indirectly, to the study of the Madden–Julian oscillation (MJO; Madden and Julian 1971, 1972). These contributions are reviewed in this chapter. We first discuss how the discovery of the MJO was influenced by Professor Yanai's observational identification of equatorial waves (section 2). We then review Professor Yanai's direct contributions to MJO research. These include the vertically integrated moist static energy budget and its development from moisture and heating profiles (section 3), cumulus momentum transport (section 4), eddy available potential and kinetic energy budgets (section 5), and tropical–extratropical interactions (section 6). The legacy of MJO research carried on by Professor Yanai's students is discussed in section 7. Concluding remarks follow in section 8.

2. The discovery of the MJO by Roland Madden and Paul Julian

Professor Yanai's early work was highly influential in the discovery of the MJO by Roland Madden and Paul Julian in the early 1970s. In a quest to explain the tropical quasi-biennial oscillation (QBO; e.g., Reed et al. 1961), Yanai and Maruyama (1966) documented the existence of short-time-scale variations in stratospheric wind at sounding stations in the west Pacific Ocean during March through July of 1958. These westward-propagating disturbances of 4–5 days and approximately $23\,\mathrm{m\,s}^{-1}$ phase speed were subsequently shown by Maruyama (1967) to be associated with the mixed Rossby–gravity waves, alternatively called Yanai waves, of equatorial shallow-water wave theory (Matsuno 1966). These disturbances were examined in more detail by Yanai et al. (1968). It was this latter paper that particularly inspired Roland Madden and Paul Julian to undertake the work that led to the MJO discovery.

The discovery of Yanai waves is discussed in more detail in chapter 3 of this monograph volume (Takayabu et al. 2016, chapter 3). In short, Yanai et al. (1968)

Corresponding author address: Eric D. Maloney, Department of Atmospheric Science, Colorado State University, 1371 Campus Delivery, Fort Collins, CO 80523-1371.
E-mail: emaloney@atmos.colostate.edu

DOI: 10.1175/AMSMONOGRAPHS-D-15-0003.1

computed power spectra and conducted an analysis of spectral coherence for winds at 17 radiosonde stations in the west Pacific during April–July 1962. Spectra were examined at vertical levels from the surface to 25 km, including an analysis of the horizontal and vertical structures of the waves as they propagated across the west Pacific. Madden and Julian were motivated by this work because of the analysis techniques and the scientific content. First, they were highly inspired by the refined spectral analysis employed by Yanai et al. (1968). Roland Madden noted that the Yanai et al. (1968) analysis was impressive "in that they used cross-spectrum analysis every which way to identify wave structures" (R. Madden 2010, personal communication). Roland Madden also mentioned that they were "very much influenced by that [analysis] and used many of the spectral techniques to come up with the 40–50 day structure" in their work of Madden and Julian (1971; 1972) that first documented the MJO.

The scientific findings contained in Yanai et al. (1968) also provided impetus for the MJO discovery. Roland Madden recalled "it was also that paper that reported results for the 1962 period that I could not reproduce from the Line Islands experiment (spring of 1967), and so we became interested in the time-varying characteristics of the tropospheric waves" (R. Madden 2010, personal communication). Motivated to study the nonstationarity of the signal, Madden and Julian conducted an initial investigation using 10 years of radiosonde data from Canton Island (3°S, 172°W) in the Pacific. Roland Madden noted that although the initial purpose of their work was to examine high-frequency tropospheric wave signals, "we got distracted from that work when we saw the large 40–50 day signal" (R. Madden 2010, personal communication). This 40–50-day spectral peak was documented by Madden and Julian (1971), and similar periodicities were also discovered at other radiosonde stations within the tropical belt (Madden and Julian 1971, 1972). The analyses of spectral coherence pioneered by Professor Yanai inspired Madden and Julian to diagnose canonical MJO structure and evolution (Fig. 4-1; Madden and Julian 1972). This remarkable figure from Madden and Julian (1972) still represents our basic understanding of MJO structure today. Roland Madden finally noted that the late 1960s and early 1970s in which the MJO was discovered "were very interesting times for me, and I know that Yanai's work was influential" (R. Madden 2010, personal communication).

Professor Yanai was one of the early users of methods for decomposing tropical variability in the wavenumber–frequency domain (Zangvil and Yanai 1981), which were developed for use in the tropical meteorology several years earlier by Professor Yanai's student Yoshikazu Hayashi at the University of Tokyo and others (Hayashi 1973; Gruber 1974; Zangvil 1975). The mixed Rossby–gravity

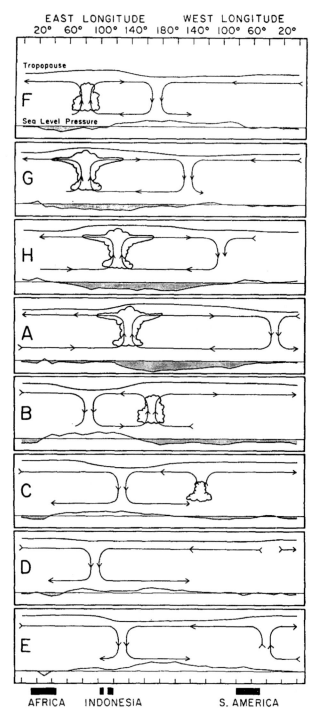

FIG. 4-1. Schematic representation of an MJO life cycle from Madden and Julian (1972).

wave or Yanai wave was the featured disturbance in the initial applications of these techniques (Hayashi 1973). Subsequent work using wavenumber–frequency decomposition of the tropical atmosphere has uncovered important dynamical insights on the MJO (Wheeler and Kiladis 1999), including the clear spectral gap between the MJO and convectively coupled Kelvin waves in wavenumber–frequency space (see chapters 3 and 6 of this monograph volume).

3. The moist static energy budget and its role in recent MJO theories

Yanai et al. (1973) developed a framework for calculating the vertically integrated moist static energy (MSE) budget starting from expressions for the apparent heat source (Q_1) and apparent moisture sink (Q_2), which has subsequently been used to provide profound insights into MJO dynamics. The more complete development of Q_1 and Q_2 in Yanai et al. (1973) follows initial introduction of these quantities in Yanai (1961), which provided an analysis of Typhoon Doris. Johnson et al. (2016, chapter 1) provide further details on this earlier development. As defined in Yanai et al. (1973) and elsewhere, the apparent heat source is

$$Q_1 = \frac{\partial \bar{s}}{\partial t} + \bar{\boldsymbol{v}} \cdot \nabla \bar{s} + \bar{\omega} \frac{\partial \bar{s}}{\partial p}$$

$$= Q_R + L(\bar{c} - \bar{e}) - \nabla \cdot \overline{(s'\boldsymbol{v}')} - \frac{\partial \overline{s'\omega'}}{\partial p}, \quad (4\text{-}1)$$

where $s = c_p T + gz$ is the dry static energy, ω is the pressure velocity, $\boldsymbol{v}$ is the horizontal wind vector, Q_R is radiative heating, L is latent heat of vaporization, c represents condensation, and e represents evaporation. Overbars represent a large-scale area average, and primes represent a deviation from that average. In (4-1), Yanai et al. (1973) excluded terms representing latent heating due to deposition, sublimation, freezing, and melting, which are generally much smaller than the other terms. Similarly, the apparent moisture sink can be defined as follows:

$$Q_2 = -L\left(\frac{\partial \bar{q}}{\partial t} + \bar{\boldsymbol{v}} \cdot \nabla \bar{q} + \bar{\omega} \frac{\partial \bar{q}}{\partial p}\right)$$

$$= L(\bar{c} - \bar{e}) + L\nabla \cdot \overline{(q'\boldsymbol{v}')} + L\frac{\partial \overline{q'\omega'}}{\partial p}, \quad (4\text{-}2)$$

where q is the water vapor mixing ratio. Subtracting (4-2) from (4-1), and then doing a mass-weighted integral over the troposphere (g^{-1} times the integral from the surface pressure to the tropopause pressure), one can show as in Yanai et al. (1973) that

$$\langle Q_1 \rangle - \langle Q_2 \rangle = \underbrace{\left\langle \frac{\partial \bar{h}}{\partial t} \right\rangle}_{\text{A}} + \underbrace{\left\langle \bar{\boldsymbol{v}} \cdot \nabla \bar{h} \right\rangle}_{\text{B}} + \underbrace{\left\langle \bar{\omega} \frac{\partial \bar{h}}{\partial p} \right\rangle}_{\text{C}}$$

$$= \langle Q_R \rangle + \text{LH} + \text{SH}, \quad (4\text{-}3)$$

where $h = c_p T + gz + Lq$ is the MSE or moist entropy, LH is the surface latent heat flux, and SH is the surface

sensible heat flux. Angled brackets represent a mass-weighted vertical integral. The right-hand side of (4-3) represents sources and sinks of moist entropy. Terms A, B, and C are the vertically integrated tendency of h, horizontal advection of h, and vertical advection of h, respectively. Yanai et al. (1973) and Yanai and Johnson (1993) noted that (4-3) is able to provide profound insight into diabatic heating processes in the tropics, and can be used to check the accuracy of tropospheric budgets given surface flux information.

Far-reaching implications of (4-3) for the MJO may not have come until much later, however. Neelin and Held (1987) used (4-3) to develop a model for tropical upper-tropospheric divergence as a proxy for deep convection. Assuming a first baroclinic mode structure to the tropical atmosphere, Neelin and Held (1987) were able to rewrite the vertical advection term in (4-3) as a quantity called "gross moist stability (GMS)" (M_h) times the vertically integrated lower-tropospheric mass convergence. In essence, M_h represents the column-integrated MSE export per unit convective activity (Raymond et al. 2009), where in Neelin and Held (1987) convective activity was represented by mass convergence in the lower half of the troposphere (or similarly, mass divergence in the upper troposphere). In Neelin and Held (1987), M_h was lowest over warm SST regions of the tropics because of higher lower-tropospheric humidity and greater moisture convergence per unit convective activity there (with the normalized upper-tropospheric export being constrained by weak tropical dry static energy gradients and thus only weakly varying across the tropics for the same mass flux profile; Charney 1963). Hence, convective activity is predicted by the Neelin and Held model to be enhanced in warm SST regions, since convection is less efficient at removing surface moist entropy sources from the column there. It has subsequently been shown that GMS is strongly regulated by vertical structure of diabatic heating (Peters and Bretherton 2006), suggesting that the first baroclinic mode structure of Neelin and Held (1987) should be relaxed. Gross moist stability was later generalized to also include horizontal advection (Raymond and Fuchs 2009).

Use of the column MSE budget to understand MJO dynamics has recently expanded. Under the assumption of weak tropical temperature gradients, which is a relatively good assumption at time scales of greater than 10 days that are characteristic of the MJO (e.g., Sobel and Bretherton 2000; Yano and Bonazzola 2009), (4-3) becomes an equation for the column-integrated moisture tendency because adiabatic cooling and diabatic heating to first order cancel to produce no net dry static energy tendency. Under such conditions, a class of balanced disturbances called moisture modes has been hypothesized to exist (Sobel et al. 2001; Raymond 2001;

Majda and Klein 2003; Maloney et al. 2010), in which gravity wave adjustment plays no role in propagation, and understanding the moisture (or MSE) budget is essential to understanding the basic maintenance and propagation of the modes. This former point distinguishes moisture modes from disturbances of equatorial shallow-water theory that are dependent on gravity for propagation, including Kelvin waves (e.g., Matsuno 1966). The strong link between tropical convection and column-integrated water vapor that has been demonstrated in observations is a key precept of moisture mode theory (e.g., Bretherton et al. 2004). Thus, (4-3) has been recently used to provide insights into the fundamental dynamics of the MJO, under the assumption that the MJO behaves like a moisture mode. This hypothesis is supported by the results of Wheeler and Kiladis (1999), which show a spectral gap between the MJO and convectively coupled Kelvin waves, and suggests that the MJO is not regulated by adjustment under gravity (see also Takayabu et al. 2016, chapter 3; chapter 6). Using (4-3) provides advantages over the moisture budget for diagnosing MJO dynamics, as it implicitly accounts for the cancellation of condensation and moisture convergence, which dominates the column-integrated MJO moisture budget.

Recent studies have used the MSE budget (4-3) to diagnose the mean state that global models must have to produce strong intraseasonal variability. In particular, models with high mean GMS tend to produce significantly weaker intraseasonal variability than those with lower GMS (e.g., Raymond and Fuchs 2009; Hannah and Maloney 2011), indicating that convection in these models is possibly too efficient at discharging column moisture, which makes it difficult to sustain moisture anomalies that are the key to MJO maintenance. In addition to the mean state diagnosis, recent studies have employed the MSE budget to diagnose fundamental internal dynamics of the MJO. In both observations (Haertel et al. 2008) and models (Hannah and Maloney 2011), shallow heating in advance of an MJO convective event has been shown to contribute to the buildup of column MSE. This is a state of negative GMS in which the net effects of shallow convection and associated divergent circulations act to moisten the column. Shallow convection has been previously hypothesized to be an important agent in moistening the column in advance of deep MJO convection (e.g., Benedict and Randall 2007). Further, horizontal MSE advection has been cited as an important regulator of eastward propagation (Maloney 2009; Maloney et al. 2010; Kiranmayi and Maloney 2011a; Andersen and Kuang 2012), including a potentially important role for intraseasonal variations in synoptic eddy activity for regulating the column MJO

moisture budget through their effects on horizontal MSE advection (Maloney 2009; Andersen and Kuang 2012). Also, recent MSE budget analyses in observations and general circulation models (GCMs) have suggested that surface flux anomalies and cloud radiative feedbacks may be important destabilization mechanisms for the MJO (Lin and Mapes 2004; Maloney 2009; Kiranmayi and Maloney 2011a), with such inferences being verified with appropriate mechanism denial experiments in general circulation models (Sobel et al. 2008, 2010; Kiranmayi and Maloney 2011b; Andersen and Kuang 2012). Finally, semiempirical MJO models have been developed that use information on the vertically integrated MSE budget from observations and GCMs to demonstrate the basic dynamics of the MJO in an idealized framework (Sobel and Maloney 2012, 2013). In summary, the budget in (4-3) derived by Yanai et al. (1973) has proven to be a powerful tool in understanding MJO dynamics in both observations and models.

4. Cumulus momentum transport

One of the exemplary aspects of Professor Yanai's work on the MJO was the gained understanding of cumulus momentum transport (CMT) during TOGA COARE and its role in maintaining the MJO low-level westerly flow. As will be discussed in more detail below, vertical transport of momentum by cumulus convection has the potential to significantly modulate the wind at different levels of the atmosphere. The strong near-surface westerly flow near and to the west of MJO convection has been hypothesized to be strengthened by such transports (e.g., Tung and Yanai 2002a). To precisely quantify the role of convective vertical momentum transport in the MJO and other convectively coupled disturbances is an extremely difficult endeavor, particularly in observational data.

Using data from the TOGA COARE inner flux array (IFA), Tung and Yanai (2002a) defined the residual ($\mathbf{X}$) of the IFA averaged momentum budget as follows:

$$\mathbf{X} = \frac{\partial \boldsymbol{v}}{\partial t} + \overline{\boldsymbol{v}} \cdot \nabla \overline{\boldsymbol{v}} + \omega \frac{\partial \overline{\boldsymbol{v}}}{\partial p} + \nabla \overline{\phi} + \lambda \mathbf{k} \times \overline{\boldsymbol{v}} = -\frac{\partial}{\partial p} \overline{\boldsymbol{v}' \omega'},$$
$$(4\text{-}4)$$

where λ is the Coriolis parameter, ϕ is the geopotential, overbars represent a large-scale area average, and primes are the deviation from this average. In (4-4), the area-averaged horizontal eddy flux divergence is assumed to be negligible. Therefore, the residual of the averaged areal budget $\mathbf{X}$ is ascribed to be due to the vertical eddy momentum flux convergence $-\partial \overline{\boldsymbol{v}' \omega'} / \partial p$ produced by convection and other unresolved motions,

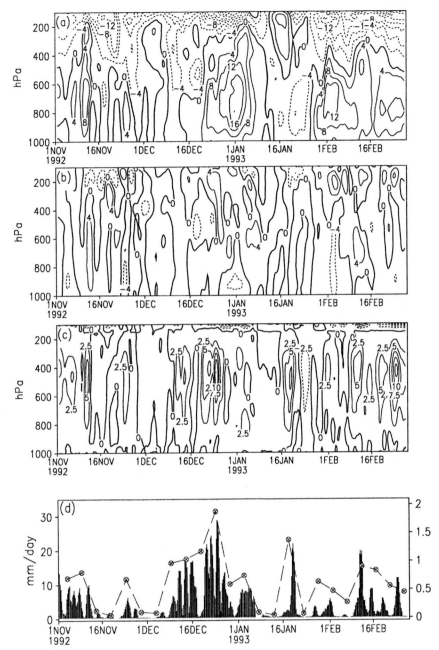

FIG. 4-2. Evolution of TOGA COARE intensive observing period (a) zonal wind, (b) meridional wind, (c) Q_1/c_p, and (d) precipitation (black circles) and a convective index derived from brightness temperature (black bars) from Tung and Yanai (2002a). The units are wind in m s^{-1} (positive solid and negative dashed contours), Q_1/c_p in K day^{-1}, and precipitation in mm day^{-1}.

thus including the effects of CMT. A strong correspondence in spectral power between brightness temperature (a proxy for deep convection) and **X** during the TOGA COARE period supports the inference that the residual represents the effects of CMT (Tung and Yanai 2002a). The results of this spectral analysis and the wavelet analysis of Tung and Yanai (2002b) also suggest

interesting multiscale interactions between the MJO and higher-frequency synoptic disturbances.

Figure 4-2 from Tung and Yanai (2002a) shows the evolution of wind fields, Q_1, and rainfall rate during the TOGA COARE period of November 1992 through February 1993. Two well-defined MJO events and associated strong low-level westerly flow transited through

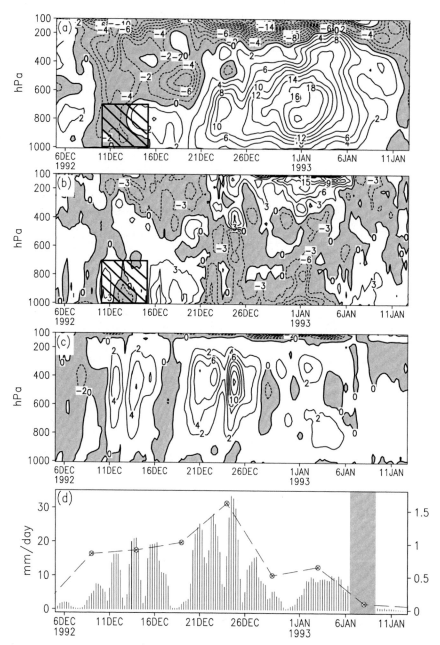

FIG. 4-3. Evolution of (a) zonal wind (b) momentum budget residual **X**, (c) Q_1/c_p, and (d) precipitation (gray circles) and a convective index derived from brightness temperature (gray bars) for part of the TOGA COARE IOP spanning the first MJO event from Tung and Yanai (2002b). Units are as in Fig. 4-2 and **X** is in m s^{-1} day^{-1}. Hatched regions in (a) and (b) represent unreliable data.

the TOGA COARE IFA. In Tung and Yanai (2002a) the time mean residual in (4-4) over the TOGA COARE period suggests that on average cumulus convection acts to decelerate the flow, and CMT is downgradient. However, the same is not true in a time-varying sense, as was also suggested by studies on individual convective features during TOGA COARE (e.g., Lewis et al. 1998; Roux 1998; Bousquet and Chong 2000).

To highlight this time variability in the sign of CMT, Fig. 4-3 from Tung and Yanai (2002b) shows the evolution of the zonal wind, momentum budget residual [(4-4)], Q_1, and precipitation during the first MJO event of the TOGA COARE. Although part of the period 6 December 1992–12 January 1993 is characterized by somewhat lower data quality, as indicated by hatching, the implied CMT in the lower troposphere during

13–21 December 1992 is of the same order of magnitude as the zonal wind acceleration, indicating a likely role of CMT for accelerating the low-level westerly flow during this MJO event. Downgradient momentum transport ensued after 21 December when the MJO westerly wind burst was mature. While of smaller amplitude, similar processes appeared to act during the second TOGA COARE MJO event of late January through early February 1993 (not shown). Tung and Yanai (2002b) argued that shallow convection or squall lines are integral to the upgradient momentum transport at the onset of the westerly wind burst associated with both of these MJO events, and such effects may have profound implications for convective parameterizations and their ability to properly simulate the MJO, since most convection parameterizations neglect the possibility of such upgradient momentum transport (e.g., Moncrieff 2004). While other studies using other datasets over longer periods (e.g., Lin et al. 2005) or different analysis techniques during the TOGA COARE period (e.g., Houze et al. 2000) have suggested a lesser role for upgradient momentum transport for supporting the flow at the westerly onset phase, the results of Tung and Yanai (2002b) have proven extremely influential for spawning a body of theoretical work on the MJO over the subsequent decade.

In the theoretical work of Majda and Biello (2004), upscale transfer of kinetic and thermal energy from wave trains of tropical synoptic-scale disturbances (e.g., superclusters; Nakazawa 1988) are necessary to maintain the large-scale MJO flow in the presence of synoptic disturbances that tilt westward with height, a paradigm also proposed by the work of Moncrieff and Klinker (1997). Superclusters refer to eastward-propagating envelopes of convection, shown by Nakazawa (1988) to most often be convectively coupled Kelvin waves. Related work by Moncrieff (2004) also documented the importance of vertically tilted mesoscale organization for supporting the large-scale MJO flow. Such ideas have been expanded in subsequent work to develop refined models of the MJO that involve multiscale interactions. Biello and Majda (2005) accounts for the evolution of convective organization during MJO events from congestus in the eastern part of the MJO convective envelope (with assumed eastward tilts with height) to supercluster-like structure in the western part (with assumed westward tilts with height) to produce a more realistic large-scale flow than in Majda and Biello (2004). Such assumptions on where particular wave types maximize relative to the MJO convective envelope and their respective tilts with height, which depend on the propagation direction, have since garnered further observational support (e.g., Kikuchi and Wang 2010). Majda and Stechmann (2009a) use a simple

dynamical model with imbedded multicloud parameterization (of Khouider and Majda 2006) that is able to simulate two-way interactions between the convectively coupled equatorial waves and the large-scale MJO flow to demonstrate that both the upgradient momentum transport observed at the initiation of the TOGA COARE westerly wind bursts and the subsequent downscale momentum transport observed by Tung and Yanai (2002b) can be captured by such a model. Work with this model was extended in Han and Khouider (2010) and Khouider et al. (2012) to show that the predominant type of convectively coupled wave produced as a function of MJO regime could be accurately captured, with the implications for upscale momentum transport for MJO dynamics also discussed.

Another important study in this line of research is that of Majda and Stechmann (2009b). This study provides a theory for both the propagation dynamics and instability mechanism of the MJO. The MJO "skeleton" is a simple Matsuno–Gill type dynamical model that without explicit synoptic-scale momentum fluxes and their effect on the MJO flow would be in a neutral state. A parameterized envelope of synoptic-scale wave activity provides the "muscle," and hence instability mechanism. The instability mechanism is assumed to involve multiscale interactions with the synoptic scale that provide upscale momentum and energy transports to maintain the large-scale MJO circulation against dissipation. Such upscale transport is parameterized in terms of the envelope of synoptic-scale wave activity. The envelope of synoptic activity serves as the heating source as well as the moisture sink through precipitation, and its temporal behavior depends on low-level moisture. With an assumed equatorially trapped meridional structure, this envelope of synoptic-scale wave activity generates the Kevin and gravest Rossby waves, which form the base function to construct the MJO. The model yields a dispersion relationship, from which a nondissipative MJO frequency is obtained. The MJO solution also includes slow $(5 \, \mathrm{m \, s^{-1}})$ propagating quadrupole vortices surrounding the envelopes of synoptic-scale wave activity, interpreted as the MJO convective envelope. This model has subsequently been used to examine its nonlinear behavior in the presence of different basic states, including a uniform SST distribution and one with a warm pool, and is able to capture other salient features of the MJO including its observed irregularity (Majda and Stechmann 2011). This model was expanded to include a frictional boundary layer that appears to improve upon the realism of the simulated MJO and aid in the selection of eastward-propagating modes (e.g., Wang and Liu 2011; Liu and Wang 2013), with the ability of synoptic-scale disturbances to destabilize the MJO dependent on where particular wave types occur relative to the MJO convective envelope (as discussed above).

Besides the theoretical work discussed above, studies with more complex models have supported the importance of CMT to the dynamics of the MJO. Deng and Wu (2010) employed a CMT parameterization that accounts for the vertical redistribution of momentum by convection as well as the effect of the perturbation pressure field generated by the interaction of the large-scale flow with convection (e.g., Wu and Yanai 1994). They demonstrated that this scheme leads to stronger, more coherent MJO convection. Specifically for the western Pacific, Deng and Wu (2011) showed that inclusion of this CMT parameterization contributed to strengthening the westerly wind phase of the MJO. Zhou et al. (2012) demonstrated that inclusion of a CMT parameterization in a version of the NCAR Community Atmosphere Model has the beneficial effect of improving the simulation of low-level westerly mean winds in the Indo-Pacific warm pool, which produces an improved simulation of intraseasonal variability. Miyakawa et al. (2012) used a global nonhydrostatic model to show that CMT helps to accelerate the westerly surface flow during MJO events, possibly reducing the eastward propagation speed of events as well as reducing momentum damping influence of the large-scale flow. Finally, "superparameterized" modeling approaches with some aspects of convective momentum transport explicitly simulated are able to produce improved intraseasonal variability (e.g., Grabowski 2002). The concept of superparameterization is discussed in extended detail in Randall et al. (2016, chapter 15).

5. Energy budgets

In a pair of papers, Yanai et al. (2000) and Chen and Yanai (2000) documented the basic characteristics of MJO events during the TOGA COARE IOP and for a longer 15-yr climatology, and then computed perturbation available potential energy (PAPE) and perturbation eddy kinetic energy (PKE) budgets to determine the dominant processes that help to maintain the large-scale MJO circulation against dissipation. In general, the TOGA COARE MJO events were found to be representative of behavior in the extended period, with a few exceptions, including the stronger variability during TOGA COARE as compared to the longer record, especially in the central Pacific near the date line. A few other interesting observations from Yanai et al. (2000) on MJO behavior during the TOGA COARE period will be discussed below.

Both Yanai et al. (2000) and Chen and Yanai (2000) computed detailed PAPE and PKE budgets using European Centre for Medium-Range Weather Forecasts (ECMWF) reanalysis products, deriving Q_1 as a residual of the thermodynamic energy equation for use in the PAPE budget. By far the largest source of PKE in the Indo-Pacific warm pool is associated with conversion of PAPE to PKE in the middle troposphere (Fig. 4-4a), with a resulting geopotential energy flux that moves energy upward and downward supporting PKE maxima in the upper and lower troposphere (not shown). PAPE generation is predominantly due to the covariance of diabatic heating and temperature perturbations (not shown), which is nearly cancelled by PKE conversion from PAPE. This cancellation is consistent with the dominant thermodynamic balance of diabatic heating and adiabatic cooling that associated with weak temperature gradient theory for the tropical atmosphere (e.g., Sobel and Bretherton 2000). The domination of Indo-Pacific warm pool MJO energy conversion processes by diabatic heating provides evidence for the idea of the convectively coupled nature of the MJO and is supported by previous and subsequent analysis (e.g., Hendon and Salby 1994; Mu and Zhang 2006; Deng and Wu 2011). Diabatic heating is also the primary energy source for large-scale motions associated with the boreal summer intraseasonal variability that occurs in the east Pacific warm pool (Maloney and Esbensen 2003). Interestingly, Yanai et al. (2000) documents a minimum in PAPE generation by diabatic heating in the Maritime Continent region, which contains a well-documented minimum in 30–60-day convective variability there (e.g., Zhang and Hendon 1997; Sobel et al. 2010).

As noted by Yanai et al. (2000), Chen and Yanai (2000), and other studies, the PKE maximum associated with the MJO actually occurs in the upper troposphere in the east Pacific just to the west of South America. This is an upper-tropospheric westerly duct region where strong tropical–extratropical interactions take place, and convective activity is a minimum (e.g., Webster and Holton 1982; Magaña and Yanai 1991). The budget analysis by Yanai et al. (2000) and Chen and Yanai (2000) shows that barotropic energy conversions and an equatorward flux of extratropical/subtropical wave energy are the dominant terms supporting the strong MJO flow in this region (Figs. 4-4b,c), which helps to explain the large PKE maximum there in the absence of diabatic heating.

The discussion section of Yanai et al. (2000) provides very interesting insights on priorities for future MJO research. For one, it is noted that a holistic view of coupling between different vertical heating modes (as defined by Q_1 and Q_2) and the large-scale MJO flow is a priority. This quest underlies many of the research areas discussed above, including the impact of higher vertical

$-\alpha\cdot\omega$ in 30–60-day band $(J\cdot kg^{-1}\cdot day^{-1})$

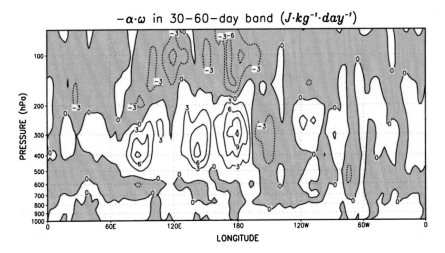

barotropic conversion in 30–60-day band $(J\cdot day^{-1}\cdot kg^{-1})$

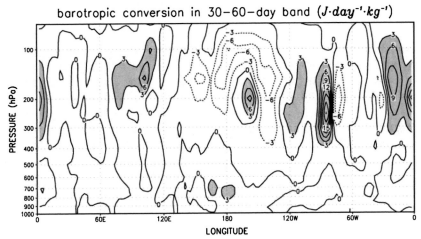

$-\nabla\cdot F_h$ in 30–60-day band $(J\cdot kg^{-1}\cdot day^{-1})$

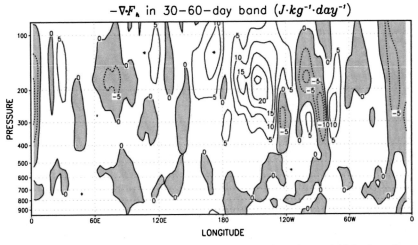

FIG. 4-4. (a) PAPE to PKE conversion, (b) barotropic conversion, and (c) horizontal convergence of wave energy flux in the 30–60-day band during the TOGA COARE experiment from Yanai et al. (2000).Units are $J\,kg^{-1}\,day^{-1}$ with negative values shaded gray.

heating modes in the moistening/drying process during an MJO life cycle (e.g., Haertel et al. 2008), and the tilted vertical structure of the MJO and the consequences for the horizontal momentum budget (e.g., Moncrieff 2004). Tung et al. (1999) suggested that three coupled modes of Q_1 and Q_2 are needed to describe heating and moistening under different large-scale convective phases of the MJO during TOGA COARE that include the convectively active phase of the MJO with heavy precipitation, the light to moderate precipitation period prior to the active phase, and the period after the MJO convective peak with strong surface wind and enhanced surface evaporation. In the same spirit, Zhang and Hagos (2009) derived two leading modes from sounding-based Q_1. The first leading mode peaks at about 400 hPa and the second at 700 hPa. Together they account for almost all structural variability of diabatic heating during TOGA COARE. These two leading modes suggest the importance of shallow as well as deep convective heating to the MJO. Yanai et al. (2000) also foreshadows the Dynamics of the MJO (DYNAMO) program during 2011–12, a field campaign designed to study MJO initiation processes (Yoneyama et al. 2013). Yanai et al. (2000) note the origination of super cloud clusters during the TOGA COARE time period near 75°E, a longitude that intersects the two DYNAMO sounding arrays. In particular, the analysis of Yanai et al. (2000) notes the possibility of superclusters near 75°E being triggered by precursor activity that first appears in upper-tropospheric wind and temperature anomalies coming from the west, possibly in association with the previous MJO event. The origin of the second MJO event during TOGA COARE can be followed back to the Western Hemisphere, at 40°W. Yanai et al. (2000) thus appear to have documented an example of "successive" MJO events, a term later coined by Matthews (2008) when describing different types of MJO initiation that also include "primary" events. The Q_1 estimation for the MJO by Yanai et al. (2000) provides a background for a model intercomparison project on the vertical structure and physical processes of the MJO, with an emphasis on its diabatic heating profiles (Zhang et al. 2013).

6. Tropical–extratropical interaction

In a series of studies, Yanai and his colleagues documented connections between equatorial wave activity and equatorward energy and momentum fluxes from the extratropics, laying the observational foundation for later work in the area of extratropical influences on equatorial waves and associated moist convection as modulated by the MJO. Zangvil and Yanai (1980)

performed time–space spectrum analyses on a 200-hPa wind dataset for the period of 1 June–31 August 1967. Using a method proposed by Yanai and Murakami (1970) that decomposes the wind field into symmetric and antisymmetric components with respect to the equator, they identified Kelvin waves (zonal wavenumber $k = 1$–2, period $\tau = 7$ days, and $k = 1$, $\tau > 20$ days), mixed Rossby–gravity waves ($k = 4$, $\tau = 5$ days), and $n = 1$ Rossby waves ($k = 2, \tau = 12$ days) from their data. They further calculated the meridional flux of wave energy due to pressure work, which propagates freely into the tropics in the east and central Pacific and Atlantic where no critical latitude exists (see also Fig. 4-6 and Knippertz 2007). Influence of this wave energy is concentrated at the zonal wavenumbers and periods of mixed Rossby–gravity waves. This analysis suggested that the mixed Rossby–gravity waves could be excited by lateral forcing from a higher latitude. The work of Zangvil and Yanai (1980) was extended by Yanai and Lu (1983) by including additional data during June–August 1972. The year 1972 differed from 1967 since mixed Rossby–gravity waves were absent and weak Kelvin and Rossby waves were present. The absence of the mixed Rossby–gravity waves was associated with a peak of meridional wave energy divergence in the tropics at characteristic wavenumbers and periods of the waves, instead of wave energy convergence as observed in 1967. Meridional wave energy convergence in the tropics, on the other hand, was observed at the wavenumbers and periods of Rossby waves. The work of Zangvil and Yanai (1980) and Yanai and Lu (1983) provided the first observational evidence for the effect of the east Pacific "westerly duct" (Webster and Holton 1982) on equatorial waves and the motivation for later theoretical explanations of laterally forced equatorial waves (Zhang and Webster 1992; Zhang 1993; Hoskins and Yang 2000). Zangvil and Yanai (1981) used satellite observations to relate mixed Rossby–gravity wave signals to tropical convection, which was one of the earliest observational studies on "convectively coupled equatorial waves." The extension of Yanai's work on tropical–extratropical interaction from synoptic to intraseasonal time scales, discussed in detail below, paved the way for later efforts that identified extratropical influences on equatorial waves in different phases of the MJO, also discussed later in this section.

Magaña and Yanai (1991) used First GARP Global Experiment (FGGE) data during the summer of 1979 to document the basic intraseasonal evolution of the subtropical circulation in the upper troposphere and tropical convection, and the nature of tropical–extratropical interactions during this time. They used an empirical orthogonal function (EOF) analysis and composite analysis to characterize the nature of such interactions. Of particular interest is that during 1979 the MJO produced substantial variations in the amplitude of the upper-tropospheric mid-Pacific trough and Mexican anticyclone as it progressed eastward. The mid-Pacific trough was amplified in the presence of enhanced convection in the far west Pacific and suppressed convection in the central Pacific, enhancing the thermal contrast between Asia and the central North Pacific. Associated upper-level convergence in the central Pacific also strengthened the trough. Related processes occurred in the Americas, where enhanced convection and associated upper-tropospheric divergence of the MJO strengthened the Mexican anticyclone. The extension of boreal summer convection associated with the MJO during 1979 to the Americas as noted in Magaña and Yanai (1991) is interesting, especially the tendency for convection near the Mexican coast to be out of phase with that to the west of 120°W. While similar signals did appear in previous composite life cycles of the MJO (e.g., Knutson and Weickmann 1987), a larger body of work has since developed that examines how the MJO influences precipitation, winds, tropical cyclone activity, and other variables in the east Pacific warm pool during boreal summer (e.g., Maloney and Hartmann 2000; Molinari and Vollaro 2000; Barlow and Salstein 2006; Jiang et al. 2012). Recent analysis using regional and global models suggests that intraseasonal variability in this region can exist in isolation from Eastern Hemisphere MJO variability, but likely phase locks given common dominant periodicities (Rydbeck et al. 2013). Magaña and Yanai (1991) were pioneers in the study of intraseasonal variability in the east Pacific region.

Fluctuations in the mid-Pacific trough on intraseasonal time scales are associated with northward propagation of angular momentum anomalies in the North Pacific, as documented in the time series of 200-hPa zonal wind anomalies (raw and bandpass filtered) at different latitude belts from Magaña and Yanai (1991; Fig. 4-5). Such fluctuations in extratropical wind anomalies associated with tropical intraseasonal variability were also noted in other previous studies (Liebmann and Hartmann 1984; Weickmann et al. 1985; Lau and Phillips 1986; Knutson and Weickmann 1987), and the global angular momentum balance associated with the MJO has in particular been a topic of much attention over the last four decades (e.g., Madden 1987; Kang and Lau 1990; Weickmann et al. 1992; Weickmann and Sardeshmukh 1994; Madden and Speth 1995; Weickmann et al. 1997).

Magaña and Yanai (1991) also argued that intraseasonal fluctuations in the mid-Pacific trough, in addition to being associated with intraseasonal fluctuations

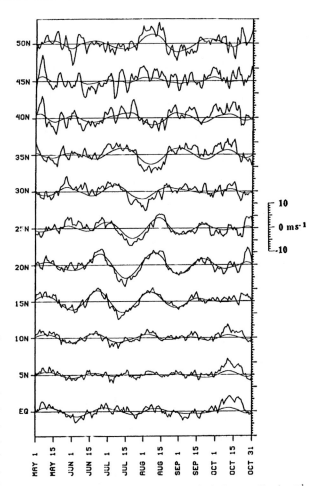

FIG. 4-5. The 200-hPa zonal mean zonal wind anomalies (m s^{-1}, +10 to −10) at 200 hPa from 50°N to 0° during 1 May to 31 October 1979. Smooth lines indicate anomalies bandpass filtered to 30–60 days. From Magaña and Yanai (1991).

of angular momentum, could also cause variations in the strength and spatial extent of the upper-tropospheric equatorial westerly duct that could admit or suppress penetration of extratropical Rossby wave energy into the tropics. The dominant pathway of such energy transport is shown in Fig. 4-6, which displays **E** vectors at 200 hPa for the total (Fig. 4-6a), the 30–60-day (Fig. 4-6b), and high-frequency (Fig. 4-6c) eddy fields. The barotropic **E** vector was defined by Magaña and Yanai (1991) using the formalism of Trenberth (1986) as

$$\mathbf{E} = \left[\frac{1}{2}(\overline{v'^2} - \overline{u'^2}), -\overline{u'v'}\right]\cos\phi.$$

Here, primes represent temporal eddies, bars represent the time mean, and ϕ is latitude. This vector represents the direction of the eddy group velocity, and $-\mathbf{E}$ is equivalent to a westerly eddy momentum flux (e.g., Hoskins et al. 1983; Trenberth 1986). Figure 4-6 indicates a notable

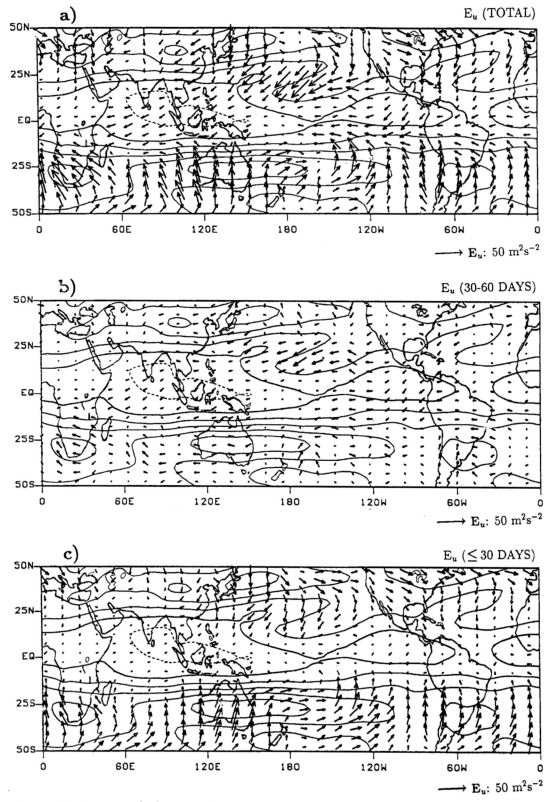

FIG. 4-6. The **E** vectors $(m^{-2}\,s^{-2})$ computed for (a) total anomaly field, (b) 30–60-day filtered fields, and (c) fields filtered to less than 30 days for the summer of 1979. The mean 200-hPa zonal wind is shown in contours with an interval of $10\,m\,s^{-1}$. From Magaña and Yanai (1991).

southwestward energy flux in the central equatorial Pacific in the area of the mid-Pacific trough. Magaña and Yanai (1991) noted that this flux is strengthened during strengthening of the mid-Pacific trough, when the westerly duct is enhanced during MJO events. Consistent with the discussion in Magaña and Yanai (1991), other studies have documented that convection in the central and east Pacific is preferentially induced in the presence of a westerly upper-tropospheric basic state when strong tropical–extratropical interactions occur (e.g., Kiladis and Weickmann 1992).

Pacific tropical–extratropical interactions and implications for convectively coupled wave activity as modulated by the MJO have been examined in more detail since the Magaña and Yanai (1991) work. Magaña and Yanai (1995) argue that a westerly or weak easterly basic state in the equatorial east Pacific upper troposphere and a favorable spatial and temporal structure and amplitude of extratropical forcing are particularly efficient at generating equatorial mixed Rossby–gravity waves. The MJO is effective at forcing variations in the strength and direction of the upper-tropospheric flow in the east Pacific that can affect mixed Rossby–gravity wave generation. Matthews and Kiladis (1999) show that during Northern Hemisphere winter in a phase when MJO convection is enhanced over the east Indian Ocean and Maritime Continent and suppressed over the South Pacific convergence zone, upper-tropospheric transients emanating from a source region in the Asian jet can penetrate more effectively into the central Pacific because of relaxation of the upper-tropospheric easterly basic state. Enhancement of high-frequency transients on 6–25-day time scales in the central Pacific results, with further rectification onto the MJO time scale (see also Meehl et al. 1996). In this context, rectification means that since precipitation is a positive definite quantity, transient activity is associated with a nonlinear precipitation signal that can project onto longer MJO time scales. Straub and Kiladis (2003a) examine the interactions of the boreal summer version of the MJO and convectively coupled mixed Rossby–gravity and Kelvin waves and demonstrate an enhancement of Kelvin wave activity to the east of the MJO convective center over the central Pacific, consistent with the hypothesis of Magaña and Yanai (1991). Straub and Kiladis (2003a) also argued for a strong mutual interaction between mixed Rossby–gravity waves and the MJO in the enhanced convective region of the MJO. Straub and Kiladis (2003b) documented a pathway by which extratropical Rossby wave trains in the Southern Hemisphere could penetrate toward the equator and initiate Kelvin wave activity in the central and east Pacific ITCZ, a process presumably enhanced during phases of the MJO when the upper-tropospheric background flow is favorable. Such a pathway is also suggested in the bottom panel of Fig. 4-6 from Magaña and Yanai (1991) and in Fig. 10 of Magaña and Yanai (1995). The importance of tropical–extratropical interactions to the MJO has been pursued over the last few decades in observational studies (Lau and Peng 1987; Hsu et al. 1990; Lau et al. 1994; Matthews et al. 1996; Straus and Lindzen 2000), theoretical work (Meehl et al. 1996; Frederiksen and Frederiksen 1997), and numerical modeling studies (Lin et al. 2007; Ray et al. 2009; Ray and Zhang 2010; Ray and Li 2013).

7. The legacy continued by Professor Yanai's students

Perhaps his students, who have pushed and continue to push the boundaries of our MJO understanding, carry on Dr. Yanai's most enduring legacy to MJO research. Their seminal contributions have continued beyond those in direct collaboration with Professor Yanai. Dr. Tatsushi Tokioka has conducted seminal research on cumulus parameterization, including development of the famous "Tokioka modification" to deep convection parameterizations, which has led to substantial improvements in MJO simulations by climate models (Tokioka et al. 1988). Dr. Tsuyoshi Nitta demonstrated the role of intraseasonal oscillations and associated westerly wind bursts to the development of the 1986/87 El Niño event (Nitta et al. 1992), and also studied teleconnections associated with boreal summer intraseasonal convective variability in the western Pacific (Nitta 1987). Dr. Masoto Murakami conducted an observational analysis of 30–40-day convective variability in Southeast Asia the western Pacific during boreal summer (Murakami 1984; Chen and Murakami 1988). Randall et al. (2016, chapter 15) discuss in more detail boreal summer intraseasonal variability in the context of superparameterization in climate models.

Dr. Steven Esbensen helped forge better understanding of the remote influence of the MJO on the Western Hemisphere, in particular the east Pacific warm pool during boreal summer (Maloney and Esbensen 2003, 2005, 2007). Study of the multiscale-scale structure of the MJO (Sui and Lau 1992), and work on air–sea interaction associated with the MJO during the TOGA COARE experiment (Sui et al. 1997; Lau and Sui 1997) were topics of study by Dr. Chung-Hsiung Sui. Dr. Victor Magaña conducted an observational analysis of 40–50-day variations in atmospheric angular momentum and length of day as a function of latitude (Magaña 1993). Dr. Xiaoqing Wu has conducted observational, cloud-resolving model, and general circulation model studies to understand the governing dynamics of the MJO, including its multiscale structure (Wu et al. 1998; Wu and LeMone 1999; Deng and Wu 2011). Dr. Baode Chen collaborated on modeling work that examined the importance of frictional convergence to the MJO (Chao and Chen 2001). Dr. Wen-Wen Tung continues seminal research on the multiscale structure of the MJO (Tung et al. 2004) and MJO predictability (Tung et al. 2011). Finally, Dr. Chih-Wen Hung has explored the relationship between the intraseasonal oscillation, Asian summer monsoon, and the mei-yu front (Hung and Hsu 2008), and also worked with Professor Yanai to examine the processes responsible for onset of the summer Australian monsoon, including the MJO (Hung and Yanai 2004). Through his students, Professor Yanai's influence will likely continue to be felt for many future generations.

8. Concluding remarks

The legacy of Professor Yanai's study of the MJO will be with us for many years to come. His pioneering analysis techniques, insightful research directions, and genuine support of young scientists have inspired a new generation of scientists in their studies of the MJO, including many of Dr. Yanai's students and also the authors of this chapter. It should also be mentioned that Professor Yanai's influence on MJO research was felt in other ways. The University of California, Los Angeles (UCLA), Tropical Meteorology and Climate Newsletter that was moderated by Professor Yanai provided a stimulating forum to discuss current MJO research in advance of formal peer review and publication. The authors took advantage of this newsletter whenever possible to introduce our latest MJO work to the tropical meteorology community.

While progress has been made in the past decades, our understanding of the MJO and our ability to simulate and predict it are still limited. Professor Yanai's contributions to MJO research—especially in the context of heating and moistening profiles, cumulus momentum transport, energy budget and conversion, and tropics–extratropics interaction—will continue to benefit future MJO studies and model development.

Acknowledgments. We thank Roland Madden, Wen-Wen Tung, and Robert Fovell for their knowledge and advice during the writing of this manuscript. We would also like to thank George Kiladis and one anonymous reviewer for their thorough and constructive reviews of the manuscript. The authors acknowledge support from the National Science Foundation Climate and Large-Scale Dynamics Program under Grants AGS-1025584, AGS-1062161, AGS-0946911, and AGS-1441916 (EDM), as well as AGS-1062202 (CZ). We also acknowledge support from the NOAA MAPP program under Grant GC09-330a (CZ) and Contracts NA12OAR4310077 (GC12-433) and NA08OAR4320893 7,14 (EDM).

REFERENCES

Andersen, J. A., and Z. Kuang, 2012: Moist static energy budget of MJO-like disturbances in the atmosphere of a zonally symmetric aquaplanet. *J. Climate,* **25,** 2782–2804, doi:10.1175/JCLI-D-11-00168.1.

Barlow, M., and D. Salstein, 2006: Summertime influence of the Madden-Julian Oscillation on daily rainfall over Mexico and Central America. *Geophys. Res. Lett.,* **33,** L21708, doi:10.1029/2006GL027738.

Benedict, J. J., and D. A. Randall, 2007: Observed characteristics of the MJO relative to maximum rainfall. *J. Atmos. Sci.,* **64,** 2332–2354, doi:10.1175/JAS3968.1.

Biello, J. A., and A. J. Majda, 2005: A new multiscale model for the Madden–Julian oscillation. *J. Atmos. Sci.,* **62,** 1694–1721, doi:10.1175/JAS3455.1.

Bousquet, O., and M. Chong, 2000: The oceanic mesoscale convective system and associated mesovortex observed 12 December 1992 during TOGA-COARE. *Quart. J. Roy. Meteor. Soc.,* **126,** 189–211, doi:10.1002/qj.49712656210.

Bretherton, C. S., M. E. Peters, and L. E. Back, 2004: Relationships between water vapor path and precipitable water over the tropical oceans. *J. Climate,* **17,** 1517–1528, doi:10.1175/1520-0442(2004)017<1517:RBWVPA>2.0.CO;2.

Chao, W. C., and B. Chen, 2001: The role of surface friction in tropical intraseasonal oscillation. *Mon. Wea. Rev.,* **129,** 896–904, doi:10.1175/1520-0493(2001)129<0896:TROSFI>2.0.CO;2.

Charney, J. G., 1963: A note on large-scale motions in the tropics. *J. Atmos. Sci.,* **20,** 607–609, doi:10.1175/1520-0469(1963)020<0607:ANOLSM>2.0.CO;2.

Chen, B., and M. Yanai, 2000: Comparison of the Madden-Julian oscillation (MJO) during the TOGA COARE IOP with a 15-year climatology. *J. Geophys. Res.,* **105,** 2139–2149, doi:10.1029/1999JD901045.

Chen, T.-C., and M. Murakami, 1988: The 30–50 day variation of convective activity over the western Pacific Ocean with emphasis on the northwestern region. *Mon. Wea. Rev.,* **116,** 892–906, doi:10.1175/1520-0493(1988)116<0892:TDVOCA>2.0.CO;2.

Deng, L., and X. Wu, 2010: Effects of convective processes on GCM simulations of the Madden–Julian oscillation. *J. Climate,* **23,** 352–377, doi:10.1175/2009JCLI3114.1.

——, and ——, 2011: Physical mechanisms for the maintenance of GCM-simulated Madden–Julian oscillation over the Indian Ocean and Pacific. *J. Climate,* **24,** 2469–2482, doi:10.1175/2010JCLI3759.1.

Frederiksen, J. S., and C. S. Frederiksen, 1997: Mechanisms of the formation of intraseasonal oscillations and Australian monsoon disturbances: The roles of convection, barotropic and baroclinic instability. *Beitr. Phys. Atmos.,* **70,** 39–56.

Grabowski, W. W., 2002: Large-scale organization of moist convection in idealized aquaplanet simulations. *Int. J. Numer. Methods Fluids,* **39,** 843–853, doi:10.1002/fld.332.

Gruber, A., 1974: The wavenumber-frequency spectra of satellite-measured brightness in the tropics. *J. Atmos. Sci.,* **31,** 1675–1680, doi:10.1175/1520-0469(1974)031<1675:TWFSOS>2.0.CO;2.

Haertel, P. T., G. N. Kiladis, A. Denno, and T. M. Rickenbach, 2008: Vertical-mode decompositions of 2-day waves and the

Madden–Julian oscillation. *J. Atmos. Sci.*, **65**, 813–833, doi:10.1175/2007JAS2314.1.

Han, Y., and B. Khouider, 2010: Convectively coupled waves in a sheared environment. *J. Atmos. Sci.*, **67**, 2913–2942, doi:10.1175/2010JAS3335.1.

Hannah, W. M., and E. D. Maloney, 2011: The role of moisture–convection feedbacks in simulating the Madden–Julian oscillation. *J. Climate*, **24**, 2754–2770, doi:10.1175/2011JCLI3803.1.

Hayashi, Y., 1973: A method of analyzing transient waves by space-time cross spectra. *J. Appl. Meteor.*, **12**, 404–408, doi:10.1175/1520-0450(1973)012<0404:AMOATW>2.0.CO;2.

Hendon, H. H., and M. L. Salby, 1994: The life cycle of the Madden–Julian oscillation. *J. Atmos. Sci.*, **51**, 2225–2237, doi:10.1175/1520-0469(1994)051<2225:TLCOTM>2.0.CO;2.

Hoskins, B. J., and G.-Y. Yang, 2000: The equatorial response to higher-latitude forcing. *J. Atmos. Sci.*, **57**, 1197–1213, doi:10.1175/1520-0469(2000)057<1197:TERTHL>2.0.CO;2.

——, I. N. James, and G. H. White, 1983: The shape, propagation and mean-flow interaction of large-scale weather systems. *J. Atmos. Sci.*, **40**, 1595–1612, doi:10.1175/1520-0469(1983)040<1595:TSPAMF>2.0.CO;2.

Houze, R. A., S. S. Chen, D. E. Kingsmill, Y. Serra, and S. E. Yuter, 2000: Convection over the Pacific warm pool in relation to the atmospheric Kelvin–Rossby wave. *J. Atmos. Sci.*, **57**, 3058–3089, doi:10.1175/1520-0469(2000)057<3058:COTPWP>2.0.CO;2.

Hsu, H.-H., B. J. Hoskins, and F.-F. Jin, 1990: The 1985/86 intra-seasonal oscillation and the role of the extratropics. *J. Atmos. Sci.*, **47**, 823–839, doi:10.1175/1520-0469(1990)047<0823:TIOATR>2.0.CO;2.

Hung, C.-W., and M. Yanai, 2004: Factors contributing to the onset of the Australian summer monsoon. *Quart. J. Roy. Meteor. Soc.*, **130**, 739–761, doi:10.1256/qj.02.191.

——, and H.-H. Hsu, 2008: The first transition of the Asian summer monsoon, intraseasonal oscillation, and Taiwan mei-yu. *J. Climate*, **21**, 1552–1568, doi:10.1175/2007JCLI1457.1.

Jiang, X., and Coauthors, 2012: Simulation of the intraseasonal oscillation over the eastern Pacific ITCZ in climate models. *Climate Dyn.*, **39**, 617–636, doi:10.1007/s00382-011-1098-x.

Johnson, R. H., P. E. Ciesielski, and T. M. Rickenbach, 2016: A further look at Q_1 and Q_2 from TOGA COARE. *Multiscale Convection-Coupled Systems in the Tropics: A Tribute to Dr. Michio Yanai, Meteor. Monogr.*, No. 56, Amer. Meteor. Soc., doi:10.1175/AMSMONOGRAPHS-D-15-0002.1.

Kang, I.-S., and K. M. Lau, 1990: Evolution of tropical circulation anomalies associated with 30-60 day oscillation of globally averaged angular momentum during northern summer. *J. Meteor. Soc. Japan*, **68**, 237–249.

Khouider, B., and A. J. Majda, 2006: A simple multicloud parame-terization for convectively coupled tropical waves. Part I: Linear analysis. *J. Atmos. Sci.*, **63**, 1308–1323, doi:10.1175/JAS3677.1.

——, Y. Han, A. J. Majda, and S. N. Stechmann, 2012: Multiscale waves in an MJO background and convective momentum transport feedback. *J. Atmos. Sci.*, **69**, 915–933, doi:10.1175/JAS-D-11-0152.1.

Kikuchi, K., and B. Wang, 2010: Spatiotemporal wavelet transform and the multiscale behavior of the Madden–Julian oscillation. *J. Climate*, **23**, 3814–3834, doi:10.1175/2010JCLI2693.1.

Kiladis, G. N., and K. M. Weickmann, 1992: Extratropical forcing of tropical Pacific convection during northern winter. *Mon. Wea. Rev.*, **120**, 1924–1939, doi:10.1175/1520-0493(1992)120<1924:EFOTPC>2.0.CO;2.

Kiranmayi, L., and E. D. Maloney, 2011a: Intraseasonal moist static energy budget in reanalysis data. *J. Geophys. Res.*, **116**, D21117, doi:10.1029/2011JD016031.

——, and ——, 2011b: Effect of SST distribution and radiative feed-backs on the simulation of intraseasonal variability in an aqua-planet GCM. *J. Meteor. Soc. Japan*, **89**, 195–210, doi:10.2151/jmsj.2011-302.

Knippertz, P., 2007: Tropical–extratropical interactions related to upper-level troughs at low latitudes. *Dyn. Atmos. Oceans*, **43**, 36–62, doi:10.1016/j.dynatmoce.2006.06.003.

Knutson, T. R., and K. M. Weickmann, 1987: 30–60 day atmo-spheric oscillations: Composite life cycles of convection and circulation anomalies. *Mon. Wea. Rev.*, **115**, 1407–1436, doi:10.1175/1520-0493(1987)115<1407:DAOCLC>2.0.CO;2.

Lau, K.-M., and T. J. Phillips, 1986: Coherent fluctuations of ex-tratropical geopotential height and tropical convection in in-traseasonal time scales. *J. Atmos. Sci.*, **43**, 1164–1181, doi:10.1175/1520-0469(1986)043<1164:CFOFGH>2.0.CO;2.

——, and L. Peng, 1987: Origin of low-frequency (intraseasonal) oscillations in the tropical atmosphere. Part I: Basic theory. *J. Atmos. Sci.*, **44**, 950–972, doi:10.1175/1520-0469(1987)044<0950:OOLFOI>2.0.CO;2.

——, and C.-H. Sui, 1997: Mechanisms of short-term sea surface temperature regulation: Observations during TOGA COARE. *J. Climate*, **10**, 465–472, doi:10.1175/1520-0442(1997)010<0465:MOSTSS>2.0.CO;2.

——, P.-J. Sheu, and I.-S. Kang, 1994: Multiscale low-frequency circu-lation modes in the global atmosphere. *J. Atmos. Sci.*, **51**, 1169–1193, doi:10.1175/1520-0469(1994)051<1169:MLFCMI>2.0.CO;2.

Lewis, S. A., M. A. LeMone, and D. P. Jorgensen, 1998: Evolution and dynamics of a late-stage squall line that occurred on 20 February 1993 during TOGA COARE. *Mon. Wea. Rev.*, **126**, 3189–3212, doi:10.1175/1520-0493(1998)126<3189:EADOAL>2.0.CO;2.

Liebmann, B., and D. L. Hartmann, 1984: An observational study of tropical–midlatitude interaction on intraseasonal time scales during winter. *J. Atmos. Sci.*, **41**, 3333–3350, doi:10.1175/1520-0469(1984)041<3333:AOSOTI>2.0.CO;2.

Lin, H., G. Brunet, and J. Derome, 2007: Intraseasonal variability in a dry atmospheric model. *J. Atmos. Sci.*, **64**, 2422–2441, doi:10.1175/2007JAS3955.1.

Lin, J.-L., and B. E. Mapes, 2004: Radiation budget of the tropical intraseasonal oscillation. *J. Atmos. Sci.*, **61**, 2050–2062, doi:10.1175/1520-0469(2004)061<2050:RBOTTI>2.0.CO;2.

——, M. Zhang, and B. Mapes, 2005: Zonal momentum budget of the Madden–Julian oscillation: The source and strength of equivalent linear damping. *J. Atmos. Sci.*, **62**, 2172–2188, doi:10.1175/JAS3471.1.

Liu, F., and B. Wang, 2013: A frictional skeleton model for the Madden–Julian oscillation. *J. Atmos. Sci.*, **70**, 3147–3156, doi:10.1175/JAS-D-12-0348.1.

Madden, R. A., 1987: Relationships between changes in the length of day and the 40- to 50-day oscillation in the tropics. *J. Geophys. Res.*, **92**, 8391–8399, doi:10.1029/JD092iD07p08391.

——, and P. R. Julian, 1971: Detection of a 40–50 day oscillation in the zonal wind in the tropical Pacific. *J. Atmos. Sci.*, **28**, 702–708, doi:10.1175/1520-0469(1971)028<0702:DOADOI>2.0.CO;2.

——, and ——, 1972: Description of global-scale circulation cells in the tropics with a 40–50 day period. *J. Atmos. Sci.*, **29**, 1109–1123, doi:10.1175/1520-0469(1972)029<1109:DOGSCC>2.0.CO;2.

——, and P. Speth, 1995: Estimates of atmospheric angular mo-mentum, friction, and mountain torques during 1987–88. *J. Atmos. Sci.*, **52**, 3681–3694, doi:10.1175/1520-0469(1995)052<3681:EOAAMF>2.0.CO;2.

Magaña, V., 1993: The 40- and 50-day oscillations in atmospheric angular momentum at various latitudes. *J. Geophys. Res.*, **98**, 10 441–10 450, doi:10.1029/93JD00343.

——, and M. Yanai, 1991: Tropical–midlatitude interaction on the time scale of 30 to 60 days during the northern summer of 1979. *J. Climate*, **4**, 180–201, doi:10.1175/1520-0442(1991)004<0180:TMIOTT>2.0.CO;2.

——, and ——, 1995: Mixed Rossby–gravity waves triggered by lateral forcing. *J. Atmos. Sci.*, **52**, 1473–1486, doi:10.1175/1520-0469(1995)052<1473:MRWTBL>2.0.CO;2.

Majda, A. J., and R. Klein, 2003: Systematic multiscale models for the tropics. *J. Atmos. Sci.*, **60**, 393–408, doi:10.1175/1520-0469(2003)060<0393:SMMFTT>2.0.CO;2.

——, and J. A. Biello, 2004: A multiscale model for tropical intraseasonal oscillations. *Proc. Natl. Acad. Sci. USA*, **101**, 4736–4741, doi:10.1073/pnas.0401034101.

——, and S. N. Stechmann, 2009a: A simple dynamical model with features of convective momentum transport. *J. Atmos. Sci.*, **66**, 373–392, doi:10.1175/2008JAS2805.1.

——, and ——, 2009b: The skeleton of tropical intraseasonal oscillations. *Proc. Natl. Acad. Sci. USA*, **106**, 8417–8422, doi:10.1073/pnas.0903367106.

——, and ——, 2011: Nonlinear dynamics and regional variations in the MJO skeleton. *J. Atmos. Sci.*, **68**, 3053–3071, doi:10.1175/JAS-D-11-053.1.

Maloney, E. D., 2009: The moist static energy budget of a composite tropical intraseasonal oscillation in a climate model. *J. Climate*, **22**, 711–729, doi:10.1175/2008JCLI2542.1.

——, and D. L. Hartmann, 2000: Modulation of eastern North Pacific hurricanes by the Madden–Julian oscillation. *J. Climate*, **13**, 1451–1460, doi:10.1175/1520-0442(2000)013<1451:MOENPH>2.0.CO;2.

——, and S. K. Esbensen, 2003: The amplification of east Pacific Madden–Julian oscillation convection and wind anomalies during June–November. *J. Climate*, **16**, 3482–3497, doi:10.1175/1520-0442(2003)016<3482:TAOEPM>2.0.CO;2.

——, and ——, 2005: A modeling study of summertime east Pacific wind-induced ocean–atmosphere exchange in the intraseasonal oscillation. *J. Climate*, **18**, 568–584, doi:10.1175/JCLI-3280.1.

——, and ——, 2007: Satellite and buoy observations of intraseasonal variability in the tropical northeast Pacific. *Mon. Wea. Rev.*, **135**, 3–19, doi:10.1175/MWR3271.1.

——, A. H. Sobel, and W. M. Hannah, 2010: Intraseasonal variability in an aquaplanet general circulation model. *J. Adv. Model. Earth Syst.*, **2** (5), doi:10.3894/JAMES.2010.2.5.

Maruyama, T., 1967: Large-scale disturbances in the equatorial lower stratosphere. *J. Meteor. Soc. Japan*, **45**, 391–408.

Matsuno, T., 1966: Quasi-geostrophic motions in the equatorial area. *J. Meteor. Soc. Japan*, **44**, 25–43.

Matthews, A. J., 2008: Primary and successive events in the Madden–Julian Oscillation. *Quart. J. Roy. Meteor. Soc.*, **134**, 439–453, doi:10.1002/qj.224.

——, and G. N. Kiladis, 1999: The tropical–extratropical interaction between high-frequency transients and the Madden–Julian oscillation. *Mon. Wea. Rev.*, **127**, 661–677, doi:10.1175/1520-0493(1999)127<0661:TTEIBH>2.0.CO;2.

——, B. J. Hoskins, J. M. Slingo, and M. Blackburn, 1996: Development of convection along the SPCZ within a Madden–Julian oscillation. *Quart. J. Roy. Meteor. Soc.*, **122**, 669–688, doi:10.1002/qj.49712253106.

Meehl, G. A., G. N. Kiladis, K. M. Weickmann, M. Wheeler, D. S. Gutzler, and G. P. Compo, 1996: Modulation of equatorial subseasonal convective episodes by tropical-extratropical interaction in the Indian and Pacific Ocean regions. *J. Geophys. Res.*, **101**, 15 033–15 049, doi:10.1029/96JD01014.

Miyakawa, T., Y. N. Takayabu, T. Nasuno, H. Miura, M. Satoh, and M. W. Moncrieff, 2012: Convective momentum transport by rainbands within a Madden–Julian oscillation in a global nonhydrostatic model with explicit deep convective processes. Part I: Methodology and general results. *J. Atmos. Sci.*, **69**, 1317–1338, doi:10.1175/JAS-D-11-024.1.

Molinari, J., and D. Vollaro, 2000: Planetary- and synoptic-scale influences on eastern Pacific tropical cyclogenesis. *Mon. Wea. Rev.*, **128**, 3296–3307, doi:10.1175/1520-0493(2000)128<3296:PASSIO>2.0.CO;2.

Moncrieff, M. W., 2004: Analytic representation of the large-scale organization of tropical convection. *J. Atmos. Sci.*, **61**, 1521–1538, doi:10.1175/1520-0469(2004)061<1521:AROTLO>2.0.CO;2.

——, and E. Klinker, 1997: Organized convective systems in the tropical western Pacific as a process in general circulation models: A TOGA COARE case study. *Quart. J. Roy. Meteor. Soc.*, **123**, 805–827, doi:10.1002/qj.49712354002.

Mu, M., and G. J. Zhang, 2006: Energetics of Madden-Julian oscillations in the National Center for Atmospheric Research Community Atmosphere Model version 3 (NCAR CAM3). *J. Geophys. Res.*, **111**, D24112, doi:10.1029/2005JD007003.

Murakami, M., 1984: Analysis of the deep convective activity over the western Pacific and Southeast Asia. Part II: Seasonal and intraseasonal variations during northern summer. *J. Meteor. Soc. Japan*, **62**, 88–108.

Nakazawa, T., 1988: Tropical super clusters within intraseasonal variations over the western Pacific. *J. Meteor. Soc. Japan*, **66**, 823–839.

Neelin, J. D., and I. M. Held, 1987: Modeling tropical convergence based on the moist static energy budget. *Mon. Wea. Rev.*, **115**, 3–12, doi:10.1175/1520-0493(1987)115<0003:MTCBOT>2.0.CO;2.

Nitta, T., 1987: Convective activities in the tropical western Pacific and their impact on the Northern Hemisphere summer circulation. *J. Meteor. Soc. Japan*, **65**, 373–390.

——, T. Mizuno, and K. Takahashi, 1992: Multi-scale convective systems during the initial phase of the 1986/87 El Niño. *J. Meteor. Soc. Japan*, **70**, 447–466.

Peters, M. E., and C. S. Bretherton, 2006: Structure of tropical variability from a vertical mode perspective. *Theor. Comput. Fluid. Dyn.*, **20**, 501–524, doi:10.1007/s00162-006-0034-x.

Randall, D., C. DeMott, C. Stan, M. Khairoutdinov, J. Benedict, R. McCrary, K. Thayer-Calder, and M. Branson, 2016: Simulations of the tropical general circulation with a multiscale global model. *Multiscale Convection-Coupled Systems in the Tropics: A Tribute to Dr. Michio Yanai*, Meteor. Monogr., No. 56, Amer. Meteor. Soc., doi:10.1175/AMSMONOGRAPHS-D-15-0016.1.

Ray, P., and C. Zhang, 2010: A case study of the mechanics of extratropical influence on the initiation of the Madden–Julian oscillation. *J. Atmos. Sci.*, **67**, 515–528, doi:10.1175/2009JAS3059.1.

——, and T. Li, 2013: Relative roles of circumnavigating waves and extratropics on the MJO and its relationship with the mean state. *J. Atmos. Sci.*, **70**, 876–893, doi:10.1175/JAS-D-12-0153.1.

——, C. Zhang, J. Dudhia, and S. S. Chen, 2009: A numerical case study on the initiation of the Madden–Julian oscillation. *J. Atmos. Sci.*, **66**, 310–331, doi:10.1175/2008JAS2701.1.

Raymond, D. J., 2001: A new model of the Madden–Julian oscillation. *J. Atmos. Sci.*, **58**, 2807–2819, doi:10.1175/1520-0469(2001)058<2807:ANMOTM>2.0.CO;2.

——, and Ž. Fuchs, 2009: Moisture modes and the Madden–Julian oscillation. *J. Climate*, **22**, 3031–3046, doi:10.1175/2008JCLI2739.1.

——, S. Sessions, A. Sobel, and Z. Fuchs, 2009: The mechanics of gross moist stability. *J. Adv. Model. Earth Syst.*, **1** (9), doi:10.3894/JAMES.2009.1.9.

Reed, R. J., W. J. Cambell, L. A. Rasmussen, and D. G. Rogers, 1961: Evidence of a downward-propagating annual wind reversal in the equatorial stratosphere. *J. Geophys. Res.*, **66**, 813–818, doi:10.1029/JZ066i003p00813.

Roux, F., 1998: The oceanic mesoscale convective system observed with airborne Doppler radars on 9 February 1993 during TOGA–COARE: Structure, evolution and budgets. *Quart. J. Roy. Meteor. Soc.*, **124**, 585–614, doi:10.1002/qj.49712454610.

Rydbeck, A. V., E. D. Maloney, S.-P. Xie, J. Hafner, and J. Shaman, 2013: Remote forcing versus local feedback of east Pacific intraseasonal variability. *J. Climate*, **26**, 3575–3596, doi:10.1175/JCLI-D-12-00499.1.

Sobel, A. H., and C. S. Bretherton, 2000: Modeling tropical precipitation in a single column. *J. Climate*, **13**, 4378–4392, doi:10.1175/1520-0442(2000)013<4378:MTPIAS>2.0.CO;2.

——, and E. D. Maloney, 2012: An idealized semi-empirical framework for modeling the Madden–Julian oscillation. *J. Atmos. Sci.*, **69**, 1691–1705, doi:10.1175/JAS-D-11-0118.1.

——, and ——, 2013: Moisture modes and the eastward propagation of the MJO. *J. Atmos. Sci.*, **70**, 187–192, doi:10.1175/JAS-D-12-0189.1.

——, J. Nilsson, and L. M. Polvani, 2001: The weak temperature gradient approximation and balanced tropical moisture waves. *J. Atmos. Sci.*, **58**, 3650–3665, doi:10.1175/1520-0469(2001)058<3650:TWTGAA>2.0.CO;2.

——, E. D. Maloney, G. Bellon, and D. M. Frierson, 2008: The role of surface heat fluxes in tropical intraseasonal oscillations. *Nat. Geosci.*, **1**, 653–657, doi:10.1038/ngeo312.

——, ——, ——, and ——, 2010: Surface fluxes and tropical intraseasonal variability: A reassessment. *J. Adv. Model. Earth Syst.*, **2** (2), doi:10.3894/JAMES.2010.2.2.

Straub, K. H., and G. N. Kiladis, 2003a: Interactions between the boreal summer intraseasonal oscillation and higher-frequency tropical wave activity. *Mon. Wea. Rev.*, **131**, 945–960, doi:10.1175/1520-0493(2003)131<0945:IBTBSI>2.0.CO;2.

——, and ——, 2003b: Extratropical forcing of convectively coupled Kelvin waves during austral winter. *J. Atmos. Sci.*, **60**, 526–543, doi:10.1175/1520-0469(2003)060<0526:EFOCCK>2.0.CO;2.

Straus, D. M., and R. S. Lindzen, 2000: Planetary-scale baroclinic instability and the MJO. *J. Atmos. Sci.*, **57**, 3609–3626, doi:10.1175/1520-0469(2000)057<3609:PSBIAT>2.0.CO;2.

Sui, C.-H., and K.-M. Lau, 1992: Multiscale phenomena in the tropical atmosphere over the western Pacific. *Mon. Wea. Rev.*, **120**, 407–430, doi:10.1175/1520-0493(1992)120<0407:MPITTA>2.0.CO;2.

——, X. Li, K.-M. Lau, and D. Adamec, 1997: Multiscale air–sea interactions during TOGA COARE. *Mon. Wea. Rev.*, **125**, 448–462, doi:10.1175/1520-0493(1997)125<0448:MASIDT>2.0.CO;2.

Takayabu, Y. N., G. N. Kiladis, and V. Magaña, 2016: Michio Yanai and tropical waves. *Multiscale Convection-Coupled Systems in the Tropics: A Tribute to Dr. Michio Yanai, Meteor. Monogr.*, No. 56, Amer. Meteor. Soc., doi:10.1175/AMSMONOGRAPHS-D-15-0019.1.

Tokioka, T., K. Yamazaki, A. Kitoh, and T. Ose, 1988: The equatorial 30–60 day oscillation and the Arakawa–Schubert penetrative cumulus parameterization. *J. Meteor. Soc. Japan*, **66**, 883–901.

Trenberth, K. E., 1986: An assessment of the impact of transient eddies on the zonal flow during a blocking episode using localized Eliassen–Palm flux diagnostics. *J. Atmos. Sci.*, **43**, 2070–2087, doi:10.1175/1520-0469(1986)043<2070:AAOTIO>2.0.CO;2.

Tung, W.-W., and M. Yanai, 2002a: Convective momentum transport observed during the TOGA COARE IOP. Part I: General features. *J. Atmos. Sci.*, **59**, 1857–1871, doi:10.1175/1520-0469(2002)059<1857:CMTODT>2.0.CO;2.

——, and ——, 2002b: Convective momentum transport observed during the TOGA COARE IOP. Part II: Case studies. *J. Atmos. Sci.*, **59**, 2535–2549, doi:10.1175/1520-0469(2002)059<2535:CMTODT>2.0.CO;2.

——, C. Lin, B. Chen, M. Yanai, and A. Arakawa, 1999: Basic modes of cumulus heating and drying observed during TOGA-COARE IOP. *Geophys. Res. Lett.*, **26**, 3117–3120, doi:10.1029/1999GL900607.

——, M. W. Moncrieff, and J.-B. Gao, 2004: A systemic analysis of multiscale deep convective variability over the tropical Pacific. *J. Climate*, **17**, 2736–2751, doi:10.1175/1520-0442(2004)017<2736:ASAOMD>2.0.CO;2.

——, J. B. Gao, J. Hu, and L. Yang, 2011: Detecting chaotic signals in heavy noise environments. *Phys. Rev.*, **83E**, 046210, doi:10.1103/PhysRevE.83.046210.

Wang, B., and F. Liu, 2011: A model for scale interaction in the Madden–Julian oscillation. *J. Atmos. Sci.*, **68**, 2524–2536, doi:10.1175/2011JAS3660.1.

Webster, P. J., and J. R. Holton, 1982: Cross-equatorial response to middle-latitude forcing in a zonally varying basic state. *J. Atmos. Sci.*, **39**, 722–733, doi:10.1175/1520-0469(1982)039<0722:CERTML>2.0.CO;2.

Weickmann, K. M., and P. D. Sardeshmukh, 1994: The atmospheric angular momentum cycle associated with a Madden–Julian oscillation. *J. Atmos. Sci.*, **51**, 3194–3208, doi:10.1175/1520-0469(1994)051<3194:TAAMCA>2.0.CO;2.

——, G. R. Lussky, and J. E. Kutzbach, 1985: Intraseasonal (30–60 day) fluctuations of outgoing longwave radiation and 250 mb streamfunction during northern winter. *Mon. Wea. Rev.*, **113**, 941–961, doi:10.1175/1520-0493(1985)113<0941:IDFOOL>2.0.CO;2.

——, S. J. S. Khalsa, and J. Eischeid, 1992: The atmospheric angular momentum cycle during the tropical Madden–Julian oscillation. *Mon. Wea. Rev.*, **120**, 2252–2263, doi:10.1175/1520-0493(1992)120<2252:TAAMCD>2.0.CO;2.

——, G. N. Kiladis, and P. D. Sardeshmukh, 1997: The dynamics of intraseasonal atmospheric angular momentum oscillations. *J. Atmos. Sci.*, **54**, 1445–1461, doi:10.1175/1520-0469(1997)054<1445:TDOIAA>2.0.CO;2.

Wheeler, M., and G. N. Kiladis, 1999: Convectively coupled equatorial waves: Analysis of clouds and temperature in the wavenumber–frequency domain. *J. Atmos. Sci.*, **56**, 374–399, doi:10.1175/1520-0469(1999)056<0374:CCEWAO>2.0.CO;2.

Wu, X., and M. Yanai, 1994: Effects of vertical wind shear on the cumulus transport of momentum: Observations and parameterization. *J. Atmos. Sci.*, **51**, 1640–1660, doi:10.1175/1520-0469(1994)051<1640:EOVWSO>2.0.CO;2.

——, and M. A. LeMone, 1999: Fine structure of cloud patterns within the intraseasonal oscillation during TOGA COARE. *Mon. Wea. Rev.*, **127**, 2503–2513, doi:10.1175/1520-0493(1999)127<2503:FSOCPW>2.0.CO;2.

——, W. W. Grabowski, and M. W. Moncrieff, 1998: Long-term behavior of cloud systems in TOGA COARE and their interactions with radiative and surface processes. Part I: Two-dimensional modeling study. *J. Atmos. Sci.*, **55**, 2693–2714, doi:10.1175/1520-0469(1998)055<2693:LTBOCS>2.0.CO;2.

Yanai, M., 1961: A detailed analysis of typhoon formation. *J. Meteor. Soc. Japan*, **39**, 187–214.

——, and T. Maruyama, 1966: Stratospheric wave disturbances propagating over the equatorial Pacific. *J. Meteor. Soc. Japan*, **44**, 291–294.

——, and M. Murakami, 1970: Spectrum analysis of symmetric and antisymmetric equatorial waves. *J. Meteor. Soc. Japan*, **48**, 331–347.

——, and M.-M. Lu, 1983: Equatorially trapped waves at the 200 mb level and their association with meridional convergence of wave energy flux. *J. Atmos. Sci.*, **40**, 2785–2803, doi:10.1175/1520-0469(1983)040<2785:ETWATM>2.0.CO;2.

——, and R. H. Johnson, 1993: Impacts of cumulus convection on thermodynamic fields. *The Representation of Cumulus Convection in Numerical Models of the Atmosphere, Meteor. Monogr.*, No. 46, Amer. Meteor. Soc., 39–62.

——, T. Maruyama, T. Nitta, and Y. Hayashi, 1968: Power spectra of large-scale disturbances over the tropical Pacific. *J. Meteor. Soc. Japan*, **46**, 308–323.

——, S. Esbensen, and J.-H. Chu, 1973: Determination of bulk properties of tropical cloud clusters from large-scale heat and moisture budgets. *J. Atmos. Sci.*, **30**, 611–627, doi:10.1175/1520-0469(1973)030<0611:DOBPOT>2.0.CO;2.

——, B. Chen, and W.-W. Tung, 2000: The Madden–Julian oscillation observed during the TOGA COARE IOP: Global view. *J. Atmos. Sci.*, **57**, 2374–2396, doi:10.1175/1520-0469(2000)057<2374:TMJOOD>2.0.CO;2.

Yano, J.-I., and M. Bonazzola, 2009: Scale analysis for large-scale tropical atmospheric dynamics. *J. Atmos. Sci.*, **66**, 159–172, doi:10.1175/2008JAS2687.1.

Yoneyama, K., C. Zhang, and C. N. Long, 2013: Tracking pulses of the Madden–Julian oscillation. *Bull. Amer. Meteor. Soc.*, **94**, 1871–1891, doi:10.1175/BAMS-D-12-00157.1.

Zangvil, A., 1975: Temporal and spatial behavior of large-scale disturbances in tropical cloudiness deduced from satellite brightness data. *Mon. Wea. Rev.*, **103**, 904–920, doi:10.1175/1520-0493(1975)103<0904:TASBOL>2.0.CO;2.

——, and M. Yanai, 1980: Upper tropospheric waves in the tropics. Part I: Dynamical analysis in the wavenumber-frequency domain. *J. Atmos. Sci.*, **37**, 283–298, doi:10.1175/1520-0469(1980)037<0283:UTWITT>2.0.CO;2.

——, and ——, 1981: Upper tropospheric waves in the tropics. Part II: Association with clouds in the wavenumber-frequency domain. *J. Atmos. Sci.*, **38**, 939–953, doi:10.1175/1520-0469(1981)038<0939:UTWITT>2.0.CO;2.

Zhang, C., 1993: Laterally forced equatorial perturbations in a linear model. Part II: Mobile forcing. *J. Atmos. Sci.*, **50**, 807–821, doi:10.1175/1520-0469(1993)050<0807:LFEPIA>2.0.CO;2.

——, and P. J. Webster, 1992: Laterally forced equatorial perturbations in a linear model. Part I: Stationary transient forcing. *J. Atmos. Sci.*, **49**, 585–607, doi:10.1175/1520-0469(1992)049<0585:LFEPIA>2.0.CO;2.

——, and H. H. Hendon, 1997: Propagating and standing components of the intraseasonal oscillation in tropical convection. *J. Atmos. Sci.*, **54**, 741–752, doi:10.1175/1520-0469(1997)054<0741:PASCOT>2.0.CO;2.

——, and S. Hagos, 2009: Bi-modal structure and variability of large-scale diabatic heating in the tropics. *J. Atmos. Sci.*, **66**, 3621–3640, doi:10.1175/2009JAS3089.1.

——, J. Gottschalck, E. D. Maloney, M. W. Moncrieff, F. Vitart, D. E. Waliser, B. Wang, and M. C. Wheeler, 2013: Cracking the MJO nut. *Geophys. Res. Lett.*, **40**, 1223–1230, doi:10.1002/grl.50244.

Zhou, L., R. B. Neale, M. Jochum, and R. Murtugudde, 2012: Improved Madden–Julian oscillations with improved physics: The impact of modified convection parameterizations. *J. Climate*, **25**, 1116–1136, doi:10.1175/2011JCLI4059.1.

Chapter 5

A Mechanism of the MJO Invoking Scale Interactions

T. N. KRISHNAMURTI, RUBY KRISHNAMURTI, ANU SIMON, AYPE THOMAS,
AND VINAY KUMAR

Department of Earth, Ocean and Atmospheric Science, Florida State University, Tallahassee, Florida

This chapter distinguishes the mechanism of tropical convective disturbances, such as a hurricane, from that of the Madden–Julian oscillation (MJO). The hurricane is maintained by organized convection around the azimuth. In a hurricane the organization of convection, the generation of eddy available potential energy, and the transformation of eddy available potential energy into eddy kinetic energy all occur on the scale of the hurricane and these are called "in-scale processes," which invoke quadratic nonlinearity. The MJO is not a hurricane type of disturbance; organized convection simply does not drive an MJO in the same manner. The maintenance of the MJO is more akin to a multibody problem where the convection is indeed organized on scales of tropical synoptic disturbances that carry a similar organization of convection and carry similar roles for the generation of eddy available potential energy and its conversion to the eddy kinetic energy for their maintenance. The maintenance of the MJO is a scale interaction problem that comes next, where pairs of synoptic-scale disturbances are shown to interact with a member of the MJO time scale, thus contributing to its maintenance. This chapter illustrates the organization of convection, synoptic-scale energetics, and nonlinear scale interactions to show the above aspects for the mechanism of the MJO.

1. Introduction

This study addresses the role of deep convection, nonlinear dynamics, and energetics for the maintenance of the Madden–Julian oscillation (MJO). The horizontal scale of the MJO is around 10 000 km. It is largely defined by zonal wavenumbers 1 and 2 (Madden and Julian 1971). This wave carries its largest amplitude over the equatorial latitudes and moves around Earth in roughly 20 to 60 days (Madden and Julian 1971; Krishnamurti and Gadgil 1985; Waliser et al. 2003). The amplitude of the zonal wind anomaly on this time scale is of the order of 1 to 3 m s^{-1}. The divergent wind on this time scale carries amplitude on the order of 1 m s^{-1}. A much cited illustration relating clouds to the MJO wave is the well-known Nakazawa diagram (Nakazawa 1988). An example is shown in Fig. 5-1. This illustration shows an active part of the MJO wave that carries a plethora of clouds. Multiple

scales of clouds are present here. The mesoconvective cloud lines, embedded within the larger-scale MJO envelope wave, propagate from east to west with typical speeds on the order of 5° to 7° longitude per day whereas the MJO moves eastward around the globe (360°) in around 40 days. This disparity in space–time scales of the clouds and mesoscales embedded in synoptic scales and the planetary-scale MJO makes it an interesting problem for scale interactions. Basically the important question we raise here is how clouds whose scale is of the order of a few kilometers communicate with the MJO, which has a scale of around 10 000 km. Two types of energy exchanges that invoke quadratic or triple product nonlinearities are important in the context of scale interactions.

In modeling multiple scales of the MJO, Biello and Majda (2005) noted the importance of westerly and easterly regimes of flows and wave trains carrying tilted organized synoptic scales. That is consistent with our findings. Scale interactions among 2-day waves and moist Kelvin waves were examined by Liu and Wang (2012a), where they noted that the convergence of eddy momentum and heat transports were sensitive to the phase of westerly winds of the lower and the upper

Corresponding author address: T. N. Krishnamurti, Department of Earth, Ocean and Atmospheric Science, Florida State University, 430 Love Building, Academic Way, Tallahassee, FL 32310.
E-mail: tkrishnamurti@fsu.edu

DOI: 10.1175/AMSMONOGRAPHS-D-15-0009.1

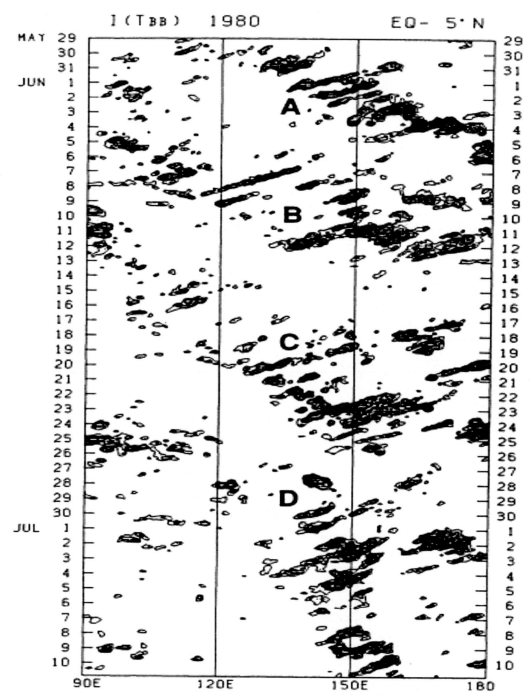

FIG. 5-1. Time–longitude section of temperature of black body (TBB; brightness temperature) index (ITBB) integrated between the equator and 5°N obtained from the 3-hourly Geostationary Meteorological Satellite (GMS) infrared data from 0000 UTC 29 May to 2100 UTC 10 Jul 1980. Symbols A to D denote the superclusters. The contour interval is 10 and shading denotes the region where values are >20 [adapted from Nakazawa (1988)].

troposphere in modulating the Kelvin wave. The role of friction was examined for the excitation of MJO by Liu and Wang (2012b), who noted a wavenumber dependence; that is, for long waves with wavenumbers less than 5, the Ekman pumping of moisture helped their growth from the moisture convergence. A formal study on scale interactions in the MJO was authored by Wang and Liu (2011), who noted that the eddy heat and moisture transports and their convergences arising from the passage of moist Kelvin waves was an important factor for

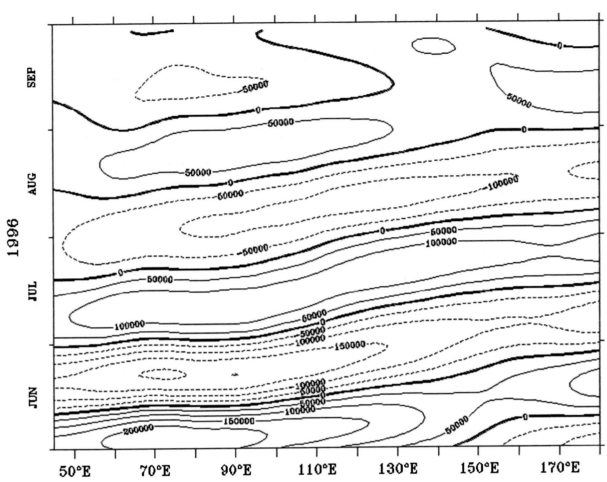

FIG. 5-2. Time–longitude section of filtered 200-hPa velocity potential ($m^2 s^{-1}$) on a Hovmöller diagram for a 122-day period during summer of 1996 in the equatorial belt 10°S–10°N.

the maintenance of the MJO. In recent years, a series of multiscale interaction mechanisms have been presented for the MJO. The upscale momentum transfer of super-clusters can drive the MJO-like circulation (Majda and Biello 2004; Biello and Majda 2005; Wang and Liu 2011; Liu and Wang 2013). The wave activity of synoptic-scale disturbance acts as a heat/moisture source/sink or vice versa—that is, an oscillator for the MJO (Majda and Stechmann 2009; Liu and Wang 2012a,b). These studies are relevant for the overall understanding of the MJO. Our study addresses the in-scale and out-of-scale ener-getics as means for the understanding of the MJO.

We address these respective interactions using energy exchange equations to illustrate the role of organized con-vection in the maintenance of the mesoscale and synoptic scale and the role of nonlinear dynamics for the exchanges of energy from the synoptic and the MJO time scales.

Figure 5-2 shows a time–longitude Hovmöller dia-gram for the 200-hPa level velocity potential, on the time scale of the MJO (20–60 days), for the summer

months of the year 1996. This is the year for which the scale interactions are computed. These are latitudinal averages between 10°S and 10°N. Positive values of the velocity potential denote regions from where divergent outflows at 200 hPa emanate. The amplitude of the di-vergent wind is largest in the Asian monsoon belt of India and East Asia as well as the western Pacific Ocean. Divergent winds are weaker over the Atlantic and east Pacific longitudes. The magnitude of the divergent wind at the 200-hPa level is between 1 and 4 m s^{-1} in the Asia–Pacific belt of the monsoon. The magnitude of the di-vergence, on the time scale of the MJO, is between 10^{-7} and $10^{-6} s^{-1}$. This weak divergence (and the compen-sating convergence of the lower troposphere) cannot account for the cloud scale or mesoconvective scales since the convergence/divergence is only on the order of 10^{-7} to $10^{-6} s^{-1}$. The passage of an active MJO at best can only initiate clouds; the subsequent growth of clouds requires a mesoscale forcing within synoptic-scale disturbances. The well-known Nakazawa diagram

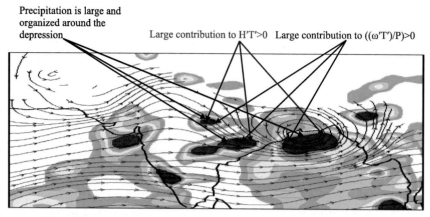

FIG. 5-3. Large precipitation elements organized around the monsoon depression carrying large values for covariances to $H'T' > 0$ and $-[(\omega'T')/P] > 0$. The heavy lines identify regions where the processes shown above the diagram are prominent. (Color bar is shown below in Fig. 5-4.)

(Fig. 5-1) shows an MJO envelope over the western Pacific Ocean. Within that envelope the synoptic-scale disturbances turn carry mesoscale convective cloud elements. Understanding this multiscale problem is central for the understanding the mechanism of the MJO.

The DYNAMO (Dynamics of the MJO) experiment in conjunction with other field experiments is aimed at collection of datasets for a better understanding of the MJO, its initiation, and its dynamics, as well as for improved prediction. There is clearly a need for such an observational experiment to better understand the role of convection in the maintenance and initiation of the MJO. The sounding radar array of DYNAMO is a vital component for such observations. Zhang et al. (2010) and Yuan and Houze (2013) are prominent contributors to the DYNAMO program. Zhang et al. (2010) have examined the heat sources for the MJO cycles and noted that both shallow convective and deep convective heating carry the MJO time scale signals over the western Pacific and the Indian Ocean regions. Lin et al. (2004) have also examined the heating profiles for the stratiform clouds during MJO passages. Yuan and Houze (2013) have examined the proportions of the total precipitation and cloud cover contributed by the MJO time scale. These are important for our overall understanding of the MJO.

The passage of the MJO from the tropical Pacific to the Atlantic has been succinctly described by Yu et al. (2011), who provide the background dynamics for the eastward motion of the MJO. The importance of moist convection for the MJO is well recognized (Hannah and Maloney 2011). That the MJO modulates tropical weather over the western Pacific has been well documented in their study. Recent studies of Zhang et al. (2010) and Ray et al. (2010) have examined important observational and modeling aspects of the MJO. They have examined the vertical distributions of heating and the structure of the MJO

environment. Their findings support the importance of deep convection and its environment. Observational and modeling studies by many authors have provided major insights into the role of convection and the structure of MJO (e.g., Kikuchi and Wang 2010; Wang et al. 2009; Waliser et al. 2009; Liu et al. 2009, among others).

2. Organization of convection in synoptic disturbances

The energetics of a hurricane is a good example on the role of the organization of convection. When one performs such an energetics (e.g., Krishnamurti et al. 2003), for a hurricane the generation of eddy available potential energy by convective heating and the conversion of eddy available potential energy into eddy kinetic energy are noted to occur on large azimuthal scales in the inner core of hurricanes. That is a clear reflection that deep convection is organized on the large azimuthal scales. The tangential winds in the inner core of a hurricane show most of the variance for azimuthal wavenumbers 1 and 2. That defines the scale of the hurricane. Although the individual elements of deep convective clouds carry scale on the order of a few kilometers each, the organization as a whole, brought about initially by shear flow instabilities, places the cloud scale into these large azimuthal scales. The above aspects of the hurricane energetics are described in Krishnamurti et al. (2003).

The MJO waves that propagate from west to east beneath these waves are tropical disturbances such as the monsoon depressions of the Indian Ocean and tropical lows and depressions of the tropical Pacific Ocean. These disturbances carry somewhat similar energetics and organization of convection. The energetics of these depressions are relevant to the coupling of synoptic and the MJO time scales that are discussed in this study.

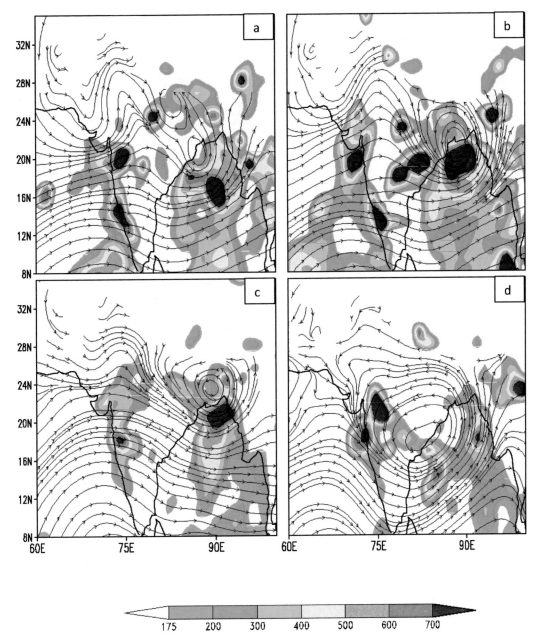

FIG. 5-4. Four cases of low pressure systems (lows) formed over the Indian region (7°–35°N, 60°–100°E) for (a) 20 Jul, (b) 21 Jul, (c) 8 Aug, and (d) 28 Aug 1996. The heavy lines identify regions where the processes shown above the diagram are prominent.

The generation and conversion of eddy available kinetic energy is largely described by covariances of heating and temperature and of vertical velocity and temperatures (Fig. 5-3); those are quadratic nonlinearities (i.e., in-scale processes in the vocabulary of scale interactions). This states that the generation and conversion, including the organization of clouds, all occur on the same scales (i.e., they are all in-scale processes). If clouds are organized along azimuthal wavenumbers L1, L2, L3, L4, etc., then all these processes will

also occur in the same repetitive scales and no communication is permitted with scales outside of a scale. The governing principle is the trigonometric selection rule (called $m = n$) for such quadratic nonlinearities.

Ideally good radar coverage would portray the organization of convection around a monsoon depression; lacking that, the mapping of daily precipitation can provide a reasonable picture of this organization. Figure 5-4 shows four examples of the 850-hPa level streamlines and 24-h rainfall totals, centered around the map time, when an intense

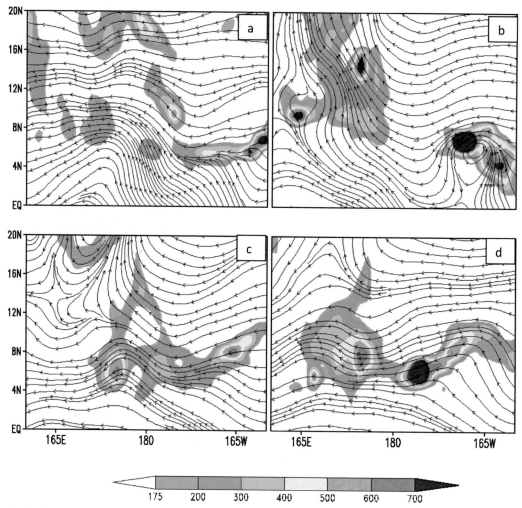

FIG. 5-5. Four cases of low pressure systems (lows) formed over the Pacific region (0°–20°N, 160°E–160°W) for (a) 22 Jul, (b) 7 Aug, (c) 29 Aug, and (d) 30 Aug 1996.

monsoon depression was present over India. These are rather typical maps of the flow field with superimposed rainfall. In all these illustrations, the organization of mesoconvective precipitating elements along the cyclonic streamlines of the monsoon depression is clearly evident. The typical rainfall intensities in these depressions are on the order of 50 mm day^{-1}. In the next section the Lorenz box energetics for these four examples are illustrated, relating the organization of convection to the eventual growth of eddy kinetic energy of the monsoon depression.

3. Lorenz box energetics on synoptic scales

The results of Lorenz box energetics are presented in this section. The synoptic-scale precipitating disturbances include tropical waves through the central Pacific Ocean and monsoon depressions over India. These were cases selected during the eastward passage of an active MJO wave (shown in Fig. 5-1) over the central Pacific and the Indian monsoon environment at the 200-hPa level. These cover the 1996 summer months. Figures 5-4 and 5-5 illustrate synoptic features for these tropical disturbances over the Indian and central Pacific regions respectively. These illustrations include the 850-hPa level streamlines and the superimposed rainfall fields. Around the quasi-circular geometry of the flow fields in these illustrations, an organization of convection is implied by these rainfall patterns. These include four separate synoptic cases for of these regions respectively. In all cases, we can see a synoptic-scale organization of the heavy rain elements. The box energetics for several of these cases are shown in Figs. 5-6a–d and 5-7a–d. These were all cases covering the 1996 season for the dates 20 July, 22 July, 6 August, and 29 August. During those dates active synoptic disturbances were present over India and the central Pacific Ocean as well. Figure 5-6 covers the results of box energetics for these afore-mentioned cases for the Indian region, and Fig. 5-7 shows the results for tropical depressions over central Pacific region. The horizontal size of these disturbances is around 3000 km, so a box was selected around these disturbances with a comparable lateral and meridional scale. The generation and energy conversion term for each case were computed and are illustrated in these panels. Boundary flux terms were not computed, so these results only confirm the internal processes. The following results stand out in all cases: There is a substantial generation of eddy available potential energy from convective heating. There is significant conversion of eddy available potential energy into eddy kinetic energy (baroclinic process) and there is a significant contribution from the local Hadley cell (i.e., energy exchange from the zonal available potential energy to zonal kinetic energy over this local domain). These salient features of the energetics are seen for the Pacific and the Indian monsoon disturbances. These results are consistent with the linear analysis of Moorthi and Arakawa (1985). The organization of convection contributes to the organization of the contributions for the covariances of convective heating and temperature and of the vertical velocity and temperature. These individual cloud elements whose horizontal scales are on the order of a few kilometers provide eddy kinetic energy for their maintenance by organizing around the scale of these disturbances.

The calculations of energy exchanges and generation and dissipation are formulated essentially following Lorenz (1967).

1) We define the principal energy quantities as follows (refer to the appendix for a list of symbols):

$$\overline{P} = \frac{c_p}{2} \int_m \frac{\gamma_d}{\gamma_d - \overline{\gamma}} \frac{[T]''^2}{\overline{T}} \, dm, \qquad (5\text{-}1)$$

where

$$dm = -\frac{1}{g} dx \, dy \, dp;$$

$$P' = \frac{c_p}{2} \int_m \frac{\gamma_d}{\gamma_d - \overline{\gamma}} \left\{ \frac{\overline{T''^2} - [T]''^2}{\overline{T}} \right\} dm; \qquad (5\text{-}2)$$

$$\overline{K} = \frac{1}{2} \int_m [(u)^2 + (v)^2] \, dm; \quad \text{and} \qquad (5\text{-}3)$$

$$K' = \frac{1}{2} \int_m \{(u - [u])^2 + (v - [v])^2\} \, dm, \qquad (5\text{-}4)$$

where $\overline{P}$, P', $\overline{K}$, and K' are, respectively, the zonal available potential energy, eddy available potential energy, zonal kinetic energy, and eddy kinetic energy over a closed domain of mass m. The $[\cdot]$ indicate a zonal average.

2) The principal energy transformation functions are the following with the angled brackets $\langle \, \rangle$ indicating an energy exchange between the quantities contained therein:

$$\langle [P] \cdot P' \rangle$$
$$= -R \int_m \left(\frac{\theta}{\overline{T}} \right) \left\{ [T'v'] \frac{\partial}{\partial y} + [T'\omega'] \frac{\partial}{\partial p} \right\} \left\{ \frac{\gamma_d}{\gamma_d - \overline{\gamma}} \frac{[T]''}{\overline{\theta}} \right\} dm,$$
$$(5\text{-}5)$$

$$\langle [P] \cdot K' \rangle = -R \int_m \frac{[T]''[\omega]''}{p} \, dm, \qquad (5\text{-}6)$$

$$\langle [K] \cdot K' \rangle = -\int_m \left\{ [u'v'] \frac{\partial}{\partial y} [u] + [u'\omega'] \frac{\partial'}{\partial p} [u] \right.$$
$$\left. + [v'v'] \frac{\partial}{\partial y} [v] + [v'\omega'] \frac{\partial}{\partial p} [v] \right\} dm, \quad \text{and}$$
$$(5\text{-}7)$$

$$\langle [P'] \cdot K' \rangle = -R \int_m \frac{\omega' T'}{p} \, dm. \qquad (5\text{-}8)$$

3) The generation of available potential energy is expressed by the relations

$$\overline{G}_i = \int_m \left(\frac{\gamma_d}{\gamma_d - \overline{\gamma}} \right) \frac{1}{\overline{T}} [T]''[H_i]'' \, dm \quad \text{and} \quad (5\text{-}9)$$

$$G'_i = \int_m \left(\frac{\gamma_d}{\gamma_d - \overline{\gamma}} \right) \frac{1}{\overline{T}} [T'H'_i] \, dm, \qquad (5\text{-}10)$$

where $\overline{G}_i$ and G'_i are, respectively, the generation terms for the zonal and the eddy available potential energy. The subscript i denotes the generation for a particular heating function H_i. In the following analysis we shall speak of four types of heating functions and their respective n contributions to the generation of available potential energy. These heating functions are shown below:

H_R = radiative warming,

H_{ST} = stable heating (i.e., large-scale condensation),

H_{CON} = convective heating (parameterized form), and

H_{SEN} = sensible heat flux from ocean and land surfaces.

4) The dissipation of kinetic energy is given by the following expressions:

$$\overline{D} = \int_m [T] \cdot [H] \, dm \quad \text{and} \qquad (5\text{-}11)$$

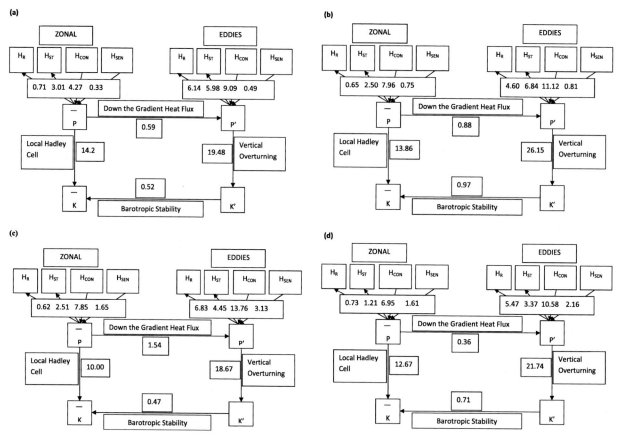

FIG. 5-6. Box energetics, following Lorenz (1967), showing some salient energy conversions and generation components (°s^{-1}). These figures pertain to the synoptic-scale disturbances for (a) 20 Jul, (b) 21 Jul, (c) 8 Aug, and (d) 28 Aug 1996 over the Indian region (7°–35°N, 60°–100°E); the symbols are explained in the table of acronyms and the text provides an explanation.

$$D' = \int_m [V' \cdot F'] \, dm, \qquad (5\text{-}12)$$

where D and D' denote the dissipation of zonal and eddy kinetic energy, respectively, $[F]$ and F' are the corresponding frictional forces per unit mass of air, and the formulation of the function F should be consistent with the momentum equations of the dynamical model in use. The formulation is same as that used in Krishnamurti (1979).

4. Scale interactions: Energy exchange from synoptic to MJO time scales

The scale interactions of the synoptic time scales (3–7 days) with the MJO time scale (20–60 days) can be addressed using energy exchanges in the frequency domain following Sheng and Hayashi (1990a,b) or from the formulations of latent heat fluxes from the boundary layer to the cloud layers (Krishnamurti et al. 2003). Both of these studies convey the same essential message on these scale interactions for the maintenance of the MJO

time scale. Here we shall first provide an interpretation of the scale interactions in the frequency domain from the latent heat perspective. Figure 5-8 shows some interesting aspects of the latent heat flux. Triple product nonlinearities exist in the current formulations of the surface similarity theory [i.e. the product of 1) a space–time varying exchange coefficient that is stability dependent, 2) the difference between the surface saturation specific humidity and the 10-m level specific humidity of air, and 3) the wind speed]. Triple product nonlinearity also exists in most formulations of diffusive fluxes of moisture (Manobianco 1988); in the planetary boundary layer [those carry products of the vertical wind shear and bulk Richardson number (stability and the inverse of wind shear)]. The Fourier-transformed moisture conservation equation carries these triple product terms. This expression tells us the gain or loss of moisture for a particular time scale from this triple product term. This formulation is revealing for the explicit scale interactions among different frequencies of the flow regimes. These mathematical details are provided in Krishnamurti et al. (2003). Using daily reanalysis datasets (ERA-40 from the

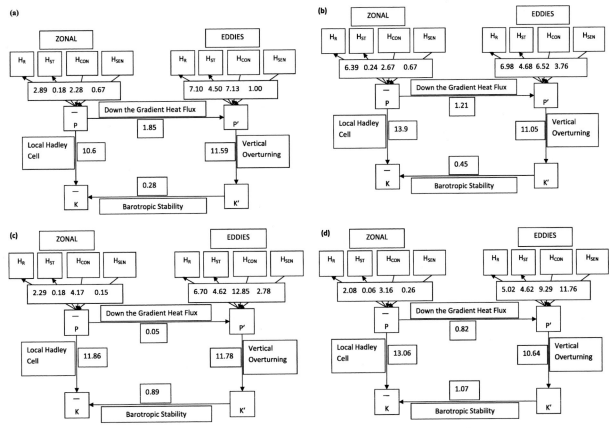

FIG. 5-7. As in Fig. 5-6, but for the Pacific region (0°–20°N, 160°E–160°W).

ECMWF) the interactions among many time scales that interacted with the MJO time scale were computed. Here the results are shown for a period covering the boreal summer season for the year 1996. This was a period with an active MJO especially over the Pacific Ocean and the Asian monsoon belt. The salient interactions (those triads that contribute to 50% or more of the total fluxes) that called for a gain of moisture on the MJO time scale arose from the interactions of the synoptic time scales (pair of frequencies) and the MJO time scale. The trigonometric selection rules for scale interactions on the frequency domain are satisfied by many pairs of synoptic time scales and the MJO time scale; an example of such is the possible interaction of the 5- and 6-day time scale with the 30-day time scale, which satisfies the rule $p = n - m$, where $p = 1/30$, $n = 1/5$, and $m = 1/6$. Of this type, there are many permissible combinations of the synoptic time scales that convey moisture to the MJO time scale. Figures 5-8a and 5-8b respectively, illustrate the total surface latent flux during the period March 1996–February 1997 and the total surface latent heat flux on the time scale of the MJO (20–60 days).

Figure 5-8c shows the contributions to the moisture flux on the MJO time scale as carries out by the salient triads. It is worth noting that a substantial proportion of the total fluxes are contributed by the salient triads. Those are shown in the next figure; the word "salient" here implies those triads that carry at least 50% of the total flux at that location. The table of numbers in Fig. 5-8d lists triplets of numbers; those are prominent triad time scales that contributed the most toward the exchange of latent heat energy from the synoptic time scales to the MJO time scale. Figures 5-8e–h illustrate the corresponding fluxes (similar to Figs. 5-8a–d, respectively) for the fluxes in the planetary boundary layer. These values of fluxes are larger for all the respective illustrations. Here again the important message is that the total moisture fluxes by the salient triads are quite significant when compared to Fig. 5-8e, which carries the total fluxes on the MJO time scale.

Figure 5-9 shows a similar illustration for the tropical western Pacific Ocean that conveys essentially the same inference as above for the Indian Ocean. The important message that is conveyed by these computations is that the vertical flux of moisture on the time scale of the MJO largely comes from selective triads that invoke scale interactions of the MJO time scale with a pair of synoptic-scale disturbances. That feature is seen largely over the

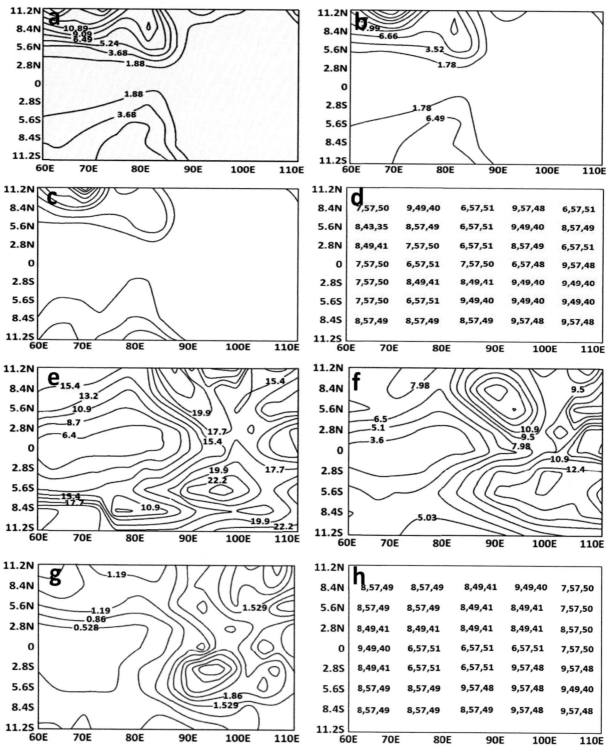

FIG. 5-8. Latent heat fluxes (W m^{-2}) over the Indian Ocean region. (a) Total latent heat fluxes on the time scale of the MJO across the constant flux layer. (b) Total fluxes of latent heat across the constant flux layer on the time scale of the MJO arising from interaction of the MJO with the synoptic time scale of 2 to 7 days. (c) Fluxes of latent heat contributed by salient (strongest contributing) triad interactions in the surface layer. (d) Salient triad interaction frequencies contributing to latent heat fluxes on the time scale of the MJO across the constant-flux layer. (e) Total latent heat fluxes on the time scale of the MJO in the planetary boundary layer (PBL) at 850 hPa. (f) Total latent heat fluxes in the PBL on the time scale of the MJO arising from interaction of the MJO time scale with the synoptic time scale of 2 to 7 days. (g) Latent heat fluxes contributed by the salient triad interactions in the PBL. (h) Salient triad interaction frequencies contributing to latent heat fluxes on the time scale of the MJO in the PBL at 850 hPa [adapted from Krishnamurti et al. (2003)].

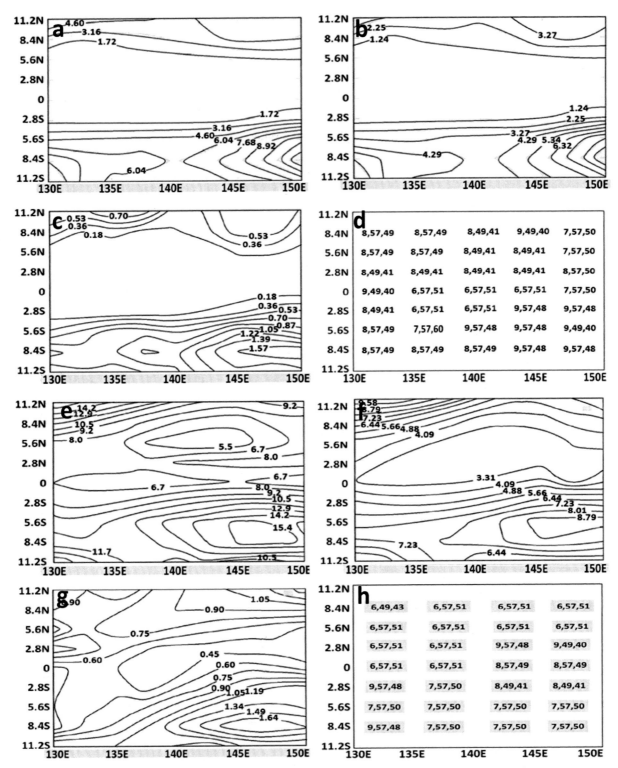

FIG. 5-9. As in Fig. 5-8, but for the west Pacific region [adapted from Krishnamurti et al. (2003)].

tropical Indian Ocean and the tropical western Pacific. We had carried out these same computations over the rest of the tropics and did not see a large contribution to such interactions of the synoptic scales with the MJO time scale.

The findings of Sheng (1986) and Sheng and Hayashi (1990a,b) addressed this problem in the frequency domain. Hayashi (1980) laid the formulation for the energy exchanges in the frequency domain. Basically this

GFDL FGGE

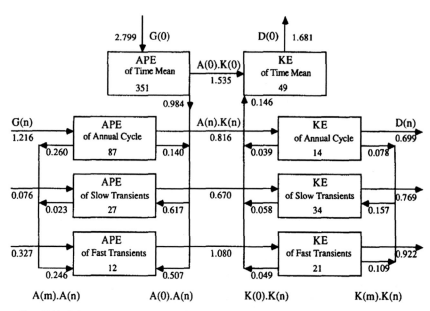

FIG. 5-10. Schematic depiction of the atmospheric energy cycle in the frequency domain estimated from the GFDL data. Numbers in boxes denote the energy storage in $10^4 \, \mathrm{J \, m^{-2}}$; fluxes of energy are in $\mathrm{W \, m^{-2}}$ [adapted from Sheng and Hayashi (1990b)].

approach follows the energetics in the wavenumber domain (Saltzman 1970). Hayashi (1980) decomposed the meteorological time series into its Fourier components in such a way that energy exchanges can be viewed through several spectral windows in the time domain. Sheng and Hayashi (1990a,b) utilized 365 days (the fundamental mode of their frequency domain) of the FGGE year dataset in their study. In the context of the MJO time scale one of their main conclusions was the energy exchanges from fast transient motions (the tropical synoptic scales to the slow transients on time scales of a month). Their study provides an important summary on the scale interactions pertinent to the synoptic scales (identified as fast transients) and the MJO time (identified as slow transients). The fast transients are a major source of energy for the slow transients. That is a kinetic to kinetic energy exchange that largely arises from the triad interactions among synoptic and the MJO time scale; generally, two members of the synoptic time scales interact with a member of the MJO time scale to bring about this exchange following the triad selection rules. Somewhat surprisingly, this study also brings to light the possibility of baroclinic energy exchange (warm air rising and relatively colder air sinking) as a source of energy for the slow transients (the MJO time scale) from the available potential energy of the slow transients. This latter feature was not confirmed in our more recent findings. There is, however, agreement on the findings that those nonlinear

advective processes that lead to the triad energy exchange terms in the energy equation are robust contributors to the maintenance of the MJO time scale.

In this study we shall further illustrate the energy exchanges from the synoptic scales to the MJO time scale following Krishnamurti et al. (2003). This also draws upon triad interactions of the latent heat fluxes of the boundary layer physics.

Sheng and Hayashi (1990b) consider three time scales, the long-term mean, which is the mean of the mean state of the 365 days and is the fundamental mode, and the annual cycle of the fast and the slow transients. Both the available potential energy and the kinetic energy exchanges in the frequency domain were calculated by them (Fig. 5-10). These results confirm the findings of the present study.

We shall not be reiterating the energy exchange equations again here, since they have appeared in Sheng and Hayashi [1990a,b; see Eqs. (9)–(12)].

The year 1996 was a robust MJO wave activity (Fig. 5-11). The eastward-moving divergent wave is best seen from the velocity potential datasets at the 200-hPa level (Krishnamurti 1971); the velocity potential on the time scale of the MJO (20–60 days) is shown here. Figure 5-11 presents alternating passages of positive and negative velocity potential anomalies shown by solid and dashed lines, respectively, on this time scale, which traverse from west to east. Superimposed on these velocity

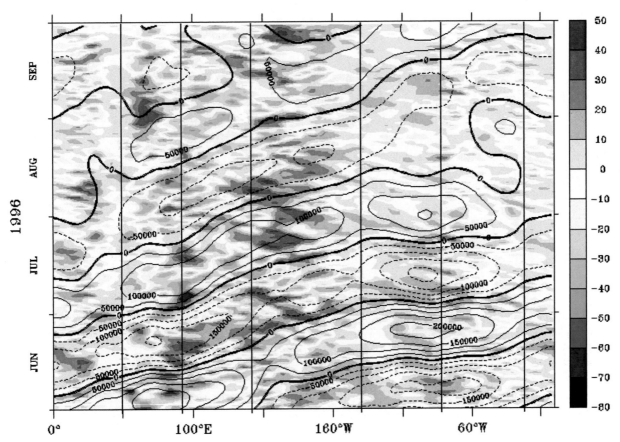

FIG. 5-11. Variability of filtered velocity potential (contours; m^2 s^{-1}) and unfiltered OLR anomaly (color shaded) for summer season of year 1996.

potential isopleths are the total outgoing longwave radiation (OLR) fields. These are all averages for the equatorial latitude belt 10°S to 10°N. This illustration does not clearly show a global relationship between the MJO and OLR (a proxy for cloud cover). For this reason, the relationship of the divergent motions and OLR was next examined over several longitudinal sectors. In Fig. 5-12 we show time series (along the abscissa) of the velocity potential and OLR over the following longitudinal belts: Africa, the Indian Ocean, western Pacific, central and eastern Pacific, and equatorial Atlantic. The correlation among these two pairs of curves is inserted. This suggests a poor relationship between OLR and the MJO waves over Africa, central and eastern Pacific, and the tropical Atlantic, implying that the relationship is stronger over the western Pacific and Indian Ocean sectors. Our study suggests that latter two regions are the most important for the scale interactions among the synoptic-scale motions (with organized convection driving the synoptic scale) and the MJO time scale via the dynamical triads. This display suggests that the MJO is not directly connected to the tropical clouds at all longitudes.

5. Conclusions

The atmospheric Kelvin wave is clearly a starting explanation for a slow eastward-moving wave with its largest amplitude residing in the equatorial latitudes. The earlier renditions of this wave come from the classical linearized shallow water equations studies that were illustrated by Matsuno (1966) and many others. Lau and Peng (1987) showed that this equatorial Kelvin wave in the absence of organized convection moves too fast at almost twice the eastward speed of the MJO wave. Using a multilevel atmospheric AGCM they showed a slowing down of the Kelvin wave by invoking wave CISK for the cumulus parameterization following Lindzen's (1974) original framework of wave CISK. That formulation essentially assumes that rising motions (on top of the boundary layer) exist along an entire active half of an MJO wave. This is equivalent to stating that MJO is quite like a hurricane and is driven by deep convection directly along its entire scale. The MJO being a planetary wave, this called for an organization of convection following the MJO along the entire active half of the MJO wave around Earth.

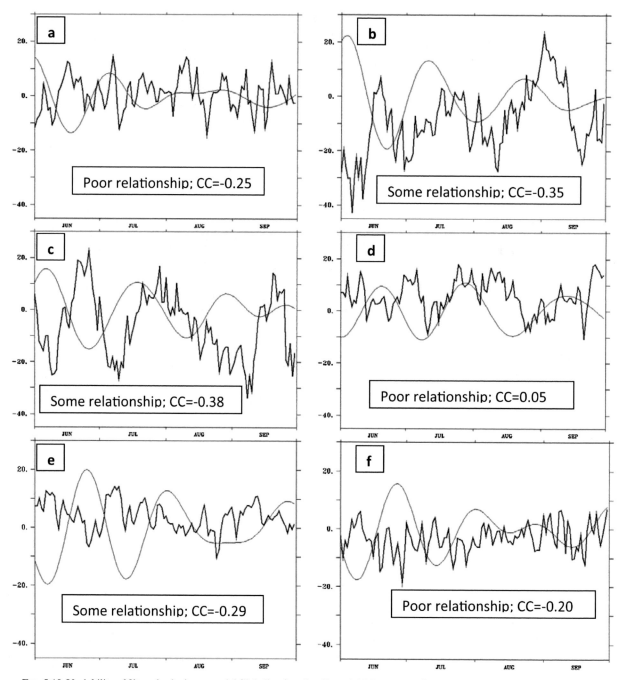

FIG. 5-12. Variability of filtered velocity potential (light lines) and unfiltered OLR anomaly (dark lines) for summer season of year 1996: (a) Africa region (10°S–10°N, 0°–50°E), (b) Indian region (10°S–10°N, 50°–90°E), (c) western Pacific region (10°S–10°N, 90°–140°E), (d) central Pacific region (10°S–10°N, 140°E–220°W), (e) eastern Pacific region (10°S–10°N, 220°–280°W), and (f) Atlantic region (10°S–10°N, 280°–340°W).

The wording "organization of convection and of dynamics" is somewhat synonymous with scale interactions that invoke quadratic and triple product nonlinearities. The trigonometric selection rules for quadratic and triple product nonlinearities in the energy equation are closely tied to the issue of organization of convection and of

dynamics respectively. If deep convection is organized around a disturbance, then the generation of eddy available potential energy, the conversion of eddy available potential energy into eddy kinetic energy, and the maintenance of kinetic energy of the disturbance all get organized around the same frequency of that disturbance. If the frequencies of that disturbance are denoted by n_i, then the selection rule for quadratic nonlinearity simply states that each member of n_i can only interact with the component frequency of n_i, all other interactions being identically equal to zero. We have addressed the energy exchanges among the synoptic scales and the MJO time scale using results from two different studies. One of these follows the approach of Hayashi (1980), which focuses on energy exchange in the frequency domain, and the other is on the latent heat fluxes of the planetary boundary layer physics (Krishnamurti et al. 2003). Both approaches invoke triple product nonlinearities and call for exchanges of energy from the synoptic scales to the MJO time scales toward the maintenance of the latter. Both of these approaches carry triads where two time scales of the synoptic scales interact with a time scale of the MJO time scale and provide energy to the latter. These transfers are facilitated by the triple product selection rules and hence we can label these as organization of dynamics.

In this paper we have shown, from formal computations of Lorenz box energetics, that during periods of presence of an active MJO, robust tropical synoptic disturbances of the central and western Pacific and the Indian monsoon environment carry organized convection that generate eddy available potential energy and a transformation of eddy available potential energy to the eddy kinetic energy on the scales of such disturbances. The maintenance of the eddy kinetic energy of such disturbances is largely attributed to convective processes for the generation of the eddy available potential energy and baroclinic processes for the transformation of the eddy available to the eddy kinetic energy.

A major scientific issue that has come up in the design of field experiments such as the DYNAMO is whether clouds organize on the space time scale of the MJO and directly drive the MJO in the same manner as organized convection drives synoptic tropical disturbances. Organized convection on the MJO time scale is seldom seen over the tropical eastern Pacific and the tropical Atlantic. The MJO wave spatially covers the global tropics as a planetary wave (Madden and Julian 1971.). Furthermore, taking one complete cycle of an MJO (20–60 days), even over the western tropical Pacific and the Asian monsoon belt we seldom see convection organized on the scale of the MJO for an entire cycle. However, whenever active synoptic scales are present we do often see well-organized convection on the scale of such disturbances. Our contention is that convection is only organized on the scale of synoptic (and mesoscale) disturbances and not on the space–time scales of the MJO. The mode of communication of the synoptic scales with the MJO is via nonlinear dynamics. This also brings up the issue of "Who modulates whom?". Nearly all the literature alludes to the modulation of tropical weather during the passage of an MJO wave. That has been verified over the tropical western Pacific and the Asian monsoon belt including northern Australia during the boreal winter season. Careful examination of weather maps over these regions always shows that tropical weather as a part of a tropical mesoscale or a synoptic-scale system. The MJO wave is very weakly divergent and carries typical magnitudes of convergence and divergence on the order of 10^{-7} to $10^{-6}\,\mathrm{s}^{-1}$. That appears to be necessary to initiate weak moistening and the start of convection; however, it is the simultaneous passage of a mesoscale or a synoptic-scale disturbance, in that region, that exploits this weak modulation of weather from the passage of the MJO wave. It is the organized convection in a mesoscale and synoptic-scale disturbance that contributes to the maintenance of the mesoscale and synoptic scale. The MJO is maintained by the synoptic scales via nonlinear energy transfers.

APPENDIX

Definition of Variables

Table 5-A1 lists the variables with their definitions.

TABLE 5-A1. Symbols and their meaning.

Symbol	Meaning of symbol
x, y, p, t	Independent variables
u, v	Horizontal velocity component
ω	Vertical velocity (dp/dt)
g	Acceleration due to gravity
P	Precipitation per unit mass of the air
R	Gas constant for air
T	Temperature of air
Θ	Potential temperature of air
c_p	Specific heat at constant pressure
$[\cdot]$	Zonal average of a variable
$-$	Areal average of a variable
γ	g/c_p
$\bar{\gamma}$	$\dfrac{Pg}{RT}\dfrac{\partial T}{\partial p}$
$''$	Departure of a variable from the meridional average
$'$	Departure of a variable from the zonal average
$\langle A \cdot B \rangle$	Energy exchange from A to B

REFERENCES

Biello, J. A., and A. J. Majda, 2005: A new multiscale model for the Madden–Julian oscillation. *J. Atmos. Sci.*, **62**, 1694–1721, doi:10.1175/JAS3455.1.

Hannah, W. M., and E. D. Maloney, 2011: The role of moisture–convection feedbacks in simulating the Madden–Julian oscillation. *J. Climate*, **24**, 2754–2770, doi:10.1175/2011JCLI3803.1.

Hayashi, Y., 1980: Estimation of nonlinear energy transfer spectra by the cross-spectral method. *J. Atmos. Sci.*, **37**, 299–307, doi:10.1175/1520-0469(1980)037<0299:EONETS>2.0.CO;2.

Kikuchi, K., and B. Wang, 2010: Spatiotemporal wavelet transform and the multiscale behavior of the Madden–Julian oscillation. *J. Climate*, **23**, 3814–3834, doi:10.1175/2010JCLI2693.1.

Krishnamurti, T. N., 1971: Tropical east–west circulations during the northern summer. *J. Atmos. Sci.*, **28**, 1342–1347, doi:10.1175/1520-0469(1971)028<1342:TEWCDT>2.0.CO;2.

——, 1979: Large-scale features of the tropical atmosphere. *Meteorology over the Tropical Oceans*, D. B. Shaw, Ed., Royal Meteorological Society, 30–56.

——, and S. Gadgil, 1985: On the structure of the 30 to 50 day mode over the globe during FGGE. *Tellus*, **37A**, 336–360, doi:10.1111/j.1600-0870.1985.tb00432.x.

——, D. R. Chakraborty, N. Cubukcu, L. Stefanova, and T. S. V. Vijaya Kumar, 2003: A mechanism of the Madden–Julian oscillation based on interactions in the frequency domain. *Quart. J. Roy. Meteor. Soc.*, **129**, 2559–2590, doi:10.1256/qj.02.151.

Lau, K.-M., and L. Peng, 1987: Origin of the low-frequency (intraseasonal) oscillation in the tropical atmosphere. Part I: Basic theory. *J. Atmos. Sci.*, **44**, 950–972, doi:10.1175/1520-0469(1987)044<0950:OOLFOI>2.0.CO;2.

Lin, J., B. Mapes, M. Zhang, and M. Newman, 2004: Stratiform precipitation, vertical heating profiles, and the Madden–Julian oscillation. *J. Atmos. Sci.*, **61**, 296–309, doi:10.1175/1520-0469(2004)061<0296:SPVHPA>2.0.CO;2.

Lindzen, R. S., 1974: Wave-CISK in the tropics. *J. Atmos. Sci.*, **31**, 156–179, doi:10.1175/1520-0469(1974)031<0156:WCITT>2.0.CO;2.

Liu, F., and B. Wang, 2012a: A model for the interaction between the 2-day waves and moist Kelvin waves. *J. Atmos. Sci.*, **69**, 611–625, doi:10.1175/JAS-D-11-0116.1.

——, and ——, 2012b: A frictional skeleton model for the Madden–Julian oscillation. *J. Atmos. Sci.*, **69**, 2749–2758, doi:10.1175/JAS-D-12-020.1.

——, and ——, 2013: Impacts of upscale heat and momentum transfer by moist Kelvin waves on the Madden–Julian oscillation: A theoretical model study. *Climate Dyn.*, **40**, 213–224, doi:10.1007/s00382-011-1281-0.

Liu, P., and Coauthors, 2009: An MJO simulated by the NICAM at 14- and 7-km resolutions. *Mon. Wea. Rev.*, **137**, 3254–3268, doi:10.1175/2009MWR2965.1.

Lorenz, E. N., 1967: The nature and theory of the general circulation of the atmosphere. Publ. 218-TP-115, WMO, 161 pp.

Madden, R. A., and P. R. Julian, 1971: Detection of a 40–50 day oscillation in the zonal wind in the tropical Pacific. *J. Atmos. Sci.*, **28**, 702–708, doi:10.1175/1520-0469(1971)028<0702:DOADOI>2.0.CO;2.

Majda, A. J., and J. A. Biello, 2004: A multiscale model for the intraseasonal oscillation. *Proc. Natl. Acad. Sci. USA*, **101**, 4736–4741, doi:10.1073/pnas.0401034101.

——, and S. N. Stechmann, 2009: The skeleton of tropical intraseasonal oscillations. *Proc. Natl. Acad. Sci. USA*, **106**, 8417–8422, doi:10.1073/pnas.0903367106.

Manobianco, J., 1988: On the observational and numerical aspects of explosive East Coast cyclogenesis. Ph.D. thesis, Department of Meteorology, Florida State University, 361 pp.

Matsuno, T., 1966: Quasigeostrophic motions in the equatorial area. *J. Meteor. Soc. Japan*, **44**, 25–43.

Moorthi, S., and A. Arakawa, 1985: Baroclinic instability with cumulus heating. *J. Atmos. Sci.*, **42**, 2007–2031, doi:10.1175/1520-0469(1985)042<2007:BIWCH>2.0.CO;2.

Nakazawa, T., 1988: Tropical super clusters within intraseasonal variations over the western Pacific. *J. Meteor. Soc. Japan*, **66**, 823–839.

Ray, P., C. Zhang, M. W. Moncrieff, J. Dudhia, J. M. Caron, L. R. Leung, and C. Bruyere, 2010: The role of the mean state on the initiation of the Madden–Julian oscillation in a nested regional climate model. *Climate Dyn.*, **36**, 161–184, doi:10.1007/s00382-010-0859-2.

Saltzman, B., 1970: Large-scale atmospheric energetic in the wave number domain. *Rev. Geophys.*, **8**, 289–302, doi:10.1029/RG008i002p00289.

Sheng, J., 1986: On the energetic of low frequency motions. Ph.D. dissertation, Florida State University, 171 pp.

——, and Y. Hayashi, 1990a: Observed and simulated energy cycles in the frequency domain. *J. Atmos. Sci.*, **47**, 1243–1254, doi:10.1175/1520-0469(1990)047<1243:OASECI>2.0.CO;2.

——, and ——, 1990b: Estimation of atmospheric energetics in the frequency domain during the FGGE year. *J. Atmos. Sci.*, **47**, 1255–1268, doi:10.1175/1520-0469(1990)047<1255:EOAEIT>2.0.CO;2.

Waliser, D. E., and Coauthors, 2003: AGCM simulations of intraseasonal variability associated with the Asian summer monsoon. *Climate Dyn.*, **21**, 423–446, doi:10.1007/s00382-003-0337-1.

——, and Coauthors, 2009: MJO simulation diagnostics. *J. Climate*, **22**, 3006–3030, doi:10.1175/2008JCLI2731.1.

Wang, B., and F. Liu, 2011: A model for scale interaction in the Madden–Julian oscillation. *J. Atmos. Sci.*, **68**, 2524–2536, doi:10.1175/2011JAS3660.1.

——, F. Huang, Z. Wu, J. Yang, X. Fu, and K. Kikuchi, 2009: Multiscale climate variability of the South China Sea monsoon: A review. *Dyn. Atmos. Oceans*, **47**, 15–37, doi:10.1016/j.dynatmoce.2008.09.004.

Yu, W., W. Han, E. D. Maloney, D. Gochis, and S.-P. Xie, 2011: Observations of eastward propagation of atmospheric intraseasonal oscillations from the Pacific to the Atlantic. *J. Geophys. Res.*, **116**, D02101, doi:10.1029/2010JD014336.

Yuan, J., and R. A. Houze Jr., 2013: Deep convective systems observed by A-Train in the tropical Indo-Pacific region affected by the MJO. *J. Atmos Sci.*, **70**, 465–486, doi:10.1175/JAS-D-12-057.1.

Zhang, C., and Coauthors, 2010: MJO signals in latent heating: Results from TRMM retrievals. *J. Atmos. Sci.*, **67**, 3488–3508, doi:10.1175/2010JAS3398.1.

Chapter 6

Monsoon Convection in the Maritime Continent: Interaction of Large-Scale Motion and Complex Terrain

CHIH-PEI CHANG

Department of Atmospheric Sciences, National Taiwan University, Taipei, Taiwan

MONG-MING LU

Central Weather Bureau, Taipei, Taiwan

HOCK LIM

Department of Physics, National University of Singapore, Singapore

ABSTRACT

The Asian monsoon is a planetary-scale circulation system powered by the release of latent heat, but important features of deep convection and rainfall distribution cannot be adequately represented by the large-scale patterns. This is mainly due to the strong influences of terrain that are important across a wide range of horizontal scales, especially over the Maritime Continent where the complex terrain has a dominant effect on the behavior of convective rainfall during the boreal winter monsoon. This chapter is a review and summary of published results on the effects on monsoon convection due to interactions between the Maritime Continent terrain and large-scale transient systems.

The Maritime Continent topographic features strongly affect both the demarcation of the boreal summer and winter monsoon regimes and the asymmetric seasonal marches during the transition seasons. In the western part of the region, the complex interactions that lead to variability in deep convection are primarily controlled by the cold surges and the synoptic-scale Borneo vortex. The Madden–Julian oscillation (MJO) reduces the frequency of weaker surges through an interference with their structure. It also influences convection, particularly on the diurnal cycle and when synoptic activities are weak. When both surges and the Borneo vortex are present, interactions between these circulations with the terrain can cause the strongest convection, which has included Typhoon Vamei (2001), which is the only observed tropical cyclone that developed within 1.5° of the equator.

The cold surges are driven by midlatitude pressure rises associated with the movement of the Siberian high. Rapid strengthening of surge northeasterly winds can be explained as the tropical response via a geostrophic adjustment process to the pressure forcing in the form of an equatorial Rossby wave group. Dispersion of meridional modes leads to a northeast–southwest orientation that allows the surge to stream downstream through the similarly oriented South China Sea. This evolution leads to a cross-equatorial return flow and a cyclonic circulation at the equator, and thus a mechanism for equatorial cyclogenesis. Although the narrow width of the southern South China Sea facilitates strengthening of the cold surge, it also severely restricts the likelihood of cyclone development so that Vamei remains to be the only typhoon observed in the equatorial South China Sea.

Climate variations from El Niño–Southern Oscillation to climate change may impact the interactions between the large-scale motion and Maritime Continent terrain because they lead to changes in the mean flow. The thermodynamic effects on the interaction between MJO and the monsoon surges and Borneo vortex over the complex terrain also need to be addressed. These and other questions such as any possible changes in the likelihood of equatorial tropical cyclogenesis as a result of climate change are all important areas for future research.

Corresponding author address: Dr. Chih-Pei Chang, Department of Meteorology, Naval Postgraduate School, 589 Dyer Road, Room 254, Monterey, CA 93943.
E-mail: cpchang@nps.edu

DOI: 10.1175/AMSMONOGRAPHS-D-15-0011.1

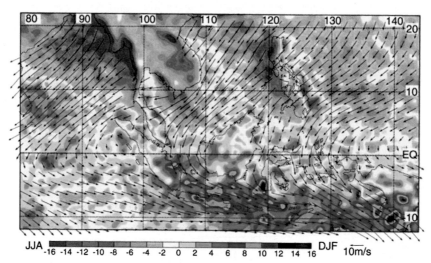

FIG. 6-1. Differences of TRMM PR rainfall and QuikSCAT winds between boreal winter and boreal summer (DJF − JJA). Warm colors are the boreal summer monsoon regime and cool colors are the boreal winter monsoon regime (from Chang et al. 2005b).

1. Monsoon rainfall pattern and complex terrain

Availability of high-resolution, satellite-derived rainfall data from the Tropical Rainfall Measuring Mission (TRMM) since November 1997 has led to many studies of the interactions of the large-scale tropical convection in the monsoon regions and the terrain. Even mesoscale terrain features have a strong influence on the large-scale distribution of monsoon rainfall (Chang et al. 2004a,b, 2005b). The traditional use of area-averaged rainfall as monsoon indexes [e.g., the all-India monsoon rainfall index (Sontakke et al. 1993) that has been used for the Indian summer monsoon since the mid-nineteenth century] was based on the perspective of a large-scale pattern of the monsoon convection and rainfall that are comparable to the scale of the monsoon circulation. However, Chang et al. (2004a, 2005b) showed that such a large-scale perspective distorts the reality of deep convection and rainfall distributions because the latter are often strongly affected by the detailed terrain features throughout the monsoon region.

Among the monsoon systems around the globe, the Asian winter monsoon has the largest meridional domain that extends from the equatorial Maritime Continent and surrounding oceans to Northern Hemisphere midlatitude Siberia (Ramage 1968; Chang et al. 2006). In the midlatitude regions, the monsoon is characterized by strong baroclinicity dominated by the cold-core Siberian high, which is the strongest surface high pressure system in the world (Ding 1994). In the equatorial region, the deep convection around the Maritime Continent is the most vigorous and extensive large-scale convection system, and its effects reach far beyond the Asian

monsoon region and can impact weather in North America (Yanai and Tomita 1998; Yang et al. 2002; Chan and Li 2004) and Europe (Neale and Slingo 2003). Strong interaction between the midlatitude and tropical components through cold surges and convection feedback to the East Asian jet stream (e.g., Chang and Lau 1982; Lau and Chang 1987; Neale and Slingo 2003) reinforces each circulation, which makes the Asian winter monsoon one of the most energetic planetary-scale circulation systems in Earth's atmosphere (Chang et al. 2006). These interactions also play key roles in the interhemispheric exchanges carried out by the Asian–Australian monsoon system (e.g., Carrera and Gyakum 2003, 2007; Guan et al. 2010).

In the Maritime Continent region, significant geographical variations also exist in the seasonal march throughout the year, which has been recognized since Braak (1921, 1929; see Ramage 1971). A main reason for these variations is the complex terrain due to islands of different sizes interspersed within the surrounding seas. The complex terrain contributes to significant local and mesoscale circulations that interacts with the large-scale monsoon circulation and the diurnal land–sea and mountain breeze circulations, particularly during El Niño events when large-scale wind conditions are weaker in this region (Moron et al. 2010; Qian et al. 2010, 2013; Robertson et al. 2011).

The importance of smaller-scale terrain influences on seasonal-scale convection can be seen in Fig. 6-1, which shows the partition of the summer and winter monsoon regimes in Southeast Asia based on the TRMM Precipitation Radar (PR) data (Simpson et al. 1996) and the QuikSCAT scatterometer winds that describe winds at

10 m above the sea surface (Liu 2002). The boreal summer and winter monsoon regimes are defined by the difference in TRMM PR rainfall and QuikSCAT winds between December–February (DJF) and June–August (JJA). Rainfall difference in warm colors means the boreal summer rainfall exceeds the boreal winter monsoon; thus it denotes the summer monsoon regime. Rainfall difference in cool colors denotes the winter monsoon regime. The general pattern reflects the mean seasonal differences that the boreal summer rainfall regime dominates north of the equator, and the boreal winter rainfall regime dominates south of the equator.

While the two regimes commingle near the equator, there is a conspicuous pattern of asymmetry in the degree of intrusion from one regime to the other regime. The boreal winter regime extends far northward into the boreal summer regime, whereas the southward extension of the boreal summer regime south of the equator is much more limited. Two effects contribute to this asymmetry: The mesoscale terrain features and the equatorward surface winds, both of which are more dominant north of the equator. Over Southeast Asia, the intrusions of the winter regime beyond 5°N occur in the following areas: east of the Philippines, northeast and northwest of Borneo in the South China Sea, east of Vietnam, the eastern coast of the Malay Peninsula, and north of Sumatra. In most of these areas the high boreal winter rainfall is due to the onshore northeasterly winter monsoon winds from the northwest Pacific and the South China Sea. These northeasterly monsoon winds are stronger than the prevailing southeasterly winds in the Southern Hemisphere because of the intense baroclinicity over the cold Asian continent during boreal winter that is significantly stronger than the baroclinicity in the Southern Hemisphere during boreal summer. There are also very few coastal areas between 5° and 10°S that face the prevailing seasonal wind, as is the case in the northern tropics.

The northeast monsoon wind is parallel to the coastline over northwest Borneo so there is little onshore wind, but the intrusion of the boreal winter regime here is also the result of the terrain effect. Over this region a quasi-stationary synoptic-scale cyclonic circulation that is associated with the heating of the island can develop in any season of the year. During boreal winter, the low-level basic-state background vorticity is also cyclonic because of both the mean northeasterly wind maximum over the South China Sea and the equatorial westerlies associated with the East Asian winter monsoon. Perturbations in this basic state can also amplify into synoptic-scale cyclonic circulations. These disturbances are often found southeast of the primary region of cold surge northeasterly winds. As a combination of these

conditions, the low-level cyclonic circulation is particularly active during winter and is a prominent feature of the boreal winter climatology (Johnson and Houze 1987, see their Fig. 10.2). Although the circulation may not be completely closed on the east side over the island, it is often referred to as the Borneo vortex and is often associated with deep convection and intense latent heat release (Cheang 1977; Lau and Chang 1987; Johnson and Houze 1987; Chang et al. 2003; Trilaksono et al. 2012). Because of the increased frequency of the Borneo vortex, the deep convection and heavy rainfall frequently extend offshore several hundred kilometers into the South China Sea during boreal winter, but much less so during boreal summer.

The effect of wind–terrain interaction for the summer monsoon regime is also clearly demonstrated in Fig. 6-1, in which the reversal of the plotted wind vectors indicates the southwest monsoon wind directions north of the equator. Here the strongest monsoonal rainfall occurs along the windward side of mountains in northwestern Philippines, northern and central Vietnam, the southwest coast of Cambodia, and the west coast of Myanmar northeast of the Bay of Bengal. It has been widely accepted that monsoon depressions in the Bay of Bengal are a major contributor to the summer monsoon rainfall in India (e.g., Krishnamurthy and Ajayamohan 2010). However, Fig. 6-1 shows that the heaviest summer monsoonal rainfall in the eastern part of the Indian monsoon region is not along the main path of the Bay of Bengal monsoon depressions that traverses the bay from southeast to northwest, but is concentrated on the northeastern part of the bay away from the Indian subcontinent. This is where the southwest monsoon winds interact with the coastal mountains near the northeastern shoreline of the bay. Thus, the wind–terrain interaction also dominates the distribution of the summer monsoon rainfall and has a stronger effect in producing the monsoon rainfall than the monsoon depressions.

2. Asymmetric seasonal transitions and terrain effects

The boreal spring [March–May (MAM)] and fall monsoon regimes [September–November (SON)] may also be delineated from the TRMM Precipitation Radar and QuikSCAT scatterometer wind data. Because these are during transitional seasons, the analysis needs be done differently from the method used to partition boreal summer and winter monsoon regimes. The spring and fall regimes are defined by Chang et al. (2005b) as the areas that have higher mean rainfall than all the other seasons. Thus, the graphic representation of these two regimes is not derived from a simple seasonal

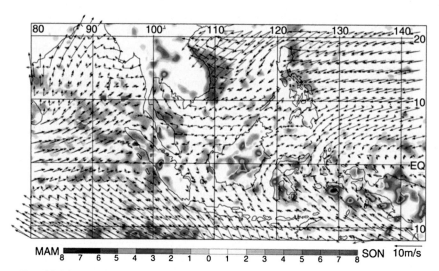

FIG. 6-2. Monsoon regimes during transition seasons deduced from TRMM PR rainfall. A grid point is identified if the rainfall during one of the two transition seasons is the maximum in the annual cycle, and the value plotted is the difference between this transition-season rainfall and the boreal winter and or boreal summer, whichever is highest. Warm colors are the boreal fall monsoon regime and cool colors are the boreal spring monsoon regime. The difference of QuikSCAT winds between the two transition seasons (SON − MAM) is plotted for the entire domain (from Chang et al. 2005b).

difference as shown in Fig. 6-1, but rather it is derived with two steps. In the first step, each grid point where the MAM rainfall (cool color) is the maximum among the four seasons is identified, and the difference between the MAM rainfall and the higher of the summer or winter rainfall is computed. The same procedure is followed at points where the SON rainfall (warm color) is the maximum of all four seasons. The resultant values are then plotted together in Fig. 6-2. The total QuikSCAT winds for both seasons are also plotted, with the MAM wind in black and the SON wind in red.

In Fig. 6-2 the SON monsoon rainfall dominates areas north of the equator and the western side of the domain. The MAM monsoon rainfall dominates areas south of the equator and the eastern side of the domain. Within the equatorial belt of 10°S–10°N, the two regimes are roughly divided by Borneo. The largest signals of the MAM monsoon rainfall are along windward coastlines in eastern Philippines, eastern Vietnam, and the northeastern Malay Peninsula, clearly indicative of the importance of local wind–terrain interaction. This is in contrast to the Southern Hemisphere MAM regime, where comparable significant monsoon rainfall does not exist. The contrast of the two transition regimes is consistent with the asymmetric seasonal march that has been noticed by many researchers. During boreal fall, the maximum convection moves southeastward from South Asia in a track that roughly follows the Southeast Asia land bridge, to reach southern Indonesia and

northern and eastern Australia during boreal winter. During boreal spring, the maximum rainfall remains mostly south of the equator without a northwestward progression that would retrace the path of the boreal fall progression (e.g., Lau and Chan 1983; Meehl 1987; Yasunari 1991; Matsumoto 1992; Matsumoto and Murakami 2002; LinHo and Wang 2002; Hung and Yanai 2004; Hung et al. 2004).

Chang et al. (2005b) and Wang and Chang (2008) hypothesized that at least a part of this spring–fall asymmetry is due to the sea level pressure (SLP) differences between land and ocean, which are driven by the different thermal memories of the ocean and atmosphere–ocean interactions. They computed the difference between boreal spring and fall SLP and showed that in the Northern Hemisphere the land SLP is higher in boreal fall and the ocean SLP is higher in boreal spring. The reverse is true in the Southern Hemisphere. This pattern is seen throughout most of the global domain other than areas of the midlatitude storm tracks. The largest SLP difference occurs over the Asian continent, leading to the stronger SON northeasterly winds in the northern South China Sea and northwestern Pacific (area A in Fig. 6-3), which causes the deep convection east of Vietnam and the Philippines in boreal fall (Fig. 6-2). The convection in these areas is much stronger than in both boreal summer and winter, even though during boreal winter the northeasterly winds are stronger. This is because the colder and drier air and the

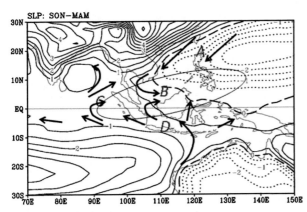

FIG. 6-3. Differences of SLP (hPa) between boreal fall and boreal spring (SON − MAM). Negative isobars are dotted and the zero line is dashed. Schematics of sea level wind differences based on the differences in the SLP pattern are indicated. The elliptic-shaped area indicates preferred belt of convergence in fall and divergence in spring. See text for details (from Chang et al. 2005b).

colder sea surface temperature (SST) make deep convection less likely to develop north of 10°N, so that strong winter cold surges actually produce drying conditions in northern and middle South China Sea (Chang et al. 2005a). In the southern South China Sea, the SLP gradient favors cyclonic flow and therefore deep convection in SON (area B in Fig. 6-3). The convection is further enhanced by the interaction of the wind with terrain on the east coast of Sumatra and west coast of Borneo.

The enhanced boreal fall convection in the eastern equatorial Indian Ocean may also be explained, at least partly, by the spring–fall SLP difference. The Bay of Bengal has higher SLP in boreal fall than in boreal spring, which suggests that in addition to the different thermal memory effect, atmosphere–ocean interaction is involved to warm the SST faster during boreal spring. One possibility is that in early spring the initially cool SST and more anticyclonic flow with weak winds cause less evaporation and more solar heating of the sea surface and downwelling in the upper ocean, so the spring SST becomes higher and the SLP becomes lower than in fall. But the land–sea redistribution of mass still contributes to lower SLP in the Bay of Bengal during boreal fall when compared to surrounding areas. The resulting difference in pressure gradient during boreal fall gives rise to cyclonic flow in the Bay of Bengal and favors increased cross-equatorial flow from the southern Indian Ocean.

The convection in and around the middle and southern South China Sea in boreal fall helps to induce southwesterly winds west of Sumatra (area C in Fig. 6-3). These southwesterly winds are enhanced by the tendency of cross-equatorial flow and the cyclonic flow in

the Bay of Bengal. Other atmospheric and oceanic factors, such as the east–west pressure gradient across the equatorial Indian Ocean, may also contribute to the development of equatorial westerly winds. These winds have two effects, both of which lead to more convection. The first is the onshore flow that causes convergence along the western coasts of northern Sumatra and the Malay Peninsula. The second is the beta effect that produces convergence in the equatorial westerlies. The increased convection may further enhance the westerlies making a positive feedback possible.

South of the equator, the SLP difference between Australia and the south Indian Ocean favors counterclockwise flow toward the equator (area D in Fig. 6-3) in boreal fall. This also enhances the clockwise cross-equatorial flow that turns westerly north of the equator, which increases the wind–terrain interactions on the west coast of land areas, such as western Borneo (area B).

The combined effect of the condition favoring southerly winds from south of the equator and northeasterly winds in the northern South China Sea and the northwestern Pacific (area A) gives rise to a broadscale belt of tendency for convergence between the equator and 20°N during boreal fall (marked by the elliptic-shaped area in Fig. 6-3). This tendency of low-level convergence favors the development of deep convection. During boreal spring, the tendency of the wind directions changes sign, and this belt becomes an area with a tendency of low-level divergence, so convection tends to be suppressed. These effects are the results of the global-scale mass redistribution between the land and ocean regions that is driven by their different thermal memories. Because of the orientation of the Asian and Australian landmasses, this redistribution facilitates the southeastward march of maximum convection from the Asian summer monsoon to the Asian winter (Australian summer) monsoon, but it deters the reverse march in boreal spring. Thus, there is an inherent tendency for the onset of the Asian summer monsoon in spring to be more abrupt and difficult to predict than the more gradual development of the Asian winter monsoon.

3. Synoptic and intraseasonal disturbances during boreal winter

During the East Asian winter monsoon, the midlatitude circulation exerts direct impacts on tropical weather through periodical and rapid strengthening of the low-level northeasterlies into the South China Sea and Maritime Continent regions following the southeastward movement of the surface high pressure (e.g.,

Ramage 1971; Chang et al. 1979, 1983; Chu and Park 1984; Lau and Chang 1987; Wu and Chan 1995; Zhang et al. 1997). The movement of the midlatitude circulation involves basically the advection of relative and planetary vorticity, but the nearly spontaneous freshening of the low-level northeasterly winds in the tropics and their large cross-isobar angles indicate a rapid progression of events on a gravity wave time scale, considerably shorter than that of the advective scale inherent in Rossby wave dynamics (Chang et al. 1983).

These processes are often referred to as "cold surges" (Ramage 1971; Compo et al. 1999; Garreaud 2001), although the rapid drop of surface temperature associated with these events rapidly diminishes over the warm waters of the South China Sea. The term "pressure surges" has also been used to identify the cold air outbreaks. This term describes well the outbreak of cold air eastward near the southern China coast (Chan and Li 2004), but it does not adequately describe the southward surges since south of about 10°N the surges are mainly manifest by the freshening of northeasterly winds rather than pressure rises. The outbreaks of cold air are almost always associated with movement of the Siberian high, which is a strong and semistationary cold-core high pressure during the boreal winter with a maximum central sea level pressure exceeding any other pressure system in Earth's atmosphere (Ding 1994). There is no clear evidence of a relationship between the Siberian high's interannual variation and circulation indexes such as El Niño–Southern Oscillation (ENSO) and North Atlantic Oscillation (NAO)/Artic Oscillation (AO), but its decadal-scale variations may be related to NAO/AO (Gong et al. 2001; Wu and Wang 2002; Gong and Ho 2004; Chang et al. 2006, 2011). Its intraseasonal variation is often forced by midlatitude wave activity and can be influenced by blocking activities both to the west in the Atlantic and Ural areas and to the east in the Pacific (Takaya and Nakamura 2005a,b; Chang and Lu 2012). It can also move away from its normal position toward the western Pacific once to several times a month with durations of a few days to one week or more (e.g., Gong and Ho 2004; Lu and Chang 2009; Park et al. 2011).

As the Siberian high moves south and eastward, the cold surge winds spread equatorward around the eastern edge of low-level anticyclones and can cover a broad longitudinal span of the entire tropical western North Pacific. However, the strongest cold surges are concentrated in the South China Sea where they can reach and cross the equator. This is because the orientation of the regional topography (Fig. 6-4) acts to restrict the flow such that the low-level northeasterly flow is channeled toward the equator. Although cold surge winds are typically dry, they are moistened by the overwater

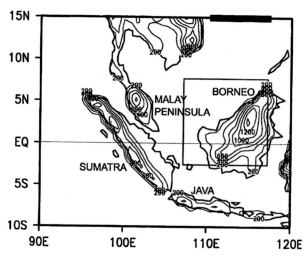

FIG. 6-4. Domain of the study and smoothed topography (m). The rectangular box is the area within which the Borneo vortex is identified. The black horizontal bar is the area for the surge index. See text for details (from Chang et al. 2005a).

trajectory (Johnson and Houze 1987; Takahashi 2012) and have been associated with enhanced upper-tropospheric outflow over the Maritime Continent, which is related to an enhanced east Asian local Hadley cell that may strengthen the East Asian jet and lead to further interactions with the midlatitude systems (Chang and Lau 1982; Lau and Chang 1987). Furthermore, the gradient of planetary vorticity together with the blocking and deflection due to topographic influences contribute to an eastward turn in the winds as they cross the equator.

The strong surges can cause heavy rainfall and are often associated with severe flooding in the equatorial zone particularly over the Malay Peninsula, Sumatra, Borneo, and other Indonesian islands (Johnson and Chang 2007; Tangang et al. 2008; S. Y. Lim et al. 2013, meeting presentation). Trilaksono et al. (2012) conducted regional model simulation of heavy precipitation events at Jakarta in West Java and found the February 2007 Jakarta flood, the worst flood event in three centuries, was associated with a strong cold surge that possessed a cold anomaly of 2°K below normal up to 1.5-km height, which is very unusual south of the equator. The cross-equatorial flow can also act to enhance the Australian monsoon trough and may contribute to tropical cyclogenesis (Holland 1984; McBride 1995).

The cold surges can even have remote influences on the tropical convection and disturbed weather away of the South China Sea. Vissa et al. (2013) documented the crucial role of the South China Sea cold surges in the development of Cyclone Sidr in the Bay of Bengal in November 2007. Cyclone Sidr was one of the most devastating tropical cyclones in the twenty-first century that caused several thousand deaths and tremendous property damage after landfall over the Bangladesh coast (Akter and Tsuboki 2012). The development of the cyclone was traced back by Vissa et al. (2013) to cold surge–induced heavy rainfall episodes at the southern Gulf of Tonkin coast. The southward progression of the surge led to the intensification of deep convection on the Vietnamese coast and interaction with Typhoon Peipah, which entered the South China Sea from the western Pacific. Typhoon Peipah transported convective cloud clusters, moisture, and westward momentum to enhance the deep convection cells over the Vietnamese coast, which then moved eastward to the Gulf of Thailand and Andaman Sea where it organized into a tropical depression that later intensified to become Cyclone Sidr in the Bay of Bengal.

In addition to the synoptic-scale cold surges, the tropical convection weather in the South China Sea and the western Maritime Continent region is also affected by local sea surface temperature (Hendon 2003; Koseki et al. 2013a) and the other large-scale disturbances including the Borneo vortex and the Madden–Julian oscillation (MJO). The MJO often has peak amplitude over the Maritime Continent during boreal winter (Madden and Julian 1972), and its convection can also strengthen the local Hadley cell and the East Asian jet (Jeong et al. 2008; He et al. 2011). The MJO causes alternating periods of large-scale active and inactive convective phases with a periodicity of 30–60 days as it propagates eastward through the region (Chang et al. 2005a, 2006; Zhang 2005; Robertson et al. 2011). Thus, the frequency of the MJO is much lower than the cold surges that typically occur more than once per month, each time lasting from a few days to one week or more. Therefore, the signal of the impact on the variation of convection weather by MJO during boreal winter is less conspicuous compared to that due to the surges. Furthermore, the complex terrain causes the large-scale structure of MJO to disintegrate (Wu and Hsu 2009) so that the concept of MJO being a large-scale precipitation envelope that smoothly propagates eastward breaks down over the Maritime Continent (Peatman et al. 2013). The MJO-associated precipitation is strongest when the synoptic systems are weak (Qian et al. 2010, 2013; S. Y. Lim et al. 2013, meeting presentation), and it is mainly identified in the diurnal cycles (Qian et al. 2010, 2013; Kanamori et al. 2013; Virts et al. 2013; Peatman et al. 2013) and over ocean (Rauniyar and Walsh 2011; Oh et al. 2012).

Around the region of Borneo a low pressure area is commonly present year-round and has been variously described as an equatorial trough or an intertropical convergence zone (ITCZ). However, the Borneo vortex is an important system of the Asian winter monsoon with a climatological property that is distinct from other vortices in the tropics. This vortex arises as a result of the interaction of the basic-state shear vorticity created by the low-level northeasterly winds over the South China Sea and the much weaker winds over the mountainous western coast of Borneo, and diabatic heating supported by abundant moisture (Chang et al. 2004a; Ooi et al. 2011; Koseki et al. 2013b). The frequency of occurrence of vortices during boreal winter in the Borneo region exceeds that of any other seasons and regions, including those that are well known for seasonal vortex activities such as the Bay of Bengal during boreal summer (Chang et al. 2004a). The variability and life cycle of the Borneo vortex has an important impact on deep convection and disturbed weather throughout the equatorial South China Sea that is comparable to those due to the cold surges (Chang et al. 2006).

The larger spatial and time scales of the MJO may lead to a conception that MJO modulates the synoptic and smaller-scale convection in this region to the degree that its wet phase may be correlated significantly to the occurrence of high-impact weather. While this depiction may provide a reasonable overview of the large-scale convection during other seasons, it is too simplistic during boreal winter because the three circulation systems differ greatly in their origin. The MJO originates over the equatorial Indian Ocean, cold surges originate from the midlatitude regions of eastern Asia, and the Borneo vortex develops locally over the southern South China Sea. These three circulation systems interact in a rather complicated way, resulting in a profound impact on the variability in deep convection over the western Maritime Continent/equatorial South China Sea region during the Asian winter monsoon.

The discussion below is based on Chang et al.'s (2005b) analysis of the interaction of the three motion systems in this region (inside rectangle in Fig. 6-4). The analysis was conducted by compositing the large-scale patterns of NCEP–NCAR reanalysis 925-hPa winds and a convection index (CI) that measures the difference between the Geostationary Meteorological Satellite (GMS) blackbody temperature and a threshold of $250°K$, with positive CI indicating a lower blackbody temperature. It will become apparent that the deep convection resulting from the three disturbance systems is strongly affected by the local terrain.

a. Mean convection and low-level kinematic fields in the western Maritime Continent region

Figure 6-5 shows the seasonal mean fields during December–February. In general, deep convection (Fig. 6-5a) is concentrated over the large islands of the region. The maximum convection occurs over Java,

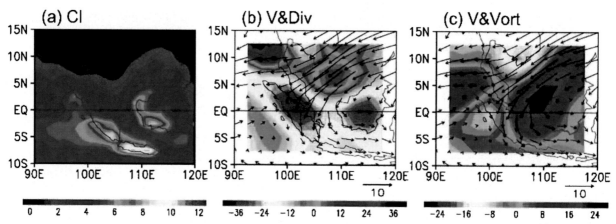

FIG. 6-5. DJF (1979/80–2000/01) mean fields of (a) CI, (b) 925-hPa winds (m s^{-1}) and divergence (shaded, 10^{-5} s^{-1}), and (c) 925-hPa winds (m s^{-1}) and vorticity (shaded, 10^{-5} s^{-1}) (from Chang et al. 2005a).

which is connected to a second convection maximum over Sumatra with an extension into the eastern Indian Ocean as part of the ITCZ south of the equator. The lack of deep convection north of 5°N is consistent with the 925-hPa divergence (Fig. 6-5b) in the northeasterly monsoon flow. This is a significant change from boreal fall when strong convection occurs off the Vietnamese coast north of 10°N as a result of the low-level convergence produced by the northeast onshore winds, as discussed in section 2. In winter the cooler and drier mean northeast winds and the lower SST in the northern and middle South China Sea suppresses deep convection until the air reaches the southern South China Sea, where it is transformed by substantial surface sensible and latent heat fluxes (Johnson and Zimmerman 1986).

The pattern of the 925-hPa winds is largely a result of the blockings and deflections of the terrain around the South China Sea. The terrain effects and the conservation of potential vorticity cause a counterclockwise turning of the cross-equatorial winds. The two primary regions of low-level convergence (Fig. 6-5b) are associated with the maxima in deep convection (Fig. 6-5a). The convergence center near Sumatra is related to the interaction of the northeasterly monsoon flow with the terrain of the Malay Peninsula. However, the convergence center over Borneo is shifted east of the primary northeasterly wind belt and is in the region of the counterclockwise turning of the winds that cross the equator. Coincident with the convergence center over Borneo is a maximum in 925-hPa relative vorticity (Fig. 6-2c). In addition to the curvature contribution to the vorticity maximum, shear vorticity results because of the interaction between the northeasterly monsoon flow and the terrain of Borneo.

Figure 6-6 shows that the Sumatra and Borneo convection centers start the season forming a V-shape

convection pattern in December, but both areas shrink and retreat southward and become separated in February. This retreat coincides with the intensification of the divergence over the South China Sea associated with the strengthening of the northeast monsoon winds. In particular, the reduction in deep convection over Sumatra is related with the increased drying influence of the strengthening northeast winds that flow toward Sumatra. The drying does not impact the convection center over Borneo as significantly since the northeast monsoon flow is nearly parallel to the Borneo coastline (Fig. 6-4), where Ekman pumping can increase moisture convergence as the strengthening northeast winds contribute to the shear and curvature vorticity of the counterclockwise circulation around Borneo. Between 2° and 4°N a substantial part of the western Borneo coastline faces northward, and as a result the blocking of strengthened northeast winds causes a direct increase in moisture convergence.

b. Cold surge and Borneo vortex

The synoptic and intraseasonal variation of deep convection in the southern South China Sea and western Maritime Continent can be described by the composites of low-level circulation and convection fields according to the activities of Borneo vortex, cold surge, and MJO. Of the total 1895 days in the 21 boreal winter seasons between 1979/80 and 2001/02, Chang et al. (2005a) identified nearly one-third as "vortex cases" and the remaining two-thirds as "no-vortex cases" based on whether one or more vortex centers appeared in the daily 0000 UTC 925-hPa streamline analysis between 5°S and 10°N and 105° and 115°E (inside rectangle in Fig. 6-4). They also identified one-fifth of the days as "surge" days when a South China Sea surge index, defined as the area-averaged northeasterly 925-hPa wind

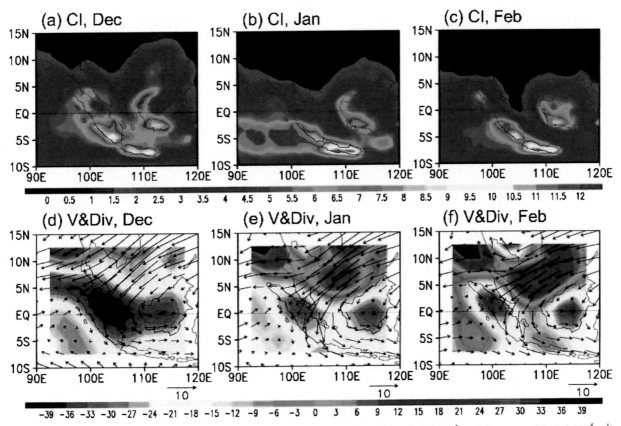

FIG. 6-6. Individual monthly means (1979/80–2000/01) for (a)–(c) CI and (d)–(f) 925-hPa winds (m s^{-1}) and divergence (shaded, 10^{-5} s^{-1}) for December in (a) and (d), January in (b) and (e), and February in (c) and (f) (from Chang et al. 2005a).

over 5°–10°N, 107°–115°E (top thick bar in Fig. 6-4), reaches 8 m s^{-1}. About 24% of the vortex days are also surge days, while only about 18% of the no-vortex days are. The frequency of both is highest in December and lowest in February.

The selection of vortex and surge days were done with unfiltered data, but in the composite of the two synoptic-scale systems the 925-hPa wind was filtered to highlight variations over the period range of 2–15 days. During no-vortex days, deep convection and low-level convergence are reduced over the southern South China Sea and enhanced to the west, southwest, and south of the South China Sea (Figs. 6-7a,c). During vortex days (Figs. 6-7b,d), the patterns are nearly identically opposite. Thus the presence of the Borneo vortex and deep convection acts to intercept transport of low-level moisture by the northeasterly monsoon flow such that convection downstream over the Malay Peninsula–Sumatra–Java region is reduced.

The composite convective index and divergence patterns for the surge and no-surge days exhibit well-recognized patterns of variability. During no-surge days, deep convection (Fig. 6-8a) and 925-hPa convergence (Fig. 6-8c) are located over Indochina, while reduced convection and low-level divergence are found over most of the remaining equatorial South China Sea region. During surge days, a near opposite pattern occurs with reduced convection (Fig. 6-8b) and low-level divergence over Indochina and enhanced convection and convergence over the remainder of the region (Fig. 6-8d), which is in agreement with previous studies of the influence of northeasterly surges on deep convection over the equatorial South China Sea (e.g., Lau and Chang 1987). During these periods, deep convection occurs in association with the blocking of the low-level winds by the terrain, which contributes to a shift of the convection pattern downstream of the maximum in 925-hPa convergence. The shift is less when the convection pattern is compared with 850-hPa convergence because there is less blocking of winds at higher elevations. The low-level convergence during surge days (Fig. 6-8d) takes on a V-shaped pattern on the windward side of the Malay Peninsula and Borneo. This pattern results from the blocking of the surge winds by the terrain (Fig. 6-4) and its location is different from the V-shaped divergence pattern associated with the monthly mean fields (Fig. 6-6), where maximum convergence centers are located over land areas.

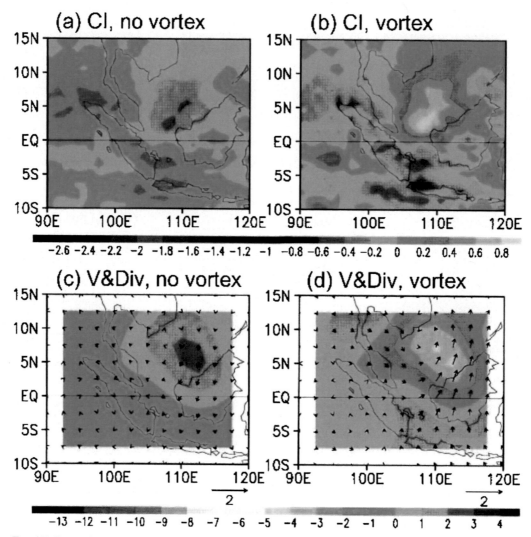

FIG. 6-7. Composite maps of CI for (a) no-vortex cases and (b) vortex cases, and composite maps of 925-hPa winds (m s^{-1}) and divergence (shaded, 10^{-5} s^{-1}) for (c) no-vortex cases and (d) vortex cases (from Chang et al. 2005a).

Figure 6-9 shows the composites when the surge and vortex classifications are considered together. When neither surge nor vortex is present (Fig. 6-9a), convection is reduced over the equatorial South China Sea and enhanced downstream of the mean northeasterly wind over the Malay Peninsula, Sumatra, and Java, where the monsoon flow that has received the ocean surface sensible and latent heat fluxes interacts with terrain. This pattern is almost reversed when a vortex is present without a cold surge (Fig. 6-9b). The presence of the Borneo vortex results in a deflection of the low-level winds and convergence to the west coast of Borneo, such that the primary area of deep convection occurs more upstream near the west coast of Borneo and over the southern South China Sea. Convection over the landmasses downstream from the surge is suppressed. The presence of a cold surge basically enhances these two

opposite patterns (Figs. 6-9c,d). The locations of convection during surge cases with and without a vortex are similar to those in the respective no-surge cases; however, the magnitude is much stronger with a surge. Convection over the southern South China Sea is strongest when both surge and vortex cases are present (Fig. 6-9d).

The enhancement of deep convection over the southern South China Sea during cases when both cold surge and Borneo vortex are present is sensitive to the strength of the cold surge (Fig. 6-10). As the intensity of the cold surge increases from weak (between 8 and 10 m s^{-1}, Figs. 6-10a,d) to moderate (10–12 m s^{-1}, Figs. 6-10b,e) to strong (greater than 12 m s^{-1}, Figs. 6-10c,f), the area covered by increased convective index values and the amplitude of the convective index increase over the southern South China Sea. The increased deep

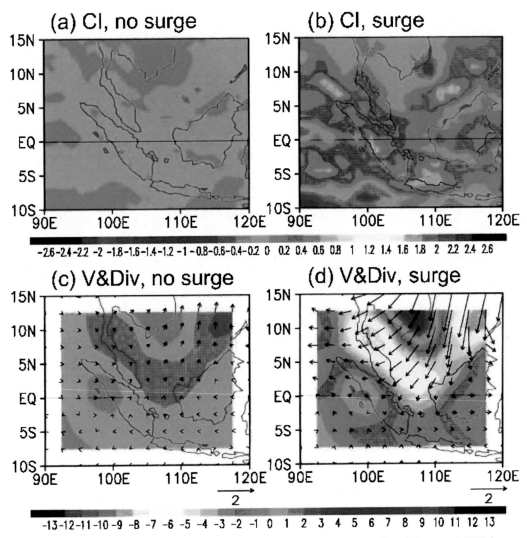

Fig. 6-8. As in Fig. 6-7, but for (a),(c) no-surge cases and (b),(d) surge cases (from Chang et al. 2005a).

convection with surge intensity results from two processes, both of which involve terrain influences. First, increased northeast winds result in increased moisture convergence near the coastal area of Borneo. Second, the northeast winds strengthen much more over sea surface than over land, resulting in increased shear vorticity that contributes to a stronger Borneo vortex.

The impact of the presence of the vortex during strong surge events (Figs. 6-10b,e and 6-10d,f) can be examined by contrasting convection, wind, and divergence patterns of these surge and vortex cases with cases of strong surges but no vortex (Fig. 6-11). Without the presence of the vortex, deep convection throughout the southern South China Sea and along the Borneo coastline is severely reduced (Figs. 6-11a,b). However, convection over the Malay Peninsula and Sumatra is increased. This is due to the lack of the counterclockwise turning of the

wind over the equatorial region (Figs. 6-11c,d), which results in reduced low-level convergence along the western Borneo coastline. Because there is no vortex to induce the clockwise turning, there is increased interaction between the northeast winds and the terrain of the Malay Peninsula and Sumatra, which contributes to increased low-level convergence and deep convection over those areas.

Another assessment of the interactions between the Borneo vortex and cold surges is to compare the location of the vortex center in the composite constructed from cases that only contain a vortex but no surge (Fig. 6-9f), all vortex cases with and without a surge (Figs. 6-5d), and the composite of vortex and surge cases (Figs. 6-9h). When a vortex is present without a surge, there is a cyclonic turning of the winds over the southern South China with no closed circulation center. In the all-vortex

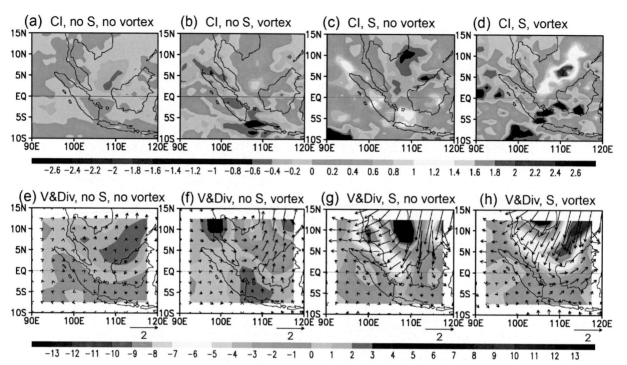

FIG. 6-9. As in Fig. 6-5, but for (a),(e) no-surge and no-vortex cases, (b),(f) no-surge and vortex cases, (c),(g) surge and no-vortex cases, and (d),(h) surge and vortex cases (from Chang et al. 2005a).

composite (Fig. 6-5d), a closed cyclonic circulation is centered over the southern South China Sea. In the vortex and surge composite (Fig. 6-9h), the center of the vortex is shifted to be oriented along the western Borneo coastline. Although the presence of the surge acts to increase the strength of the vortex, the surge results in a shift of the vortex center from being located over the southern South China Sea (Fig. 6-9b) to be near the Borneo landmass (Fig. 6-9d).

An extreme case of the interaction between a strong cold surge and a Borneo vortex is the rare formation of Typhoon Vamei near the equator on 26 December 2001 (Chang et al. 2003). In addition to the vanishing Coriolis parameter, an important reason for such a formation to be extremely rare is the typical displacement of the vortex center toward Borneo under strong surge conditions. Since a tropical cyclone needs to develop over ocean, a vortex with much of the cyclonic circulation that lies over land cannot intensify into a tropical cyclone regardless of the strength of the surge. This interesting case of equatorial tropical cyclone formation will be discussed in section 4.

c. The effects of the Madden–Julian oscillation

Chang et al. (2005b) used a singular-value decomposition (SVD) on 30–60-day filtered 850-hPa winds and OLR to identify MJO activity over the Maritime Continent

region. In their analysis the periods of MJO presence are separated from those of no-MJO presence. Within a period of MJO presence, four phases—dry, wet, and two transitions—defined by the two leading SVD modes are shown in Fig. 6-12. It is important to note the different meanings in the common usage of the term MJO. In the operational forecast community, the wet phase has often been called an MJO, but here the wet phase is only one of the four phases of an MJO period. Similarly, a dry phase is not the same as a period of no MJO, even though both are characterized by suppressed convection.

The convection pattern of Chang et al.'s MJO phase 1 (dry phase, Fig. 6-12a) resembles that of the Wheeler and Hendon (2004) MJO Real-time Multivariate (RMM) phases 1 and 2, in that the enhanced convection appears over the eastern Indian Ocean and the suppressed convection appears over the Maritime Continent. The convection pattern of Chang et al. MJO phase 3 (wet phase, Fig. 6-12c) resembles that of the RMM phases 5 and 6, both of which have enhanced convection over the Maritime Continent. However, there is an important difference between the two MJO index schemes. The Wheeler and Hendon index is based on the empirical orthogonal functions of the combined fields of 850- and 200-hPa zonal winds and OLR. The index does not contain meridional wind information, which is essential in the structure of synoptic-scale weather

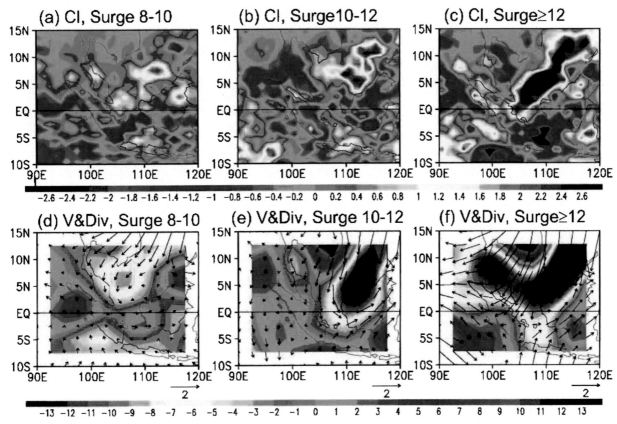

FIG. 6-10. Composite maps of (a)–(c) CI and (d)–(f) 925-hPa winds (m s^{-1}) and divergence (shaded, 10^{-5} s^{-1}) for surge and vortex cases when the surge is weak [in (a) and (d); surge index between 8 and 10 m s^{-1}, 76 cases]; moderate [in (b) and (e); surge index between 10 and 12 m s^{-1}, 41 cases]; and strong [in (c) and (f); surge index greater than 12 m s^{-1}, 34 cases] (from Chang et al. 2005a).

systems. The Chang et al. MJO phases use the 850-hPa total wind and are therefore directly applicable for studying the interactions between MJO and cold surges and Borneo vortex, both of which have strong meridional wind component. It will become clear that as an MJO moves over the Maritime Continent, it is the interaction of its meridional wind components with the synoptic systems that reveals the most discernible effects of the interactions.

During the dry phase (Fig. 6-12a), convection over the Maritime Continent is reduced and equatorial easterly anomalies exist between 80° and 150°E. During the dry-to-wet phase (Fig. 6-12b), the region is in a transition from the reduced convection regime of the MJO, which has moved eastward, to the approaching active convective regime of the MJO. Increased convection and low-level westerlies exist immediately west and south of the Malay Peninsula, Sumatra, and Java. During the wet phase (Fig. 6-12c), convection is enhanced over the eastern portion of the Maritime Continent and low-level westerly anomalies exist throughout the region. Furthermore, the Australian monsoon trough is very well defined and 850-hPa northeasterly winds increase throughout the South China Sea. During the wet-to-dry

transition phase (Fig. 6-12d), the active convection regime has moved to the equatorial western Pacific and the reduced convection regime is approaching the Maritime Continent from the west. Low-level equatorial westerly (easterly) anomalies are found to the east (west) of the Maritime Continent.

In addition to affecting the convection over the Maritime Continent, the circulation of the MJO also interacts with the synoptic-scale disturbances. In particular, Rossby-type responses that are manifest in the subtropical circulations flanking the MJO-induced enhanced or reduced convection are associated with large-scale meridional wind patterns. These meridional winds may act to reinforce or weaken a cold surge event. The net effect of the MJO is to suppress cold surges (Table 6-1). When MJO is present, the chance of a cold surge during the dry and dry-to-west phases (15%) is almost one-half those during the wet and wet-to-dry phases (28%) and during periods when no MJO is present (29%). This contrast is consistent with the anomalous 30–60-day 850-hPa wind patterns (Fig. 6-12), in that there are anomalous southerly winds during the dry-to-west phases over the South China Sea that apparently inhibit the development of cold

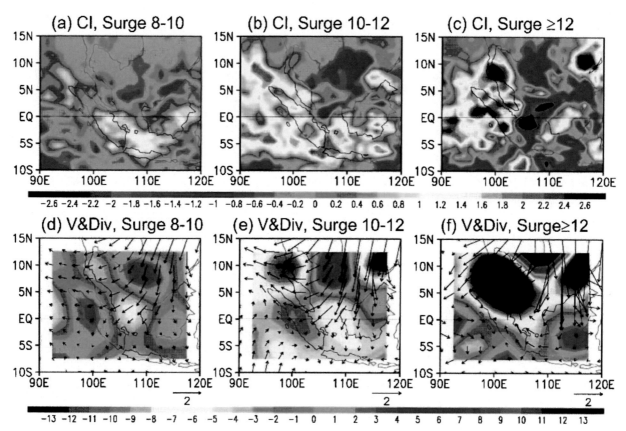

FIG. 6-11. As in Fig. 6-8, but for surge and no-vortex cases when the surge is (a),(d) weak (136 cases), (b),(e) moderate (62 cases), and (c),(f) strong (31 cases) (from Chang et al. 2005a).

surges. Chang et al. (2005b) further showed that it is mainly the weak surges (surge index between 8 and 10 m s^{-1}) that are inhibited.

The MJO has an even larger suppression effect on the Borneo vortex, with nearly twice as many vortex cases occur during no-MJO periods than MJO periods (Table 6-2). On the other hand, the distribution of the number of vortex cases varies only modestly among the four phases when MJO is present. During the MJO dry phase, the composite Borneo vortex case (Fig. 6-13a) is similar to the composite of vortex cases with no surge (Figs. 6-9b,f). There is a broad area of counterclockwise turning of the low-level winds over the southern South China Sea with no closed circulation. Convection is increased over the southern South China Sea and reduced over the equatorial Southern Hemisphere, Malay Peninsula, and Sumatra. This similarity indicates that the MJO dry phase may inhibit cold surges so the Borneo vortex most likely occurs without a surge during this phase. This is consistent with the anomalous 30–60-day subtropical ridge over the western North Pacific with southerly anomalies over the South China Sea in Fig. 6-12a. During the dry-to-wet phase, the deep convection over the equatorial South China Sea and the low-level winds

associated with the Borneo vortex (Fig. 6-13b) become more organized. Northeast winds appear over the southern South China Sea, but their magnitude is much less than that in the vortex and surge composites (Fig. 6-9d). Overall, the vortex composite patterns associated with both the dry and dry-to-wet phases are not very different from composites of vortex-only and vortex–surge cases, suggesting that the MJO primarily influences the Borneo vortex by inhibiting cold surges and reducing their enhancement effect on the Borneo vortex.

Although the MJO wet phase is associated with a higher cold surge frequency than the dry phase, the cyclonic circulation of the vortex seems to be more linked to cyclonic horizontal shear associated with equatorial westerly winds rather than northeasterly winds that extend through the southern South China Sea. These increased equatorial westerlies are associated with the enhanced MJO-scale convection over the eastern portion of the Maritime Continent (Fig. 6-12c). Consequently, the center of the vortex is located over the southern South China Sea (Fig. 6-13c). During the wet-to-dry phase (Fig. 6-13d), the vortex is very weak with cyclonic shear present only over the extreme western South China Sea and Malay Peninsula. Southwesterly

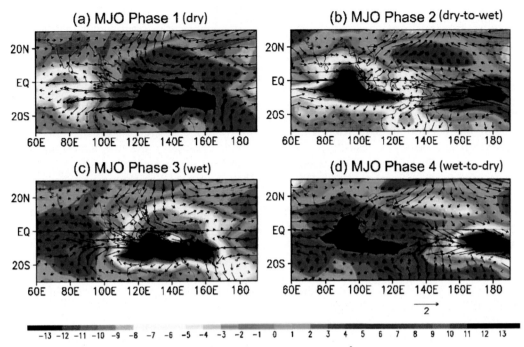

FIG. 6-12. Composite 850-hPa winds (m s^{-1}) and anomalous OLR (W m^{-2}) for the four phases of MJO based on time coefficients of an SVD analysis of the winds and OLR. The terms dry and wet refer to increased or reduced convection over the Maritime Continent (adapted from Chang et al. 2005a).

anomalies exist over the primary region of the southern South China Sea, and reduced deep convection spreads northeastward along the west coast of Borneo. The relationships among the MJO, cold surges, and the Borneo vortex are summarized in Fig. 6-14. It is clear that the presence of the MJO is associated with fewer numbers of vortex cases, and the occurrence of vortex cases during periods of weak surges is most reduced during the MJO. Therefore, while the presence of a surge acts to increase the strength of the Borneo vortex, the frequency of surges is reduced when the MJO is present. Often the MJO-scale circulation pattern directly opposes the cold surge wind pattern. Therefore, weak surges may be more inhibited during periods of strong MJO. Primarily because of the impact of the MJO on cold surge intensity and frequency, 66% of the vortex cases occur during no-MJO periods. The Borneo vortex is least likely to occur when the inactive convective portion of the MJO extends to the Maritime Continent with large-scale low-level diffluence that acts to restrict the impact of cold surges on convection in the southern

South China Sea. This complex relationship among MJO, cold surges, and the Borneo vortex and the effects of the topography contribute to the variability in convection patterns over a variety of space and time scales.

4. Equatorial cyclogenesis and the formation of Typhoon Vamei (2001)

The monsoon heavy rainfall events often produce heavy floods and major economic and life losses in the Maritime Continent area, particularly during boreal winter when cold surges and synoptic disturbances including the Borneo vortex are most active (Johnson and Chang 2007). However, tropical cyclogenesis has never been observed over the warm surface of the equatorial South China Sea until the development of Typhoon Vamei in December 2001. The formation of Typhoon Vamei was especially noteworthy because it formed at 1.5°N in the southern tip of the South China Sea at 0000 UTC 27 December 2001, a latitude that most previous literatures and textbooks (e.g., Gray 1968; Anthes

TABLE 6-1. The number of surge cases with respect to the MJO and the phase of the MJO.

	No MJO days	MJO days	MJO phases 1–2	MJO phases 3–4
No. of no-surge days	845	670	381	289
No. of surge days	242	138	56	82
Percentage of surge days	29%	21%	15%	28%

TABLE 6-2. The distribution of vortex cases (days) with respect to MJO periods and phases of the MJO.

	No MJO	MJO	MJO phase 1	MJO phase 2	MJO phase 3	MJO phase 4
No vortex	675	542	148	163	157	124
Vortex	412	216	57	69	50	40

1982; McBride 1995) ruled out for development because of the smallness of the Coriolis parameter. Thus an equatorial development needs a basic-flow relative vorticity that reaches certain minimum magnitude. Even though the Borneo vortex is a climatological quasi-stationary feature during the Asian winter monsoon, it is of synoptic spatial and time scales. The only possible mechanism to provide the required large-scale background vorticity will have to come from the sustained cold surges with periods up to a week or more.

The equatorial cyclogenesis of Typhoon Vamei was very different from drifting vortices that have been observed to cross the equator since the beginning of the satellite era (e.g., Chang and Maas 1976) because such vortices were generated away from the equator and their intensity while crossing the equator was usually weak. It was also different from twin tropical cyclones that saddle the equator (e.g., Keen 1982; Lander 1990), where the centers of these tropical cyclones were all sufficiently away from the equator for the Coriolis parameter to play a dominant role. The cyclogeneses induced by cross-equatorial cold surge discussed by Love (1985) occurred at subtropical latitudes (13°–16°).

a. Dynamics of surge-forced equatorial cyclogenesis in the absence of terrain

Two decades before the observation of Vamei, Lim and Chang (1981) proposed an equatorial cyclogenesis process to study the dynamics of a midlatitude cold surge into the tropics. They used a barotropic divergent model and argued that even though the tropical motions responding to surges are likely to be associated with cumulus convection, the shallow-water system may be regarded as a two-level approximation to a baroclinic system because in the free atmosphere the two are equivalent for a given vertical structure (Matsuno 1966). They chose a scale height for their shallow-water model such that the model waves have phase speeds close to those observed. This approximation is justified because Chang (1977) has shown that in a baroclinic model with a balance between diabatic heating and damping, the equatorial wave behavior is altered in such a way that the vertical wavelength of even the higher-frequency waves tends to be comparable to the depth of the troposphere.

Lim and Chang's (1981) shallow-water model on an equatorial beta plane with no mean flow is

$$\left.\begin{array}{c} \dfrac{\partial u'}{\partial t} + g\dfrac{\partial h'}{\partial x} - \beta y v' = 0 \\[2mm] \dfrac{\partial v'}{\partial t} + g\dfrac{\partial h'}{\partial y} + \beta y u' = 0 \\[2mm] \dfrac{\partial h'}{\partial t} + H\left(\dfrac{\partial u'}{\partial t} + \dfrac{\partial v'}{\partial t}\right) = \Phi' \end{array}\right\}.$$

Here u', v', and h' are the perturbation zonal and horizontal velocity and height, respectively; g, β, and H are the constant gravitational acceleration, equatorial Coriolis parameter gradient, and mean height, respectively. The variable Φ' represents a time-dependent mass forcing that is used to emulate pressure surge from the tropics and is expressed in a Fourier series form:

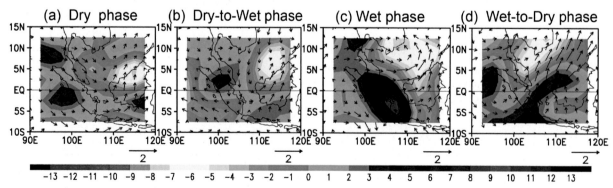

FIG. 6-13. Composite maps of convective indices and 925-hPa winds (m s^{-1}) for MJO and vortex cases when the MJO is in (a) dry, (b) dry-to-wet, (c) wet, and (d) wet-to-dry phases (adapted from Chang et al. 2005a.)

$$\Phi' = \left. \begin{array}{ll} \sum_{m=0}^{\infty} [\Phi_c(m,\zeta)\cos kx + \Phi_s(m,\zeta)\sin kx]\dfrac{t^2}{2\tau^3}e^{-t/\tau} & \text{for} \quad t \geqq 0 \\[2mm] 0 & \text{for} \quad t < 0 \end{array} \right\}.$$

Here Φ_c and Φ_s are the expansion coefficient of the space-dependent part of the forcing function in a Fourier series in x direction, m is the number of waves, ζ is the nondimensionalized y coordinate, k is the x wavenumber, τ is a time constant; various parameters in this equation are used to specify the spatial and time scale of the forcing. The time-dependent part is normalized in the sense that total time integration gives a constant value of unity. Lim and Chang (1981) studied various time scales of large-spatial-scale forcing cases by specifying the forcing to have a Gaussian spatial distribution in the form of

$$\Phi' = \exp\left\{ -\frac{1}{2}\left[\left(\frac{\zeta - a}{\sigma}\right)^2 + \left(\frac{x}{\lambda}\right)^2 \right] \right\} \left(\frac{t^2}{2\tau^3}e^{-t/\tau} \right).$$

Here a is the nondimensional latitude of the center of forcing; λ and σ are the nondimensional half-width of the x and y profile of the forcing function, respectively; and the ζ profile is expanded in terms of the Hermite solutions $D_n(\zeta)$:

$$\exp\left[-\frac{1}{2}\left(\frac{\zeta - a}{\sigma} \right)^2 \right] = \sum_{n=0}^{\infty} p_n D_n(\zeta),$$

with the expansion coefficient p_n, governed by the following recurrent relation:

$$p_n = \frac{a}{n(1+\sigma^2)}p_{n-1} + \frac{\sigma^2 - 1}{2n(1+\sigma^2)}p_{n-2},$$

with $p_{-1} = 0$ and

$$p_0 = \left(\frac{2}{1+\sigma^2} \right)^{1/2}\sigma\exp\left[-\frac{a^2}{2(1+\sigma^2)} \right].$$

Lim and Chang considered a forcing centered at 30°N with $\tau = 1$ day, $\sigma = 1$ (corresponding to ~11° latitude), $\lambda = R/5$ where R is the radius of Earth, $a = 2.5$, and $c = 30\,\mathrm{m\,s^{-1}}$, which implies a phase speed of ~10 m s^{-1} for the gravest Rossby mode ($n = 1$). The result computing up to zonal wavenumber 29 and Hermite mode 29, which provide better than 99.9% accuracy, is shown in Fig. 6-15.

The time evolution of velocity and geopotential depicts the geostrophic adjustment process after the forcing is turned on. On day 2, a high pressure center is building up over the area of forcing with a slightly subgeostrophic anticyclonic outflow. The equatorward wind speed is stronger because of the smaller Coriolis parameter. It becomes noticeably cross isobaric south of 15°S and flows cross the equator, backing sharply from northeasterly to westerly in the process. The flow strengthens and becomes more geostrophic on day 3. The backing of the cross-equatorial flow becomes more pronounced resulting in a distinct east-northeast–west-southwest equatorial shear line located southeast of the high. On day 4, a vortex (area a in Fig. 6-15) develops on this shear line, and south of it a southwesterly cross-equatorial flow is established. Two days later, the equatorial vortex drifts southwestward and begins to develop around it a small high pressure area just south of the equator. The midlatitude anticyclone now shows a marked northeast–southwest tilt with a ridgeline extending toward the equator. Parallel to this ridgeline a band of low pressure appears to the east. The southwesterly cross-equatorial flow strengthens and makes a distinct wave (area d) as it swings back south to merge with the eastward-propagating equatorial westerlies. A region of counterclockwise rotation is formed at the equator to the west of area d, nearly enclosing the band of low pressure. Thus an equatorial cyclogenesis appears to take place as a result of the forcing at 30°N.

By day 9, the northeast–southwest tilt of the midlatitude anticyclone becomes very pronounced. A streak of strong northeasterly winds surges from about 25°N to near the equator. Between this northeasterly streak and the southwesterly cross-equatorial flow, a very marked shear line is developed and extends from just south of the equator up to about 15°N (area c). This shear line also roughly divides a general area of high pressure to its west from a band of low pressure to its east, where the counterclockwise rotation becomes more conspicuous with closed circulation centered just north of the equator. This large area of complex motions drifts slowly westward with the midlatitude anticyclone. The equatorial wave (area d) and westerlies (area b) propagate farther to the east and become detached from this area of major response.

The identity of the response can be analyzed by a mode-by-mode computation. Figures 6-16a–f show the velocity field of all wave modes of $n = -1$ and 0 and the Rossby modes of $n = 1, 2, 3,$ and 4, respectively, while Fig. 6-16g shows the combined velocity field of the

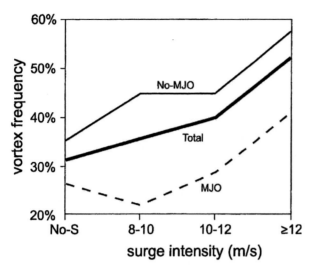

FIG. 6-14. The percentage of days containing a vortex relative to the surge intensity, for all (total) days, no-MJO days, and MJO days. The surge intensity is the average 925-hPa northerly wind along 15°N between 110° and 117.5°E (from Chang et al. 2005a).

Rossby modes $n = 1$ to 9, all for day 9. Comparing Fig. 6-16 and Fig. 6-15, it is clear that the main response that drifts slowly westward is a congregation of Rossby waves, while the eastward-propagating equatorial wave (Fig. 6-16d) and westerlies (Fig. 6-16b) are manifestations of the mixed Rossby–gravity and Kelvin wave groups, respectively.

The development of the main response may be interpreted in terms of wave-group behavior of the Rossby waves of different Hermite modes. For each Hermite mode (n), Rossby waves of different zonal wavenumber combine to form a wave group with a zonal width comparable to the longitudinal extent of the forcing. The wave groups of the lower Hermite modes, $n = 1$–3 have local wavenumber near 3, while those of higher modes, $n = 4$–7, have local wavenumber near 5. All of these wave groups have westward group velocity, but the speed is smaller for the higher modes, as can be expected from the characteristics of the Rossby wave dispersive relationship. The wave group $n = 1$ therefore moves fastest to the west, followed successively by wave groups of increasing n. As time goes on, the wave groups of different n drift farther and farther apart. Since equatorial motions are mainly made up of contributions from lower modes while midlatitude motions are mainly made up of contributions from the higher modes, the development of a northeast–southwest tilt in the midlatitude anticyclone and the equatorial disturbances may be seen as a natural consequence of the gradual westward displacement of the lower mode wave groups relative to the higher mode wave groups. The strong northeasterly streak is built up with contributions from many Hermite modes; modes 1 and 2 make up the far

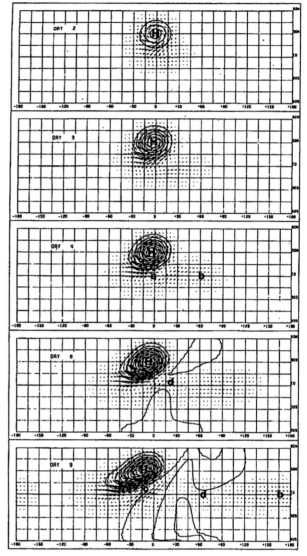

FIG. 6-15. The velocity and geopotential (solid lines) fields for days 2, 3, 4, 6, and 9 in the equatorial beta-plane model after a forcing centered at 30°N is turned on with $\tau = 1$ day. See text for details (from Lim and Chang 1981).

southwest end of the streak, while modes 4 and 5 make up the northeast part. The southwesterly cross-equatorial flow is made up mostly of contributions from modes 2–6. As the lower modes move farther and farther westward, their influences on the eastern part of the disturbance diminish and the motions there gradually develop higher mode characteristics.

Many features in Lim and Chang's (1981) Rossby mode of response are reminiscent of the northeasterly surges during the Asian winter monsoon. Following a pressure surge in the midlatitudes, a belt of strong winds rapidly builds up, sweeping down from about 20°N to the equator and veering gradually from north-northeasterly

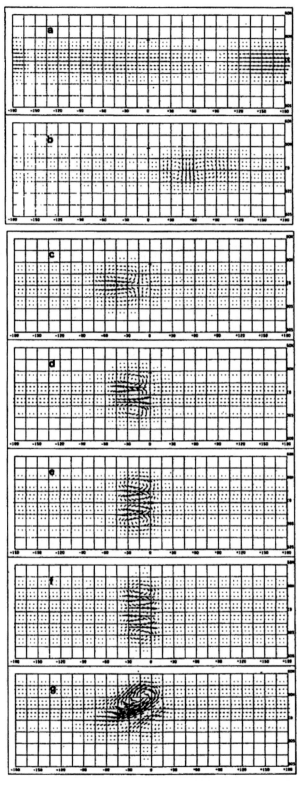

FIG. 6-16. The wave groups of different Hermite modes at day 9: (a) Kelvin wave group ($n = -1$) and (b) mixed Rossby–gravity wave group of $n = 0$; (c)–(f) Rossby wave groups of $n = 1$–4; (g) the combined velocity field of Rossby wave groups of $n = 1$–9. For the same speed, the vectors in (a) and (b) are drawn 5 times longer than those in (c)–(g) (from Lim and Chang 1981).

to east-northeasterly. This wind belt appears to correspond to the wind surges observed over the South China Sea that often develop within 24 h after a steep pressure rise occurs in southern China. Although the flow pattern is basically a feature of the Rossby-type motion, its development however, is not through the usual mechanism of vorticity advection but through a process of separation of the Rossby mode response from the gravity mode responses. In a way, this may be regard as a geostrophic adjustment process on an equatorial beta plane. The velocity scale therefore is the gravity wave speed and not the advective speed, which is consistent with the rapid propagation of the wind surge over the South China Sea.

The northeasterly wind belt predicted by the theory, however, is too broad and penetrates farther westward than observed. More realistic predictions could possibly be made had the effects of terrain been considered. Since wind surges over the South China Sea occur mainly in the lowest 1 km of the atmosphere, frictional dissipation could significantly reduce the wind speed over land and thereby leave a narrower wind belt over the ocean. The westward extent of the surge also would be curtailed by the mountain ranges of Indochina and the Malay Peninsula as the barrier effect would be significant for the shallow layer of strong winds.

An important implication of Lim and Chang's (1981) theory is that the counterclockwise rotation developed at the equator southeast of the main anticyclone is an inherent property of the Rossby wave group response to the cold surge even in the absence of terrain. It is consistent with the mean winter condition in the general vicinity of Borneo and southern South China Sea, and the enhancement of the Borneo vortex following surges discussed in section 3. The terrain effect of the Borneo Island is no doubt a fundamental reason for the tendency of development of low pressure and deep cumulus convection in this region, but the geostrophic adjustment and potential vorticity conservation following a cross-equatorial surge from the north can independently cause an equatorial cyclogenesis.

b. The development of equatorial Typhoon Vamei

The life cycle of the cyclone is shown in Fig. 6-17, which is based on the best track and intensity published by JTWC. The storm made landfall over southeast Johor on the southern tip of peninsular Malaysia, about 50 km northeast of Singapore, at 0830 UTC 27 December 2001, 12 h after the formation. It then weakened rapidly to a tropical depression. It continued in its west-northwest track across southern Johor and the Malacca Straits, and made landfall again in Sumatra. Upon entering the Bay of Bengal, the storm regenerated and continued its northwest track before dissipating in the central Bay of

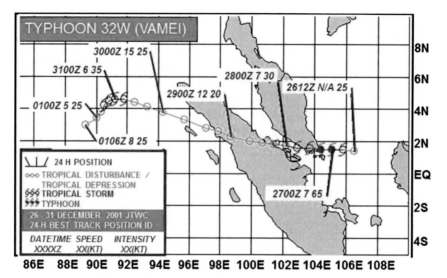

FIG. 6-17. Best track and intensity of Vamei from 1200 UTC 26 Dec 2001 to 0000 UTC 1 Jan 2002 (diagram courtesy of JTWC).

Bengal on 31 December 2001. During the short period of 12 h as a typhoon and another 12 h as a tropical storm, Vamei caused damage to two U.S. Navy ships, including a carrier, and flooding and mudslides in southern peninsular Malaysia's Johor and Pahang states. More than 17 000 people were evacuated and five lives were lost.

The detection of Vamei as a typhoon-strength tropical cyclone was somewhat accidental (Chang and Wong 2008). Vamei was initially classified as a tropical storm by the Japan Meteorological Agency with estimated winds of 21 m s^{-1}. It was later upgraded it to the typhoon category by the JTWC in Hawaii when shipboard observations of sustained winds of 39 m s^{-1} and gusts up to 54 m s^{-1} from several U.S. Navy ships within the small eyewall were reported.

Figure 6-18 shows the MODIS satellite image on 27 December 2001. Vamei's circulation center can be estimated to be just north of 1°N, but an eye is not observable under the clouds. Even though the size of the typhoon is quite small, the spiral cloud bands emanating out from around the center clearly indicate that the storm circulation was on both sides of the equator. Figure 6-19 shows the Japanese GMS image at 0232 UTC of the same day. Feeder bands from both sides of the equator spiral into the center of Vamei, where a small eye is visible. An eye was also observed in TRMM and SSM/I images within the preceding 2 h. The size of the eye estimated from different sensors ranges from 28 to 50 km in diameter. Vamei's small size as it formed in the southern end of the South China Sea and the short duration before its initial landfall make it difficult to observe its highest wind speed from ground-based

observations or to estimate its intensity from satellite images. Without the chance passage by the U.S. ship (USS) *Carl Vinson* carrier group through its eyewall, JTWC may not be able to operationally upgrade the intensity of the storm to a typhoon.

The synoptic development preceding the formation is shown in Fig. 6-20, which is a time sequence of the 1° × 1° Navy Operational Global Atmospheric Prediction System (NOGAPS) 850-hPa wind and vorticity (Chang et al. 2003). Starting from 19 December 2001, a cold surge developed rapidly over the South China Sea, while the center of the Borneo vortex was located near 3°N on the northwest coast (not shown). The strength and persistence of this surge was helped by the strong meridional gradient of sea level pressure in the equatorial South China Sea during December 2001, which was above normal according to the report by the Bureau of Meteorology, Northern Territory region, published in 2002. Figure 6-20 depicts the southwestward movement of the vortex from along the Borneo coast toward the equator. By 21 December, the center of the vortex had moved off the coast over water, where the open sea region in the southern end of the South China Sea narrows to about 500 km, with Borneo to the east and the Malay Peninsula and Sumatra to the west. This overwater location continued for several days. While the vortex center remained in the narrow equatorial sea region, the strong northeasterly surge persisted, and was slightly deflected to the northwest of the vortex.

As discussed in section 3, this near "trapping" of the Borneo vortex by a sustained surge is unusual because normally the vortex center would be pushed eastward by the strengthening surge that streaks southwestward in

FIG. 6-18. MODIS satellite image of 27 Dec 2001 showing Typhoon Vamei near Singapore
(from Chang et al. 2003).

the middle of the South China Sea. Figure 6-21 shows the location of counterclockwise circulation centers during the 51 boreal winters of 1951/52–2001/02 based on a streamline analysis of the NCEP–NCAR reanalysis 925-hPa wind. The distribution of the frequencies of vortex centers lasting for 96 h or more in four subregions near the northwestern coast of Borneo is superimposed on the diagram. There were only six occurrences of a Borneo vortex center that stayed continuously for 96 h in the southwestern box that is in the narrow equatorial region. This unusual trapping caused the cross-equatorial flow wrapped around the vortex and provided a background area of cyclonic relative vorticity with a magnitude of $>1 \times 10^{-5}\,\mathrm{s}^{-1}$, which is comparable to that of the Coriolis parameter 5° away from the equator, a latitude where formation of tropical cyclones is not considered unusual.

Cold surges are frequent events in the South China Sea during the Asian winter monsoon, but the surge preceding the development of Typhoon Vamei was especially intense and long lasting. Chang et al. (2003) reported that the period of 0000 UTC 19–0000 UTC 25 December 2001 was the most sustained and intense surge period among the approximately 15 surge episodes in the three winters that QuickSCAT wind data became available. Furthermore, none of the other 14 surge periods coincided with the occurrence of a Borneo

vortex that migrated into, and stayed in, the narrow open sea region (the southwest box in Fig. 6-21).

Evidence of the strength of Vamei and its relationship with the cold surge can also be confirmed from

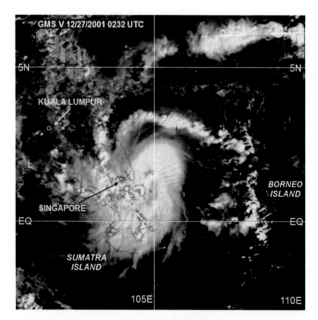

FIG. 6-19. Japanese GMS image at 0232 UTC 27 Dec 2001 (from Chang and Wong 2008).

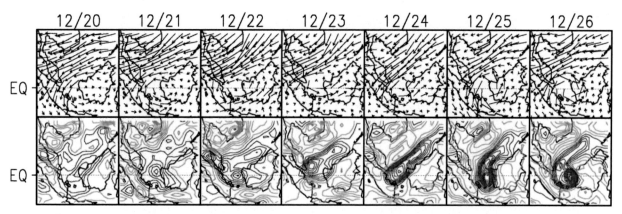

FIG. 6-20. NOGAPS 1° × 1°850-hPa wind and vorticity (red positive, green negative) at 0000 UTC 20–26 Dec 2001 (from Chang et al. 2003).

QuickSCAT satellite scatterometer wind data. Figure 6-22 shows that the QuickSCAT wind direction and speed at 2232 UTC 26 December 2001 captured both the signal of Vamei as it develops to typhoon strength and the remnant of the continuing surge wind upstream in the northern South China Sea. In the southern perimeter, the wind speed at 10-m height has already reached above $27 \, \mathrm{m \, s}^{-1}$ over an area of about 1° latitude × 1° longitude. The northern spiral band extends to about 6°N and is detached from the cold surge wind belt farther north.

Chang and Wong (2008) examined the results of Lim and Chang's cold surge theory and found a number of interesting features that resemble the characteristics of the Vamei development process. Figures 6-23a and 6-23b show the theoretical solutions three days apart (day 3 and day 6 in Fig. 6-15), with the location of the high center on day 6 (Fig. 6-23b) shifted eastward by $0.4L$, where L is the approximate radius of the high pressure cell, to align the high centers that otherwise will show a westward propagation. Figures 6-23c and 6-23d are the NOGAPS 850-hPa wind analysis for 19 December and 22 December 2007, respectively.

As discussed earlier, Fig. 6-23a resembles a typical cold surge event that follows the southeastward movement of an East Asian surface high center with the development of a northeast–southwest tilt. This tilt is due to the dispersion of equatorial beta-plane Rossby waves in which the lower meridional modes have larger amplitudes closer to the equator and therefore propagate westward more quickly. As the northeasterly wind strengthens south of the high center, it streams southward, and after crossing the equator, it turns eastward between the equator and 15°S leading to a counterclockwise curvature flow at the equator. Figure 6-23b shows the solution three days later, in which the northeast–southwest tilt becomes even more pronounced. To the southeast of the northeasterly surge

streak, southwesterly cross-equatorial winds enhance the counterclockwise circulation belt over the equator, which is now northeast–southwest oriented. The counterclockwise flow in both Figs. 6-23a and 6-23b are highlighted by color streamlines. Figures 6-23c and 6-23d show the NOGAPS 850-hPa wind analysis at the beginning of the actual cold surge (Fig. 6-23c, 0000 UTC 19 December), and 3 days later (Fig. 6-23d, 0000 UTC 22 December). Color streamlines are also sketched on Figs. 6-23c and 6-23d to highlight the counterclockwise rotation across the equator

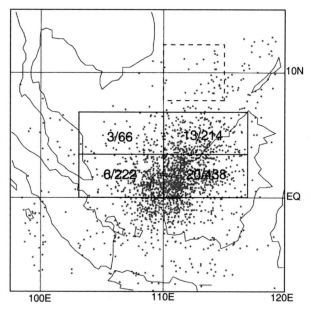

FIG. 6-21. Location of NCEP–NCAR 925-hPa counterclockwise circulation centers during 51 boreal winters. The four subregional boxes are enclosed by equator–7°N, and 103°–117°E. The internal partitions are 3°N and 110°E. The first number in each box indicates the frequency of persistent 925-hPa cyclonic circulation center that lasted for 96 h or more, based on the 1951/52–2001/02 DJF NCEP–NCAR 2.5° × 2.5° reanalysis. The second number is the total number of days a center is identified (from Chang et al. 2003).

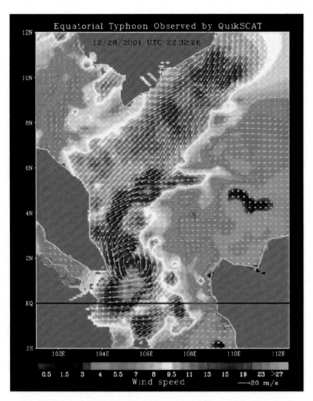

FIG. 6-22. QuickSCAT wind direction and speed (color shading and arrow length) at 2232 UTC 26 Dec 2001 showing the typhoon strength of Vamei and the remnant of the continuing surge wind upstream in the northern South China Sea (diagram courtesy of Jet Propulsion Laboratory/NASA).

that is produced by the cold surge flow. Comparing Figs. 6-23a,b and 6-23c,d, there is a strong resemblance in the overall geometry of the development of the cross-equatorial counterclockwise circulation over the 3-day span.

As pointed out in section 4b, the surge wind belt produced by the theory is too broad and penetrates too far westward. The observed intense surge wind belt is confined within the width of the South China Sea of about 750 km, which is approximately one-half of the width in the equatorial beta-plane solution in Fig. 6-23b. If the effects of topography are included, the strongest surge winds in the theory would be confined over the South China Sea, and the westward extent of the surge would be constrained by the barrier effect of the Indochina and the Malay Peninsula mountain ranges. These effects will also add a forcing to induce the flow to turn equatorward in addition to that required by the conservation of potential vorticity.

The theory and the actual development may be made to resemble each other by scaling the east–west dimension of Figs. 6-23a and 23b to one-half of the original size ($L = 15°$ longitude instead of 30° in Fig. 6-23), or treating the highlighted rectangular area in Fig. 6-23a as

being comparable to the domain of the NOGAPS plots in Figs. 6-23c and 6-23d. In addition, the shifting of the high center in Fig. 6-23b eastward by $0.4L$ in Fig. 6-23a may be accounted for by the slower propagation due to reduced zonal scale and two factors in real-world cold surge events: The eastward movement of the East Asian surface high center due to the midlatitude westerly mean flow, and the fixed location of the surge belt that is restricted geographically by the South China Sea.

c. Why is tropical cyclone development in the equatorial South China Sea so rare?

The interaction between the winter monsoon circulation and the complex terrain and the moisture from the warm ocean surface provided the latent heat and favorable vorticity conditions for development of Typhoon Vamei. However, the strong cold surge and Borneo vortex that led to the development of Vamei are commonly observed major systems of the Asian winter monsoon, and abundant low-level warm and moist air is constantly present over the South China Sea during every season. The most interesting question is not necessarily just how or why Typhoon Vamei could form so close to the equator. Rather, the question is why more typhoon formations have not been observed in the equatorial South China Sea (Chang and Wong 2008). Indeed, the many numerical simulations of Typhoon Vamei (e.g., Koh 2006; Chambers and Li 2007; Juneng et al. 2007; Tangang et al. 2007; Loh et al. 2011) have not addressed this key question. Numerical model simulations should not be considered completely successful only by simulation of the development of Vamei, but also by how successfully the model simulates the *nondevelopment of the numerous other cases of cold surge and Borneo vortex* that occur in every boreal winter.

An important reason tropical cyclones did not develop in these other cases is due to the terrain factor—the narrowness of the equatorial South China Sea. This factor plays two counteracting roles that combine to make the occurrence of the typhoon formation possible but rare. One role is the channeling and strengthening of the cross-equatorial surge winds that help produce the background cyclonic vorticity at the equator. On the other hand, the open water region of approximately 5° longitude is only sufficient to accommodate the diameter of a small tropical cyclone. It is too small for most synoptic disturbances to remain over the water for more than a day or so, particularly during strong surge events when a Borneo vortex is likely displaced to the east. In the unusual case of Typhoon Vamei, the durations of the intense cold surge and the Borneo vortex remained over water significantly longer than normal, which allowed the interaction to continue for nearly a

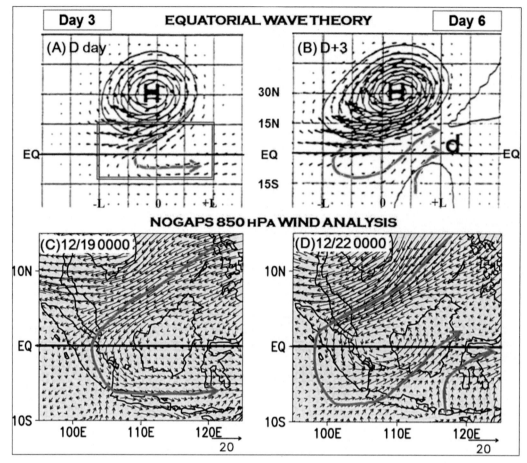

FIG. 6-23. Comparison of Lim and Chang's (1981) (a),(b) barotropic equatorial beta-plane cold surge theory and (c),(d) observed NOGAPS 850-hPa wind analysis, each for two time periods separated by three days. Because the narrow width of the South China Sea confined the width of the intense surge belt to about one-half of that in the terrain-free equatorial beta-plane solution, a comparison of the theory and the actual development may be made by scaling the east–west dimension of (a) and (b) to one-half of the original size ($L = 15°$ longitude instead of $30°$), or treating the highlighted rectangular area in (a) as being comparable to the domain of the NOGAPS plots. The location of the high center in (b) was also shifted eastward by $0.4L$ to account for the reduced zonal scale, the typical eastward movement of the East Asian surface high center, and the constraining effect of the surge belt by the terrain around the South China Sea. See text for details (adapted from Lim and Chang 1981; Chang et al. 2003; Chang and Wong 2008).

week until the storm was formed. Based on an analysis of the NCEP–NCAR reanalyses during the boreal winters of 1951/52–2001/02, Chang et al. (2003) estimated that the probability of an equatorial development from similar conditions to be about once in a century or longer. Although this estimate appears consistent with historical records, it is not known whether there might be other near-equatorial developments not observed during the presatellite era.

5. Summary

The Asian monsoon system, which is defined by the seasonal reversal of winds and the associated maximum rainfall seasons, is driven by the continent–ocean thermal contrast. It derives its energy mainly from the latent heat release in deep convection that produces the heavy rainfall. Even though the monsoon circulation has a very large spatial scale, the convection and rainfall patterns are heavily influenced by the complex terrain on various scales. Indeed, mesoscale and local terrain features can strongly distort the monsoon rainfall distribution, which cannot be considered from just a large-scale perspective. This terrain effect is especially prominent during boreal winter, when the center of deep convection is located in the deep tropics around the Maritime Continent region, where islands of different sizes are interspersed within the surrounding oceans.

Over this region, the equator serves as a general demarcation between the Asian summer (to its north) and winter (to its south) monsoon rainfall regimes. However, the seasonal convection is locally dominated by interactions between the complex terrain and the annual reversal of the large-scale surface winds. These interactions cause the summer and winter monsoon rainfall regimes to intertwine across the equator. In particular, the boreal winter regime extends far northward along the eastern flanks of the major island groups and landmasses. However, there is no complementary extension of the boreal summer regime into southern latitudes.

The seasonal march is also asymmetric during the transitional seasons, with the maximum convection following a gradual southeastward progression from the Asian summer monsoon to the Asian winter monsoon, but a sudden transition in the reverse. The mass redistribution between land and ocean areas during the transition season contributes to this asymmetric march, which can be traced to the orientations of the Asia and Australia landmasses. These orientations and various local terrain features produce sea level pressure patterns that lead to asymmetric wind–terrain interactions throughout the region, and a low-level divergence asymmetry that promotes the southward march of convection during boreal fall but opposes the northward march during boreal spring.

The deep convection and heavy rainfall in the Maritime Continent region during the Asian winter monsoon are among the most energetic tropical convection systems. In the western part of this region and over the equatorial South China Sea, the convection is strongly affected by disturbances ranging from synoptic to intraseasonal time scales. The strongest variability of convection and heavy rainfall are due to the effects of the northeasterly cold surges and the Borneo vortex and their interactions with the terrain. MJO plays a secondary role in lessening these effects when the synoptic-scale systems are of modest intensity.

The effects of a northeasterly surge occur mainly through the interaction with local topography and the dynamic response to the change in latitude. These effects contribute to the turning of the winds and localized patterns of deep convection. In days without a Borneo vortex, deep convection tends to be suppressed over the South China Sea and Borneo and enhanced downstream over the landmasses on the western and southern peripheries of the equatorial South China Sea. The pattern is reversed in days with a vortex. The simultaneous presence of a cold surge enhances this contrast. The surge also interacts with the Borneo vortex, in that the vortex is strengthened and its center shifts from over the South China Sea to over the western coast of Borneo. The frequency of cold surges and vortex days is reduced during periods when the MJO is present. Often the

MJO-related circulation patterns oppose the synoptic-scale cold surge and vortex circulations. Thus, a primary impact of the MJO is to inhibit weak cold surge events, which then produces a secondary impact on the Borneo vortex via interactions between the cold surge winds and the vortex.

The strongest motion system that developed as a result of the interaction of a cold surge with a Borneo vortex is Typhoon Vamei off Singapore on 26 December 2001. With its center at 1.5°N and its circulation encompassing both sides of the equator, Typhoon Vamei was the first observed equatorial tropical cyclogenesis. The development depended on a basic flow relative vorticity that is apparently provided by an abnormally strong and sustained cold surge prior to the formation.

The mechanism for a surge-induced equatorial cyclogenesis may be explained via the linear equatorial wave theory. In the absence of terrain, a mass source at 30°N that simulates the effect of the sea level pressure rise due to the migration of the Siberian high can generate the cold-surge-like northeasterly winds over the South China Sea. After an initial period of gravity wave–type motions with strong northerly winds, the main tropical response takes the form of a Rossby wave group. A pronounced northeast–southwest tilt in this Rossby wave group develops because of the faster westward group velocity of the lower meridional modes relative to the higher meridional modes. The dispersion of the Rossby waves gives rise to a cyclonic shear zone near the equator where the Borneo vortex is often observed. These responses suggest that the equatorward strengthening of northeasterly surge winds and the cyclonic circulation can develop without the effect of terrain and latent heating, and thus may be considered the mechanism to generate and maintain the basic-state relative vorticity in the region where Typhoon Vamei formed.

Whereas cold surges and Borneo vortices are common occurrences in the South China Sea during boreal winter, equatorial formation of tropical cyclones are extremely rare because of the terrain effects that are both necessary for and hinder development. The narrowness of the South China Sea helps to channel and strengthen the cross-equatorial surge winds to produce the background cyclonic vorticity at the equator. However, it is difficult for a preexisting Borneo vortex to remain over the water long enough to develop into a tropical cyclone. The formation of Typhoon Vamei was due to unusually persistent durations of an intense cold surge and a Borneo vortex circulation that remained over open water. The probability of a recurrence of these conditions based on the climatology of cold surge and Borneo vortex behavior has been estimated to be about once in a century.

The Maritime Continent is a region that is strongly influenced by ENSO. Even though the greatest influence occurs in the dry boreal summer and boreal fall transitional season (Haylock and McBride 2001; Hendon 2003; Chang et al. 2004b), various local impacts from ENSO are also found during boreal winter (Chang et al. 2004b; Robertson et al. 2011). How the interactions of the transient motion systems are affected by ENSO, such as whether some effects may be through diurnal cycle (Qian et al. 2010, 2013), are unanswered questions. Effects of longer-term variations, from decadal to climate change scales, on these interactions are also mostly unknown. The skills of current climate models are typically low for the Maritime Continent area, with the simulated monsoon–ENSO connection generally weaker than observed, and no clear model consensus on monsoon rainfall over the second half of the twentieth century (Jourdain et al. 2013).

Another important area for future research concerns the role of the MJO. The interaction between the MJO and cold surges and the Borneo vortex discussed in this chapter relate only to the dynamic structure of the motion systems. Not much is known at this time on the thermodynamic interaction between MJO and these synoptic-scale systems under strong terrain effects over a broad area, and on how the surges and Borneo vortex may affect the MJO. The recent field program of the Cooperative Indian Ocean Experiment on Intraseasonal Variability (CINDY)/DYNAMO (Yoneyama et al. 2013; Zhang 2013) was held in the Indian Ocean to study the initiation of MJO, where the open ocean and small island topography are very different from the complex and steep terrain over the Maritime Continent. Field observation experiments, such as the planned Years of Maritime Continent (YMC), and high-resolution modeling studies will help the understanding of the interaction between MJO and the cold surges and Borneo vortex.

Regarding the rare development of Typhoon Vamei (2001), many mesoscale numerical weather prediction models were able to simulate the equatorial cyclogenesis based on initial conditions that are not obviously dissimilar from those observed on many other December or January days. Numerical experiments to simulate both Typhoon Vamei and the nondevelopment of these other days are needed to elucidate the reasons that Typhoon Vamei was the exception. Longer-term climate model integrations may give an indication as to whether there may be a shortening of the century-scale return period of the equatorial tropical cyclogenesis under climate change scenarios.

Acknowledgments. This review was based on research in the past decade supported by the National Science Foundation, the Office of Naval Research, and the National Science Council of Taiwan, R.O.C. We wish to thank Professors Russ Elsberry and Bob Haney and three anonymous reviewers for their helpful comments.

REFERENCES

Akter, N., and K. Tsuboki, 2012: Numerical simulation of Cyclone Sidr using a cloud-resolving model: Characteristics and formation process of an outer rainband. *Mon. Wea. Rev.*, **140**, 789–810, doi:10.1175/2011MWR3643.1.

Anthes, R. A., 1982: *Tropical Cyclones: Their Evolution, Structure and Effects.* Amer. Meteor. Soc., 208 pp.

Braak, C., 1921: *Het Klimaat van Nederlandsch-Indië. Deel I: Algemeene Hoofdstukken, Verh. Magn. Meteor. Obs. Batavia, Indones.*, No. 8, Javasche Boekhandel & Drukkerij, 787 pp.

——, 1929: *Het Klimaat van Nederlandsch-Indië. Deel II: Local Klimatologie, Verh. Magn. Meteor. Obs. Batavia, Indones.*, No. 8, Javasche Boekhandel & Drukkerij, 802 pp.

Carrera, M. L., and J. R. Gyakum, 2003: Significant events of interhemispheric atmospheric mass exchange: Composite structure and evolution. *J. Climate*, **16**, 4061–4078, doi:10.1175/1520-0442(2003)016<4061:SEOIAM>2.0.CO;2.

——, and ——, 2007: Southeast Asian pressure surges and significant events of atmospheric mass loss from the Northern Hemisphere, and a case study analysis. *J. Climate*, **20**, 4678–4701, doi:10.1175/JCLI4266.1.

Chambers, C. R. S., and T. Li, 2007: Simulation of a near-equatorial typhoon Vamei (2001). *Meteor. Atmos. Phys.*, **98**, 67–80, doi:10.1007/s00703-006-0229-0.

Chan, J. C. L., and C. Y. Li, 2004: The East Asian winter monsoon. *East Asian Monsoon*, C.-P. Chang, Ed., World Scientific Series on Asia-Pacific Weather and Climate, Vol. 2, World Scientific, 54–106.

Chang, C.-P., 1977: Viscous internal gravity waves and low-frequency oscillations in the tropics. *J. Atmos. Sci.*, **34**, 901–912, doi:10.1175/1520-0469(1977)034<0901:VIGWAL>2.0.CO;2.

——, and A. E. Maas Jr., 1976: A case of cross-equatorial displacement of a vortex. *Mon. Wea. Rev.*, **104**, 653–655, doi:10.1175/1520-0493(1976)104<0653:ACOCDO>2.0.CO;2.

——, and K. M. Lau, 1982: Short-term planetary-scale interactions over the tropics and midlatitude during northern winter. Part I: Contrasts between active and inactive periods. *Mon. Wea. Rev.*, **110**, 933–946, doi:10.1175/1520-0493(1982)110<0933:STPSIO>2.0.CO;2.

——, and T. S. Wong, 2008: Rare typhoon development near the equator. *Recent Progress in Atmospheric Sciences: Applications to the Asia-Pacific Region*, K. N. Liou, M. D. Chou, and H. H. Hsu, Eds., World Scientific, 172–181.

——, and M.-M. Lu, 2012: Intraseasonal predictability of Siberian high and East Asian winter monsoon and its interdecadal variability. *J. Climate*, **25**, 1773–1778, doi:10.1175/JCLI-D-11-00500.1.

——, J. E. Erickson, and K. M. Lau, 1979: Northeasterly cold surges and near-equatorial disturbances over the Winter MONEX area during December 1974. Part I: Synoptic aspects. *Mon. Wea. Rev.*, **107**, 812–829, doi:10.1175/1520-0493(1979)107<0812:NCSANE>2.0.CO;2.

——, J. E. Millard, and G. T. J. Chen, 1983: Gravitational character of cold surges during Winter MONEX. *Mon. Wea. Rev.*, **111**, 293–307, doi:10.1175/1520-0493(1983)111<0293:GCOCSD>2.0.CO;2.

——, C. H. Liu, and H. C. Kuo, 2003: Typhoon Vamei: An equatorial tropical cyclone formation. *Geophys. Res. Lett.*, **30**, 1150, doi:10.1029/2002GL016365.

——, P. A. Harr, J. McBride, and H.-H. Hsu, 2004a: Maritime Continent monsoon: Annual cycle and boreal winter variability. *East Asian Monsoon*, C.-P. Chang, Ed., World Scientific, 107–150.

——, Z. Wang, J. Ju, and T. Li, 2004b: On the relationship between western Maritime Continent monsoon rainfall and ENSO during northern winter. *J. Climate*, **17**, 665–672, doi:10.1175/1520-0442(2004)017<0665:OTRBWM>2.0.CO;2.

——, P. A. Harr, and H. J. Chen, 2005a: Synoptic disturbances over the equatorial South China Sea and western Maritime Continent during boreal winter. *Mon. Wea. Rev.*, **133**, 489–503, doi:10.1175/MWR-2868.1.

——, Z. Wang, J. McBride, and C. H. Liu, 2005b: Annual cycle of Southeast Asia—Maritime Continent rainfall and the asymmetric monsoon transition. *J. Climate*, **18**, 287–301, doi:10.1175/JCLI-3257.1.

——, ——, and H. Hendon, 2006: The Asian winter monsoon. *The Asian Monsoon*, B. Wang, Ed., Praxis, 89–127.

——, M. M. Lu, and B. Wang, 2011: The Asian winter monsoon. *The Global Monsoon System: Research and Forecast*, 2nd ed. C. P. Chang et al., Eds., World Scientific Series on Asia-Pacific Weather and Climate, Vol. 5, World Scientific, 99–110.

Cheang, B. K., 1977: Synoptic features and structures of some equatorial vortices over the South China Sea in the Malaysian region during the winter monsoon of December 1973. *Pure Appl. Geophys.*, **115**, 1303–1333, doi:10.1007/BF00874411.

Chu, P.-S., and S.-U. Park, 1984: Regional circulation characteristics associated with a cold surge event over East Asia during winter MONEX. *Mon. Wea. Rev.*, **112**, 955–965, doi:10.1175/1520-0493(1984)112<0955:RCCAWA>2.0.CO;2.

Compo, G. P., G. N. Kiladis, and P. J. Webster, 1999: The horizontal and vertical structure of East Asian winter monsoon pressure surges. *Quart. J. Roy. Meteor. Soc.*, **125**, 29–54, doi:10.1002/qj.49712555304.

Ding, Y. H., 1994: *Monsoons over China*. Kluwer Academic, 419 pp.

Garreaud, R. D., 2001: Subtropical cold surges: Regional aspects and global distribution. *Int. J. Climatol.*, **21**, 1181–1197, doi:10.1002/joc.687.

Gong, D.-Y., and C.-H. Ho, 2004: Intra-seasonal variability of wintertime temperature over East Asia. *Int. J. Climatol.*, **24**, 131–144, doi:10.1002/joc.1006.

——, S. W. Wang, and J. H. Zhu, 2001: East Asian winter monsoon and Arctic Oscillation. *Geophys. Res. Lett.*, **28**, 2073–2076, doi:10.1029/2000GL012311.

Gray, W. M., 1968: Global view of tropical disturbances and storms. *Mon. Wea. Rev.*, **96**, 669–700, doi:10.1175/1520-0493(1968)096<0669:GVOTOO>2.0.CO;2.

Guan, Z., C. Lu, S. Mei, and J. Cong, 2010: Seasonality of inter-annual inter-hemispheric oscillations over the past five decades. *Adv. Atmos. Sci.*, **27**, 1043–1050, doi:10.1007/s00376-009-9126-z.

Haylock, M., and J. McBride, 2001: Spatial coherence and predictability of Indonesian wet season rainfall. *J. Climate*, **14**, 3882–3887, doi:10.1175/1520-0442(2001)014<3882:SCAPOI>2.0.CO;2.

He, J., H. Lin, and Z. Wu, 2011: Another look at influences of the Madden-Julian oscillation on the wintertime East Asian weather. *J. Geophys. Res.*, **116**, D03109, doi:10.1029/2010JD014787.

Hendon, H. H., 2003: Indonesian rainfall variability: Impacts of ENSO and local air–sea interaction. *J. Climate*, **16**, 1775–1790, doi:10.1175/1520-0442(2003)016<1775:IRVIOE>2.0.CO;2.

Holland, G. J., 1984: On the climatology and structure of tropical cyclones in the Australian/southwest Pacific region: II. Hurricanes. *Aust. Meteor. Mag.*, **32**, 17–31.

Hung, C.-W., and M. Yanai, 2004: Factors contributing to the onset of the Australian summer monsoon. *Quart. J. Roy. Meteor. Soc.*, **130**, 739–758, doi:10.1256/qj.02.191.

——, X. Liu, and M. Yanai, 2004: Symmetry and asymmetry of the Asian and Australian summer monsoons. *J. Climate*, **17**, 2413–2426, doi:10.1175/1520-0442(2004)017<2413:SAAOTA>2.0.CO;2.

Jeong, J.-H., B.-M. Kim, C.-H. Ho, and Y.-H. Noh, 2008: Systematic variation in wintertime precipitation in East Asia by MJO-induced extratropical vertical motion. *J. Climate*, **21**, 788–801, doi:10.1175/2007JCLI1801.1.

Johnson, R. H., and J. R. Zimmerman, 1986: Modification of the boundary layer over the South China Sea during a Winter MONEX cold surge event. *Mon. Wea. Rev.*, **114**, 2004–2015, doi:10.1175/1520-0493(1986)114<2004:MOTBLO>2.0.CO;2.

——, and R. A. Houze Jr., 1987: Precipitating cloud systems of the Asian monsoon. *Monsoon Meteorology*, C.-P. Chang and T. N. Krishnamurti, Eds., Oxford University Press, 298–353.

——, and C.-P. Chang, 2007: Winter MONEX: A quarter century and beyond. *Bull. Amer. Meteor. Soc.*, **88**, 385–388, doi:10.1175/BAMS-88-3-385.

Jourdain, N. C., A. S. Gupta, A. S. Taschetto, C. C. Ummenhofer, A. F. Moise, and K. Ashok, 2013: The Indo-Australian monsoon and its relationship to ENSO and IOD in reanalysis data and the CMIP3/CMIP5 simulations. *Climate Dyn.*, **41**, 3073–3102, doi:10.1007/s00382-013-1676-1.

Juneng, L., F. T. Tangang, C. J. Reason, S. Moten, and W. A. W. Hassan, 2007: Simulation of tropical cyclone Vamei (2001) using the PSU/NCAR MM5 model. *Meteor. Atmos. Phys.*, **97**, 273–290, doi:10.1007/s00703-007-0259-2.

Kanamori, H., T. Yasunari, and K. Kuraji, 2013: Modulation of the diurnal cycle of rainfall associated with the MJO observed by a dense hourly rain gauge network at Sarawak, Borneo. *J. Climate*, **26**, 4858–4875, doi:10.1175/JCLI-D-12-00158.1.

Keen, R. A., 1982: The role of cross-equatorial tropical cyclone pairs in the Southern Oscillation. *Mon. Wea. Rev.*, **110**, 1405–1416, doi:10.1175/1520-0493(1982)110<1405:TROCET>2.0.CO;2.

Koh, T.-Y., 2006: Numerical weather prediction research in Singapore and case study of Tropical Cyclone Vamei. *Extended Abstracts, Winter MONEX: A Quarter Century and Beyond, WMO Asian Monsoon Workshop,* Kuala Lumpur, Malaysia, WMO, 3.1.2. [Available online at http://www.ims.nus.edu.sg/activities/npde/files/tykoh.pdf.]

Koseki, S., T.-Y. Koh, and C.-K. Teo, 2013a: Effects of the cold tongue in the South China Sea on the monsoon, diurnal cycle and rainfall in the Maritime Continent. *Quart. J. Roy. Meteor. Soc.*, **139**, 1566–1582, doi:10.1002/qj.2052.

——, ——, and ——, 2013b: Borneo vortex and meso-scale convective rainfall. *Atmos. Chem. Phys. Discuss.*, **13**, 21 079–21 124, doi:10.5194/acpd-13-21079-2013.

Krishnamurthy, V., and R. S. Ajayamohan, 2010: Composite structure of monsoon low pressure systems and its relation to Indian rainfall. *J. Climate*, **23**, 4285–4305, doi:10.1175/2010JCLI2953.1.

Lander, M. A., 1990: Evolution of the cloud pattern during the formation of tropical cyclone twins symmetrical with respect to the equator. *Mon. Wea. Rev.*, **118**, 1194–1202, doi:10.1175/1520-0493(1990)118<1194:EOTCPD>2.0.CO;2.

Lau, K.-M., and P. H. Chan, 1983: Short-term climate variability and atmospheric teleconnections from satellite-observed outgoing longwave radiation. Part II: Lagged correlations. *J. Atmos. Sci.*, **40**, 2751–2767, doi:10.1175/1520-0469(1983)040<2751:STCVAA>2.0.CO;2.

——, and C.-P. Chang, 1987: Planetary scale aspects of winter monsoon and teleconnections. *Monsoon Meteorology*, C.-P. Chang and T. N. Krishnamurti, Eds., Oxford University Press, 161–202.

Lim, H., and C.-P. Chang, 1981: A theory for mid-latitude forcing of tropical motions during winter monsoons. *J. Atmos. Sci.*, **38**, 2377–2392, doi:10.1175/1520-0469(1981)038<2377:ATFMFO>2.0.CO;2.

LinHo, and B. Wang, 2002: The time–space structure of the Asian–Pacific summer monsoon: A fast annual cycle view. *J. Climate*, **15**, 2001–2019, doi:10.1175/1520-0442(2002)015<2001:TTSSOT>2.0.CO;2.

Liu, W. T., 2002: Progress in scatterometer application. *J. Oceanogr.*, **58**, 121–136, doi:10.1023/A:1015832919110.

Loh, W. T., L. Jueng, and F. T. Tangang, 2011: Sensitivity of Typhoon Vamei (2001) simulation to planetary boundary layer parameterization using PSU/NCAR MM5. *Pure Appl. Geophys.*, **168**, 1799–1811, doi:10.1007/s00024-010-0176-z.

Love, G., 1985: Cross-equatorial influence of winter hemisphere subtropical cold surges. *Mon. Wea. Rev.*, **113**, 1487–1498, doi:10.1175/1520-0493(1985)113<1487:CEIOWH>2.0.CO;2.

Lu, M. M., and C. P. Chang, 2009: Unusual late-season cold surges during the 2005 Asian winter monsoon: Roles of Atlantic blocking and the central Asian anticyclone. *J. Climate*, **22**, 5205–5217, doi:10.1175/2009JCLI2935.1.

Madden, R. A., and P. R. Julian, 1972: Description of global-scale circulation cells in the tropics with a 40–50 day period. *J. Atmos. Sci.*, **29**, 1109–1123, doi:10.1175/1520-0469(1972)029<1109:DOGSCC>2.0.CO;2.

Matsumoto, J., 1992: The seasonal changes in Asian and Australian monsoon regions. *J. Meteor. Soc. Japan*, **70**, 257–273.

——, and T. Murakami, 2002: Seasonal migration of monsoons between the Northern and Southern Hemisphere as revealed from equatorially symmetric and asymmetric OLR data. *J. Meteor. Soc. Japan*, **80**, 419–437, doi:10.2151/jmsj.80.419.

Matsuno, T., 1966: Quasigeostrophic motions in the equatorial area. *J. Meteor. Soc. Japan*, **44**, 25–43.

McBride, J. L., 1995: Tropical cyclone formation. *Global Perspective on Tropical Cyclones*, R. L. Elsberry, Ed., WMO Tech. Doc. 693, 63–105.

Meehl, G. A., 1987: The annual cycle and interannual variability in the tropical Pacific and Indian Ocean region. *Mon. Wea. Rev.*, **115**, 27–50, doi:10.1175/1520-0493(1987)115<0027:TACAIV>2.0.CO;2.

Moron, V., A. W. Robertson, and J.-H. Qian, 2010: Local versus large-scale characteristics of monsoon onset and post-onset rainfall over Indonesia. *Climate Dyn.*, **34**, 281–299, doi:10.1007/s00382-009-0547-2.

Neale, R., and J. Slingo, 2003: The Maritime Continent and its role in the global climate: A GCM study. *J. Climate*, **16**, 834–848, doi:10.1175/1520-0442(2003)016<0834:TMCAIR>2.0.CO;2.

Oh, J.-H., K.-Y. Kim, and G.-H. Lim, 2012: Impact of MJO on the diurnal cycle of rainfall over the western Maritime Continent in the austral summer. *Climate Dyn.*, **38**, 1167–1180, doi:10.1007/s00382-011-1237-4.

Ooi, S. H., A. A. Samah, and P. Braesicke, 2011: A case study of the Borneo vortex genesis and its interactions with the global circulation. *J. Geophys. Res.*, **116**, D21116, doi:10.1029/2011JD015991.

Park, T.-W., C.-H. Ho, and S. Yang, 2011: Relationship between the Arctic Oscillation and cold surges over East Asia. *J. Climate*, **24**, 68–83, doi:10.1175/2010JCLI3529.1.

Peatman, S. C., A. J. Mathews, and D. P. Stevens, 2013: Propagation of the Madden–Julian Oscillation through the Maritime Continent and scale interaction with the diurnal cycle of precipitation. *Quart. J. Roy. Meteor. Soc.*, **140**, 814–825, doi:10.1002/qj.2161.

Qian, J.-H., A. W. Robertson, and V. Moron, 2010: Interactions between ENSO, monsoon and diurnal cycle in rainfall variability over Java, Indonesia. *J. Atmos. Sci.*, **67**, 3509–3523, doi:10.1175/2010JAS3348.1.

——, ——, and ——, 2013: Diurnal cycle in different weather regimes and rainfall variability over Borneo associated with ENSO. *J. Climate*, **26**, 1772–1790, doi:10.1175/JCLI-D-12-00178.1.

Ramage, C. S., 1968: Role of a tropical "maritime continent" in the atmospheric circulation. *Mon. Wea. Rev.*, **96**, 365–369, doi:10.1175/1520-0493(1968)096<0365:ROATMC>2.0.CO;2.

——, 1971: *Monsoon Meteorology*. Academic Press, 296 pp.

Rauniyar, S. P., and K. J. E. Walsh, 2011: Scale interaction of the diurnal cycle of rainfall over the Maritime Continent and Australia: Influence of the MJO. *J. Climate*, **24**, 325–348, doi:10.1175/2010JCLI3673.1.

Robertson, A. W., V. Moron, J.-H. Qian, C.-P. Chang, F. Tangang, E. Aldrian, T. Y. Koh, and L. Juneng, 2011: The Maritime Continent monsoon. *The Global Monsoon System: Research and Forecast*, 2nd ed. C. P. Chang et al., Eds., World Scientific Series on Asia-Pacific Weather and Climate, Vol. 5, World Scientific, 85–98.

Simpson, J., C. Kummerow, W.-K. Tao, and R. F. Adler, 1996: On the Tropical Rainfall Measuring Mission. *Meteor. Atmos. Phys.*, **60**, 19–36, doi:10.1007/BF01029783.

Sontakke, N. A., G. B. Pant, and N. Singh, 1993: Construction of all-India summer monsoon rainfall series for the period 1844–1991. *J. Climate*, **6**, 1807–1811, doi:10.1175/1520-0442(1993)006<1807:COAISM>2.0.CO;2.

Takahashi, H. G., 2012: Orographic low-level clouds of Southeast Asia during the cold surges of the winter monsoon. *Atmos. Res.*, **131**, 22–33.

Takaya, K., and H. Nakamura, 2005a: Geographical dependence of upper-level blocking formation associated with intraseasonal amplification of the Siberian high. *J. Atmos. Sci.*, **62**, 4441–4449, doi:10.1175/JAS3628.1.

——, and ——, 2005b: Mechanisms of intraseasonal amplification of the cold Siberian high. *J. Atmos. Sci.*, **62**, 4423–4440, doi:10.1175/JAS3629.1.

Tangang, F. T., L. Juneng, and C. J. Reason, 2007: MM5 simulated evolution and structure of Typhoon Vamei (2001). *Solid Earth, Ocean Science & Atmospheric Science*, Y.-T. Chen, Ed., Advances in Geosciences, Vol. 9, World Scientific, 191–207.

——, E. Salimun, P. N. Vinayachandran, Y. K. Seng, C. J. C. Reason, S. K. Behera, and T. Yasunari, 2008: On the roles of the northeast cold surge, the Borneo vortex, the Madden-Julian Oscillation, and the Indian Ocean Dipole during the extreme 2006/2007 flood in southern peninsular Malaysia. *Geophys. Res. Lett.*, **35**, L14S07, doi:10.1029/2008GL033429.

Trilaksono, N. J., S. Otsuka, and S. Yoden, 2012: A time-lagged ensemble simulation on the modulation of precipitation over West Java in January–February 2007. *Mon. Wea. Rev.*, **140**, 601–616, doi:10.1175/MWR-D-11-00094.1.

Virts, K. S., J. M. Wallace, M. L. Hutchins, and R. H. Holzworth, 2013: Diurnal lightning variability over the Maritime Continent: Impact of low-level winds, cloudiness, and the MJO. *J. Atmos. Sci.*, **70**, 3128–3146, doi:10.1175/JAS-D-13-021.1.

Vissa, N. K., A. N. V. Satyanarayana, and B. Prasad Kumar, 2013: Impact of South China Sea cold surges and Typhoon Peipah on initiating Cyclone Sidr in the Bay of Bengal. *Pure Appl. Geophys.*, **170**, 2369–2381, doi:10.1007/s00024-013-0671-0.

Wang, Z., and C.-P. Chang, 2008: Mechanism of the asymmetric monsoon transition as simulated in an AGCM. *J. Climate*, **21**, 1829–1836, doi:10.1175/2007JCLI1920.1.

Wheeler, M. C., and H. H. Hendon, 2004: An all-season Real-time Multivariate MJO index: Development of an index for monitoring and prediction. *Mon. Wea. Rev.*, **132**, 1917–1932, doi:10.1175/1520-0493(2004)132<1917:AARMMI>2.0.CO;2.

Wu, B., and J. Wang, 2002: Possible impacts of winter Arctic Oscillation on Siberian high, the East Asian winter monsoon and sea-ice extent. *Adv. Atmos. Sci.*, **19**, 297–320, doi:10.1007/s00376-002-0024-x.

Wu, C.-H., and H.-H. Hsu, 2009: Topographic influence on the MJO in the Maritime Continent. *J. Climate*, **22**, 5433–5448, doi:10.1175/2009JCLI2825.1.

Wu, M. C., and J. C. L. Chan, 1995: Surface features of winter monsoon surges over south China. *Mon. Wea. Rev.*, **123**, 662–680, doi:10.1175/1520-0493(1995)123<0662:SFOWMS>2.0.CO;2.

Yanai, M., and T. Tomita, 1998: Seasonal and interannual variability of atmospheric heat sources and moisture sinks as determined from NCEP–NCAR reanalysis. *J. Climate*, **11**, 463–482, doi:10.1175/1520-0442(1998)011<0463:SAIVOA>2.0.CO;2.

Yang, S., K.-M. Lau, and K.-M. Kim, 2002: Variations of the East Asian jet stream and Asian–Pacific–American winter climate anomalies. *J. Climate*, **15**, 306–325, doi:10.1175/1520-0442(2002)015<0306:VOTEAJ>2.0.CO;2.

Yasunari, T., 1991: The monsoon year—A new concept of the climatic year in the tropics. *Bull. Amer. Meteor. Soc.*, **72**, 1331–1338, doi:10.1175/1520-0477(1991)072<1331:TMYNCO>2.0.CO;2.

Yoneyama, K., C. Zhang, and C. N. Long, 2013: Tracking pulses of the Madden–Julian oscillation. *Bull. Amer. Meteor. Soc.*, **94**, 1871–1891, doi:10.1175/BAMS-D-12-00157.1.

Zhang, C., 2005: Madden-Julian Oscillation. *Rev. Geophys.*, **43**, RG2003, doi:10.1029/2004RG000158.

——, 2013: The CINDY/DYNAMO field campaign: Advancing our understanding and prediction of MJO initiation. *Extended Abstracts, Fifth WMO Int. Workshop on Monsoons*, Macao, China, WMO, 294–298.

Zhang, Y., K. R. Sperber, and J. S. Boyle, 1997: Climatology and interannual variation of the East Asian winter monsoon: Results from the 1979–95 NCEP/NCAR reanalysis. *Mon. Wea. Rev.*, **125**, 2605–2619, doi:10.1175/1520-0493(1997)125<2605:CAIVOT>2.0.CO;2.

Chapter 7

Impacts of the Tibetan Plateau on Asian Climate

GUOXIONG WU AND YIMIN LIU

State Key Laboratory of Numerical Modeling for Atmospheric Sciences and Geophysical Fluid Dynamics (LASG), Institute of Atmospheric Physics, Chinese Academy of Sciences, Beijing, China

ABSTRACT

Professor Yanai is remembered in our hearts as an esteemed friend. Based on his accomplishments in tropical meteorology and with his flashes of insight he led his group at the University of California, Los Angeles, in the 1980s and 1990s to explore the thermal features of the Tibetan Plateau (TP) and its relation to the Asian monsoon, and he brought forward the TP meteorology established by Ye Duzheng et al. in 1957 to a new stage. In cherishing the memory of Professor Yanai and his great contribution to the TP meteorology, the authors review their recent study on the impacts of the TP and contribute this chapter as an extension of their chapter titled "Effects of the Tibetan Plateau" published by Yanai and Wu in 2006 in the book *The Asian Monsoon.*

The influence of a large-scale orography on climate depends not only on the mechanical and thermal forcing it exerts on the atmosphere, but also on the background atmospheric circulation. In winter the TP possesses two leading heating modes resulting from the relevant dominant atmospheric circulations, in particular the North Atlantic Oscillation and the North Pacific Oscillation. The prevailing effect of the mechanical forcing of the TP in wintertime generates a dipole type of circulation, in which the anticyclonic gyre in the middle and high latitudes contributes to the warm inland area to the west, and the cold seashore area to the east, of northeast Asia, whereas the cyclonic gyre in low latitudes contributes to the formation of a prolonged dry season over central and southern Asia and moist climate over southeastern Asia. Such a dipole circulation also generates a unique persistent rainfall in early spring (PRES) over southern China.

In 1980s, Yanai and his colleagues analyzed the in situ observation and found that the constant potential temperature boundary layer over the TP can reach about 300 hPa before the summer monsoon onset. This study supports these findings, and demonstrates that such a boundary layer structure is a consequence of the atmospheric thermal adaptation to the surface sensible heating, which vanishes quickly with increasing height. The overshooting of rising air, which is induced by surface sensible heating, then can form a layer of constant potential temperature with a thickness of several kilometers.

The thermal forcing of the TP on the lower tropospheric circulation looks like a sensible heat–driven air pump (SHAP). It is the surface sensible heating on the sloping sides of the plateau that the SHAP can effectively influence the Asian monsoon circulation. In spring the SHAP contributes to the seasonal abrupt change of the Asian circulation and anchors the earliest Asian summer monsoon onset over the eastern Bay of Bengal. In summer, this pumping, together with the thermal forcing over the Iranian Plateau, produces bimodality in the South Asian high activity in the upper troposphere, which is closely related to the climate anomaly patterns over South and East Asia. Because the isentropic surfaces in the middle and lower troposphere intersect with the TP, in summertime the plateau becomes a strong negative vorticity source of the atmosphere and affects the surrounding climate and even the Northern Hemispheric circulation via Rossby wave energy dispersion. Future prospects in related TP studies are also addressed.

Corresponding author address: Guoxiong Wu, Institute of Atmospheric Physics, LASG, 40 HuayanBeili, Chaoyang District, Beijing, 9804, China.
E-mail: gxwu@lasg.iap.ac.cn

DOI: 10.1175/AMSMONOGRAPHS-D-15-0018.1

1. Introduction

Monsoons are generally considered an atmospheric response to seasonal changes in land–sea thermal contrast induced by the annual cycle of the solar zenith angle (Wallace and Hobbs 1977; Holton 2004). The Asian summer monsoon (ASM) is the strongest element of the global monsoon system (Flohn 1957; Chang 2004; Wang 2006). In addition to land–sea thermal contrast, it is affected by large-scale mountain ranges such as the Tibetan Plateau (TP) (Yeh et al. 1957; Ye and Gao 1979; Yanai et al. 1992; Yanai and Wu 2006). The TP, also called the Qinghai-Xizang Plateau in China, extends over the area of 27°–45°N, 70°–105°E, covering a region about one-quarter the size of the entire Chinese territory. Its mean elevation is more than 4000 m above sea level, with the peak of Mount Everest at 8844 m (near 300 hPa) standing on its southern fringe. In winter the TP together with the Iranian Plateau serves as a giant wall across almost the whole Eurasian continent that blocks cold outbreaks from the north and confines the winter monsoon to eastern and southern Asia (Chang et al. 2006).

Yanai and Wu (2006) gave a thorough review of the past studies about the effects of the TP. The review starts from the research in the 1950s on the jet stream and the warm South Asian high, and the early progress of TP research. The review then goes over studies concerning the mechanical effects of the TP on large-scale motion, the winter cold surge, and the summer negative vorticity source over the TP. The review also covers the importance of the thermal influences of the TP on the seasonal circulation transition and Asian monsoon onset based on different datasets and numerical experiments (Ye and Gao 1979; Tao and Chen 1987; Wu et al. 1997, 2002; X. Liu et al. 2001; Y. Liu et al. 2001; Liu et al. 2002; Mao et al. 2002a,b; Wang and LinHo 2002). This paper is an effort to review the evaluation of the heating source on the TP through the analyses of spatial and temporal distributions of $\langle Q_1 \rangle$ and $\langle Q_2 \rangle$ in the notation used by Yanai et al. (1973) and based on observations from the First GARP Global Experiment (FGGE) (December 1978–November 1979) and the Qinghai–Xizang Plateau Meteorology Experiment (QXPMEX), conducted from May to August 1979 by Chinese meteorologists (Zhang et al. 1988).

Since then, great efforts have been made to understand the mechanism concerning how the TP forcing, either mechanical or thermodynamical, can affect the regional as well as global climate. Some of the results are summarized in this study, and can be considered as a complement to the review of Yanai and Wu (2006). Both the diagnosis and numerical experiments are used to get new insights into our understanding.

The remainder of this chapter is organized as follows. In section 2, there is an analysis of the relative importance of mechanical and thermal forcing induced by large-scale mountains. The diabatic heating characteristics of atmosphere over TP, the heating variability, and its impacts in winter are given in section 3. Section 4 focuses on how the mechanical forcing of the TP in late winter and early spring contributes to the occurrence of the persistent rainfall in early spring (PRES) over southern China and the early Asian monsoon onset. Section 5 discusses how large-scale topography affects the atmospheric vertical and horizontal circulations in summer based on idealized sensitivity experiments. Section 6 describes how the summertime continental forcing over Eurasia and the local forcing due to the TP work together to intensify the monsoon in East Asia and the dry climate in the Middle Asia, and how another plateau, the Iranian Plateau (IP), influences the Asian monsoon formation. Section 7 describes how the climbing and deflecting flow induced by topography influence the ASM and its configuration. A brief description of the change of the thermal forcing over the TP and a discussion on its impacts on the summer pattern of the eastern Asian precipitation are presented in section 8. Perspectives on future study are given in section 9.

2. Relative importance of mechanical and thermal forcing induced by large-scale mountains

The response of the atmospheric circulation to a thermal forcing could be considered as the response to a topography with a so-called equivalent mountain height H_Q (Held 1983), which is inversely proportional to the intensity of the basic flow u. For stationary waves forced by a zonal wavenumber-1 heating, Chen (2001) calculated H_Q corresponding to different zonal flows and found that H_Q is about 1 km for $u = 20\,\mathrm{m\,s^{-1}}$ and about 40 km for $u = 2\,\mathrm{m\,s^{-1}}$. For $u < 10\,\mathrm{m\,s^{-1}}$, the height of the equivalent topography is much larger than that of the real topography. Figure 7-1 shows the latitude–height cross section along 90°E of the climate-mean zonal wind in winter and summer months. In winter, the westerly jet of more than $40\,\mathrm{m\,s^{-1}}$ is just over the TP, whereas in summer the TP is located between easterlies to its south and westerlies to its north. These imply that mechanical forcing of the TP should be dominant in winter, whereas its heating is more important than topography in forcing the summer stationary waves in the subtropics.

a. Mechanical forcing

Figure 7-2 summarizes the linear atmospheric response to pure large-scale mechanical forcing (Wu 1984). For a

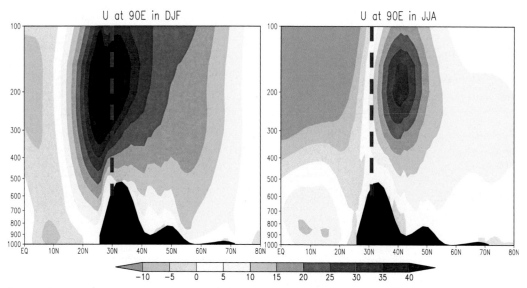

FIG. 7-1. Latitude–height cross section along 90°E of the climate-mean zonal wind (m s^{-1}) in (left) winter and (right) summer. The vertical dashed purple line shows the location of the TP.

barotropic atmosphere without friction, westerly flow climbing over a smaller (planetary scale) mountain will produce a ridge (trough) over the mountain ridge. The existence of friction always causes a ridge (trough) on the windward (lee) side of a mountain. For a baroclinic atmosphere without friction, westerly flow climbing over a smaller-scale mountain will produce a ridge over the mountain with its amplitude quickly decreasing

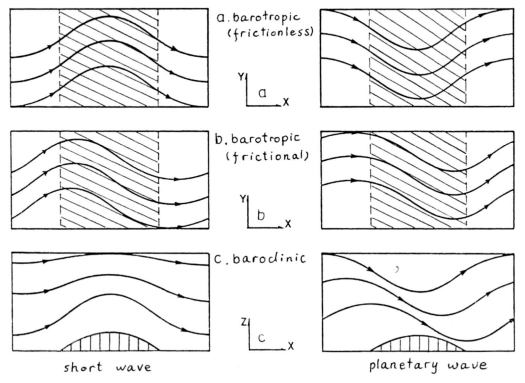

FIG. 7-2. Response of the atmosphere to large-scale mechanical forcing for different mountain scales under different atmospheric conditions for (left) short and (right) planetary waves: (a) frictionless barotropic, (b) frictional barotropic, and (c) baroclinic. The oblique lines in (a) and (b) and the vertical lines in (c) indicate where mountains are located [from Wu (1984)].

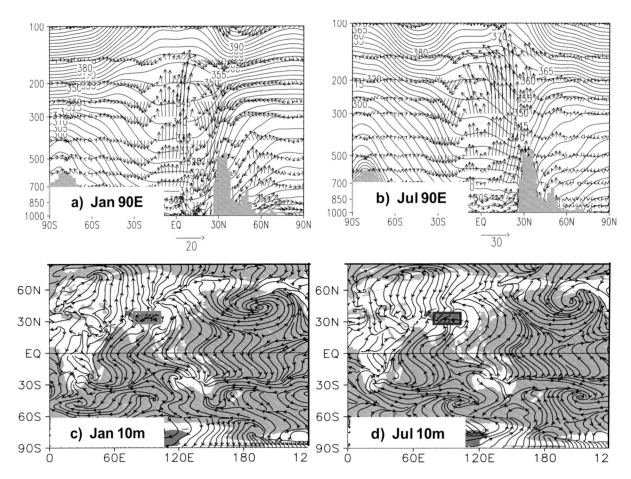

FIG. 7-3. TP-SHAP: The cross sections along 90°E of the (a) January and (b) July mean potential temperature (K) and circulation (vectors, m s^{-1}) projected on the cross sections. The deviations from the annual mean of the (c) January and (d) July mean streamlines at 10 m above the surface calculated from the NCEPII reanalysis for the period 1979–98. The blue box in (c) and the red box in (d) indicate the location of the TP [from Wu et al. (2007)].

vertically, whereas when climbing over a planetary-scale mountain, the flow will generate a westward tilting planetary mountain wave with wave energy propagating upward (Charney and Drazin 1961). On the other hand, the nonlinear atmospheric response to pure mechanical forcing can produce either climbing or deflecting flows depending on its height. Subjected to the constrains of energy and angular momentum conservations, a critical mountain height H_c, which is from about several hundred to one thousand meters, can be derived (Wu 1984). If the height of a mountain is higher than H_c, airflow impinging upon a mountain will be deflected around the mountain, whereas for a mountain lower than H_c, airflow can climb over and produce air descent on the lee side.

b. Thermal forcing: TP sensible heat–driven air pump

Figures 7-3a and 7-3b show the cross sections along 90°E of the monthly mean potential temperature and wind vector projected on the cross sections for January

and July, respectively, in which the vertical velocity w has been multiplied by 10^3. In January, cold temperatures over continental areas and warm temperatures over ocean are prominent. Atmospheric cooling indicated by air descent ($\mathbf{V} \cdot \nabla \theta < 0$) prevails in the free troposphere over continents. The strongest cooling is over the TP and its southern slope, with the strong heating and ascent over the equator in contrast. During summer, the warmest center of potential temperature is just above the TP. This is in agreement with the existence of the warm temperature center in July over the plateau (Yanai et al. 1992). Strong ascent prevails in the area from east of the TP to eastern China (figure not shown) and to its south from the TP to the Bay of Bengal (BOB) (Fig. 7-3b), penetrating the isentropic surfaces upward almost perpendicularly and indicating the existence of a strong heating source over the area in summer. Figures 7-3c and 7-3d show, respectively, the deviations from the annual mean of the January-mean and July-mean streamlines 10 m above the surface calculated from the NCEP-II reanalysis for the period 1979–98. It is evident that deviating air currents flow from the winter hemisphere into the summer hemisphere, diverging out of continents and converging toward oceans in midlatitude areas in the winter hemisphere, but converging toward the continents and diverging out of oceans in the summer hemisphere. A remarkable circulation reversal appears over the Asian–Australian (A–A) monsoon area. In January the surface air diverges out of the TP region toward South Africa and Australia, whereas in July the surface air is converged into the TP region from South Africa and Australia, characterizing the seasonal reversal of the A–A monsoon flows.

In summary, cold and dry air in winter descends over the TP then diverges toward the surrounding areas, whereas in summer moist and warm air converges from the surroundings, particularly from the Indian and western Pacific Oceans, toward the TP region where it is uplifted to the free atmosphere to form cloud and monsoon precipitation. The whole process repeats from year to year and acts like a huge air pump. Since the work done by the air pump is basically driven by the surface sensible heating of the TP, it was named the TP sensible heat–driven air pump (TP-SHAP).

c. TP-forced stationary waves and the associated climate

Figure 7-4 depicts the winter- and summer-mean distributions of potential temperatures and the stream field composed of zonal deviation winds at 850 hPa. In winter months the topography retards the impinging midlatitude winds and deflects the westerly from zonal to circumcolumnar flow (Fig. 7-4a). The deviation streamlines depict an asymmetric dipole shape with an anticyclone to the north of the TP and a cyclone to its south. The streamlines then converge over eastern China and go into its eastern pole but come out of its western pole and diverge over central Asia. The anticyclonic deviation circulation gyre in high latitudes transports warm air northward to its west but cold air southward to its east. As a result, the isotherms in the high latitudes of Asia tilt from northwest to southeast, and the temperature at 130°E is colder than that at 50°E by 10 K at 40°N and 14 K at 50°N. On the other hand, the cyclonic deviation circulation gyre in low latitudes transports dry air southward to South Asia subcontinent, but moist air northward to the Indochina Peninsula and southern China. As a result, a prolonged dry season in South Asia and persistent rainy season in Southeast Asia and southern China are observed before the Asian monsoon onset. In the summer months, the strong TP heating excites a huge cyclonic deviation circulation over East

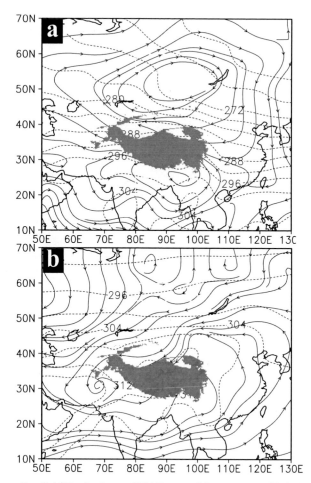

FIG. 7-4. Distributions at 850-hPa potential temperature (dashed lines; K) and the stream fields (lines with arrows) composed of wind deviations from the corresponding zonal means based on the NCEP–NCAR reanalysis for 1968–97. (a) Winter average between December and February. (b) Summer average between June and August [from Wu et al. (2007)].

Asia, and the strong pumping of the TP-SHAP makes the surrounding flows converge into the TP area. Therefore, the summer pattern of the deviation stream field at 850 hPa resembles a huge cyclonic spiral and the TP-SHAP looks like a spiral pump. In fact, the summer TP is an important genesis location of vortices that can propagate eastward and result in torrential rain along the Yangtze River.

3. Thermal characteristics over the TP in winter and their implications

There are three kinds of atmospheric diabatic heating: diffusive sensible heating including surface sensible heating; condensation latent heating, including both large-scale condensation heating and deep convection condensation heating; and radiative heating, which

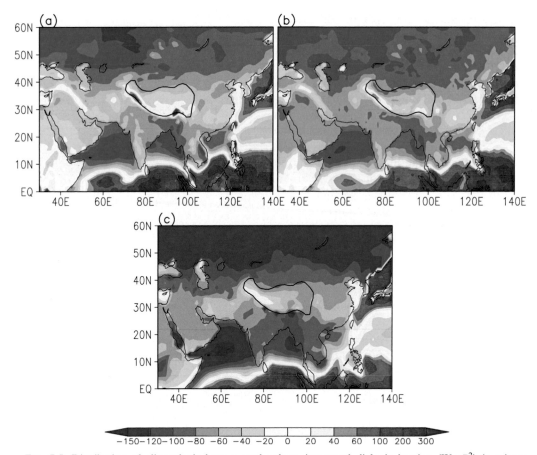

FIG. 7-5. Distribution of climatological-mean total column-integrated diabatic heating $(W\,m^{-2})$ in winter (December–February; the 3000 m orographic height is depicted by the bold black contour): (a) ERA-40, (b) JRA-25, and (c) NCEP2 [from Yu et al. (2011a)].

includes positive shortwave radiative heating and negative longwave radiative cooling. Usually the heating data in reanalyses are not as reliable as those of the temperature and wind. However, such heating data from ECMWF and NCEP have been gauged by the extent of their dynamical consistency with the large-scale circulation and it was shown that they are reasonably consistent estimates (Nigam 1994, 1997). Here we will use the above three kinds of heating to investigate the thermal status over the TP.

a. Climatology of the TP thermal status

The diabatic heating characteristics of the atmosphere over the TP in wintertime was analyzed by using three kinds of reanalysis data of diabatic heating. The strong column-integrated heating is observed over the southwestern flank of the TP, equatorial Indian Ocean, and western Pacific as well as the storm track regions over the northwestern Pacific in winter (Fig. 7-5). They are predominantly due to condensational latent heat release that accompanies the precipitation when westerly climbs

up the TP, the ITCZ rains and the storm precipitation (Figs. 7-6a,b), and to strong surface sensible heating (Figs. 7-6c,d) from warmer ocean to the east coast of China. There is diabatic cooling over the Asian continent and tropical Indian Ocean and South China Sea (Fig. 7-5). The stronger cooling in high latitudes results from the remarkable surface cooling (Figs. 7-6c,d), and weaker solar radiation (Figs. 7-6g,h), whereas the cooling over the north Indian Ocean is mainly caused by the stronger longwave radiation (Figs. 7-6e,f). Although there are some quantitative differences among different reanalysis data, three kinds of diabatic heating data basically yield similar climatological distributions (Yu et al. 2011a).

b. Interannual variation of wintertime diabatic heating and its impacts

EOF analysis shows that the leading mode (EOF1) of interannual variation of the total column-integrated diabatic heating in January over the TP shows the most remarkable variation of the diabatic heating concentrates

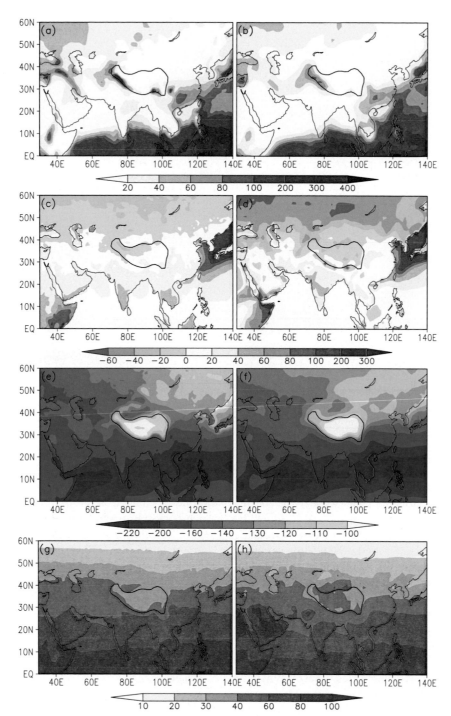

FIG. 7-6. Distribution of climatological mean column-integrated diabatic heating component in winter (W m^{-2}): (a),(b) latent heating, (c),(d) sensible heating, (e),(f) longwave radiative heating, and (g),(h) shortwave radiative heating, from (left) JRA-25 and (right) NCEP2 [from Yu et al. (2011a)].

on its west and southeast (Fig. 7-7), a pattern fairly close to that of the climate-mean total heating as shown in Fig. 7-5. The three reanalysis datasets produce similar spatial and temporal variation of the heating. Data

diagnosis and numerical modeling indicate that the interannual variation of this heating is closely related with the corresponding upstream westerly, just to the southwest side of the TP. Because circulation data are more

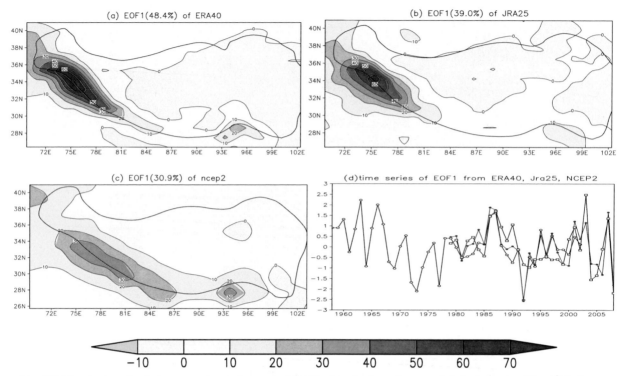

FIG. 7-7. Distribution of main interannual variation mode (EOF1) of total column-integrated diabatic heating (W m^{-2}) in January over the TP (the orographic height of 3000 m is depicted by the black contour) from analysis of (a) ERA-40 (1958–2002), (b) JRA-25 (1979–2008), and (c) NCEP (1979–2008) [from Yu et al. (2011b)]. (d) The corresponding PCs; black, red, and blue curves are for ERA-40, JRA-25, and NCEP, respectively.

reliable compared with diabatic heating data, a westerly index (the West Index) representing the TP heating status is thus defined as the mean westerly at 700 hPa averaged over the upstream region (20°–30°N, 40°–70°E) (Yu et al. 2011b). The West Index is then used to identify the circulation associated with the TP heating variation. Figure 7-8 demonstrates the regression of the West Index with circulations in different layers. Different from summer, there appear consistent cyclonic (or anticyclonic) anomalous circulations over the TP from the lower troposphere to the upper troposphere when the column-integrated diabatic heating over the TP is above (or below) normal, demonstrating an equivalent-barotropic abnormal structure (Figs. 7-8a–c). When the upstream westerly to the southwest of the TP is getting stronger, cyclonic anomalous circulation in the northwest of TP develops due to topography blocking, the detouring flow around the south side of TP becomes stronger, and the cyclonic anomalous circulation around TP also develops (Yu et al. 2011b). Consequently, these lead to the increase of precipitation, latent heating, and total heating over the west and southeast of the TP. It becomes apparent that the abnormality of diabatic heating in winter over the TP is the result of atmospheric circulation abnormality and the mechanical effect of TP.

It is demonstrated that the heating over the TP in winter is a local response to a specific abnormal mode of atmospheric circulation in the Northern Hemisphere, which is closely related with the teleconnection wave pattern oriented from northwest to southeast in the Northern Hemisphere. Nevertheless, the diabatic heating anomalies over the Tibetan Plateau was reported to be closely related to the extreme heavy snow and cold event in January 2008 over southern China as shown in Fig. 7-8d (Bao et al. 2010; Li et al. 2011).

4. TP forcing in spring and Asian summer monsoon onset

a. Persistent rains in early spring over southern China

Besides the well-known Asian monsoon, another less-known unique feature of Asian rainfall is the occurrence of persistent rains in early spring (PRES) over southern China. From late February to early May (pentads 12–26) before the onset of the Asian monsoon, persistent rains occur over a large area south of the Yangtze River. However, the formation mechanism is still unclear. Tian and Yasunari (1998) proposed a mechanism of time lag in the spring warming between land and sea to explain

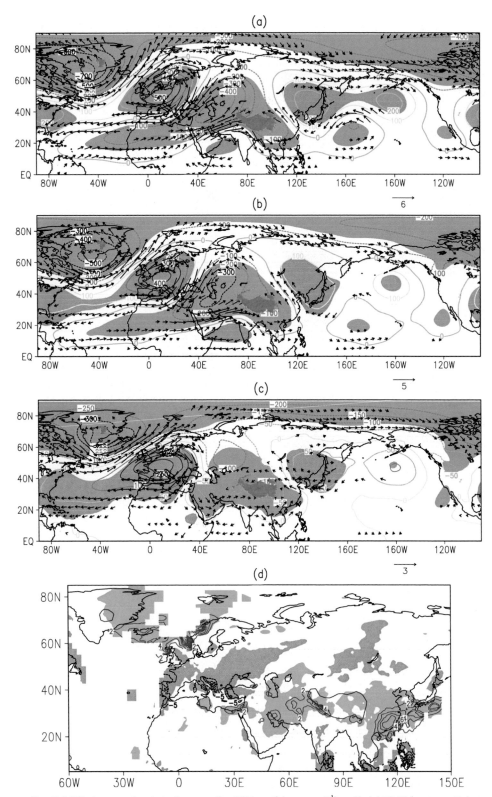

FIG. 7-8. Wind vectors (statistically exceeding 95% confidence; m s^{-1}) and height field (contours; Pa) at (a) 200, (b) 500, and (c) 850 hPa. (d) Land precipitation regressed on the West Index from ERA-40 [contours with red (blue) are positive (negative); shading indicates correlation coefficients statistically exceeding 95% confidence. In (a)–(c) orange (blue) shading shows positive (negative) correlation; in (d) green (red) shading shows positive (negative) correlation [from Yu et al. (2011b)].

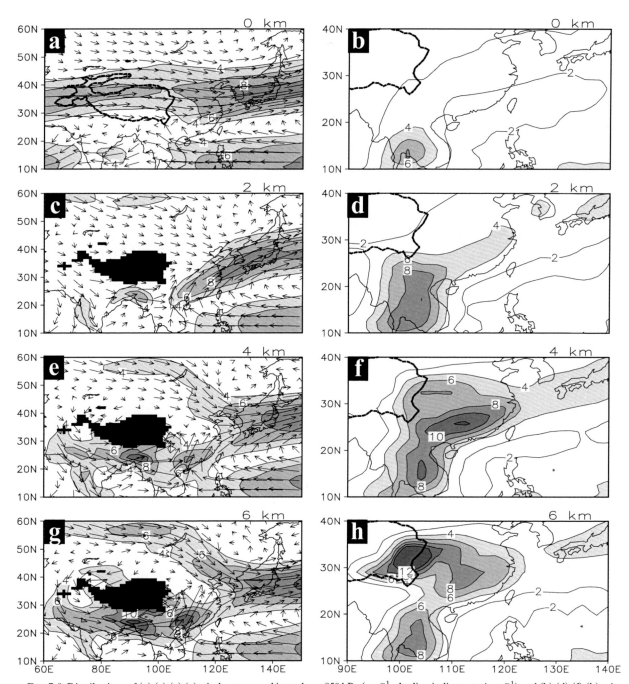

FIG. 7-9. Distributions of (a),(c),(e),(g) wind vectors and isotachs at 850 hPa (m s^{-1}; shading indicates >4 m s^{-1}) and (b),(d),(f),(h) rain (mm day^{-1}) in the perpetual spring sensitivity experiments with (top to bottom) different TP elevations and averaged over 30 months. The black shading in the left panels and the bolder solid curve in the right panels are the main part of TP. The TP maximum elevation is 0 km in (a),(b), 2 km in (c),(d), 4 km in (e),(f), and 6 km in (g),(h) [from Wan and Wu (2007)].

the formation of the PRES. However, a recent study of Wan and Wu (2007) shows that such a heating time-lag mechanism also exists between Mexico and the western North Atlantic in the same period, but there is no PRES in North America. It turns out that the time-lag mechanism is a necessary condition but not sufficient for the occurrence of the PRES.

As demonstrated in Fig. 7-4a, before the monsoon onset the TP can generate a dipole-type stationary circulation pattern. The deflected flow brings cold air from the north and moist air from south, which then converge over eastern China. This may contribute to the formation of PRES. To verify this hypothesis, a series of numerical experiments were conducted by Wan and Wu (2007) and

Wu et al. (2007). To focus on physical mechanisms, perpetual April integrations are designed in the following experiments. The solar angle is set to 10 April. Figure 7-9 shows the wind vector at 850 hPa and rainfall in the experiments when the elevation of the TP is leveled at 0, 2, 4, and 6 km, respectively (Wu et al. 2007).

In the "no TP" experiment, the westerly jet in Eurasia does not split, and a zonally oriented subtropical anticyclone belt dominates in the middle to low latitudes (Fig. 7-9a). Significant rainfall appears only in the southwest of the Indochina Peninsula, in association with the easterly perturbation forced by land–sea thermal contrast (Fig. 7-9b). After the elevation of the TP reaches 2 km, the westerly jet starts to split into its northern and southern branches (Fig. 7-9c). They meet downstream of the TP, forming a strong Asian jet. The belt of subtropical anticyclone breaks, and a prominent anticyclonic center appears over the western Pacific. Therefore the distribution of rainfall takes the shape of PRES (Fig. 7-9d). In the 4-km TP experiment, the split westerly jets on both north and south sides of the TP are strengthened, the circulation pattern (Fig. 7-9e) is already close to the simulation counterpart in the control simulation in which the seasonal variation is included (Wu et al. 2007), and the PRES appears (Fig. 7-9f). In the 6-km TP experiment, the wintertime TP dipole circulation becomes stronger compared to the control simulation. As a result, the PRES is weakened, and the main rainfall center moves to the eastern TP (Fig. 7-9h) where the convergent entrance pole of the TP dipole in the wintertime deviation streamline pattern is located.

The above experiments prove that PRES are formed efficiently due to the deflecting effects of the TP. When the deflected flow—the cold air from the north and the warm and moist air from the south—meet downstream of the TP PRES are formed. The experiments further demonstrate that the intensity and location of the PRES are to some extent influenced by the elevation of the TP.

b. Asian summer monsoon onset

Wu and Zhang (1998) showed that it is due to mechanical as well as thermal forcing of the TP that the onset of the Asian summer monsoon is composed of three consequential stages. The earliest is over the region from the eastern BOB to the western Indochina Peninsula in early May. The Bay of Bengal monsoon onset creates favorable conditions for the South China Sea (SCS) monsoon onset in mid-May (Liu et al. 2002b). This leads to great changes in both large-scale circulation and diabatic heating over Asia. Finally, the onset of the Indian monsoon appears in early June. Since then, many studies have been conducted to determine the underlying mechanism. A numerical simulation based on idealized land–sea distribution developed by Liang et al. (2005) proved that TP heating in late spring greatly intensifies the southern branch of the wintertime dipole of the 850-hPa stream field. This intensified southern branch of the dipole enhances the southerlies to the southeast of Tibet and brings heavy rainfall and more latent heating over the eastern Bay of Bengal and to its east, resulting in Asian summer monsoon onset over the eastern BOB (Fig. 7-10a). It also produces prevailing northerlies to the southwest of Tibet, resulting in less rainfall and more sensible heating over the Indian subcontinent. If the TP is moved westward by 30° in longitude to the north of the Arabian Sea, then the onset site of the Asian monsoon is also moved westward by 30° (Fig. 7-10b). It is evident that the TP anchors the onset site of the Asian summer monsoon to a place over the eastern BOB, just to the southeast of the TP. It is also found that stronger surface sensible heating on the TP in spring leads to an earlier seasonal transition in eastern Asia (Duan and Wu 2005; Mao and Duan 2005).

Commonly, in the lower troposphere the monsoon onset is preceded by the development of a monsoon onset vortex (MOV) (Krishnamurti 1981). The northward moving of the onset vortex can lead to the break of the ridgeline of subtropical anticyclone over the eastern BOB region in the lower troposphere and activate the India–Burma trough, resulting in the BOB summer monsoon onset. Case studies (Wu et al. 2011, 2012a) demonstrate that the formation of the BOB monsoon onset vortex is a consequence of in situ air–sea interaction modulated by the land–sea thermal contrast in South Asia and TP forcing in spring, which can be interpreted schematically by Fig. 7-11: in spring the dominant cold northwesterly over India that is induced by the TP forcing generates strong surface sensible heating and cyclonic circulation. The resultant southwesterly over the northwestern BOB together with the seasonal near-equator westerly forms a steady anticyclone circulation to the north of the BOB and cyclone circulation to its south (Fig. 7-11a) and produces offshore ocean current as well as upwelling in the western BOB. Warm surface water is transported eastward and accumulated on the eastern BOB (Fig. 7-11b). More importantly, the strong downward shortwave radiation over the northern BOB under the anticyclone condition heats the surface water with a thin thermocline of less than 20 m and the SST rises fast over the central and northeastern BOB, forming a strong spring BOB warm pool there (Figs. 7-11b,c). On the southern rim of the warm pool, stronger surface sensible heating from ocean to atmosphere is produced because of the development of the near-equator westerly (Fig. 7-11c), where temperature and heating are positively correlated and

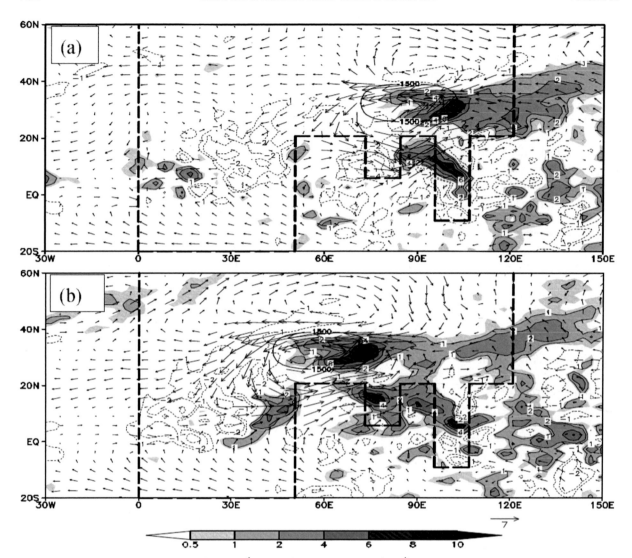

FIG. 7-10. Differences in precipitation (mm day^{-1}; shading) and wind vectors (m s^{-1}) at 850 hPa with and without TP experiments during the Asian summer monsoon onset time. The virtual TP is centered at (a) 33°N, 90°E and (b) 33°N, 60°E by using the GOALS-SAMIL model. The heavy dashed line denotes the continent boundary. [Adapted from Liang et al. (2005).]

atmospheric available energy is generated, thus favoring of the development of monsoon onset vortex over the southern BOB.

In spring, the zonal westerly just crosses the TP. Then because of the strong zonal advection the TP heating in late spring generates cold (warm) air temperatures upstream (downstream) of the TP. This can stimulate the summer-type meridional temperature gradient in the upper troposphere (cold in south and warm in north) to the east of the TP and enhance the winter-type meridional temperature gradient in the upper troposphere (cold in north and warm in south) to its west, providing a favorable background to the east of the BOB for summer monsoon onset to follow (Wu and Zhang 1998). More importantly, the significant latent

heat release due to the BOB monsoon generates a stationary Rossby wave train in the middle and upper troposphere with an anticyclone circulation over southern China, bringing colder air southward to the southern coast of China to meet the northward transported warm and moist air, which is brought about by transient eddies exited by the BOB monsoon latent heating as well (Liu et al. 2002). Consequently, the SCS summer monsoon onset commences.

The huge latent heating released by the BOB and SCS monsoon stimulates the unstable development of the South Asian high (SAH) in the upper troposphere, which extends westward with a remarkable divergence developing on its southwest just above the southeastern coast of Arabian Sea, forming a local and strong upper

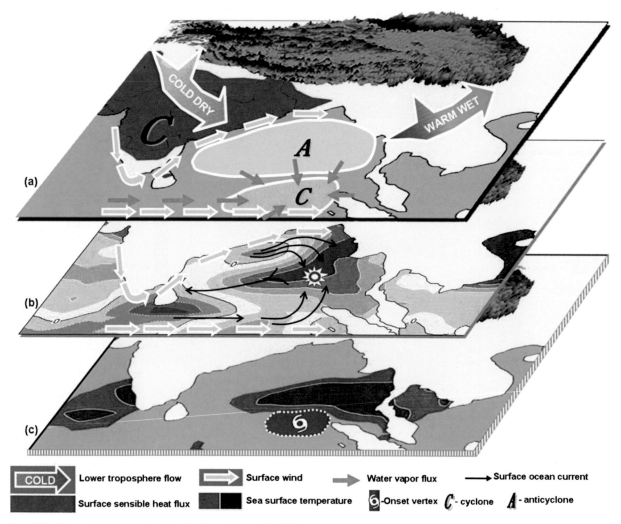

FIG. 7-11. Schematic diagram showing the formation of the BOB monsoon onset vortex as a consequence of in situ air–sea interaction modulated by the land–sea thermal contrast in South Asia and TP forcing in spring. See the text for details [from Wu et al. (2012a)].

layer pumping. Indian summer monsoon is induced at last (Zhang et al. 2014).

It becomes clear that the TP forcing in late spring can control the progression of the Asian summer monsoon onset.

5. Topography impacts on regional circulation and ASM

a. Topography impacts on regional circulation

Figure 7-3 demonstrates that the TP-SHAP plays significant roles in regulating the Asian monsoon climate. However, it is not clear whether the removal of surface sensible heating especially on the sloping surfaces will suppress the convergence (divergence) of the surface air flows from (into) the surrounding areas in the lower layers. To illustrate the significance of the sensible

heating over the sloping surfaces in the operation of the TP air pump, a series of aquaplanet experiments have been conducted based on an atmospheric general circulation model FGOALS_s developed at LASG, Institute of Atmospheric Physics (IAP). For the present purpose, aquaplanet experiments are designed with an idealized topography being introduced into the aquaplanet in which Earth's surface in the model is covered merely by ocean. There are four experiments (Fig. 7-12) (Wu et al. 2007):

1) ALLSH: All of the virtual TP surfaces possess surface sensible heating;
2) SLPSH: Only the sloping TP surface possesses surface sensible heating;
3) TOPSH: Only the top TP surface possesses surface sensible heating; and
4) NOSH: None of the TP surfaces possesses surface sensible heating.

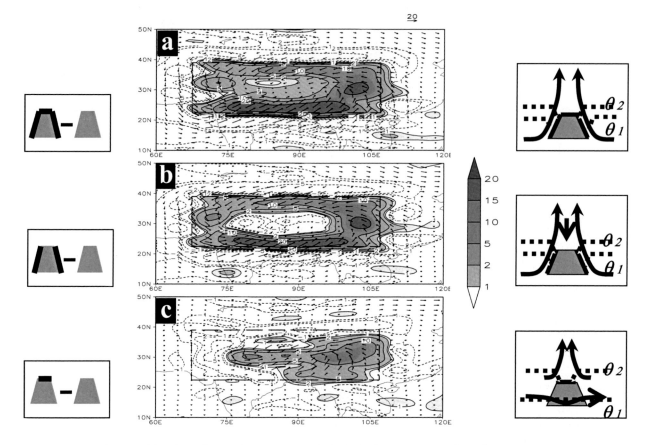

FIG. 7-12. Distributions of the wind difference (vectors) and vertical velocity ($-\omega$, shading, 10^{-2} Pa s^{-1}) on the $\sigma = 0.991$ surface from the perpetual July experiments (a) ALLSH-NOSH, (b) SLPSH-NOSH, and (c) TOPSH-NOSH, for (left) experiment designs and (right) mechanism interpretations [from Wu et al. (2007)]. The dashed line denotes the continent boundary [from Wu et al. (2012b)].

In constructing all the "no sensible heat" cases, the diffusive heating term in the atmospheric thermodynamic equation is set to zero while the surface sensible heating from the surface continues to release. Therefore, the surface energy balance is maintained. Then the differences between the three pairs (i.e., ALLSH–NOSH, SLPSH–NOSH, and TOPSH–NOSH) can be considered, respectively, as the impacts on the circulation of the TP due to all surface sensible heating, sloping-surface sensible heating alone, and top-surface sensible heating alone. Their differences in horizontal wind and vertical motion at the lower model level $\sigma = 0.991$ are presented in Fig. 7-12. Coastlines are added into the figure only for reference and comparison. In the all-surface sensible heating case (Fig. 7-12a), surface air from the Arabian Sea, Indian subcontinent, BOB, Indochina Peninsula, and the other neighboring areas are converged into the TP area and ascend over the TP. The enhanced upward movement of more than -0.2 Pa s^{-1} is on its southern and eastern slopes where the release of latent heating further intensifies

the ascent. The sloping-surface sensible heating produces similar circulations (Fig. 7-12b) as in the case with the surface sensible heating impose on all the TP surfaces (Fig. 7-12a) except over the platform: because of the absence of the in situ surface sensible heating, air descends over its top in association with radiation cooling. On the other hand, the mere top-surface sensible heating produces convergence flow only above the platform where the elevation is already higher than 3 km (Fig. 7-12c). There is no convergence at the lower elevations, and the main ascent appears over the eastern TP as a stationary Rossby wave response to the top surface heating. The results shown above can be explained by the right panels in Fig. 7-12. In the presence of surface sensible heating on the sloping surfaces (Figs. 7-12a,b), the heated air particles at the sloping surface penetrate the isentropic surfaces and slide upward. The air in the lower elevation layers in the surrounding areas is therefore pulled into the plateau region, forming strong rising motion and even heavy rainfall over the TP. On the contrary, in the top-heating-only case (Fig. 7-12c), although the platform heating can result in air convergence above the plateau it cannot pull air from below. This is because when an air particle is traveling in the lower layer and impinging on the TP, it has to stay at the same isentropic surface because of the absence of diabatic heating from the sloping lateral surface of the TP. Therefore, the air particle just goes around the TP at a rather horizontally located θ surface and no apparent ascent occurs. Therefore, there are no significant impacts on monsoon rainfall. It becomes evident that such an air pump is driven by the surface sensible heating on the TP (the TP-SHAP).

b. Topography impacts on ASM

The general circulation model FGOALS_s was used again to conduct idealized experiments for understanding the roles of large-scale mountains in the Asian summer monsoon. Three experiments, Exp TRO, Exp TP, and Exp IPTP, are designed to investigate the influence on Asian monsoon of the TP and Iranian Plateau in which idealized land–sea distribution and orography are used (Wu et al. 2012b):

(a) Exp TRO, the main Eurasian continent located over 20°–90°N, 0°–120°E and three square-shaped tropical lands over 35°S–20°N, 0°–50°E; 5°–20°N, 75°–85°E; and 9°S–20°N, 95°–105°E are integrated to form the "Afro-Eurasian continent."
(b) Exp TP, an idealized TP with the following ellipsoidal shape is placed on the continent in Exp TRO:

$$h(\lambda, \varphi) = h_{\max} \cos\left(\frac{\pi}{2} \frac{\lambda - \lambda_0}{\lambda_d}\right) \cos\left(\frac{\pi}{2} \frac{\varphi - \varphi_0}{\varphi_d}\right),$$

where $(\lambda_0, \varphi_0), (\lambda_d, \varphi_d)$, and $h_{\max}$ are taken at (32.5°N, 87.5°E), (8.25°N, 25.0°E), and 5 km.
(c) Exp IPTP, an idealized IP, where $(\lambda_0, \varphi_0), (\lambda_d, \varphi_d)$, and $h_{\max}$ are set at (32.5°N, 53.4°E), (8.25°N, 22.5°E) and 3 km, is added to Exp TP.

Figure 7-13a shows the simulated distributions of precipitation and the wind field in July at the near-surface level $\sigma = 0.991$ in Exp TP. Figure 7-13b shows the differences between the experiments with and without the TP. The elevated mountain heating produces not only lower-troposphere cyclonic circulation but also strong rainfall ($>16\,\mathrm{mm\,day}^{-1}$) on the southeastern slope of the TP, partly because the local sensible heating generates a geostrophic Rossby wave with air ascent to the east and descent to the west, and partly because over the southeastern TP the upstream warm, moist air that is transported along the southwesterly from the Indian Ocean is pumped upward. The TP acts to enhance the coupling between the lower and upper tropospheric circulations and between the subtropical

and tropical monsoon circulations (Fig. 7-14), resulting in an intensification of the East Asian summer monsoon and a weakening of the South Asian summer monsoon (Fig. 7-13b). Linking the Iranian Plateau to the TP substantially reduces the precipitation over Africa and increases the precipitation over the Arabian Sea and the northern Indian subcontinent, effectively contributing to the development of the South Asian summer monsoon (Figs. 7-13c–e).

6. Thermal control of the Asian summer monsoon

The relative impacts of various land–sea distributions (LSDs) and mountains on Asian monsoon extent and intensity have been assessed by Xu et al. (2009, 2010a,b) based on a series of AGCM simulations. Their results indicate that the presence of a midlatitude zonal LSD induces a strong zonal pressure gradient between the continent and ocean, which in turn results in the formation of an East Asian subtropical monsoon. The presence of the Asian mountains results in a stronger South Asian summer monsoon (SASM), as well as East Asian summer monsoon. The results further show that the LSD plays a more fundamental role than topography in determining the extent of Asian and African monsoons, whereas the tropical zonal LSD and Asian mountains both play a crucial role for establishing summer monsoon convection over the South Asian region.

a. Influence of land–sea thermal contrast on the ASM

The influence of land–sea thermal contrast and plateau forcing on the ASM was also investigated by Wu et al. (2012c) through employing the GCM FGOALS_s. The model is integrated with prescribed, seasonally varying sea surface temperature (SST) and sea ice. The controlled climate integration is referred to as the CON experiment. Although the ASM intensity possesses a stronger bias compared with observation induced by stronger cross-equatorial flow, the modeled precipitation (Fig. 7-15a) in general captures the main features of the ASM compared with observations (Fig. 7-15b), performing reasonably well in simulating the maximum centers over the western coast of India, the Bay of Bengal, and the southeastern slopes of the TP.

Since the monsoon is traditionally considered an atmospheric response to the seasonal land–sea thermal contrast, it is reasonable to infer that the precipitation forced not by orography but by the land–sea distribution alone could be considered as a monsoon prototype. A no-mountain experiment NMT is thus designed, being the same as the CON run except that all the mountains worldwide are removed. The modeled precipitation

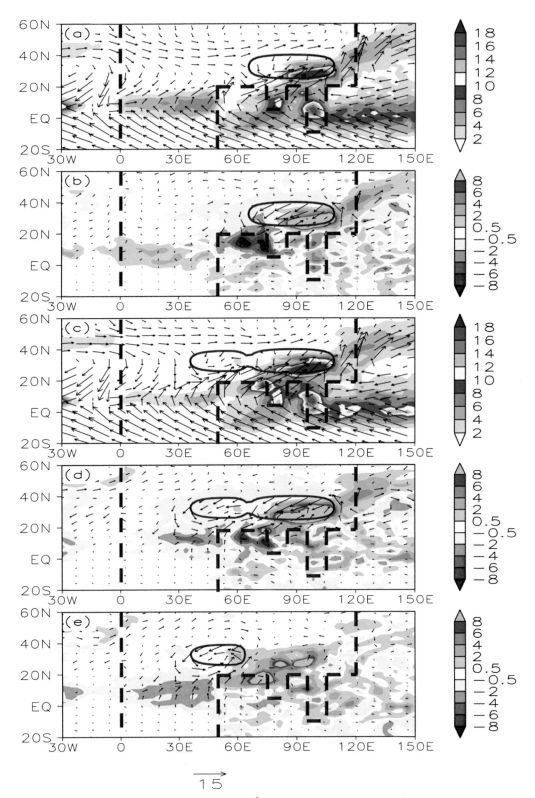

FIG. 7-13. Wind vectors at $\sigma = 0.991$ (arrows, m s^{-1}; the unit vector is shown at the bottom of the figure) and precipitation (shading; mm day^{-1}) in July (a) in Exp TP, and the difference between (b) Exp TP and Exp TRO; (c) in Exp IPTP, and (d) the difference between Exp IPTP and Exp TRO; and (e) the difference between Exp IPTP and Exp TP. The heavy red curve denotes the 700-m orographic contour; the heavy dashed line denotes the continent boundary [from Wu et al. (2012b)].

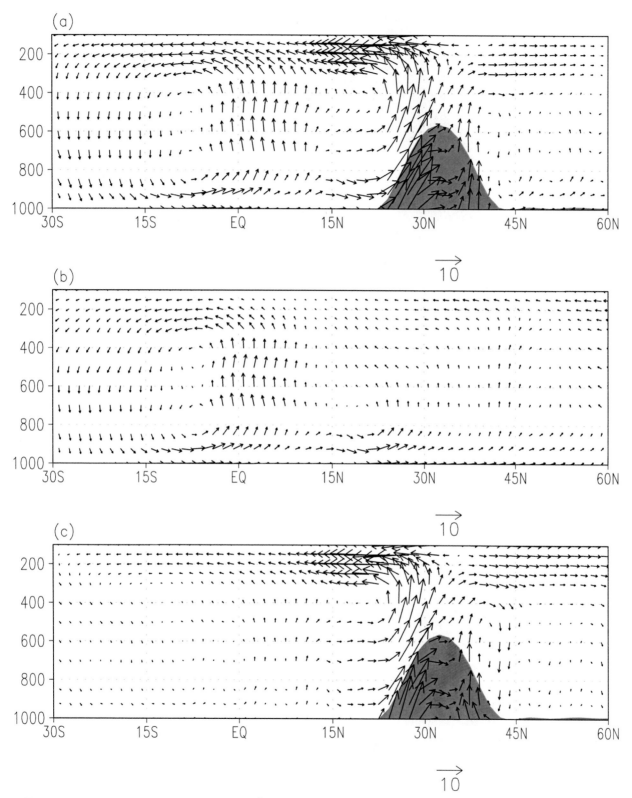

FIG. 7-14. Mean meridional circulation (v, $-\omega$; m s^{-1}) averaged over the eastern continent domain between 90° and 120°E in (a) Exp IPTP and (b) Exp TRO, and (c) their difference. The shading in (a),(c) denotes the TP shape across 87.5°E. The vertical pressure velocity ω has been amplified by -60 for better visibility [from Wu et al. (2012b)].

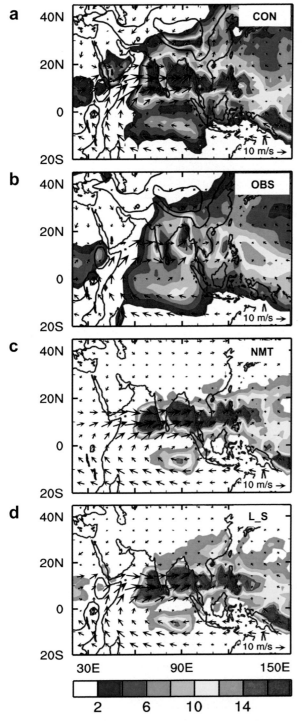

FIG. 7-15. Impacts of land–sea thermal contrast on the Asian summer monsoon showing the summer precipitation rate (color shading, mm day^{-1}) and 850-hPa winds (vectors, m s^{-1}) for (a) the control experiment CON; (b) observations averaged over the period 1979–2009 from Global Precipitation Climatology Project (GPCP) for precipitation and from NCEP–DOE AMIP-II Reanalysis (R-2) for winds; (c) experiment NMT in which the global surface elevations are set to zero; and (d) experiment L_S in which only the elevations of the Iranian Plateau (IP) and the Tibetan Plateau (TP) are set to zero. Contours in (a) and (b) over continental areas indicate elevations higher than 1500 and 3000 m [from Wu et al. (2012c)].

(Fig. 7-15c) is confined to south of 20°N, with the maximum centers (>18 mm day^{-1}) located between 10° and 15°N, as in the control run. Remarkable changes compared with the control run are seen in the subtropical area: the SASM north of 20°N and the East Asian summer monsoon (EASM) are substantially reduced. In an experiment in which only the IP and TP (Exp IPTP) is removed, the simulated precipitation pattern (Fig. 7-15d) is similar to that in NMT (Fig. 7-15c). Because our main concern is how the extensive Asian mountains IPTP influence the ASM, Fig. 7-15d could be considered a component of the ASM that is induced by land–sea thermal contrast alone. The experiment is thus termed the L_S experiment.

b. Influence of IPTP mechanical insulation on ASM

The differences (DIFF) in circulation and precipitation between CON (Fig. 7-15a) and L_S (Fig. 7-15d), as shown in Fig. 7-16a, are forced by mechanisms other than land–sea thermal forcing. Such mechanisms are required to 1) produce a cyclonic circulation at 850 hPa over the subtropical continent between 20° and 40°N, circumambulating the IPTP; 2) reduce precipitation over tropical oceans and the northwestern Pacific; and 3) increase precipitation mainly over the Asian continent, with maximum centers over India, the northern BOB, the southern slopes of the TP, and eastern Asia.

The absence of precipitation over northern India in the L_S experiment might reflect the removal of the "IPTP insulator," which might result in the southward advection of dry, cold air from the subtropics and a lack of tropical convective instability and rainfall as proposed by Boos and Kuang (2010, hereinafter BK10). Were this the case, merely adding the IPTP (but not allowing its surface sensible heating to heat the atmosphere) into the L_S experiment, which is defined as the IPTP_M experiment, would be sufficient to produce the monsoon rainfall in the northern South Asia. However, the results in Fig. 7-16b indicate that this is not the case. In the IPTP_M experiment, the patterns of both precipitation and circulation at 850 hPa are similar to those in the L_S experiment. Similarly, if we merely add IP and TP separately into the L_S experiment (i.e., the IP_M and TP_M experiments, respectively), the resultant precipitation and circulation distributions (Figs. 7-16c and 7-16d, respectively) are also similar to those in the L_S experiment. These results demonstrate that in summer, mechanical insulation of the IP and TP has a minor influence on the generation of the ASM, as it cannot produce the required compensating rainfall and precipitation patterns as shown in Fig. 7-16a.

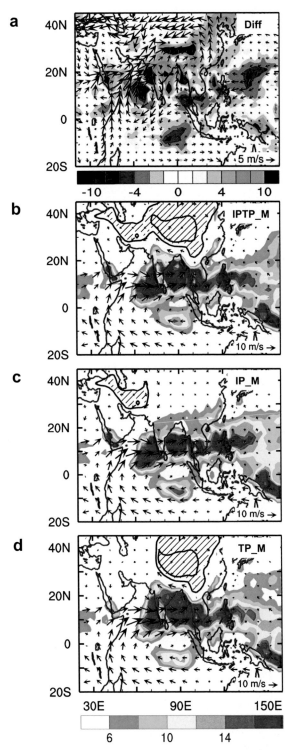

FIG. 7-16. As in Fig. 7-15 but for impacts of mountain mechanical forcing: (a) the difference (DIFF) between the CON and L_S experiments, indicating the compensating rainfall and circulation required to make up the total monsoon; (b) experiment IPTP_M in which the IP and TP mechanical forcing exists; (c) experiment IP_M in which the IP's mechanical forcing exists; and (d) experiment TP_M in which the TP's mechanical forcing exists. Thick black contours surrounding gray-hatched regions [except in (a)] indicate elevations higher than 1500 and 3000 m [from Wu et al. (2012c)].

c. Influence of IPTP thermal forcing on the ASM

Three sets of experiments were designed with surface sensible heating on the IP (IP_SH), TP (TP_SH), and IPTP (IPTP_SH), respectively, in order to study the influence of orographically elevated thermal forcing on the ASM. In IP_SH (Fig. 7-17a), the IP thermal forcing generates a cyclonic circulation encircling the IP, similar to the western parts of the compensating circulation in Fig. 7-16a. The forcing also results in reduced precipitation, mainly over the tropical Indian Ocean and the northwest Pacific, and increased precipitation over the Asian continent west of 100°E (especially over Pakistan, northern India, and the southwestern slopes of the TP), a pattern similar to that of the compensating precipitation west of 100°E, indicating the important role of the IP in generating the northern South Asian summer monsoon (SASM).

In the TP_SH experiment (Fig. 7-17b), TP thermal forcing also generates a cyclonic circulation encircling the TP. Correspondingly, reduced precipitation occurs west of 80°E; in contrast, increased precipitation occurs east of 80°E, especially over the BOB, the southern slopes of the TP, and East Asia. They are similar to the compensating precipitation and circulation patterns in the region east of 80°E (Fig. 7-16a), indicating that TP thermal forcing plays a dominant role in the generation of the EASM and the eastern part of the SASM.

In the IPTP_SH experiment (Fig. 7-17c), the elevated IPTP heating results in reduced precipitation in tropical oceans, and increased precipitation over the Asian continent to the north. The heating also generates a cyclonic circulation at 850 hPa over the Asian subtropical continental areas, with relatively isolated centers over the IP and TP. The results shown in Fig. 7-17c are basically equivalent to the linear addition of the results in Figs. 7-17a and 7-17b, indicating the important but contrasting roles of IP and TP thermal forcing in different parts of the ASM. More significantly, the precipitation and circulation patterns generated by IPTP thermal forcing (Fig. 7-17c) are close to those required to compensate the ASM (Fig. 7-16a). This result demonstrates that in addition to land–sea thermal contrast, the thermal forcing of large mountain ranges in Asia is an important factor in producing the ASM, especially over continental areas.

7. Topography-induced climbing and deflecting flow and configuration of ASM

a. Influence of climbing versus deflecting topographic effects on the ASM

More than 85% of the total atmospheric water vapor, as measured by specific humidity, generally resides in a layer below 3 km above sea level. In order for monsoon

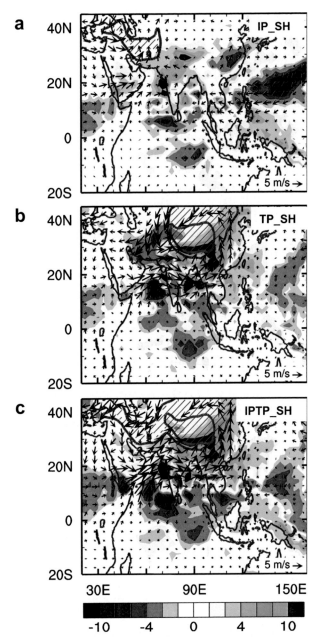

FIG. 7-17. As in Fig. 7-15, but for impacts of mountain thermal forcing generated by the elevated surface sensible heating of (a) the Iranian Plateau (IP_SH), (b) the Tibetan Plateau (TP_SH), and (c) the IP and TP (IPTP_SH). Thick red contours surrounding red-hatched regions indicate elevations higher than 1500 and 3000 m [from Wu et al. (2012c)].

clouds and precipitation to form, lower-tropospheric water vapor must be lifted by vertical motions forced either internally or externally. One of the internal forcings is the type of cold and/or warm fronts. This mechanism is important in middle and high latitudes, especially in winter, but is not important in the tropics in summer because the air temperature in the tropics is relatively uniform.

The mechanical forcing of mountains is an important external forcing: airflow impinging upon mountains is either deflected to produce encircling flow or lifted to produce climbing flow. Consequently, clouds and precipitation are generated around mountains. However, if a mountain is higher than several hundred meters, the conservation constraint of angular momentum and energy means that the airflow passes around the mountain rather than rising over it (Wu 1984).

Thermal forcing can also generate atmospheric ascent, because large-scale atmospheric potential temperature (θ) increases with height. According to the steady-state thermodynamic equation,

$$\mathbf{V} \cdot \nabla\theta = Q, \qquad (7\text{-}1)$$

where $\mathbf{V}$ is air velocity. In regions of heating ($Q > 0$), air should penetrate isentropic surfaces upward. There are several types of atmospheric heating. Shortwave radiation is weakly absorbed directly by the atmosphere. In the absence of cloud, longwave radiation can easily escape into space. Condensation heating normally occurs above the cloud base. Surface sensible heating, however, can increase the near-surface entropy and result in the development of convective instability and trigger atmospheric ascent, and it is effective in generating atmospheric ascent in the lower troposphere.

If surface sensible heating occurs on a mountain slope, and if the mountain is high enough, large amounts of moisture in lower layers are readily transported to the free atmosphere. The TP in summer is a heat source for the atmosphere and has a strong influence on weather and climate. When a moist and warm southwesterly approaches the TP, the air becomes heated, starts to penetrate isentropic surfaces, and slides upward along its sloping surface.

Figure 7-18 shows the distribution of precipitation and streamlines at the $\sigma = 0.89$ surface, which is about 1 km from the surface. In the CON experiment (Fig. 7-18a), when the water conveyer belt originating from the Southern Hemisphere meanders eastward through the South Asian subcontinent, the effects of land–sea thermal forcing mean that severe precipitation centers are formed along 15°N. The rest of the water vapor is transported to sustain the East Asian monsoon, although some swerves northward over northern India and the BOB. The pumping effect of TP-SHAP results in the convergence of air toward the TP. The upward streamlines are perpendicular to the TP contours, eventually forming a cyclonic circulation at the southeastern corner of the TP. Consequently, heavy monsoon rainfall occurs over northern India and western China, with a maximum center ($>18\,\mathrm{mm\,day}^{-1}$) appearing over the

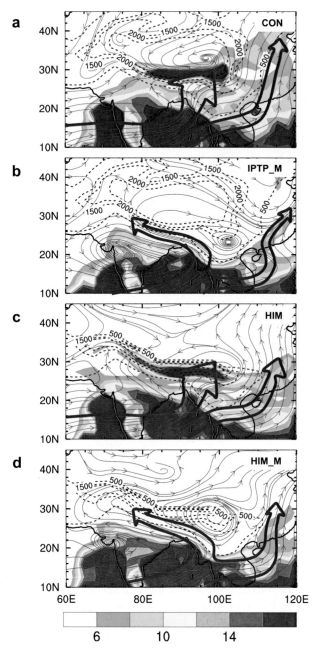

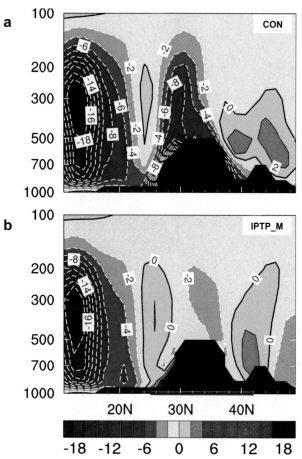

FIG. 7-18. Mechanisms and relative contributions of the climbing and deflecting effects of mountains, showing the summer precipitation rate (color shading, mm day^{-1}) and streamlines at the $\sigma = 0.89$ level for the (a) CON, (b) IPTP_M, (c) HIM, and (d) HIM_M experiments. Dashed contours enclose elevations higher than 1500 and 3000 m with red and black colors, respectively, indicating with and without surface sensible heating of the mountains. Dark blue open arrows denote the main atmospheric flows impinging on the TP, either climbing up the plateau in (a) and (c) or moving around the plateau, parallel to orographic contours, in (b) and (d). [Adapted from Wu et al. (2012c).]

FIG. 7-19. Structure of the South Asian summer monsoon, showing 80°–90°E longitudinally averaged vertical–meridional cross sections of pressure vertical velocity (contour interval, 2×10^{-2} Pa s^{-1}) for experiments (a) CON and (b) IPTP_M. The black mass indicates the topography [from Wu et al. (2012c)].

southeastern slopes of the TP. The condensation heating of this rainfall center generates cyclonic circulation in the lower layer and further intensifies the EASM.

In the IPTP_M experiment (Fig. 7-18b), in contrast, when the water vapor flux from the main water conveyer belt approaches the TP, it is not heated and the airflow remains at the same isentropic surface [Eq. (7-1)]. Consequently, the streamlines do not climb up the TP; instead, they move around the mountains, parallel to orographic contours. Thus, no monsoon develops over northern India and the TP, and the EASM is substantially weakened. These results indicate that the thermal forcing of large-scale mountains plays a dominant role in the generation of the northern and eastern parts of the SASM and the EASM.

Boos and Kuang claimed that the impact of TP thermal forcing is less important than the thermal insulation of the Himalayas in the formation of the SASM (BK10;

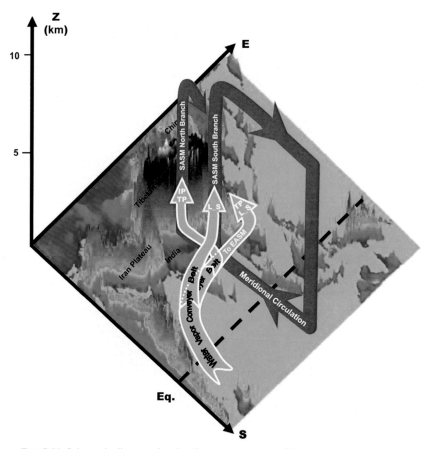

FIG. 7-20. Schematic diagram showing the gross structure of the Asian summer monsoon. For the southern branch, water vapor along the conveyer belt is lifted up due mainly to land–sea (L_S) thermal contrast in the tropics; for the northern branch, the water vapor is drawn away from the conveyer belt northward toward the foothills and slopes of the TP, and is uplifted to produce heavy precipitation that is controlled mainly by IPTP-SHAP; the rest of the water vapor is transported northeastward to sustain the East Asian summer monsoon, which is controlled by the land–sea thermal contrast as well as thermal forcing of the TP [adapted from Wu et al. (2012c)].

Boos and Kuang 2013). Their conclusion was based on a numerical experiment in which the main part of TP was removed while only its southern rim (the Himalayas) is maintained and the resultant SASM was similar to that in the control experiment with the whole TP presented. They proposed that the Himalayas block cold, dry air from the north and the high surface entropy over India can thus induce moist convection and drive the SASM. Wu et al. (2012c) followed BK10's experiment design based on the FGOALS model and carried a similar numerical experiment (HIM). The modeled SASM rainfall as shown in Fig. 7-18c is also similar to that in the CON run (Fig. 7-18a) as reported by BK10. However, in a parallel experiment HIM_M, which is the same as HIM but without surface sensible heating from the Himalaya to the atmosphere, the northern branch of the SASM disappeared completely (Fig. 7-18d) as in the

IPTP_M run (Fig. 7-18b). Since in the HIM_M experiment the insulation effect is kept as in HIM, the results demonstrate that the insulation effect of the TP is insignificant in the formation of the SASM and the thermal pumping of the IPTP plays a fundamental role. Notice that in all the numerical experiments in which topography is removed (e.g., Kasahara and Washington 1971; Hahn and Manabe 1975; BK10), there was no cold and dry advection in summer from the subtropics to northern India. This is because in summer the subtropical area is warmer than in the tropics. Furthermore, the high surface entropy over northern India depends on high surface temperature as well as high humidity. It is due to the pumping effect of the IPTP that the abandoned moisture over ocean can be transported into inland India and result in high surface entropy there. In this regard the SASM is also thermally controlled.

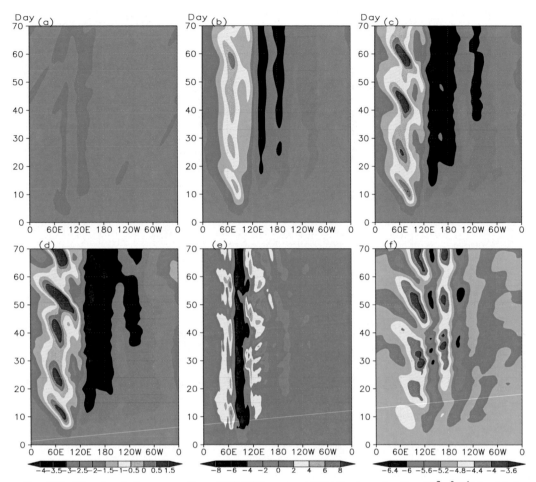

FIG. 7-21. Time–longitude cross section of 25°–35°N averaged 200-hPa streamfunction ($10^7 \, m^2 \, s^{-1}$) for (a) the O_TP experiment (simulation only with the TP orography), (b) the OQ_TP experiment (with the TP orography and observed heating above), (c) the O1.6Q_TP experiment (with the TP orography and 1.6 times heating above), and (d) the O2Q_TP experiment (with the TP orography and 2 times heating above). Panels (e),(f) are similar to (a) but for ω at 400 hPa ($10^{-2} \, Pa \, s^{-1}$) and for the 40°–60°N averaged 250-hPa streamfunction in the O1.6Q_TP experiment, respectively. The left color bar underneath the panels is for the streamfunction in (a)–(d), the middle bar for ω, and the right bar for the streamfunction in (f) [from Liu et al. (2007b)].

b. Structure of the ASM

A striking feature of the above experiments is the insensitivity of the southern part of the SASM to IPTP forcing: actually in all experiments (Figs. 7-15, 7-16, and 7-18), the intensity and spatial distribution of precipitation south of 20°N show little change compared with the control while the configuration or thermal status of the Tibetan/Iranian Plateau shows a marked change. Figures 7-19a and 7-19b show 80°–90°E longitudinally averaged latitude–height cross sections from the CON and IPTP_M experiments, demonstrating that the vertical velocity is divided into a southern branch near 15°N and a northern branch at about 25°N. In the CON run (Fig. 7-19a), strong rising associated with the southern SASM is located over the northern Indian Ocean. Ascending air is also dominant

above the TP, with maxima located near the surface, indicating the importance of surface sensible heating in generating orographic ascent. In the IPTP_M experiment (Fig. 7-19b), the lack of surface heating on the TP results in two remarkable sinking centers over its slopes; thus, the northern branch of the SASM disappears over northern India. However, the intensity and location of the southern branch is largely unchanged. In fact, in all the experiments the southern SASM branch remains steady, with a center ($>18 \times 10^{-2} \, Pa \, s^{-1}$) at about 400 hPa, locked to the south of the coastline. The insensitivity of the southern branch of the SASM to orographic change indicates that the land–sea thermal contrast plays a dominant role in its generation and variation.

The above discussion is summarized schematically in Fig. 7-20. The meridional circulation of the SASM can

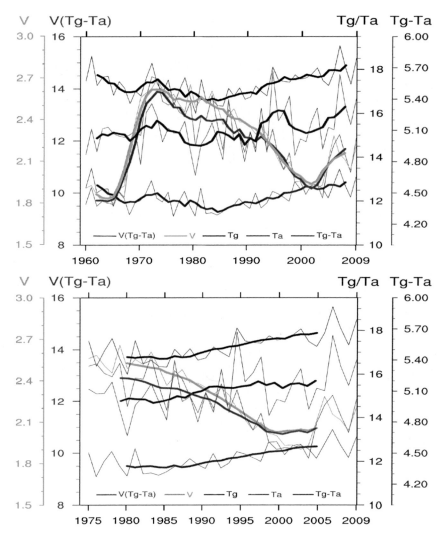

FIG. 7-22. Evolutions of JJA means averaged over the Tibetan Plateau stations of surface soil temperature T_g, surface air temperature T_a, difference between T_g and T_a, surface wind speed V, and the parameter of surface sensible heat flux PSH = $V(T_g - T_a)$. Values shown are (a) for 1960–2009 and for the corresponding 5-yr running mean and (b) for 1975–2009 and the corresponding 11-yr running mean. Units are °C for T_g, T_a, and $(T_g - T_a)$; m s^{-1} for V; and °C m^{-1} s^{-1} for PSH [from Liu et al. (2012)].

be divided into southern and northern branches. Its southern branch is located in the tropics: water vapor that originates from the Southern Hemisphere and is transported along the zonally oriented "water vapor conveyer belt" is lifted upward due to the land–sea thermal contrast, forming monsoon precipitation there. The northern branch occurs along the southern margin of the IPTP in the subtropics. When the conveyer belt approaches the TP, part of its water vapor is hauled away and turned northward, then lifted upward by the IPTP-SHAP, resulting in heavy precipitation in the monsoon trough over northern India and along the foothills and slopes of the TP. The rest of the water vapor along the water conveyer belt is transported

northeastward to sustain the EASM, which is controlled by the land–sea thermal contrast and thermal forcing of the TP. These results highlight the dominant roles of the land–sea thermal contrast and IPTP thermal forcing in influencing the ASM.

8. Variation of the thermal forcing over the TP and its impacts

The real atmosphere goes through periods in which the monsoon variability is suppressed or enhanced depending on the intensities of the TP and tropical heating and the vertical extent of the heating in the TP region.

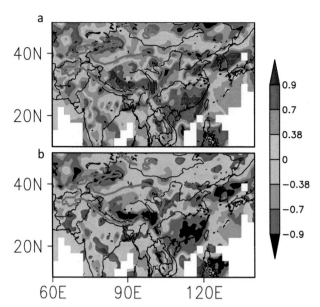

FIG. 7-23. Distributions of the correlation coefficient for 1980–2004 and between the June–August (JJA) precipitation and the TP-averaged (a) T_a and (b) PSH = $V(T_g - T_a)$. The source data were processed using 11-yr running mean before the correlation calculation was performed. The heavy gray contours are the main body of the TP [adapted from Liu et al. (2012)].

a. Biweekly oscillation of the South Asian anticyclone

Severe weather events in China and in the Asian monsoon regions have been found to have a close relationship with the nature of the Tibetan upper tropospheric anticyclone, the South Asian anticyclone (SAA). A quasi-periodic oscillation of the SAA was revealed in the spectral analysis performed by Tao and Zhu (1964) and Tao and Ding (1981) with periods of 10–16 days. Liu et al. (2007b) found that heating over the plateau leads to a potential vorticity (PV) minimum and that if it is sufficiently strong, say 1.5 or 2 times the observed average, the flow is unstable, producing a quasi-biweekly oscillation (Figs. 7-21c,d). During this oscillation, the Tibetan anticyclone changes from a single center over the southwestern side of the plateau to a split–double structure with centers over China and the Middle East and back again. These characteristics are similar to observed variability in the region. Associated with this quasi-biweekly oscillation are significant variations in the strength of the ascent over the plateau (Fig. 7-21e) and the Rossby wave pattern over the North Pacific (Fig. 7-21f).

The origin of the variability is instability associated with the zonally extended potential vorticity PV minimum on a θ surface, as proposed by Hsu and Plumb (2000). This minimum is due to the tendency to reduce the PV above the heating over the plateau and to advection by the consequent anticyclone of high PV around from the east and low PV to the west. The deep

convection to the south and southeast of the plateau tends to suppress the quasi-biweekly oscillation because the low PV produced above it acts to reduce the meridional PV gradient reversal. The occurrence of the oscillation depends on the relative magnitude of the heating in the two regions (Liu et al. 2007a,b).

b. Decadal variation of the TP thermal forcing and the rainfall in East China

Duan et al. (2006) and Duan and Wu (2008) reported that temporal change in the annual-mean surface atmospheric temperature T_a over the plateau increased at a rate of $0.4°C\,decade^{-1}$ during the period of 1980–2003. Data analysis based on station observations reveals that many meteorological variables averaged over the TP are closely correlated, and their trends during the past decades are well correlated with the rainfall trend of the Asian summer monsoon. Liu et al. (2012) and Zhu et al. (2012) further diagnosed and confirmed the existence of a weakening decadal trend in TP thermal forcing from the mid-1970s to the end of the 1990s, characterized by weakened surface sensible heat flux in spring and summer (Fig. 7-22). They also indicated that the weakening trend in surface sensible heating is highly correlated with surface wind speed V. This weakening trend in thermal forcing is significantly correlated with decreasing summer precipitation over northern South Asia and northern China and increasing precipitation over northwestern China, southern China, and the Korean peninsula during the same period (Fig. 7-23).

However, such a correlation does not necessarily imply causality. An atmospheric general circulation model, the HadAM3, was employed to elucidate the causality between the weakening TP forcing (Fig. 7-24) and the change in the Asian summer monsoon rainfall. Results demonstrate that a weakening in surface sensible heating over the TP results in reduced summer precipitation in the plateau region and a reduction in the associated latent heat release in summer (Fig. 7-24a). These changes in turn result in the weakening of the near-surface cyclonic circulation surrounding the plateau and the subtropical anticyclone over the subtropical western North Pacific (Fig. 7-24b), in agreement with the results obtained from the idealized TP experiment (Fig. 7-13b) but with opposite polarities. The southerly that normally dominates East Asia, ranging from the South China Sea to northern China, then weakens, resulting in a weaker equilibrated Sverdrup balance between positive vorticity generation and latent heat release. Consequently, the convergence of water vapor transport is confined to southern China, forming a unique anomaly pattern in monsoon rainfall (Fig. 7-24c), "south wet and north dry."

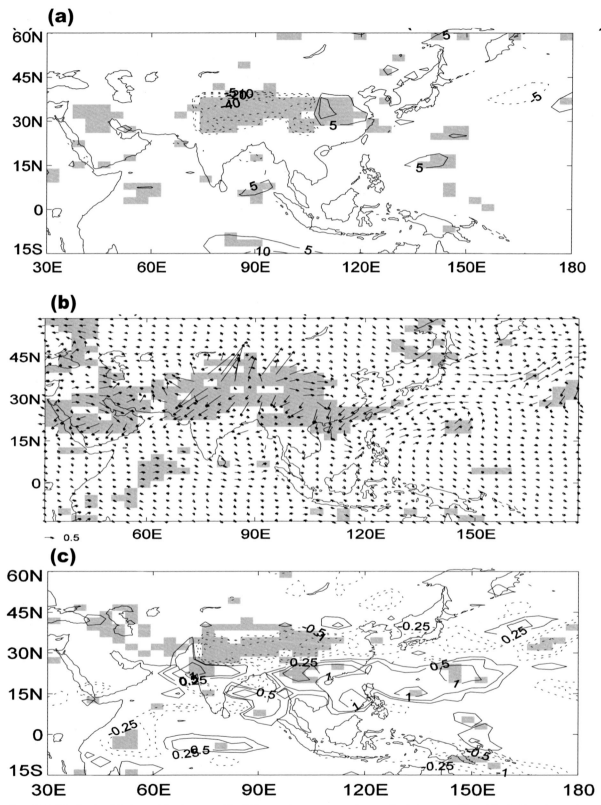

FIG. 7-24. Weakening TP thermal forcing: monthly difference between experiment albedo increase (Exp Albedo-I) and experiment albedo decrease (Exp Albedo-D) in terms of the JJA-mean (a) surface upward sensible and latent heat flux anomalies (W m^{-2}), (b) surface wind (m s^{-1}), and (c) precipitation (mm day^{-1}). Shading indicates regions where anomalies are significant at the 95% confidence level (t test) [adapted from Liu et al. (2012)].

It is interesting to see that the summertime surface sensible heating over the TP has strengthened since the beginning of this century (Fig. 7-22). A similar result was also found by Si and Ding (2013) based on the 72 stations over the central and eastern TP. This then implies that the wet south and dry north rainfall anomaly pattern over eastern China may come to an end and an opposite pattern may appear in the near future.

9. Future perspective

In nature, the influences of the IPTP orography and its surface sensible heating cannot be separated. The significance of the dominance of IPTP thermal forcing in influencing the ASM lies in the fact that, over the modern-day orography, the thermal status of the IPTP varies due to natural and anthropogenic factors. By focusing on changes in the thermal status of the IPTP, the dominance of thermal controls on the ASM may provide us with a tangible way of identifying climate trends in the Asian summer monsoon in a warming world, and of improving weather forecasts, climate predictions, and projections in areas affected by the Asian monsoon. The challenges could be to 1) understand the modulation of the TP on the air–sea interaction and 2) quantify the impact of the TP.

The modulation of the TP on air–sea interaction is a newly recognized monsoon dynamic. How such modulation influences the seasonal and interannual variability of the ASM and the ENSO–monsoon relationship must be further studied.

Quantifying the TP impact is crucial for the further understanding and predictions of the Asian monsoon. It is necessary to increase in situ observation stations, improve the quality of the satellite data over the plateau, improve the description of physical processes in numerical models—boundary processes, cloud, radiation, convection, etc.,—over the plateau. Work will focus on identifying and explaining potentially significant teleconnections, such as the influence of Tibetan snow cover on the Asian monsoon and Northern Hemispheric conditions. Employing existing theories in physics, chemistry, and mathematics to reveal the complex and relevant multisphere interaction and to get new insights into the impacts of the TP on climate will be significant for enhancing our understanding the roles of the TP in regional and global climate and improving the fidelity of climate model and prediction.

Acknowledgments. Material used here is based on cooperative studies among the authors and Drs. Buwen Dong, Anmin Duan, Qiong Zhang, Weiping Li, Xin Liu, Xiaoyun Liang, Tongmei Wang, Zaizhi Wang, Jingjing Yu, Jieli Hong, and Yue Guan.

REFERENCES

Bao, Q., G. Wu, Y. Liu, Y. Jing, Z. Wang, and T. Zhou, 2010: An introduction to the coupled model FGOALS1.1-s and its performance in East Asia. *Adv. Atmos. Sci.*, **27**, 1131–1142, doi:10.1007/s00376-010-9177-1.

Boos, W. R., and Z. Kuang, 2010: Dominant control of the South Asian monsoon by orographic insulation versus plateau heating. *Nature*, **463**, 218–222, doi:10.1038/nature08707.

——, and ——, 2013: Sensitivity of the South Asian monsoon to elevated and non-elevated heating. *Sci. Rep.*, **3**, 1192, doi:10.1038/srep01192.

Chang, C.-P., Ed., 2004: *East Asian Monsoon.* World Scientific, 564 pp.

——, Z. Wang, and H. Hendon, 2006: The Asian winter monsoon. *The Asian Monsoon*, B. Wang et al., Eds., Springer, 89–127.

Charney, J. G., and P. G. Drazin, 1961: Propagation of planetary-scale disturbances from the lower into the upper atmosphere. *J. Geophys. Res.*, **66**, 83–109, doi:10.1029/JZ066i001p00083.

Chen, P., 2001: Thermally forced stationary waves in a quasigeostrophic system. *J. Atmos. Sci.*, **58**, 1585–1594, doi:10.1175/1520-0469(2001)058<1585:TFSWIA>2.0.CO;2.

Duan, A. M., and G. X. Wu, 2005: Role of the Tibetan Plateau thermal forcing in the summer climate patterns over subtropical Asia. *Climate Dyn.*, **24**, 793–807, doi:10.1007/s00382-004-0488-8.

——, and ——, 2008: Weakening trend in the atmospheric heat source over the Tibetan Plateau during recent decades. Part I: Observations. *J. Climate*, **21**, 3149–3164, doi:10.1175/2007JCLI1912.1.

——, ——, Q. Zhang, and Y. Liu, 2006: New proofs of the recent climate warming over the Tibetan Plateau as a result of the increasing greenhouse gases emissions. *Chin. Sci. Bull.*, **51**, 1396–1400, doi:10.1007/s11434-006-1396-6.

Flohn, H., 1957: Large-scale aspects of the "summer monsoon" in South and East Asia. *J. Meteor. Soc. Japan*, 75th anniversary volume, 180–186.

Hahn, D. G., and S. Manabe, 1975: The role of mountains in the South Asian monsoon circulation. *J. Atmos. Sci.*, **32**, 1515–1541, doi:10.1175/1520-0469(1975)032<1515:TROMIT>2.0.CO;2.

Held, I. M., 1983: Stationary and quasi-stationary eddies in the extratropical troposphere: Theory. *Large-Scale Dynamical Processes in the Atmosphere*, B. Hoskins and R. Pearse, Eds., Academic Press, 127–168.

Holton, J. R., 2004: *An Introduction to Dynamic Meteorology.* Elsevier Academic, 535 pp.

Hsu, C. J., and R. A. Plumb, 2000: Nonaxisymmetric thermally driven circulations and upper-tropospheric monsoon dynamics. *J. Atmos. Sci.*, **57**, 1255–1276, doi:10.1175/1520-0469(2000)057<1255:NTDCAU>2.0.CO;2.

Kasahara, A., and W. M. Washington, 1971: General circulation experiments with a six-layer NCAR model, including orography, cloudiness, and surface temperature calculations. *J. Atmos. Sci.*, **28**, 657–701, doi:10.1175/1520-0469(1971)028<0657:GCEWAS>2.0.CO;2.

Krishnamurti, T. N., 1981: Cooling of the Arabian Sea and the onset-vortex during 1979. Recent progress in equatorial oceanography: A report of the final meeting of SCOR Working Group 47, Scientific Committee on Oceanic Research, 1–12.

Li, L. F., Y. M. Liu, and C. Y. Bo, 2011: Impacts of diabatic heating anomalies on an extreme snow event over South China in January 2008 (in Chinese). *Climatic Environ. Res.*, **16**, 126–136.

Liang, X. Y., Y. M. Liu, and G. X. Wu, 2005: Effect of Tibetan Plateau on the site of onset and intensity of the Asian summer monsoon (in Chinese). *Acta Meteor. Sin.*, **63**, 799–805.

Liu, X., G. X. Wu, W. P. Li, and Y. M. Liu, 2001: Thermal adaptation of the large-scale circulation to the summer heating over the Tibetan Plateau (in Chinese). *Prog. Nat. Sci.*, **11**, 207–214.

——, W. P. Li, and G. X. Wu, 2002: Interannual variations of the diabatic heating over the Tibetan Plateau and the Northern Hemispheric circulation in summer. *Acta Meteor. Sin.*, **60**, 267–277.

Liu, Y. M., G. X. Wu, H. Liu, and P. Liu, 2001: Condensation heating of the Asian summer monsoon and the subtropical anticyclone in the Eastern Hemisphere. *Climate Dyn.*, **17**, 327–338, doi:10.1007/s003820000117.

——, J. C. L. Chan, J. Y. Mao, and G. X. Wu, 2002: The role of Bay of Bengal convection in the onset of the 1998 South China Sea summer monsoon. *Mon. Wea. Rev.*, **130**, 2731–2744, doi:10.1175/1520-0493(2002)130<2731:TROBOB>2.0.CO;2.

——, Q. Bao, A. M. Duan, Z. A. Qian, and G. X. Wu, 2007a: Recent progress in the study in China of the impact of Tibetan Plateau on the climate. *Adv. Atmos. Sci.*, **24**, 1060–1076, doi:10.1007/s00376-007-1060-3.

——, B. J. Hoskins, and M. Blackburn, 2007b: Impact of Tibetan topography and heating on the summer flow over Asia. *J. Meteor. Soc. Japan*, **85B**, 1–19, doi:10.2151/jmsj.85B.1.

——, G. X. Wu, J. L. Hong, B. Dong, A. Duan, Q. Bao, and L. J. Zhou, 2012: Revisiting Asian monsoon formation and change associated with Tibetan Plateau forcing: II. Change. *Climate Dyn.*, **39**, 1183–1195, doi:10.1007/s00382-012-1335-y.

Mao, J. Y., and A. M. Duan, 2005: The predictability of the reversal of the ridge of the subtropical anticyclone and the onset of the summer Asian monsoon (in Chinese). *Impacts of the Land–Sea Thermal Contrast on the Climate in China*, Y. Liu and Z. Qian, Eds., China Meteorological Press, 173–177.

——, G. Wu, and Y. Liu, 2002a: Study on modal variation of subtropical high and its mechanism during seasonal transition. Part I: Climatological features of subtropical high structure (in Chinese). *Acta Meteor. Sin.*, **60**, 400–408.

——, ——, and ——, 2002b: Study on modal variation of subtropical high and its mechanism during seasonal transition. Part II: Seasonal transition index over Asian monsoon region (in Chinese). *Acta Meteor. Sin.*, **60**, 409–420.

Nigam, S., 1994: On the dynamical basis for the Asian summer monsoon rainfall–El Niño relationship. *J. Climate*, **7**, 1750–1771, doi:10.1175/1520-0442(1994)007<1750:OTDBFT>2.0.CO;2.

——, 1997: The annual warm to cold phase transition in the eastern equatorial Pacific: Diagnosis of the role of stratus cloud-top cooling. *J. Climate*, **10**, 2447–2467, doi:10.1175/1520-0442(1997)010<2447:TAWTCP>2.0.CO;2.

Si, D., and Y.-H. Ding, 2013: Decadal change in the correlation pattern between the Tibetan Plateau winter snow and the East Asian summer precipitation during 1979–2011. *J. Climate*, **26**, 7622–7634, doi:10.1175/JCLI-D-12-00587.1.

Tao, S.-Y., and F. K. Zhu, 1964: The variation of 100-mb circulation over South Asia in summer and its association with march and withdrawal of the West Pacific subtropical high (in Chinese). *Acta Meteor Sin.*, **34**, 385–395.

——, and Y.-H. Ding, 1981: Observational evidence of the influence of the Qinghai-Xizang (Tibet) Plateau on the occurrence of heavy rain and severe convective storms in China. *Bull. Amer. Meteor. Soc.*, **62**, 23–30, doi:10.1175/1520-0477(1981)062<0023:OEOTIO>2.0.CO;2.

——, and L. Chen, 1987: A review of recent research on the East Asian summer monsoon in China. *Monsoon Meteorology*, Oxford University Press, 60–92.

Tian, S. F., and T. Yasunari, 1998: Climatological aspects and mechanism of spring persistent rains over central China. *J. Meteor. Soc. Japan*, **76**, 57–71.

Wallace, J. M., and P. V. Hobbs, 1977: *Atmospheric Science: An Introductory Survey*. Academic Press, 467 pp.

Wan, R. J., and G. X. Wu, 2007: Mechanism of the spring persistent rains over southeastern China. *Sci. China*, **50D**, 130–144, doi:10.1007/s11430-007-2069-2.

Wang, B., 2006: *The Asian Monsoon*. Springer, 787 pp.

——, and LinHo, 2002: Rainy seasons of the Asian-Pacific monsoon. *J. Climate*, **15**, 386–398, doi:10.1175/1520-0442(2002)015<0386:RSOTAP>2.0.CO;2.

Wu, G. X., 1984: The nonlinear response of the atmosphere to large-scale mechanical and thermal forcing. *J. Atmos. Sci.*, **41**, 2456–2476, doi:10.1175/1520-0469(1984)041<2456:TNROTA>2.0.CO;2.

——, and Y. S. Zhang, 1998: Tibetan Plateau forcing and the timing of the monsoon onset over South Asia and the South China Sea. *Mon. Wea. Rev.*, **126**, 913–927, doi:10.1175/1520-0493(1998)126<0913:TPFATT>2.0.CO;2.

——, W. Li, H. Guo, H. Liu, J. Xue, and Z. Wang, 1997: Sensible heat driven air-pump over the Tibetan Plateau and its impacts on the Asian summer monsoon (in Chinese). *Collection in Memory of Zhao Jiuzhang*, Y. Duzheng, Ed., Chinese Science Press, 116–126.

——, S. Lan, Y. Liu, L. Hui, S. Sun, and W. Li, 2002: Impacts of land surface processes on summer climate. *Selected Papers of the Fourth Conference on East Asia and Western Pacific Meteorology and Climate*, C. P. Chang et al., Eds., World Scientific, 64–76.

——, and Coauthors, 2007: The influence of mechanical and thermal forcing by the Tibetan Plateau on Asian climate. *J. Hydrometeor.*, **8**, 770–789, doi:10.1175/JHM609.1.

——, Y. Guan, T. M. Wang, Y. M. Liu, J. H. Yan, and J. Y. Mao, 2011: Vortex genesis over the Bay of Bengal in spring and its role in the onset of the Asian summer monsoon. *Sci. China Earth Sci.*, **54**, 1–9, doi:10.1007/s11430-010-4125-6.

——, ——, Y. M. Liu, J. H. Yan, and J. Y. Mao, 2012a: Air–sea interaction and formation of the Asian summer monsoon onset vortex over the Bay of Bengal. *Climate Dyn.*, **38**, 261–279, doi:10.1007/s00382-010-0978-9.

——, Y. Liu, B. Dong, X. Liang, A. Duan, Q. Bao, and J. Yu, 2012b: Revisiting Asian Monsoon formation and change associated with Tibetan Plateau forcing: I. Formation. *Climate Dyn.*, **39**, 1169–1181, doi:10.1007/s00382-012-1334-z.

——, Y. M. Liu, B. He, Q. Bao, A. M. Duan, and F. F. Jin, 2012c: Thermal controls on the Asian summer monsoon. *Nat. Sci. Rep.*, **2**, 404, doi:10.1038/srep00404.

Xu, Z. F., C. B. Fu, and Y. F. Qian, 2009: The relative roles of land–sea distribution and orography in the Asian monsoon intensity. *J. Atmos. Sci.*, **66**, 2714–2729, doi:10.1175/2009JAS3053.1.

——, Y. F. Qian, and C. B. Fu, 2010a: The role of land–sea distribution and orography in the Asian monsoon. Part I: Land–sea distribution. *Adv. Atmos. Sci.*, **27**, 403–420, doi:10.1007/s00376-009-9005-7.

——, ——, and ——, 2010b: The role of land–sea distribution and orography in the Asian monsoon. Part II: Orography. *Adv. Atmos. Sci.*, **27**, 528–542, doi:10.1007/s00376-009-9045-z.

Yanai, M., and G.-X. Wu, 2006: Effects of the Tibetan Plateau. *The Asian Monsoon*, B. Wang et al., Eds., Springer, 513–549.

——, S. Esbensen, and J.-H. Chu, 1973: Determination of bulk properties of tropical cloud clusters from large-scale heat and moisture budgets. *J. Atmos. Sci.*, **30**, 611–627, doi:10.1175/1520-0469(1973)030<0611:DOBPOT>2.0.CO;2.

——, C. F. Li, and Z. S. Song, 1992: Seasonal heating of the Tibetan Plateau and its effects on the evolution of the Asian summer monsoon. *J. Meteor. Soc. Japan*, **70**, 319–351.

Ye, D. Z., and Y. X. Gao, 1979: *Meteorology of the Qinghai-Xizang Plateau.* Chinese Science Press, 278 pp.

Yeh, T. C., S.-W. Lo, and P.-C. Chu, 1957: The wind structure and heat balance in the lower troposphere over Tibetan Plateau and its surrounding. *Acta Meteor. Sin.*, **28**, 108–121.

Yu, J. J., Y. M. Liu, and G. X. Wu, 2011a: An analysis of the diabatic heating characteristic of atmosphere over the Tibetan Plateau in winter I: Climatology. *Acta Meteor. Sin.*, **69**, 79–88.

——, ——, and ——, 2011b: An analysis of the diabatic heating characteristic of atmosphere over the Tibetan Plateau in winter II: Interannal variation. *Acta Meteor. Sin.*, **69**, 89–98.

Zhang, J. J., and Coauthors, 1988: *Advances in the Qinghai-Xizang Plateau Meteorology–The Qinghai-Xizang Meteorology Experiment (QXPMEX 1979) and Research.* Chinese Science Press, 268 pp.

Zhang, Y. N., G. X. Wu, Y. M. Liu, and Y. Guan, 2014: The effects of asymmetric potential vorticity forcing on the instability of South Asia high and Indian summer monsoon onset. *Sci. China Earth Sci.*, **57**, 337–350, doi:10.1007/s11430-013-4664-8.

Zhu, X. Y., Y. M. Liu, and G. X. Wu, 2012: An assessment of summer sensible heat flux on the Tibetan Plateau from eight data sets. *Sci. China Earth Sci.*, **55**, 779–786, doi:10.1007/s11430-012-4379-2.

Chapter 8

Multiscale Temporal Mean Features of Perturbation Kinetic Energy and Its Budget in the Tropics: Review and Computation

BAODE CHEN

Key Laboratory of Numerical Modeling for Tropical Cyclone, China Meteorological Administration, and Shanghai Typhoon Institute of China Meteorological Administration, Shanghai, China

WEN-WEN TUNG

Department of Earth, Atmospheric, and Planetary Sciences, Purdue University, West Lafayette, Indiana

MICHIO YANAI

Department of Atmospheric and Oceanic Sciences, University of California, Los Angeles, Los Angeles, California

ABSTRACT

The authors examined the maintenance mechanisms of perturbation kinetic energy (PKE) in the tropical regions for multiple time scales by computing and analyzing its budget equation. The emphasis has been placed on the mean features of synoptic and subseasonal variabilities using a 33-yr dataset. From analysis of the contributions from u-wind and v-wind components, the PKE maximum in the Indian Ocean is attributed less to synoptic variability and more to intraseasonal variability in which the Madden–Julian oscillation (MJO) dominates; however, there is strong evidence of seasonal variability affiliated with the Asian monsoon systems. The ones in the eastern Pacific and Atlantic Oceans are closely related to both intraseasonal and synoptic variability that result from the strong MJO and the relatively large amplitude of equatorial waves.

The maintenance of the PKE budget mainly depends on the structure of time mean horizontal flows, the location of convection, and the transport of PKE from the extratropics. In the regions with strong convective activities, such as the eastern Indian Ocean to the western Pacific, the production of PKE occurs between 700 and 200 hPa at the expense of perturbation available potential energy (PAPE), which is generated by convective heating. This gain in PKE is largely offset by divergence of the geopotential component of vertical energy flux; that is, it is redistributed to the upper- and lower-atmospheric layers by the pressure field. Strong PKE generation through the horizontal convergence of the extratropical energy flux takes place in the upper troposphere over the eastern Pacific and Atlantic Ocean, and is largely balanced by a PKE loss due to barotropic conversion, which is determined solely by the sign of longitudinal stretching deformation. However, over the Indian Ocean, there is a net PKE loss due to divergence of energy flux, which is compensated by PKE gain through the shear generation.

---

1. Introduction

The perturbation kinetic energy (PKE) is extensively used to measure and study transient wave activity in the tropics (e.g., Webster and Chang 1988; Arkin and Webster 1985; Chen and Yanai 2000; Yanai et al. 2000). Murakami and Unninayar (1977) showed that the region of maximum PKE along the equator in January and February of 1971 was located in the vicinity of the equatorial westerlies. Using National Meteorological Center (NMC) operational tropical objective analyses from 1968 to 1979, Arkin and Webster (1985) indicated that the maximum PKE at 200 hPa in the tropics is collocated with a zone of equatorial westerlies where

Corresponding author address: Dr. Baode Chen, Shanghai Typhoon Institute, 166 Puxi Rd., Shanghai 200030, China.
E-mail: baode@mail.typhoon.gov.cn

DOI: 10.1175/AMSMONOGRAPHS-D-15-0017.1

convective activity is minimum. Furthermore, Webster and Yang (1989) extended the analysis of Arkin and Webster by considering the vertical distribution of the mean PKE and its relationship with the structure of mean zonal wind field. Their study and others (e.g., Liebmann 1987; Webster and Chang 1988) showed quite similar results: the tropics can be divided into two regimes conditioned on the direction of the time-mean upper-level zonal wind at the equator, and that PKE maximum is located within the westerlies.

It needs to be pointed out that the PKE utilized in aforementioned studies was defined as the time mean kinetic energy of the motions on having time scales of less than one month. In consequence, the intraseasonal variability containing the Madden–Julian oscillation [MJO; Madden and Julian (1971), (1972), (1994); Zhang (2005); see also chapter 4 (Maloney and Zhang 2016) and chapter 5 (Krishnamurti et al. 2016) in this monograph], which is among the most prominent large-scale motions in the tropical atmosphere, had been underestimated. Yanai et al. (2000) examined the PKE in the intraseasonal time scale and its budget for the TOGA COARE intensive observation period (IOP; hereafter COARE IOP) by integrating the power spectra and cospectra over the period range of 30–60 days. They showed that, in addition to two PKE maxima over the equatorial eastern Pacific and Atlantic Oceans where the westerlies were predominant in the upper troposphere, a large PKE center associated with strong convective activity over the warm pool region was located above the 200-hPa level. It was found that the interaction between convection and large-scale circulation, via production and conversion of perturbation available potential energy (PAPE) in the middle troposphere, plays a major role in the maintenance and growth of the MJO over the Indian Ocean–western Pacific warm pool, where the wave energy flux is clearly radiating upward and downward from the convective source region. In the central-eastern Pacific, where deep cumulus convection is suppressed, there are strong equatorward fluxes of wave energy from the subtropics of both hemispheres, causing horizontal convergence of wave energy flux in the equatorial upper troposphere. The results obtained from the climatological data indicate that the conclusions derived from the COARE IOP cases by Yanai et al. (2000) can be generalized to a great degree (Chen and Yanai 2000).

The most conspicuous feature of the tropical PKE is the collocation of PKE maximum with the equatorial westerlies in the upper troposphere. Various theories and hypotheses have been proposed to explain this feature. Charney (1969) studied the meridional propagation of large-scale wave disturbances into the tropics and suggested that disturbances propagating into the tropics will tend to be confined in the upper troposphere and lower stratosphere where zonal winds are weak easterly or westerly. Bennett and Young (1971) discussed in detail the effects of horizontal shear on the meridional propagation of disturbances. They pointed out that 1) disturbances with large eastward phase speeds cannot propagate into the tropics, 2) disturbances whose phase speeds coincide with the mean flow somewhere are absorbed at the critical latitude, and 3) disturbances with greater westward phase speeds than the mean flow may freely propagate into the tropics. Several studies (e.g., Webster and Holton 1982; Magaña and Yanai 1991; Zhang and Webster 1992) further emphasized the interaction of tropical and extratropical activity, and indicated that the regional maximum in PKE is a result of the equatorward propagation of disturbances into the tropics through the westerly duct (i.e., the convergence of equatorward wave energy flux). On the other hand, Webster and Chang (1988) and Chang and Webster (1990, 1995) have offered another explanation, that is, the PKE maximum may be associated with the accumulation of wave energy flux emanating from the tropical source region through nonuniform zonal flow. In addition, Wang and Xie (1996) suggested that a westerly vertical shear provides a favorable condition for the trapping of Rossby and Yanai waves in the upper-troposphere waves and may be partially responsible for the PKE maximum in the upper-tropospheric westerly duct.

The PKE budget analysis has further been used to diagnose the energetics of the intraseasonal variability and MJO in model simulations. Mu and Zhang (2006) examined the PKE budget in the modified NCAR CAM3, and pointed out that different mechanisms are responsible for the PKE production at different locations. Deng and Wu (2011) computed the PKE budget to delineate the physical processes that led to improved MJO simulations by a general circulation model. These studies demonstrated that the PKE budget could be a powerful tool to evaluate model performance and to investigate the physical processes resulting in the development and maintenance of the simulated tropical disturbances.

This chapter reviews and expands on the seminal work by Yanai et al. (2000). The primary objective is to examine the maintenance mechanisms of PKE in various synoptic and subseasonal time scales (i.e., 2–15, 17–24, and 30–60 days) by evaluating the PKE budget equation using a long period dataset, and, for the sake of model evaluation, to provide a climatological mean distribution of the PKE budget. It is also intended to serve as a reference for the PKE and PAPE budget computations, specifically the computation of energy transformation and generation functions using cross-spectrum analysis. The formula of the generation function calls for the apparent heat source (Q_1; Yanai et al. 1973), which is the budget residual of the thermodynamic equation; therefore, we describe its computation to demonstrate the principle of computing budget equations in the spherical-pressure coordinates.

The data and analysis procedures are briefly described in section 2. Section 3 examines spatial distribution of the tropical PKE in different time scales. Sections 4, 5, and 6 present results from the terms of shear generation, conversion of PAPE, and energy flux in the PKE budget equation, respectively. In addition, an analysis for the PKE budget residual is also included in section 6. The summary and discussions are given in section 7. Finally, methods for computing cross-spectra and budget equation residuals are described in the appendixes.

2. Data and analysis procedures

The primary dataset used in this study is the European Centre for Medium-Range Weather Forecasts (ECMWF) interim reanalysis (ERA-Interim) which is the latest global atmospheric reanalysis produced by the ECMWF to replace ERA-40. The ERA-Interim was generated by a frozen global data assimilation system along with an observational database and overcame several difficult data assimilation problems that were mostly related to the use of satellite data (Dee et al. 2011). The ERA-Interim provided an improved representation of the hydrological cycle, a more realistic stratospheric circulation, and better temporal consistency on a range of time scales. An advantage of using reanalysis is that the dataset is not subject to changes in the operational analyses. Furthermore, in the tropics, especially in the areas with sparse observations, both operational analysis and reanalyses depend greatly on the first guess supplied by the forecast model, and consequently depend critically on the diabatic heating distribution produced by the physical parameterization of the model. Comparison between the precipitation from the ERA-Interim, and independent surface- and satellite-based observations indicate that the ERA-Interim has substantial improvement in the precipitation due to improvements in the cloud and convection schemes and that the analyzed circulation is probably near reality.

The ERA-Interim dataset used in the study covers a 33-yr period from 1979 to 2011 and was averaged into daily mean data with a grid mesh of $1.5° \times 1.5°$ at 28 pressure levels. The domain of study is 50°S–50°N and 0°–360°.

Following Yanai et al. (2000), the PKE is defined as $\overline{k} = (\overline{u'^2} + \overline{v'^2})/2$, where the overbar and prime represent the time mean over a defined period and departure from the time mean, respectively. The equation governing PKE under the quasi-static (hydrostatic) approximation can be written as

$$\frac{\partial \overline{k}}{\partial t} = -\overline{\mathbf{v}'\mathbf{v}'} \cdot \nabla \cdot \overline{\mathbf{v}} - \overline{\mathbf{v}'\omega'} \cdot \frac{\partial \overline{\mathbf{v}}}{\partial p} - \overline{\alpha'\omega'}$$

$$- \nabla \cdot \mathbf{F}_h - \frac{\partial F_p}{\partial p} + \overline{\mathbf{v}' \cdot \mathbf{f}_r}. \qquad (8\text{-}1)$$

In Eq. (8-1), $\mathbf{F}_h \equiv \overline{\phi'\mathbf{v}'} + \overline{k}\,\overline{\mathbf{v}} + \overline{\mathbf{v}'^2\mathbf{v}'}/2$, and $F_p \equiv \overline{\phi'\omega'} + \overline{k}\,\overline{\omega} + \overline{\mathbf{v}'^2\omega'}/2$, where $\mathbf{v}$ is the horizontal velocity, $\omega = dp/dt$ (the vertical p velocity), p is the pressure, α is the specific volume, ∇ is the isobaric gradient operator, ϕ is the geopotential, and $\mathbf{f}_r$ is the frictional force per unit mass.

The right-hand side terms of Eq. (8-1) represent, respectively, the production of PKE through barotropic conversion process (the shear generation terms), the conversion from the perturbation available potential energy (more details in section 5) through $(\alpha, -\omega)$ correlation, the horizontal and vertical convergence of wave energy flux $(\mathbf{F}_h, F_p)$, and the work done by frictional force. Furthermore, the energy flux $(\mathbf{F}_h, F_p)$ is separated into two components: $(\mathbf{F}_h, F_p) = (\mathbf{F}_h, F_p)_{gh} + (\mathbf{F}_h, F_p)_{PKE}$, where $(\mathbf{F}_h, F_p)_{gh} = (\overline{\phi'\mathbf{v}'}, \overline{\phi'\omega'})$ is the geopotential component that contributes to redistribution of energy by pressure, and $(\mathbf{F}_h, F_p)_{PKE} = (\overline{k}\,\overline{\mathbf{v}} + \overline{\mathbf{v}'^2\mathbf{v}'}/2, \overline{k}\,\overline{\omega} + \overline{\mathbf{v}'^2\omega'}/2)$ mainly represents PKE flux.

The terms mentioned above (i.e., various energy transformation functions) involve covariance between two quantities. By utilizing cross-spectrum analysis technique, the contribution from variability at a certain frequency to the total covariance (i.e., cospectrum) can be obtained. Therefore, Eq. (8-1) can be calculated at different frequencies. In this study the lag-correlation method is used in the spectral analysis to conserve variance and covariance (see details in appendix A).

We used a moving block bootstrap method to estimate the covariance terms. Since we are interested in synoptic to intraseasonal time scales, the data were grouped into 189 successive overlapping (by 63 days) segments with a length of 183 days. The calculation was carried out for each segment and the final result was obtained by averaging over all available segments of the 33-yr record. The mean of each segment was first removed. In the lag-correlation method the maximum lag number (M) is set to 60 days, then the variability can be resolved at periods of $2M/k$, ($k = 0, 1, 2, \ldots, M$) (i.e., the mean, 120 days, 60 days, 40 days, 30 days, etc.). The variance and covariance were grouped into the contributions from three period bands: 30–60 days (intraseasonal time scale), 17–24 days (interim time scale), and 2–15 days (synoptic time scale).

The selection of these time scales is ultimately subjective. The multiscale tropical convection–coupled disturbances are known to have broadband time spectra and often scale-invariant characteristics below the

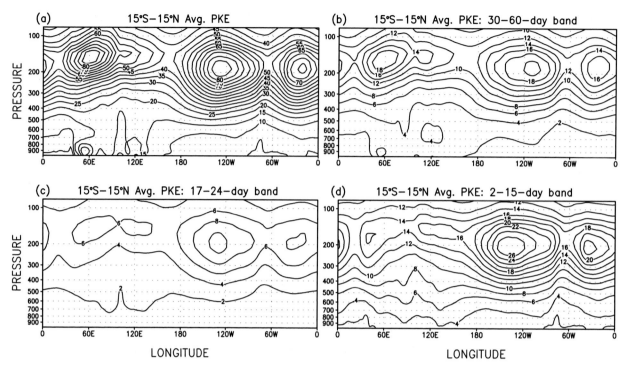

FIG. 8-1. Longitude–height cross sections of PKE in the (a) total period range, (b) 30–60-day band, (c) 17–24-day band, and (d) 2–15-day band (J kg^{-1}).

intraseasonal time scale (e.g., Tung and Yanai 2002a; Tung et al. 2004). As exemplified by the high-resolution spatial–temporal modes of tropical convection signals extracted from satellite infrared brightness temperature throughout 1984–2006 in Tung et al. (2014), ~(30–60)-day spectral maxima are prevailing features in the time spectra across various modes, accompanied with a power-law decaying range down to below 10 days until the diurnal cycle (e.g., Fig. 5 in their paper). The 2–15-day scale was chosen for known equatorial waves active in this range (e.g., review by Kiladis et al. 2009) and its significance in short- to extended-range synoptic weather forecasting. The interim 17–24-day range is thus named accordingly.

3. Distribution of PKE

Figures 8-1a–d show the PKE averaged between 15°S and 15°N as functions of longitude and height (pressure) (hereafter x–p section) for the total and the three period bands. In Fig. 8-1a, the maximum centers in total PKE are located above the tropical Indian, the eastern Pacific, and Atlantic Oceans; they occupy the mid- to upper troposphere between 400 and 100 hPa. The two maxima in the Western Hemisphere are stronger than the one in the Eastern Hemisphere and reside in a lower altitude. The spatial distribution of PKE in the three period bands bears a resemblance to that of total PKE, but

naturally with smaller magnitudes. It is noted that, in the intraseasonal time scale, the maximum in the Eastern Hemisphere shows nearly equal intensity to those in the Western Hemisphere (Fig. 8-1b). However, in the synoptic time scale, the former appears weaker than the latter (Fig. 8-1d). Moreover, the PKE maxima in the intraseasonal time scale are relatively higher [~(150–200) hPa] than those in the other two time scales (~200 hPa). In this time scale, the maximum center around 120°W is more eastward displaced than other period bands. Compared with the PKE in the other two period bands, that in the interim 17–24-day band has very small contribution to the total PKE.

To interpret the PKE in the total PKE (Fig. 8-1a), one must consider the mean PKE (zero frequency) and a portion of the seasonal PKE (120-day period) in addition to the three period bands (see section 2 and appendix A). In general, the mean PKE accounts for 10% of the total PKE, and the 120-day PKE accounts for 15% (neither is shown here). Particularly, the intensity of seasonal PKE is evidently quite strong in the Indian Ocean, with the 120-day PKE reaching up to 25%–30% of the total PKE maximum center. This large seasonal PKE is considered to be associated with the Asian monsoon system, which migrates annually between the Northern and Southern Hemispheres in accordance with the apparent movement of the sun (e.g., Webster et al. 1998; Chao and Chen 2001). It may also be related to the seasonality of the

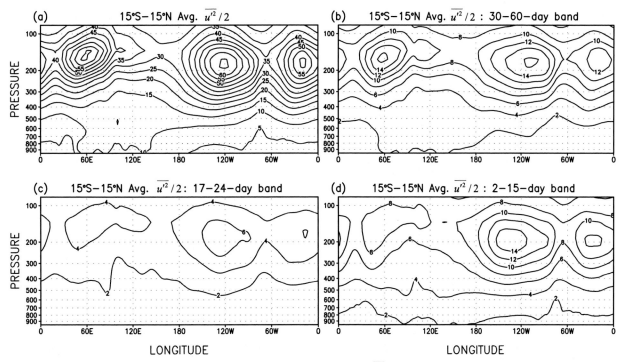

FIG. 8-2. As in Fig. 8-1, but for $\overline{u'^2}/2$.

tropical intraseasonal variability (e.g., reviews by Madden and Julian 1994; Lau and Waliser 2012).

Figures 8-2 and 8-3 show $\overline{u'^2}/2$ and $\overline{v'^2}/2$, the respective contributions of zonal and meridional wind components to the PKE in Fig. 8-1. As seen in Figs. 8-2a and 8-3a, and compared with Fig. 8-1a, the contribution from the $\overline{u'^2}/2$ to the total PKE is larger than that from the $\overline{v'^2}/2$. In particular, much greater contribution from $\overline{u'^2}/2$ to the

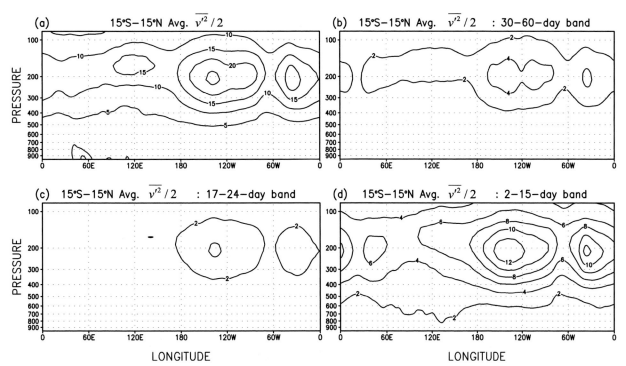

FIG. 8-3. As in Fig. 8-1, but for $\overline{v'^2}/2$.

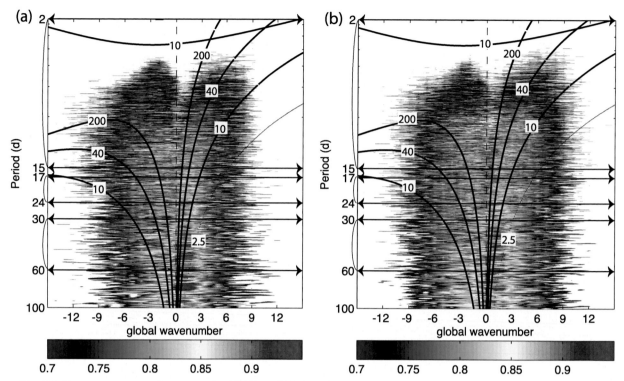

FIG. 8-4. Frequency–wavenumber spectra for (a) 850- and (b) 200-hPa zonal wind, calculated using the fast Fourier transform after a Hanning window was applied. Logarithmic powers normalized by the domain maximum are shown if ≥0.7. The ordinates are scaled logarithmically. The dispersion curves shown are meridional mode, $n = 1$ inertia–gravity waves, equatorial Rossby waves, and equatorial Kelvin waves. The associated equivalent depths are in m. The three period bands examined in the paper are marked: 30–60, 17–24, and 2–15 days.

PKE maximum over the Indian Ocean is observed. The contribution to total $\overline{u'^2}/2$ by the intraseasonal band is overall comparable to that by the synoptic band (Figs. 8-2b and 8-2d); however, much larger contribution from seasonal (not shown) and intraseasonal variability (Fig. 8-2a) can be seen over the Indian Ocean. From Figs. 8-3b and 8-3c, contribution to the total $\overline{v'^2}/2$ from variability of the 2–15-day band is much more significant in the eastern Pacific and Atlantic Oceans, indicating the presence of relatively large-amplitudes Rossby waves that are considered to be associated with an equatorward propagation of wave energy from higher latitudes by many authors (e.g., Kiladis 1998; Straub and Kiladis 2003).

Moreover, comparing Fig. 8-2b with Fig. 8-3b, in the 30–60-day band $\overline{u'^2}/2$ is much stronger than $\overline{v'^2}/2$; comparing Fig. 8-2b with Fig. 8-1b, about ¾ of PKE in 30–60-day band comes from the contribution of $\overline{u'^2}/2$, suggesting that the MJO plays a key role in the PKE distribution because of the large MJO amplitude in u-wind component especially at its "dry" phase across the Western Hemisphere (e.g., review by Madden and Julian 1994). This is further corroborated by Fig. 8-4, which shows the frequency–wavenumber power spectra of the zonal wind at 850 and 200 hPa. The spectra are

overlaid with dispersion curves as in previous work by Hayashi (1982), Takayabu (1994), and Wheeler and Kiladis (1999); here, they were calculated by assuming a static base state, with the marked equivalent depths for each equatorially trapped shallow-water wave type (Matsuno 1966; Lindzen and Matsuno 1968). These spectra indicate that the eastward-moving MJO is the most dominant intraseasonal signal in both wind fields. By examining Fig. 8-2d and Fig. 8-3d together, it is shown that the contribution to PKE in the 2–15-day period band from $\overline{v'^2}/2$ is comparable to that from $\overline{u'^2}/2$ in terms of magnitude. The relatively large contribution of $\overline{v'^2}/2$ to PKE in the 2–15-day period might be related to strong wave activities in particular over the eastern Pacific and Atlantic Oceans. A pronounced feature in this region is the upper-tropospheric "westerly duct" in the northern fall through spring that allows extratropical Rossby waves to propagate through (Webster and Holton 1982; Tomas and Webster 1994; Yang and Hoskins 1996). In the 17–24-day period, both $\overline{u'^2}/2$ and $\overline{v'^2}/2$ are much smaller than those in the other two period bands (Figs. 8-1c, 8-2c, and 8-3c).

In summary, two PKE maxima in the eastern Pacific and Atlantic Oceans mostly comprise the variance of two time scales: the 30–60-day intraseasonal time scale

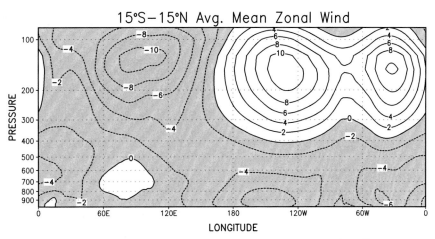

FIG. 8-5. Longitude–height cross section of mean zonal wind (m s^{-1}).

and the 2–15-day synoptic time scale. The variance from both time scales, as well as from the 120-day seasonal variation, contribute to the PKE maxima in the Indian Ocean. On the intraseasonal time scale in all three regions, the dominant contributor to the PKE is the zonal wind perturbations. Moreover, the synoptic time-scale activities are stronger in the eastern Pacific and Atlantic Oceans than those of the Indo-Pacific warm-pool sector, and are marked with comparable contributions from both zonal and meridional wind perturbations.

Figure 8-5 shows the longitude–height section of zonal wind averaged between 15°S and 15°N. The total PKE maximum centers over the eastern Pacific and Atlantic Oceans in Fig. 8-1a are collocated with the equatorial westerlies in the upper troposphere, while large values of PKE and easterlies in the upper levels coincide in the regions from the Indian Ocean to the western Pacific where the most active convection exists. The collocation in the eastern Pacific and Atlantic Oceans was

recognized a long time ago and well documented (e.g., Murakami and Unninayar 1977; Arkin and Webster 1985; Webster and Chang 1988), but recognition of the one in the Eastern Hemisphere is relatively new (Yanai et al. 2000).

Figure 8-6 is the sum of all terms on the right-hand side of Eq. (8-1) except friction forcing for all period ranges. Practically, the time tendency of PKE is zero for a long time mean, so the residual represents the negative friction forcing (i.e., $-\overline{\mathbf{v}' \cdot \mathbf{f}_r}$) plus the computation error. It can be seen that the $\overline{\mathbf{v}' \cdot \mathbf{f}_r}$ is negative everywhere and especially in the boundary layer, although areas with small positive values are found over the oceans around 700 hPa and on the upwind side of the Andes and the Ethiopian Highlands at around 200 hPa. These positive signals are nontrivial; they may well be associated with momentum transports of convection and convectively coupled waves [e.g., review by Sui and Yanai (1986); Tung and Yanai (2002a,b); and

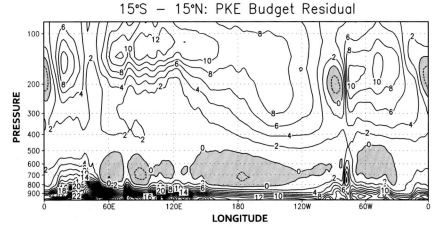

FIG. 8-6. Longitude–height cross section of PKE budget residual in total period range (J kg^{-1} day^{-1}).

fundamental theories in chapter 9 (Khouider and Majda 2016) and chapter 10 (Majda and Stechmann 2016)] as well as mountain-induced internal gravity waves (e.g., Durran 1990). Therefore, the residual indicates that our budget calculation is rather accurate and reliable as well.

In the subsequent sections, we will discuss processes that maintain the PKE distribution by calculating and analyzing the PKE budget equation. Since the interim band contributes minimally to the total PKE, we mainly focus on the discussions of total, intraseasonal, and synoptic period bands. The interim band is mentioned only when there are sufficiently significant findings.

4. Horizontal shear generation

To gain a better understanding of the role played by the horizontal shear generation or barotropic conversion in the maintenance of PKE, it is helpful to rewrite this term as

$$\underbrace{-\overline{v'v'} \cdot \mathbf{V} \cdot \overline{\mathbf{v}}}_{SG} = \underbrace{-\overline{u'u'}\frac{\partial \overline{u}}{\partial x}}_{SG_1} - \underbrace{\overline{v'v'}\frac{\partial \overline{v}}{\partial y}}_{SG_2} - \underbrace{\overline{u'v'}\left(\frac{\partial \overline{u}}{\partial y}+\frac{\partial \overline{v}}{\partial x}\right)}_{SG_3} .$$

That is, SG (horizontal shear generation of PKE) is divided into the PKE generation through the work done by stretching deformation of mean horizontal flows (SG_1 and SG_2), and by shearing deformation (SG_3).

Figures 8-7a–c show the longitude–height sections of SG for total, intraseasonal, and synoptic period ranges. From Fig. 8-7a (total period range), three large positive generation values in the upper troposphere can be found around the eastern coast of Africa (40°–60°E), the western coast of America (130°–80°W), and the central Atlantic Ocean (30°W–10°E). Moreover, the negative values cover Africa, America, and a broad region from the central Indian Ocean to the eastern Pacific Ocean. In the intraseasonal and synoptic period ranges (Figs. 8-7b and 8-7c), both distributions of SG are quite similar to Fig. 8-7a but overall SG in the intraseasonal time scale is stronger than that in the synoptic time scale.

Terms SG_1, SG_2, and SG_3 are shown in Figs. 8-8a–c, Figs. 8-9a–c, and Figs. 8-10a–c for the total, intraseasonal, and synoptic period range, respectively. Among the three terms, SG_1 is the dominant one in all period ranges and determines net effect of the horizontal shear generation (SG). As a result, SG_1 shows a very similar structure as SG. In all period ranges SG_2 shows an opposite sign with SG_1 and offsets a portion of the contribution of SG_1 although the offsetting values are different in various period ranges. Finally, SG_3 is rather small and negligible in comparison with SG_1.

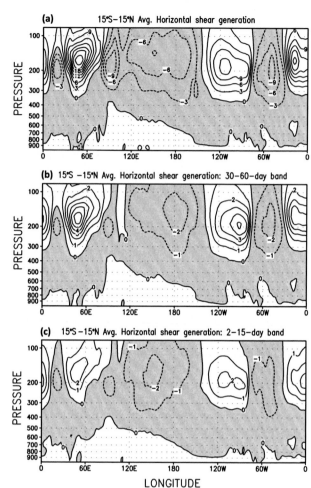

FIG. 8-7. Longitude–height cross sections of horizontal shear generation (SG) in the (a) total period range, (b) 30–60-day band, (c) 17–24-day band, and (d) 2–15-day band (J kg^{-1} day^{-1}).

The sign of SG, which indicates generation or destruction of PKE, is mainly determined by that of SG_1, which is, in turn, decided solely by the sign of longitudinal stretching deformation $\partial \overline{u}/\partial x$ due to $\overline{u'u'} > 0$; that is, in the region of $\partial \overline{u}/\partial x < 0$ ($\partial \overline{u}/\partial x > 0$), PKE is generated (destroyed). The longitude–height section of $\partial \overline{u}/\partial x$ is illustrated in Fig. 8-11. From Fig. 8-11 together with Figs. 8-5, 8-7, and 8-8, it can be seen that, through work done by longitudinal stretching of the mean flow, the generation of PKE occurs to the east of the equatorial westerly and to the west of the easterly where $\partial \overline{u}/\partial x < 0$, and the destruction occurs to the west of the westerly and to the east of the easterly where $\partial \overline{u}/\partial x > 0$. From the wave propagation view, the term SG_1 is related to the accumulation of wave energy that accompanies wave action flux in a spatially varying mean flow. Webster and Chang (1988) and Chang and Webster (1990) theoretically explored the phenomenon of energy accumulation of transient equatorially

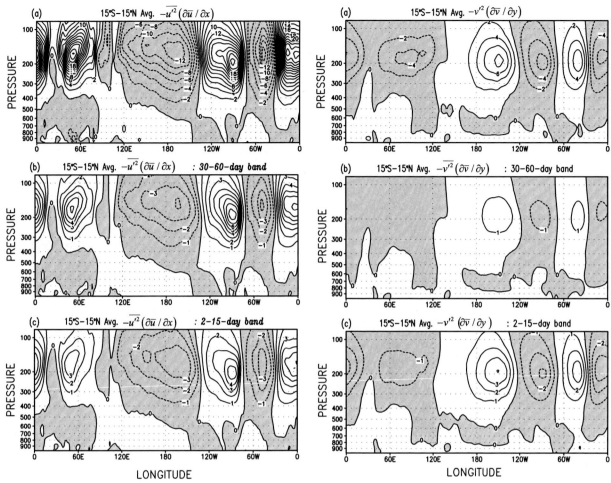

FIG. 8-8. As in Fig. 8-7, but for SG$_1$.   FIG. 8-9. As in Fig. 8-7, but for SG$_2$.

trapped modes, and pointed out that regions where $\partial \overline{u}/\partial x > 0$ ($\partial \overline{u}/\partial x < 0$) are regions of wave action flux (i.e., wave energy density) divergence (convergence). Thus, the accumulation of wave energy occurs to the east of the maximum westerly wind that enhances the intensity of disturbances (i.e., the generation of PKE).

In the period ranges discussed, PKE generation from vertical shear of the mean flow [i.e., $-\overline{\mathbf{v}'\omega'} \cdot (\partial\overline{\mathbf{v}}/\partial p)$] (not shown) is very small and its contribution is negligible.

5. Conversion of perturbation available potential energy

Figures 8-12a–c illustrate the longitude–height cross sections of the total covariance of specific volume (α) and vertical motion ($-\omega$) and the partial covariance in the 30–60-day, and 2–15-day period ranges, respectively. The term $-\overline{\alpha'\omega'} > 0$ represents conversion from PAPE to PKE through rising (sinking) motion of warm (cold) air parcels (baroclinic conversion). In the four period ranges,

the large conversion from PAPE to PKE occurs almost exclusively in the deep tropospheric layer between 500 and 200 hPa. In the total period range (Fig. 8-12a), the generation of PKE takes place in almost all the tropical domain shown, with the largest values in a region from the Indian Ocean across the western Pacific warm pool beyond the date line and in the continents of South America and Africa where the most active cumulus convection is observed. In the intraseasonal time scale, the most significant conversion from PAPE to PKE is confined to the Indo-Pacific warm-pool region, which is in agreement with a well-known and well-documented fact that the most active convection associated with the MJO exists only over a region from the central Indian Ocean to the central Pacific Ocean (see review by Zhang 2005). In the synoptic time scale, in addition to the large and broad generation of PKE existing between the Indian and Pacific Oceans, two centers are found in the continents of Africa and South America suggesting a strong convection activity in the synoptic time scale.

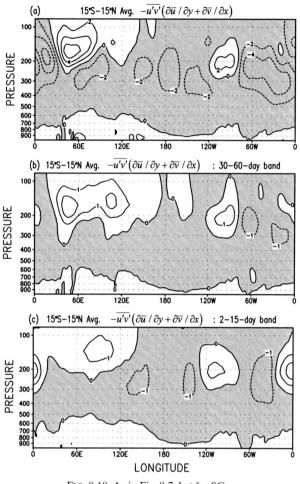

FIG. 8-10. As in Fig. 8-7, but for SG$_3$.

Following Yanai et al. (2000), the PAPE is defined as $\bar{e} = \overline{\alpha'^2}/2S$, where $S = -\bar{\alpha}\partial\bar{\theta}/\partial p$ is the static stability factor and θ is the potential temperature; the approximate equation for the time change of PAPE is

$$\frac{\partial \bar{e}}{\partial t} = -\frac{1}{S}\overline{\alpha'\mathbf{v}'}\cdot\nabla\bar{\alpha} + \overline{\alpha'\omega'} - \nabla\cdot\mathbf{E}_h - \frac{\partial E_p}{\partial p} + \frac{R}{c_p Sp}\overline{\alpha'Q_1'},$$

(8-2)

with $\mathbf{E}_h \equiv \bar{e}\,\bar{\mathbf{v}} + \overline{\mathbf{v}'\alpha'^2/2S}$ and $E_p \equiv \bar{e}\,\bar{\omega} + \overline{\alpha'^2\omega'/2S}$, R is the gas constant of dry air, c_p is the specific heat capacity at constant pressure, and Q_1 is the diabatic heating. The right-hand side of Eq. (8-2) represent, respectively, the conversion of mean available potential energy through horizontal heat flux due to perturbation, the conversion from PAPE into PKE through the $(\alpha, -\omega)$ correlation [cf. Eq. (8-1)], the horizontal and vertical convergence of fluxes of PAPE fluxes ($\mathbf{E}_h$, E_p), and production of PAPE through the (α, Q_1) correlation. Here, Q_1 can be evaluated as the residual of the large-scale thermodynamic equation budget (e.g., Yanai et al. 1973; Yanai and Johnson 1993; see appendix B for computation). Figures 8-13a–c show the term $(R/c_p Sp)\overline{\alpha'Q_1'}$ for the four period ranges. Compared with Figs. 8-12a–c, the terms $(R/c_p Sp)\overline{\alpha'Q_1'}$ and $-\overline{\alpha'\omega'}$ bear a remarkable resemblance to each other with regard to the distributions and the magnitudes, indicating that the latent-heating release and transport of heat in deep convection is a primary source of PAPE, and is almost solely converted into PKE through the correlation of (α, ω). This particular PKE generation and conversion were first noted in the Marshall Island area by Nitta (1970, 1972), and confirmed by Wallace (1971) and Kung and Merritt (1974).

Figures 8-12 and 8-13 indicate that the maximum heating and the conversion from PAPE to PKE take place above midtroposphere between 400 and 300 hPa for all frequency bands. In the tropics, apart from deep convection, a great amount of stratiform cloud exists, including mesoscale anvils, as well as those not

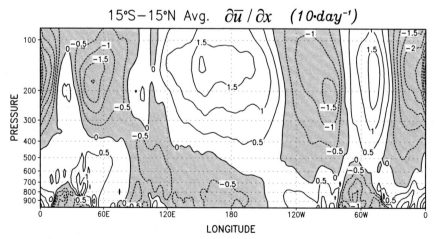

FIG. 8-11. Longitude–height cross section of the longitudinal stretching deformation of mean zonal wind ($\partial\bar{u}/\partial x$; 10 day^{-1}).

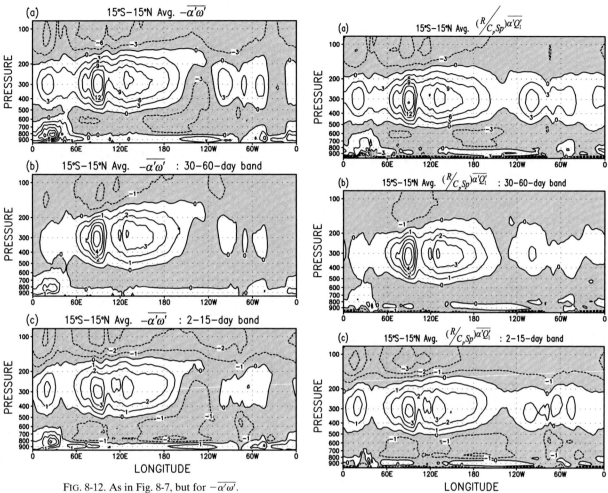

FIG. 8-12. As in Fig. 8-7, but for $-\overline{\alpha'\omega'}$.

FIG. 8-13. As in Fig. 8-7, but for $(R/c_p Sp)\overline{\alpha'Q_1'}$.

associated with convective systems (e.g., Houze 1982; Schumacher and Houze 2003). Lin et al. (2004) attributed the top heaviness of the heating profile of the MJO to stratiform precipitation. For other time scales, further investigation is warranted.

6. Energy flux and its convergence

a. Energy flux

Figures 8-14a–c show $(F_x, F_p)_{gh}$ (as in the geopotential component of wave energy flux) along the equatorial band between 15°S and 15°N for the total and two period ranges (see appendix C for the procedure to produce the plot). The most outstanding feature recognized in Fig. 8-14a is that the wave energy flux emanates upward and downward from a region (250–400 hPa) of wave energy flux over the warm pool from the eastern Indian Ocean to the date line. In addition, two regions between 100 and 400 hPa are identified off the coast of South America (~90°W) and East Africa (~30°E) where a

very strong westward horizontal transport exists besides the vertical emanations. Compared with Fig. 8-12, it can be found that the regions from where wave energy fluxes emanate away in Fig. 8-14a are collocated with the areas where the large production of PKE from PAPE occurs. In the 30–60-day band (Fig. 8-14b), $(F_x, F_p)_{gh}$ shows noticeable upward and downward emanations in the Indo-Pacific sector where the deep convective activity associated with the MJO resides, and dominant westward components between 100 and 400 hPa from South America. Moreover, $(F_x, F_p)_{gh}$ in the 2–15-day band (Fig. 8-14c) possesses the largest contribution to the total flux and shows a similar structure to those of the total period and the 30–60-day band.

The PKE components of wave energy flux along the 15°S–15°N latitude band [i.e., $(F_x, F_p)_{PKE}$] are illustrated in Figs. 8-15a–c for the total and two period ranges. It is striking that, for all bands, the regions of large eastward

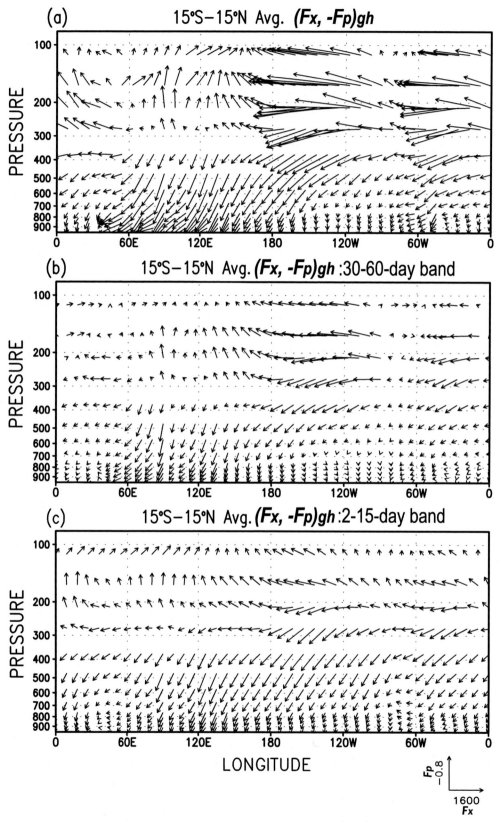

FIG. 8-14. As in Fig. 8-7, but for $(F_x, F_p)_{gh}$ (J m s^{-1} kg^{-1}).

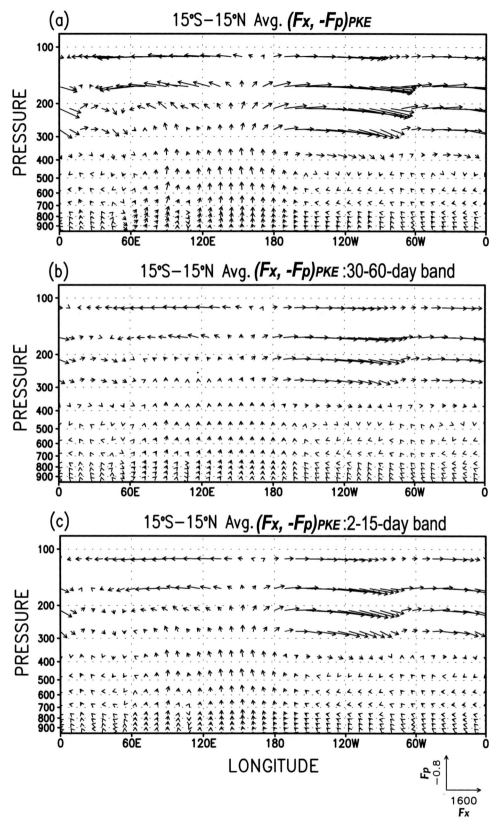

FIG. 8-15. As in Fig. 8-14, but for $(F_x, F_p)_{\text{PKE}}$.

horizontal flux between 400 and 100 hPa are from the date line to the American west coast, as well as from the Atlantic Ocean to the west coast of Africa with the largest convergence around South America. Vertical overturning features are also seen in these two regions, in particular in the 2–15-day period. From the total wave energy flux [i.e., $(F_x, F_p)_{gh} + (F_x, F_p)_{PKE}$] (Figs. 8-16a–c), the potential and PKE components of wave energy flux are partially canceled out over the Western Hemisphere. The potential component dominates the total wave energy flux and largely determines the distribution of the energy wave flux for all but the 2–15-day band in which the overturning features seen in the PKE component over the eastern Pacific and Atlantic Ocean are carried into the total wave energy flux.

Horizontal maps showing the total horizontal wave energy flux [i.e., (F_x, F_y)] in a vertical layer between 200 and 225 hPa are displayed in Figs. 8-17a–d for the total, intraseasonal, interim, and synoptic period ranges. It is noted that as time scales reduce from Figs. 8-17a to 8-17d, the prevailing energy flow patterns morph from being largely rotational in Fig. 8-17a to being convergent in Fig. 8-17d, likely reflecting the distinction between the geostrophic (Rossby) mode and the inertia–gravity mode of atmospheric motions. This gradual transition is more obvious in the Eastern Hemisphere than the Western Hemisphere where background westerlies permit extratropical Rossby waves to propagate through (cf. Fig. 8-5). In the total and 30–60-day period ranges (Figs. 8-17a and 8-17b), two rotational centers straddle the equator over the Indian Ocean. From the Pacific to the Atlantic regions, the energy flux alternates between northward and southward cross-equatorial transports. The vast area over the western Pacific between 10°S and 10°N is occupied by a northward flux, while the eastern Pacific has a southward flux. There is a southward flux from the extratropics to the Maritime Continent at around 120°E, adjacent to a rotational center between 120° and 150°E. Along the equator, the energy flux is clearly eastward over the Indian Ocean; two likely centers of energy flux convergence are ~120°E and ~150°W.

In the shorter period ranges (Figs. 8-17c,d), the energy flux over the equatorial Indian Ocean is westward, the opposite of that in Figs. 8-17a,b. Here, the energy flux converges over Africa and South America on the equator. Another center of convergence is evident over the Maritime Continent at ~120°E in Fig. 8-17c. Near the date line, the center of divergence is seen in both Fig. 8-17c (~150°W) and Fig. 8-17d (~180°). The eastern Pacific exhibits the most pronounced cross-equatorial southward energy fluxes delivered from the Northern Hemispheric extratropics.

In all time scales, but most significantly in the 2–15-day range, the northward flux from the southern extratropics and the southward flux from the northern extratropics meet near the equatorial regions from the central to eastern Pacific Ocean and the equatorial Atlantic Ocean, where the mean westerlies prevail showing the propagation of wave energy into the tropics from higher latitudes. From horizontal maps of $(F_x, F_y)_{gh}$ and $(F_x, F_y)_{PKE}$ (not shown), generally, the two west–east components [i.e., $(F_x)_{gh}$ and $(F_x)_{PKE}$] mostly offset each other for all four period ranges with relatively larger values in the geopotential part. But, the two north–south components [i.e., $(F_y)_{gh}$ and $(F_y)_{PKE}$] have the same signs in most regions where the PKE part is dominant, leading to an enhancement in the north–south components of total wave energy flux, particularly, the northward flux from the southern extratropics and the southward flux from the northern extratropics in the sector from the central to eastern Pacific Ocean and Atlantic Ocean.

It should be pointed out that there is an apparent inconsistency between the zonal component of the total fluxes at ~200 hPa in Fig. 8-16a and those around the equator in Fig. 8-17a, especially between 0°–120°E and 120°W–0°. As shown in Fig. 8-17a, the total fluxes in these areas have a very strong east–west component beyond 10°S and 10°N in the upper troposphere. They end up dominating the local sign of the averaged F_x in Fig. 8-16a. The averaged fluxes in Figs. 8-16b and 8-16c are not as much affected by the moderately strong zonal component in the higher latitudes, however.

b. Convergence of energy flux

The convergence of the geopotential and the PKE components of the energy fluxes is further divided into horizontal and vertical parts [i.e., $-\nabla \cdot (\mathbf{F}_h)_{gh}$, $-\nabla \cdot (\mathbf{F}_h)_{PKE}$, $-\partial(F_p)_{gh}/\partial p$, and $-\partial(F_p)_{PKE}/\partial p$]. Figures 8-18a–c illustrate the longitude–height cross sections of $-\nabla \cdot (\mathbf{F}_h)_{gh}$ averaged between 15°S and 15°N. In Fig. 8-18a, (total period band) wave energy flux convergence [$-\nabla \cdot (\mathbf{F}_h)_{gh} > 0$] exists in most areas from the western Indian Ocean across the Pacific Ocean to the eastern Pacific in the upper troposphere above 600 hPa. In addition, two fairly narrow but strong convergence centers cover most of Africa and South America. Two strong negative extrema are located at both coasts of Africa (around 5° and 45°E), and a relatively weak negative center is close to the western coast of South America (i.e., around 90°W). In the 30–60-day band (Fig. 8-18b), the patterns are quite similar to those in the total period band, but with a much stronger negative center off the west coast of South America (~90°W). In the 17–24-day band (not shown), strong convergence is found in

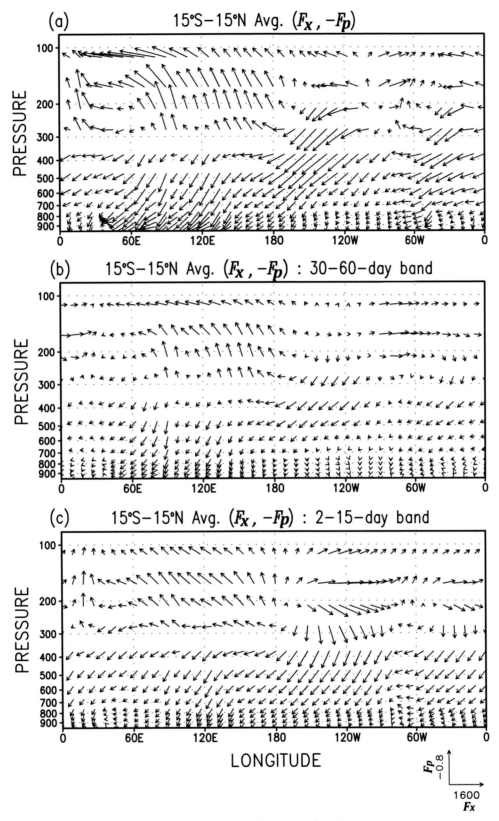

FIG. 8-16. As in Fig. 8-14, but for (F_x, F_p).

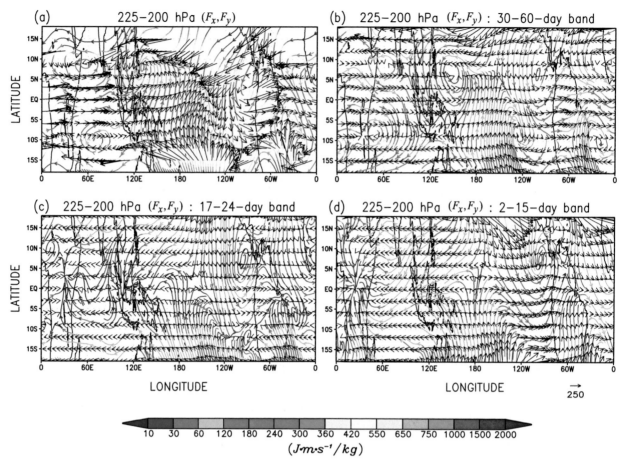

FIG. 8-17. Horizontal maps of (F_x, F_y) and its magnitudes (shaded) in the (a) total period range, (b) 30–60-day band, (c) 17–24-day band, and (d) 2–15-day band (J m s^{-1} kg^{-1}).

the sector from the central Pacific to the middle Atlantic Oceans. In the 2–15-day band (Fig. 8-18c), except for a weak negative center in the eastern coast of Africa, strong convergence $[-\nabla \cdot (\mathbf{F}_h)_{\mathrm{gh}} > 0]$ almost covers the entire globe in the middle to high troposphere.

The longitude–height cross sections of $-\nabla \cdot (\mathbf{F}_h)_{\mathrm{PKE}}$ are shown in Figs. 8-19a–c for the total and two period ranges. Compared with the cross sections for $-\nabla \cdot (\mathbf{F}_h)_{\mathrm{gh}}$ (i.e., Figs. 8-18a–c), in the regions between the Indian Ocean to the western Pacific, $-\nabla \cdot (\mathbf{F}_h)_{\mathrm{PKE}}$ is nearly compensated by that of geopotential component for all period bands. On the other hand, fairly strong convergence [i.e., $-\nabla \cdot (\mathbf{F}_h)_{\mathrm{PKE}} > 0$] takes place in the regions from the central to eastern Pacific, particularly, with the largest values in the 2–15-day band, which considerably enhances the convergence of the total wave energy flux there. In the other regions, $-\nabla \cdot (\mathbf{F}_h)_{\mathrm{PKE}}$ has an opposite sign to those of the geopotential component with relatively smaller values, indicating that $-\nabla \cdot (\mathbf{F}_h)_{\mathrm{gh}}$ mainly determines the distributions of $-\nabla \cdot \mathbf{F}_h$ in these regions.

Figures 8-20a–c illustrate the longitude–height cross sections of $-\nabla \cdot \mathbf{F}_h$ along the equator. In Fig. 8-20a (total period band) wave energy flux convergence ($-\nabla \cdot \mathbf{F}_h > 0$) exists in the areas from the western Indian Ocean to the eastern Pacific Ocean in the upper troposphere above 600 hPa; and the maximum of convergence is located at the central to eastern Pacific Ocean around 160°W. In addition, two other convergence centers can be identified off the eastern coast of America (~40°W) and over Africa. Strong divergence of wave energy flux ($-\nabla \cdot \mathbf{F}_h < 0$) occurs around the Arabic Sea and the western coast of Africa, and a weak divergence center is also seen in the eastern Pacific. In the 30–60-day band (Fig. 8-20b) the largest convergence of energy flux takes place in the two areas from the central to eastern Pacific Ocean and from the eastern coast of America to the central Atlantic Ocean, and the divergence is sporadically distributed with three centers in the central Atlantic Ocean, the eastern coast of Africa, and the eastern Pacific Ocean.

In 2–15-day band (Fig. 8-20c), the convergence of energy flux exists in most of the domain except around

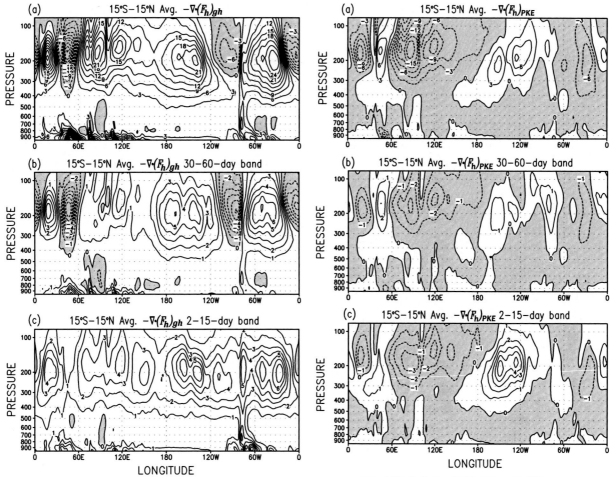

FIG. 8-18. As in Fig. 8-7, but for $-\mathbf{V} \cdot (\mathbf{F}_h)_{\mathrm{gh}}$.

FIG. 8-19. As in Fig. 8-7, but for $-\mathbf{V} \cdot (\mathbf{F}_h)_{\mathrm{PKE}}$.

the central Indian Ocean where there is small divergence. The largest convergence center is located at the eastern Pacific Ocean, and it extends westward to the western Pacific Ocean and eastward across the Atlantic Ocean to the eastern coast of Africa with two centers at 30°W and 20°E, respectively. The two convergence centers are found in the eastern Pacific Ocean and the central Atlantic Ocean in the 17–24-day band (not shown), and the small divergence occurs in other regions. Apart from the maximum of convergence in the regions from the central to eastern Pacific Ocean around 160°W in the upper troposphere above 600 hPa, which is strengthened by the contributions from both geopotential and PKE components, the distributions of $-\mathbf{V} \cdot \mathbf{F}_h$ is mainly decided by the divergence and convergence of geopotential component of energy wave flux for all period bands.

Figures 8-21a–c show the longitude–height cross sections of $-\partial (F_p)_{\mathrm{gh}}/\partial p$ along the equator for the total and two period bands. The most outstanding features

are that the predominant divergence occurs in the middle troposphere of 400–200 hPa for the whole longitudinal domain, and that the maximum divergence is collocated with the most intensive convection. In the 30–60-day period band (Fig. 8-21b), the strongest divergence exists in the sector between the Indian and Pacific Oceans where the eastward propagation of convection associated with the MJO is found. On the other hand, in the 2–15-day band (Fig. 8-21c) the strong divergence covers all tropical longitudes in the middle troposphere and is sandwiched by two positive areas in the lower and upper troposphere, although local variation is evident. The longitude–height cross sections of $-\partial F_p/\partial p$ {i.e., $-[\partial (F_p)_{\mathrm{gh}}/\partial p + \partial (F_p)_{\mathrm{PKE}}/\partial p]$} are displayed in Figs. 8-22a–c. Compared with Figs. 8-21a–c [i.e., $-\partial (F_p)_{\mathrm{gh}}/\partial p$], the contribution of geopotential component largely determines the distribution of total vertical divergence and convergence of wave energy flux, and $-\partial (F_p)_{\mathrm{PKE}}/\partial p$ (not shown) is very small and its contribution is negligible.

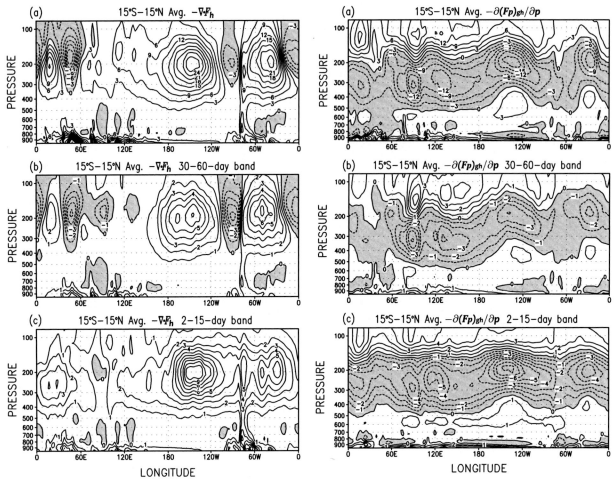

FIG. 8-20. As in Fig. 8-7, but for $-\mathbf{V} \cdot \mathbf{F}_h$.

FIG. 8-21. As in Fig. 8-7, but for $-\partial(F_p)_{\text{gh}}/\partial p$.

7. Summary and discussion

In this chapter we reviewed and demonstrated the diagnostic utility of PKE in the tropical regions for various time scales. We calculated all terms of the PKE budget equation except the one related to frictional force by applying cross spectrum analysis techniques to the computation of variance and covariance. This approach is designed in such a way that the budget of PKE can be analyzed in a desired frequency or period domain. Our emphasis has been placed on the time mean features of subseasonal variability based on a 33-yr dataset. The computational methods are described in appendixes A and B for the readers' reference.

In the course of our demonstration, three large PKE centers were found in the upper troposphere over the Indian Ocean, the eastern Pacific, and Atlantic Oceans; these are consistent with previous findings. From the analysis of the contributions from zonal and meridional perturbation wind components, the PKE maximum center in the Indian Ocean was shown to be dominated by the zonal component and is associated with synoptic variability and intraseasonal variability in which the MJO dominates. There was also evidence of strong seasonal variability, likely associated with the Asian monsoon systems and the seasonality of the tropical intraseasonal variability. The PKE maxima in the eastern Pacific and Atlantic Oceans appeared closely related to both intraseasonal and synoptic variability that result from the strong MJO and relatively large amplitude of tropical waves. These synoptic waves have comparable contributions from both zonal and meridional wind perturbations. In addition, the collocation of PKE maximum with the equatorial westerlies in the upper troposphere is evident over the eastern Pacific and Atlantic Oceans.

From an energetics point of view, to make the PKE budget balance, gain of PKE through certain processes must be compensated by the loss of the energy through

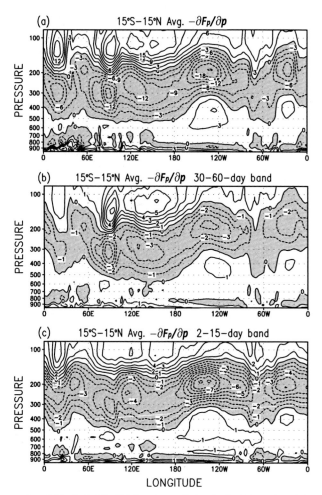

FIG. 8-22. As in Fig. 8-7, but for $-\partial F_p/\partial p$. in the (a) total period range, (b) 30–60-day band, (c) 17–24-day band, and (d) 2–15-day band (J kg^{-1} day^{-1}).

other processes. This energy balance is achieved, from a dynamical point of view, by adjusting the atmosphere to a new structure to be compatible with the energetics constraints through the dynamics and thermodynamics operating within the atmosphere. Our subsequent analysis revealed that the maintenance of the PKE budget mainly depends on the distribution of time mean horizontal flows, the location of convection, and the transport of PKE from the extratropics.

The most important conclusions can be summarized as follows for all period bands discussed. In the regions with strong convective activity, such as the Indian Ocean to the western Pacific, the gain of PKE occurs between 700 and 200 hPa through a positive correlation of $(\alpha, -\omega)$ at the expense of PAPE generated by convective heating through positive (α, Q_1) correlation. Such a gain is largely offset by the divergence of the geopotential component of vertical energy flux;

that is, it is redistributed to the upper and lower atmospheric layers by the pressure field. Strong PKE generation through the horizontal convergence of the extratropical energy flux takes place in the upper troposphere over the eastern Pacific and Atlantic Ocean, and is largely balanced by PKE loss due to the work done by horizontal shear, which is decided solely by the sign of longitudinal stretching deformation $\partial \overline{u}/\partial x$. However, over the Indian Ocean, there is a net PKE loss due to divergence of energy flux, which is compensated by PKE gain through the shear generation.

Finally, our calculation is performed by integrating the variance or covariance over a predefined period range. Notable differences can be found for the variability for different periods or frequencies. However, the results presented here can only serve as a general and static picture for a particular variability such as the MJO, because the detailed features of spectra are somewhat obscured by strong noise in the spectrum (e.g., weather noise and that resulted from budget calculations). To obtain the details of the PKE budget associated with a particular tropical system, proper prefiltering or significant tests must be performed. Moreover, for this chapter to be a first primer of PKE analysis, our focus is only placed on the time mean feature. More detailed statistical features, such as those conditioned on the phase of interannual variability, are of great interest for future research.

Acknowledgments. This work is based on an unpublished brief manuscript BC and WWT coauthored with the late Professor Michio Yanai. The authors BC and WWT are grateful for the constructive reviews and insightful comments by Drs. Jian-Wen Bao, Winston C. Chao, Robert G. Fovell, Changhai Liu, and one anonymous reviewer. BC was supported by the National Natural Science Foundation of China (Grant 41175094). WWT was partially supported by U.S. NSF Grants CMMI-0826119 and CMMI-1031958.

APPENDIX A

An Application of Cross-Spectrum Analysis to the Calculation of the Energy Transformation Functions

The terms in the PKE budget in Eq. (8-1) (i.e., various energy transformation functions) involve covariance between two quantities. Following the classic cross-spectrum analysis of equally spaced finite records (e.g., in early studies of tropical waves by Maruyama 1967, 1968 and Yanai et al. 1968), we consider two time series

$x_i (i = 1, 2, \ldots, N)$ and $y_i (i = 1, 2, \ldots, N)$, in which time means [i.e., $\bar{x} = (1/N)\sum_{i=1}^{N} x_i$ and $\bar{y} = (1/N)\sum_{i=1}^{N} y_i$] are removed. We first estimate the cross-covariance function by $C_{xy}(l) = [1/(N-1)]\sum_{i=1}^{N-l} x_i y_{i+l}$ where $l = 0, 1, 2, \ldots, M (\leq N/2)$ and M is the maximum of lag number that determines the frequency resolution (or interval) by $\Delta f = 1/(2M\Delta t) = f_N/M$ (Δt is the time interval and f_N is the Nyquist frequency). Then we construct the symmetric (or even) and the antisymmetric (or odd) parts {i.e., $C_+(l) = (1/2)[C_{xy}(l) + C_{yx}(l)]$ and $C_-(l) = (1/2)[C_{xy}(l) - C_{yx}(l)]$}. The cospectrum is obtained by the cosine transform of the symmetric component, that is, $K_{xy}(k) = 4\Delta t[\sum_{l=0}^{M} C_+(l)\cos(k\pi l/M)\delta_l]$ for $k = 0, 1, 2, \ldots, M$; for $l = 0, M, \delta_l = 1/2$, otherwise $\delta_l = 1$. The quadrature spectrum is the sine transform of the antisymmetric component {i.e., $Q_{xy}(k) = 4\Delta t[\sum_{l=0}^{M} C_-(l)\sin(k\pi l/M)\delta_l]$, $k = 0, 1, 2, \ldots, M$}. It can be shown that the covariance is $(1/N)\sum_{i=1}^{N} x_i y_i = \sum_{k=0}^{M}[K_{xy}(k) + Q_{xy}(k)]$, and the variance is $(1/N)\sum_{i=1}^{N} x_i^2 = \sum_{k=0}^{M}[K_{xx}(k)]$. Therefore, the terms $K_{xy}(k) + Q_{xy}(k)$ and $K_{xx}(k)$ can be identified as the contributions of perturbations with frequency of $k/(2M\Delta t)$ to the total covariance and variance, respectively; and the contribution from a particular frequency band can be obtained by aggregating those components within that frequency band. Using this technique we can calculate the variance and covariance terms, such as $\overline{u'^2}$, $\overline{v'^2}$, $\overline{\omega'T'}$, etc., for a certain period band, and analyze the PKE budget for that particular period band.

APPENDIX B

Example of Budget Computation Using Finite-Difference Approximation in the Spherical-Pressure Coordinates: Q_1

In research we frequently perform budget evaluations given state variables and equations of conservation laws. When the computation is executed with observational data, the results are always tainted with measurement and computational errors. To suppress the latter type of errors, the finite-difference approximation of these equations cannot be arbitrary. As a tribute to late Prof. Yanai, we use the apparent heat source, Q_1, which is the budget residual of the thermodynamic equation, as an example to show the appropriate finite-difference approximation in the spherical-pressure coordinates. The readers can code the approximation for the PKE in Eq. (8-1) and obtain the time-mean budget residual (i.e., $\overline{\mathbf{v}' \cdot \mathbf{f}_r'}$) (Fig. 8-6) with the information provided in appendixes A and B.

Following Yanai et al. (1973) and Yanai and Johnson (1993), the apparent heat source, Q_1, is the budget residual of the thermodynamic equation:

$$Q_1 \equiv c_p \left(\frac{p}{p_0}\right)^{\kappa} \left(\frac{\partial \bar{\theta}}{\partial t} + \bar{\mathbf{v}} \cdot \nabla \bar{\theta} + \bar{\omega} \frac{\partial \bar{\theta}}{\partial p}\right)$$
$$= Q_R + L(\bar{c} - \bar{e}) - \nabla \cdot \overline{s'\mathbf{v}'} - \frac{\partial \overline{s'\omega'}}{\partial p},$$

where $p_0 = 1000 \, \text{hPa}$, $\kappa = R/c_p$, Q_R is the radiative heating rate, s is the dry static energy, and c and e are the rates of condensation and evaporation (of cloud water) per unit mass of air, respectively. We have assumed that the Reynolds conditions and their consequences hold with sufficient accuracy. Note that, unlike in the main text or appendix A, which deal with time mean and its perturbations, the overbar here denotes *the running horizontal average with respect to a large-scale area* and the prime denotes *the deviation from the average*. That is, the scale separation takes place in space. The variables u, v, ω, and θ resolved at the grid points of the reanalysis are regarded as "large scale" and are marked with overbars in the first equation. They are used as the direct input to calculate the residual Q_1, which is then interpreted according the second equation as the total effect of radiative heating, latent heat released by net condensation, and the convergence of fluxes of sensible heat due to subgrid-scale eddies such as cumulus convection and turbulence.

The finite-difference scheme of the thermodynamic equation is casted in a vertically staggered grid so that Q_1 is computed between two standard isobaric levels of the reanalysis data. Furthermore, its advective form is consistent with its flux form. In the discrete spherical-pressure coordinate system (λ, φ, p, and t), where λ is the longitude and φ is the latitude, let i be the longitude index, j be the latitude index, k be the pressure index, and n be the time index. The distance between two longitudinal grid points depends on their latitudinal location j, that is, $\Delta x_j = a\cos\varphi_j\Delta\lambda$, with a the earth's radius. The distance between two latitudinal grid points is a constant, that is, $\Delta y = a\Delta\varphi$.

Following is a pseudocode for the finite-difference approximations for the terms in the first equation, at an arbitrary grid point with indices (i, j, k, n):

$$\left(\frac{\partial \bar{\theta}}{\partial t}\right)_{i,j,k,n} \approx \frac{1}{2}\left(\frac{\bar{\theta}_{i,j,k,n+1} - \bar{\theta}_{i,j,k,n-1}}{2\Delta t}\right.$$
$$\left. + \frac{\bar{\theta}_{i,j,k+1,n+1} - \bar{\theta}_{i,j,k+1,n-1}}{2\Delta t}\right),$$

$$\overline{\mathbf{v}} \cdot \nabla \overline{\theta}_{i,j,k,n} \approx \frac{1}{2} \left\{ \frac{1}{2} \frac{1}{a \cos\varphi_j} \left[(\overline{u}_{i+1,j,k,n} + \overline{u}_{i-1,j,k,n}) \frac{\overline{\theta}_{i+1,j,k,n} - \overline{\theta}_{i-1,j,k,n}}{2\Delta\lambda} \right. \right.$$

$$\left. + (\overline{v}_{i,j+1,k,n} \cos\varphi_{j+1} + \overline{v}_{i,j-1,k,n} \cos\varphi_{j-1}) \frac{\overline{\theta}_{i,j+1,k,n} - \overline{\theta}_{i,j-1,k,n}}{2\Delta\varphi} \right]$$

$$+ \frac{1}{2} \frac{1}{a \cos\varphi_j} \left[(\overline{u}_{i+1,j,k+1,n} + \overline{u}_{i-1,j,k+1,n}) \frac{\overline{\theta}_{i+1,j,k+1,n} - \overline{\theta}_{i-1,j,k+1,n}}{2\Delta\lambda} \right.$$

$$\left. \left. + (\overline{v}_{i,j+1,k+1,n} \cos\varphi_{j+1} + \overline{v}_{i,j-1,k+1,n} \cos\varphi_{j-1}) \frac{\overline{\theta}_{i,j+1,k+1,n} - \overline{\theta}_{i,j-1,k+1,n}}{2\Delta\varphi} \right] \right\}, \quad \text{and}$$

$$\overline{\omega} \frac{\partial \overline{\theta}}{\partial p}_{i,j,k,n} \approx \frac{1}{2} (\overline{\omega}_{i,j,k,n} + \overline{\omega}_{i,j,k+1,n}) \frac{\overline{\theta}_{i,j,k,n} - \overline{\theta}_{i,j,k+1,n}}{\Delta p}.$$

If the reanalysis has a horizontal mesh of $1.5° \times 1.5°$, 25-hPa vertical intervals, and 6-hourly time resolution, then $\Delta\lambda = 1.5° \times \pi/180° = \Delta\varphi$, $\Delta p = 2500$ Pa, and $\Delta t = 6$ h $= 21\,600$ s. Last, it is conventional to present Q_1 in terms of Q_1/c_p (K day^{-1}). At this point, readers are referred to Yanai and Tomita (1998) and Tung et al. (1999) for more aspects of computing heat and moisture budget residuals and the related analyses.

APPENDIX C

Determining the Vertical Scale of a Longitude–Height Vector Plot

A map's scale indicates the relationship between a certain distance on the map and the real distance on the earth's surface. A longitude–height plot usually has different scales between horizontal and vertical directions. When making a longitude–height vector plot, in which the magnitudes of horizontal and vertical components are represented by certain horizontal and vertical lengths on the plot, two scales must be identical. For the horizontal direction, let L be a length on the plot and X be a distance on the ground; for the vertical direction, let H be a vertical length on the plot and Z be a real vertical distance, thus the scales for both directions are $R_X = L/X$ and $R_Z = H/Z$, respectively. To make a (u, w) vector plot, the proportion, $R_X u_p : R_Z w_p = u : w$, must be maintained, where u_p and w_p are the scaled u and w. Therefore, the (u_p, w_p) should be used to make the plot with $(u_p, w_p) = (u, R_X w/R_Z)$ or $(u_p, w_p) = (R_Z u/R_X, w)$.

REFERENCES

Arkin, P. A., and P. J. Webster, 1985: Annual and interannual variability of tropical–extratropical interaction: An empirical study. *Mon. Wea. Rev.*, **113**, 1510–1523, doi:10.1175/1520-0493(1985)113<1510:AAIVOT>2.0.CO;2.

Bennett, J. R., and J. A. Young, 1971: The influence of latitudinal wind shear upon large-scale wave propagation into the tropics. *Mon. Wea. Rev.*, **99**, 202–214, doi:10.1175/1520-0493(1971)099<0202:TIOLWS>2.3.CO;2.

Chang, H. R., and P. J. Webster, 1990: Energy accumulation and emanation at low latitudes. Part II: Nonlinear response to strong episodic equatorial forcing. *J. Atmos. Sci.*, **47**, 2624–2644, doi:10.1175/1520-0469(1990)047<2624:EAAEAL>2.0.CO;2.

——, and ——, 1995: Energy accumulation and emanation at low latitudes. Part III: Forward and backward accumulation. *J. Atmos. Sci.*, **52**, 2384–2403, doi:10.1175/1520-0469(1995)052<2384:EAAEAL>2.0.CO;2.

Chao, W. C., and B. D. Chen, 2001: On the origin of monsoon. *J. Atmos. Sci.*, **58**, 3497–3507, doi:10.1175/1520-0469(2001)058<3497:TOOM>2.0.CO;2.

Charney, J. G., 1969: A further note on large-scale motions in the tropics. *J. Atmos. Sci.*, **26**, 182–185, doi:10.1175/1520-0469(1969)026<0182:AFNOLS>2.0.CO;2.

Chen, B. D., and M. Yanai, 2000: Comparison of the Madden–Julian oscillation (MJO) during the TOGA COARE IOP with a 15-year climatology. *J. Geophys. Res.*, **105**, 2139–2149, doi:10.1029/1999JD901045.

Dee, D., and Coauthors, 2011: The ERA-Interim reanalysis: Configuration and performance of the data assimilation system. *Quart. J. Roy. Meteor. Soc.*, **137**, 553–597, doi:10.1002/qj.828.

Deng, L., and X. Wu, 2011: Physical mechanisms for the maintenance of GCM-simulated Madden–Julian oscillation over the Indian Ocean and Pacific. *J. Climate*, **24**, 2469–2482, doi:10.1175/2010JCLI3759.1.

Durran, D. R., 1990: Mountain waves and downslope winds. *Atmospheric Processes over Complex Terrain, Meteor. Monogr.*, No. 45, Amer. Meteor. Soc., 59–82.

Hayashi, Y., 1982: Space-time spectral analysis and its applications to atmospheric waves. *J. Meteor. Soc. Japan*, **27**, 156–171.

Houze, R. A., Jr., 1982: Cloud clusters and large-scale vertical motions in the tropics. *J. Meteor. Soc. Japan*, **60**, 396–410.

Khouider, B., and A. J. Majda, 2016: Models for multiscale interactions. Part I: A multicloud model parameterization. *Multiscale Convection-Coupled Systems in the Tropics: A Tribute to Dr. Michio Yanai, Meteor. Monogr.*, No. 56, Amer. Meteor. Soc., doi:10.1175/AMSMONOGRAPHS-D-15-0004.1.

Kiladis, G. N., 1998: Observations of Rossby waves linked to convection over the eastern tropical Pacific. *J. Atmos. Sci.*, **55**, 321–339, doi:10.1175/1520-0469(1998)055<0321:OORWLT>2.0.CO;2.

——, M. C. Wheeler, P. T. Haertel, K. H. Straub, and P. E. Roundy, 2009: Convectively coupled equatorial waves. *Rev. Geophys.*, **47**, RG2003, doi:10.1029/2008RG000266.

Krishnamurti, T. N., R. Krishnamurti, A. Simon, A. Thomas, and V. Kumar, 2016: A mechanism of the MJO invoking scale interactions. *Multiscale Convection-Coupled Systems in the Tropics: A Tribute to Dr. Michio Yanai, Meteor. Monogr.*, No. 56, Amer. Meteor. Soc., doi:10.1175/AMSMONOGRAPHS-D-15-0009.1.

Kung, E. C., and L. P. Merritt, 1974: Kinetic energy source in large-scale tropical disturbances over the Marshall Island area. *Mon. Wea. Rev.*, **102**, 489–502, doi:10.1175/1520-0493(1974)102<0489:KESILS>2.0.CO;2.

Lau, W. K. M., and D. E. Waliser, Eds., 2012: *Intraseasonal Variability of the Atmosphere-Ocean Climate System.* 2nd ed. Springer, 613 pp.

Liebmann, B., 1987: Observed relationship between large-scale tropical convection and the tropical circulation on subseasonal time scales during Northern Hemisphere winter. *J. Atmos. Sci.*, **44**, 2543–2561, doi:10.1175/1520-0469(1987)044<2543:ORBLST>2.0.CO;2.

Lin, J., B. Mapes, M. Zhang, and M. Newman, 2004: Stratiform precipitation, vertical heating profiles, and the Madden–Julian oscillation. *J. Atmos. Sci.*, **61**, 296–309, doi:10.1175/1520-0469(2004)061<0296:SPVHPA>2.0.CO;2.

Lindzen, R. S., and T. Matsuno, 1968: On the nature of large scale wave disturbances in the equatorial lower stratosphere. *J. Meteor. Soc. Japan*, **46**, 215–221.

Madden, R. A., and P. R. Julian, 1971: Detection of a 40–50 day oscillation in the zonal wind in the tropical Pacific. *J. Atmos. Sci.*, **28**, 702–708, doi:10.1175/1520-0469(1971)028<0702:DOADOI>2.0.CO;2.

——, and ——, 1972: Description of global-scale circulation cells in the tropics with a 40–50 day period. *J. Atmos. Sci.*, **29**, 1109–1123, doi:10.1175/1520-0469(1972)029<1109:DOGSCC>2.0.CO;2.

——, and ——, 1994: Observations of the 40–50-day tropical oscillation—A review. *Mon. Wea. Rev.*, **122**, 814–837, doi:10.1175/1520-0493(1994)122<0814:OOTDTO>2.0.CO;2.

Magaña, V., and M. Yanai, 1991: Tropical–midlatitude interaction on the timescales of 30 to 60 days during the northern summer of 1979. *J. Climate*, **4**, 180–201, doi:10.1175/1520-0442(1991)004<0180:TMIOTT>2.0.CO;2.

Majda, A. J., and S. N. Stechmann, 2016: Models for multiscale interactions. Part II: Madden–Julian oscillation, moisture, and convective momentum transport. *Multiscale Convection-Coupled Systems in the Tropics: A Tribute to Dr. Michio Yanai, Meteor. Monogr.*, No. 56, Amer. Meteor. Soc., doi:10.1175/AMSMONOGRAPHS-D-15-0005.1.

Maloney, E. D., and C. Zhang, 2016: Dr. Yanai's contributions to the discovery and science of the MJO. *Multiscale Convection-Coupled Systems in the Tropics: A Tribute to Dr. Michio Yanai, Meteor. Monogr.*, No. 56, Amer. Meteor. Soc., doi:10.1175/AMSMONOGRAPHS-D-15-0003.1.

Maruyama, T., 1967: Large-scale disturbances in the equatorial lower stratosphere. *J. Meteor. Soc. Japan*, **45**, 391–408.

——, 1968: Time sequence of power spectra of disturbances in the equatorial lower stratosphere in relation to the quasi-biennial oscillation. *J. Meteor. Soc. Japan*, **46**, 327–342.

Matsuno, T., 1966: Quasi-geostrophic motions in the equatorial area. *J. Meteor. Soc. Japan*, **44**, 25–43.

Mu, M., and G. J. Zhang, 2006: Energetics of Madden–Julian oscillations in the National Center for Atmospheric Research Community Atmosphere Model version 3 (NCAR CAM3). *J. Geophys. Res.*, **111**, D24112, doi:10.1029/2005JD007003.

Murakami, T., and M. S. Unninayar, 1977: Atmospheric circulation during December 1970 through February 1971. *Mon. Wea. Rev.*, **105**, 1024–1038, doi:10.1175/1520-0493(1977)105<1024:ACDDTF>2.0.CO;2.

Nitta, T., 1970: A study of generation and conversion of eddy available potential energy in the tropics. *J. Meteor. Soc. Japan*, **48**, 524–528.

——, 1972: Energy budget of wave disturbances over the Marshall Islands during the years of 1956 and 1958. *J. Meteor. Soc. Japan*, **50**, 71–84.

Schumacher, C., and R. A. Houze, 2003: Stratiform rain in the tropics as seen by the TRMM precipitation radar. *J. Climate*, **16**, 1739–1756, doi:10.1175/1520-0442(2003)016<1739:SRITTA>2.0.CO;2.

Straub, K. H., and G. N. Kiladis, 2003: Extratropical forcing of convectively coupled Kelvin waves during austral winter. *J. Atmos. Sci.*, **60**, 526–543, doi:10.1175/1520-0469(2003)060<0526:EFOCCK>2.0.CO;2.

Sui, C.-H., and M. Yanai, 1986: Cumulus ensemble effects on the large-scale vorticity and momentum fields of GATE. Part I: Observational evidence. *J. Atmos. Sci.*, **43**, 1618–1642, doi:10.1175/1520-0469(1986)043<1618:CEEOTL>2.0.CO;2.

Takayabu, Y.-N., 1994: Large-scale cloud disturbances associated with equatorial waves. Part I: Spectral features of the cloud disturbances. *J. Meteor. Soc. Japan*, **72**, 433–449.

Tomas, R., and P. J. Webster, 1994: Horizontal and vertical structure of cross-equatorial wave propagation. *J. Atmos. Sci.*, **51**, 1417–1430, doi:10.1175/1520-0469(1994)051<1417:HAVSOC>2.0.CO;2.

Tung, W.-w., and M. Yanai, 2002a: Convective momentum transport observed during the TOGA COARE IOP. Part II: Case studies. *J. Atmos. Sci.*, **59**, 2535–2549, doi:10.1175/1520-0469(2002)059<2535:CMTODT>2.0.CO;2.

——, and ——, 2002b: Convective momentum transport observed during the TOGA COARE IOP. Part I: General features. *J. Atmos. Sci.*, **59**, 1857–1871, doi:10.1175/1520-0469(2002)059<1857:CMTODT>2.0.CO;2.

——, C. Lin, B. Chen, M. Yanai, and A. Arakawa, 1999: Basic modes of cumulus heating and drying observed during TOGA-COARE IOP. *Geophys. Res. Lett.*, **26**, 3117–3120, doi:10.1029/1999GL900607.

——, M. W. Moncrieff, and J. B. Gao, 2004: A systematic analysis of multiscale deep convective variability over the tropical Pacific. *J. Climate*, **17**, 2736–2751, doi:10.1175/1520-0442(2004)017<2736:ASAOMD>2.0.CO;2.

——, D. Giannakis, and A. J. Majda, 2014: Symmetric and anti-symmetric convective signals in the Madden–Julian Oscillation. Part I: Basic modes in infrared brightness temperature. *J. Atmos. Sci.*, **71**, 3302–3326, doi:10.1175/JAS-D-13-0122.1

Wallace, J. M., 1971: Spectral studies of tropospheric wave disturbances in the tropical western Pacific. *Rev. Geophys. Space Phys.*, **9**, 557–612, doi:10.1029/RG009i003p00557.

Wang, B., and X. Xie, 1996: Low-frequency equatorial waves in vertically sheared zonal flow. Part I: Stable waves. *J. Atmos. Sci.*, **53**, 449–467, doi:10.1175/1520-0469(1996)053<0449:LFEWIV>2.0.CO;2.

Webster, P. J., and J. R. Holton, 1982: Cross-equatorial response to middle-latitude forcing in a zonally varying basic state. *J. Atmos. Sci.*, **39**, 722–733, doi:10.1175/1520-0469(1982)039<0722:CERTML>2.0.CO;2.

——, and H. R. Chang, 1988: Equatorial energy accumulation and emanation regions: Impacts of a zonally varying basic state. *J. Atmos. Sci.*, **45**, 803–829, doi:10.1175/1520-0469(1988)045<0803:EEAAER>2.0.CO;2.

——, and S. Yang, 1989: The three-dimensional structure of perturbation kinetic energy and its relationship to the zonal wind field. *J. Climate*, **2**, 1210–1222, doi:10.1175/1520-0442(1989)002<1210:TTDSOP>2.0.CO;2.

——, V. O. Magaña, T. N. Palmer, J. Shukla, R. A. Tomas, M. Yanai, and T. Yasunari, 1998: Monsoons: Processes, predictability, and the prospects for prediction. *J. Geophys. Res.*, **103**, 14 451–14 510, doi:10.1029/97JC02719.

Wheeler, M., and G. N. Kiladis, 1999: Convectively coupled equatorial waves: Analysis of clouds and temperature in the wavenumber-frequency domain. *J. Atmos. Sci.*, **56**, 374–399, doi:10.1175/1520-0469(1999)056<0374:CCEWAO>2.0.CO;2.

Yanai, M., and R. H. Johnson, 1993: Impacts of cumulus convection on thermodynamic fields. *The Representation of Cumulus Convection in Numerical Models*, K. A. Emanuel and D. J. Raymond, Eds., Amer. Meteor. Soc., 39–62.

——, and T. Tomita, 1998: Seasonal and interannual variability of atmospheric heat sources and moisture sinks as determined from NCEP–NCAR reanalysis. *J. Climate*, **11**, 463–482, doi:10.1175/1520-0442(1998)011<0463:SAIVOA>2.0.CO;2.

——, T. Maruyama, T. Nitta, and Y. Hayashi, 1968: Power spectra of large-scale disturbances over the tropical Pacific. *J. Meteor. Soc. Japan*, **46**, 308–323.

——, S. Esbensen, and J.-H. Chu, 1973: Determination of bulk properties of tropical cloud clusters from large-scale heat and moisture budgets. *J. Atmos. Sci.*, **30**, 611–627, doi:10.1175/1520-0469(1973)030<0611:DOBPOT>2.0.CO;2.

——, B. D. Chen, and W.-w. Tung, 2000: The Madden–Julian oscillation observed during the TOGA COARE IOP: Global view. *J. Atmos. Sci.*, **57**, 2374–2396, doi:10.1175/1520-0469(2000)057<2374:TMJOOD>2.0.CO;2.

Yang, G. Y., and B. J. Hoskins, 1996: Propagation of Rossby waves of nonzero frequency. *J. Atmos. Sci.*, **53**, 2365–2378, doi:10.1175/1520-0469(1996)053<2365:PORWON>2.0.CO;2.

Zhang, C., 2005: Madden-Julian Oscillation. *Rev. Geophys.*, **43**, RG2003, doi:10.1029/2004RG000158.

——, and P. J. Webster, 1992: Laterally forced equatorial perturbations in a linear model. Part I: Stationary transient forcing. *J. Atmos. Sci.*, **49**, 585–607, doi:10.1175/1520-0469(1992)049<0585:LFEPIA>2.0.CO;2.

Chapter 9

Models for Multiscale Interactions. Part I: A Multicloud Model Parameterization

BOUALEM KHOUIDER

Department of Mathematics and Statistics, University of Victoria, Victoria, British Columbia, Canada

ANDREW J. MAJDA

Center for Atmosphere Ocean Science, and Department of Mathematics, Courant Institute, New York University, New York, New York

ABSTRACT

In this chapter, a model parameterization for organized tropical convection and convectively coupled tropical waves is presented. The model is based on the main three cloud types, congestus, deep, and stratiform, that are observed to play an important role in the dynamics and morphology of tropical convective systems. The model is based on the self-similarity across scales of tropical convective systems and uses physically sound theory about the mutual interactions between the three cloud types and the environment. Both linear analysis and numerical simulations of convectively coupled waves and the Madden–Julian oscillation are discussed.

1. Introduction

Convection in the tropics is organized on a hierarchy of scales ranging from the convective cell of a few kilometers to planetary-scale disturbances such as the Madden–Julian oscillation (MJO) (Nakazawa 1974). Cloud clusters and superclusters occur on the meso- and synoptic scales and often appear embedded in each other and within the MJO envelope. Analysis of outgoing longwave radiation cross correlated with the reanalysis products helped identify the synoptic-scale superclusters as the moist analogs of the equatorial shallow-water waves of Matsuno (1966) but with a severely reduced phase speed (Takayabu 1994; Wheeler and Kiladis 1999) and a front-to-rear vertical tilt in zonal wind, temperature, heating, and humidity profiles (Wheeler et al. 2000; Straub and Kiladis 2002). These moisture-coupled waves are often referred to as convectively coupled waves (CCWs; Takayabu 1994; Wheeler and Kiladis 1999; Wheeler et al. 2000; Kiladis et al.

2009). Convectively coupled Kelvin waves associated with the deepest baroclinic mode dominate the spectral variability on the synoptic scales and propagate, along the equator, at speeds ranging from 12 to $20 \, \mathrm{m \, s^{-1}}$, unlike their dry counterparts that travel at $50 \, \mathrm{m \, s^{-1}}$ (Kiladis et al. 2009).

For almost a decade, simple primitive equation models involving a single baroclinic vertical mode, forced by deep convection, have been used with some relative success for theoretical and numerical studies of convectively coupled waves (Emanuel 1987; Mapes 1993; Neelin and Yu 1994; Yano et al. 1995, 1998; Majda and Shefter 2001b; Fuchs and Raymond 2002; Majda and Khouider 2002). They were somewhat able to reproduce scale-selective instability, at the synoptic scale, of Kelvin waves with a propagation speed in the observed range. However, these one-baroclinic-mode models failed badly to reproduce the important[1] front-to-rear tilt in

Corresponding author address: Dr. Boualem Khouider, Mathematics and Statistics University of Victoria, P.O. Box 3045 STN CSC, Victoria, BC V8W 3P4, Canada.
E-mail: khouider@math.uvic.ca

---

[1] As we will see in the next chapter (Majda and Stechmann 2016), the vertical tilt is important for momentum transport from convective systems toward larger scales.

DOI: 10.1175/AMSMONOGRAPHS-D-15-0004.1

the vertical and many other important features of CCWs. Moreover, these early models are based on two major theories for the destabilization of large-scale waves by convection, namely, they are divided into convergence-driven models and quasi-equilibrium models. Convergence models date back to the work of Charney and Eliassen (1964) and Ooyama (1964) followed by Yamasaki (1969), Hayashi (1971), and Lindzen (1974). The convergence models, also called convective instability of the second kind (CISK) models, sustain convection through large reservoirs of convectively available potential energy (CAPE) driven by low-level moisture convergence. Such models exhibit extreme sensitivity to grid-scale variability, and linearized stability analysis reveals the undesirable feature of catastrophic instability with growth rates increasing with the wavenumber (Yano et al. 1998; Majda and Shefter 2001b). In the quasi-equilibrium thinking, first introduced by Arakawa and Shubert (1974), one assumes a large-scale quasi-equilibrium state where CAPE is nearly constant and deep convection acts as an energy regulator in restoring quickly the equilibrium by consuming any excess of CAPE. The triggering and the amplification of convection in quasi-equilibrium models rely on surface fluxes. Indeed, such quasi-equilibrium models are linearly (Neelin and Yu 1994) and even nonlinearly stable (Frierson et al. 2004). The most popular mechanism used in concert with the quasi-equilibrium models to create instability is wind-induced surface heat exchange (WISHE) (Emanuel 1987; Emanuel et al. 1994). Both WISHE and CISK theories were initially proposed for hurricanes (Zehnder 2001; Craig and Gray 1996). There is no basic observational evidence for their validity for CCWs (Straub and Kiladis 2003). The phenomenon of phase speed reduction is associated, in these one-baroclinic-mode models, solely with a reduction in the background stratification due to moisture coupling of the waves; but the observed backward vertical tilt is suggestive of a strong projection of the CCWs onto shallower vertical modes, with much slower gravity wave speeds, in addition to the prominence of the fastest first baroclinic mode.

Recent analysis of observations over the Indo-Pacific warm pool in the tropics reveals the ubiquity of three cloud types above the boundary layer: shallow congestus clouds, stratiform clouds, and deep penetrative cumulus clouds (Lin and Johnson 1996; Johnson et al. 1999). Furthermore, recent analysis of convectively coupled waves on large scales reveals a similar multicloud convective structure with leading shallow congestus cloud decks that moisten and precondition the lower troposphere followed by deep

convection and finally trailing decks of stratiform precipitation; this structure applies to the eastward-propagating convectively coupled Kelvin waves (Wheeler and Kiladis 1999; Straub and Kiladis 2002) and westward-propagating 2-day waves (Haertel and Kiladis 2004) that reside on equatorial synoptic scales of the order of 1000 to 3000 km in the lower troposphere as well as the planetary-scale Madden–Julian oscillation (Kiladis et al. 2005; Dunkerton and Crum 1995). An inherently multiscale theory for the Madden–Julian oscillation with qualitative agreement with observations that is based on these three cloud types is proposed by Majda and Biello (2004) and Biello and Majda (2005).

Furthermore, despite the observational evidence, none of the models with a single vertical mode mentioned earlier account for the multimode nature of tropical convection and the importance of the different cloud types; they are forced by a heating profile based solely on the deep, penetrative clouds. Parameterizations with two convective heating modes (systematically representing a deep, convective mode and a stratiform mode) have first appeared in the work of Mapes (2000). Majda and Shefter (2001a) proposed and analyzed a simpler version of Mapes' model based on a systematic Galerkin projection of the primitive equations onto the first two linear, baroclinic modes yielding a set of two shallow-water systems. The first baroclinic system is heated by deep convective clouds, while the second baroclinic system is heated aloft and cooled below by stratiform anvils. Linear stability analysis of this model convective parameterization revealed a mechanism of stratiform instability independent of WISHE (Majda and Shefter 2001a; Majda et al. 2004). Numerical simulations carried out in Majda et al. (2004) revealed the resemblance of many features of the moist gravity waves for the Majda and Shefter (2001a) model and the real-world convective superclusters (Straub and Kiladis 2002; Majda et al. 2004). One visible shortcoming of the Majda and Shefter (MS) model, however, is its short-cutting of the role of the congestus heating and the systematic interaction of the cloud types with moisture. In fact, the Majda and Shefter model requires the use of WISHE in order to produce realistic amplification of convectively coupled gravity waves during nonlinear simulations (Majda et al. 2004).

2. The multicloud model

In a seminal paper, the authors (Khouider and Majda 2006b) proposed a new model convective parameterization within the framework of the MS model.

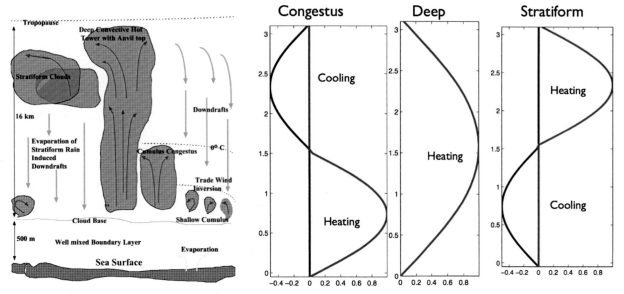

FIG. 9-1. (left) Schematic of the three tropical cloud types interacting with the well-mixed planetary boundary layer above the sea surface through convective updrafts and downdrafts: the trade wind inversion, the freezing level with temperature 0°C, and tropopause layers are shown. (right) Vertical profiles of heating and cooling fields associated with the three cloud types. [Figure 1 from Khouider and Majda (2008b). ©2008 American Meteorological Society. Reprinted with permission.]

In addition to the deep convective and stratiform clouds, the new model carries cumulus congestus clouds that serve to heat the second baroclinic mode from below and cool it from above as in actual congestus cloud decks.

To help the visualization, a cartoon of the cloud types is shown in Fig. 9-1 together with the associated heating profiles. The first baroclinic mode is heated by deep convection, as in the simple one-baroclinic-mode models discussed above, and the second baroclinic mode is heated and cooled by both congestus and stratiform clouds. Congestus clouds lead tropical convective systems and precondition the environment prior to deep convection. They heat the lower troposphere through condensation and induce a cooling of the upper troposphere through detrainment at the cloud top and by blocking longwave radiation emanating from the surface. Stratiform clouds, on the other hand, are observed to trail behind deep convection, and they heat the upper troposphere as deep clouds enter their freezing phase and cool the lower troposphere because of the evaporation of stratiform rain that falls into the already dry environment.

The minimal dynamical core for the multicloud model consists of two coupled (linear) shallow water systems, forced and coupled through the heating rates associated with the three cloud types:

$$\frac{\partial \mathbf{v}_j}{\partial t} + \beta y \mathbf{v}_j^{\perp} - \nabla \theta_j = -C_d(u_0)\mathbf{v}_j - \frac{1}{\tau_W}\mathbf{v}_j, \quad j = 1, 2,$$

$$\frac{\partial \theta_1}{\partial t} - \mathrm{div}\mathbf{v}_1 = \frac{\pi}{2\sqrt{2}}P + S_1,$$

$$\frac{\partial \theta_2}{\partial t} - \frac{1}{4}\mathrm{div}\mathbf{v}_2 = \frac{\pi}{2\sqrt{2}}(-H_s + H_c) + S_2, \qquad (9\text{-}1)$$

where the two modes $j = 1$ and 2 are coupled through the nonlinear source terms. These equations are derived through a systematic Galerkin projection of the equatorial beta-plane primitive equations onto the first and second baroclinic modes that are directly forced by the heating rates H_c, H_d, H_s associated with the three cloud types. Following Khouider and Majda (2008b), the total precipitation is set to $P = H_d + \xi_s H_s + \xi_c H_c$, and it includes contributions from deep convection, stratiform, and congestus clouds. Here, ξ_s and ξ_c are parameters, with values between 0 and 1, representing the relative contributions of stratiform and congestus clouds to surface precipitation. In this framework, the horizontal velocity and potential temperature are given in terms of their first and second baroclinic components $\mathbf{v}_j$ and θ_j, $j = 1, 2$, respectively, via the Galerkin expansions:

$$\mathbf{v} = \sqrt{2}\mathbf{v}_1 \cos(z) + \sqrt{2}\mathbf{v}_2 \cos(2z),$$

$$\theta = \sqrt{2}\theta_1 \sin(z) + 2\sqrt{2}\theta_2 \sin(2z). \qquad (9\text{-}2)$$

TABLE 9-1. Parameters and symbols used in the multicloud model. While the typical values of some parameters are given here, the interested reader may consult the original papers as those values change considerably from case to case. The primes represent deviations from the RCE solution.

Symbol	Description	Value
C_d	Turbulent boundary layer drag coefficient	0.001
τ_W	Rayleigh damping time scale	75 days
τ_R	Newtonian cooling time scale	50 days
β	Gradient of Coriolis force at the equator	$2.28 \times 10^{-11}\,\mathrm{m^{-1}\,s^{-1}}$
u	Turbulence velocity scale	$2\,\mathrm{m\,s^{-1}}$
ξ_s, ξ_c	Contribution of respectively stratiform and congestus heating to total precipitation	Typically 05 and 1.25, respectively
$\tilde{Q}$	Background moisture stratification	0.9
$\tilde{\alpha}$	Coefficient of second baroclinic nonlinear convergence	0.1
$\tilde{\lambda}$	Relative contribution of second baroclinic moisture convergence, linear term	0.8
α_s	Stratiform heating adjustment coefficient	0.25 (varies)
α_c	Congestus heating adjustment coefficient	0.1 (varies)
a_1	Coefficient of θ_{eb} in Q_d	0.5 (varies)
a_2	Coefficient of q in Q_d	$1 - a_1$ (varies)
a_0	Relative strength of buoyancy fluctuation in Q_d	2 (varies)
γ_2	Relative contribution of θ_2 in Q_d	0.1 (varies)
a_0'	Relative strength of buoyancy fluctuation in Q_c	2 (varies)
γ_2'	Relative contribution of θ_2 in Q_c	0.1 (varies)
μ	Contribution of stratiform rain evaporation to downdraft	0.25 (varies)
Λ	Moisture switch function	1 if $\theta_{eb} - \theta_{em} \geq 20\,\mathrm{K}$ 0 if $\theta_{eb} - \theta_{em} \leq 10\,\mathrm{K}$ $0.1(\theta_{eb} - \theta_{em}) - 1$ if $10\,\mathrm{K} \leq \theta_{eb} - \theta_{em} \leq 20\,\mathrm{K}$
$\overline{\Lambda}$	Moisture switch function at RCE	$0.1(\overline{\theta}_{eb} - \overline{\theta}_{em}) - 1$ (varies)
$Q_d = \max\left(\left\{\overline{Q} + \dfrac{1}{\tau_{\mathrm{conv}}}[a_1\theta_{eb}' + a_2q' - a_0(\theta_1' + \gamma_2\theta_2')]\right\}, 0\right)$	Deep convective heating potential	Variable
$Q_c = \max\left(\left\{\overline{Q} + \dfrac{\alpha_c}{\tau_{\mathrm{conv}}}[\theta_{eb}' - a_0'(\theta_1' + \gamma_2'\theta_2')]\right\}, 0\right)$	Congestus heating potential	Variable
$D_0 = m_0 \max\left(1 + \mu\dfrac{H_s - H_c}{\overline{Q}}, 0\right)(\theta_{eb} - \theta_{em})$	Downdraft fluxes	Variable
$m_0 = \dfrac{H_T Q_{R,1}^0}{\{1 + \mu[\alpha_s(1 - \overline{\Lambda}) - \overline{\Lambda}\alpha_c]\}(\overline{\theta}_{eb} - \overline{\theta}_{em})}$	Downdraft mass flux scale	Set by RCE solution
$\overline{Q} = (1 - \overline{\Lambda})Q_{R,1}^0$	Convective heating potential at RCE	(Varies)

In (9-1), $\mathbf{v}^{\perp} = \mathbf{k} \times \mathbf{v}$, where $\mathbf{k}$ is the upward vertical unit vector, and $\mathbf{V}$ and div are the gradient and divergence operators, respectively. The nonlinear advection terms are neglected for simplicity; they are believed to play a secondary role in the presence of convective heating. The other source terms represent radiative cooling rates: $S_j = -Q_{R,j}^0 - (\tau_R)^{-1}\theta_j$, $j = 1, 2$. Here and elsewhere in the text, the parameters and variables that are not properly defined in the text are described in Table 9-1.

In addition to this dynamical core [(9-1)], there are equations for the boundary layer equivalent potential temperature θ_{eb} and the vertically integrated moisture content q:

$$\frac{\partial \theta_{eb}}{\partial t} = \frac{1}{h_b}(E - D)$$

$$\frac{\partial q}{\partial t} + \mathrm{div}[(\mathbf{v}_1 + \tilde{\alpha}\mathbf{v}_2)q] + \tilde{Q}\,\mathrm{div}(\mathbf{v}_1 + \tilde{\lambda}\mathbf{v}_2) = -P + \frac{1}{H_T}D,$$

(9-3)

Here, $h_b \approx 500$ m is the height of the moist boundary layer, and $H_T = 16$ km is the tropospheric height, while $\tilde{Q}$, $\tilde{\lambda}$, and $\tilde{\alpha}$ are parameters associated with a prescribed moisture background and perturbation vertical profiles. According to the first equation in (9-3), θ_{eb} changes in response to the downdrafts D and the sea surface evaporation E. When setting the closure for the forcing terms in (9-3), conservation of vertically integrated moist static energy is used as a design principle. The moisture equation in (9-3) is derived through a systematic vertical averaging of the water vapor conservation equation (Khouider and Majda 2006b,a). The parameter $\tilde{\lambda}$ measures the strength of moisture convergence due to the second baroclinic mode, and it plays an important role in the dynamics of convectively coupled tropical waves (Khouider and Majda 2006b).

The main nonlinearities of the model are in the source terms H_c, H_d, H_s, and D. The stratiform and congestus heating rates H_s and H_c satisfy relaxation-type equations:

$$\frac{\partial H_s}{\partial t} = \frac{1}{\tau_s}(\alpha_s H_d - H_s), \qquad (9\text{-}4)$$

and

$$\frac{\partial H_c}{\partial t} = \frac{1}{\tau_c}\left(\alpha_c \frac{\Lambda - \Lambda^*}{1 - \Lambda^*} Q_c - H_c\right). \qquad (9\text{-}5)$$

The deep convective heating H_d and the downdrafts D are given diagnostically by

$$H_d = \frac{1 - \Lambda}{1 - \Lambda^*} Q_d \quad \text{and} \quad D = \Lambda D_0. \qquad (9\text{-}6)$$

The "moisture switch" function Λ controls the transition between congestus to deep convection, and it depends on the difference between boundary layer and midtropospheric equivalent potential temperatures $\theta_{eb} - \theta_{em}$. The diagnostic functions Q_d, Q_c, D_0, and Λ all involve nonlinear switches, and they are described in detail in Khouider and Majda (2008b) (see Table 9-1). These source terms take slightly different forms in different versions of the multicloud model such as Khouider and Majda (2006b, 2008b).

3. Convectively coupled equatorial waves in the multicloud model

In this section, we exhibit some typical solutions of the multicloud model equations (9-1)–(9-5) in the form of linear waves and long-time nonlinear simulations.

For the linear waves, the base state is a state of radiative–convective equilibrium (RCE). RCE is a homogeneous (in space and time) steady-state solution for the governing equations (9-1)–(9-5) around which waves can grow and oscillate. In the multicloud model, an RCE solution is determined by fixing three important climatological parameters: the longwave radiative cooling rate $Q^0_{R,1}$, the discrepancy between the boundary layer equivalent potential temperature and its saturation value $\theta^*_{eb} - \bar{\theta}_{eb}$, and the discrepancy between the boundary layer and midtropospheric equivalent potential temperatures $\bar{\theta}_{eb} - \bar{\theta}_{em}$. They are fixed according to climatological values that are recorded in the tropics (Khouider and Majda 2006b, 2008b): $Q^0_{R,1} = 1$ K day^{-1} and $\theta^*_{eb} - \bar{\theta}_{eb} = 10$ K, while $10 \leq \bar{\theta}_{eb} - \bar{\theta}_{em} \leq 20$ K or higher/lower according to whether we want to study a case of a dry or a moist middle troposphere.

The PDE system [(9-1)–(9-5)] is then linearized around the relevant RCE solution and linear solutions are sought on the form $U(x, t) = \hat{U} \exp[i(kx - \omega t)]$, where U is the vector of diagnostic variables. Here, k is the zonal wavenumber, and $\omega = \omega(k)$ is the generalized frequency determined as an eigenvalue of the corresponding linear system for a fixed value of k, while $\hat{U}(k)$ is the associated eigenvector. The real part of $\omega(k)$ defines the phase speed of the wave solution: $c(k) = \Re[\omega(k)]/k$, while the imaginary part represents its growth rate.

More details on the construction of climatologically sound RCE solutions for the multicloud model are found in Khouider and Majda (2006b,a, 2008b). Before proceeding to the study of linear wave solution to the PDE system, it is important to select RCE's that are stable to small perturbation. This is easily achieved by looking at solutions of the linear system when $k = 0$, which essentially represents solutions to the zonally averaged linear solutions. As shown in Khouider and Majda (2006b,a), this system exhibits interesting bifurcation behavior with respect to the model parameters. A typical stability diagram of the background RCE solution with respect to the parameters α_c, α_2, and $\bar{\theta}_{eb} - \bar{\theta}_{em}$ is shown in Fig. 9-2. Such diagrams are used as guidelines to select the appropriate parameters for the multicloud model (Khouider and Majda 2006b,a, 2008b). Parameter values for which the homogeneous RCE is unstable cannot possibly support waves since the associated background itself will grow and change considerably during the integration of the model. Further linear wave analysis associated with an unstable background is meaningless. The associated parameter values or range of values are thus automatically discarded.

However, the important characteristic of the multicloud model is that, in the appropriate parameter regime, it exhibits a scale-selective instability of wave solutions that have several key physical and dynamical features resembling observed convectively coupled equatorial waves. In Figs. 9-3 and 9-4, we show linear

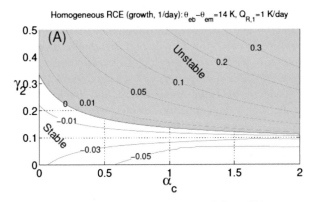

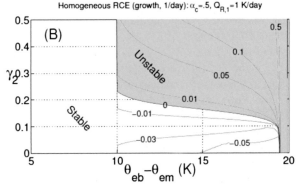

FIG. 9-2. Bifurcation diagram for the homogeneous state RCE (a) in the γ_2–α_c plane for fixed $\bar{\theta}_{eb} - \bar{\theta}_{em} = 14$ K and (b) in the γ_2–$\bar{\theta}_{eb} - \bar{\theta}_{em}$ plane for fixed $\alpha_c = 0.5$. The regions of positive maximum growth among the $k = 0$ modes are shaded and a few contours are plotted. [Figure 2 from Khouider and Majda (2006b). ©2006 American Meteorological Society. Reprinted with permission.]

and nonlinear solutions for the multicloud equations obtained in the case of a simple flow above the equator, where the beta effect is ignored. When the full (nonlinear) multicloud equations are integrated numerically, for a long enough time, with an initial condition consisting of the RCE solution plus a small random perturbation, the solution goes to a statistical steady state that exhibits the wavelike disturbances that have the same features as their linear equivalents, including a reduced phase speed of roughly $17\,\mathrm{m\,s}^{-1}$ and a front-to-rear vertical tilt in wind, temperature, and heating field. In particular, note that the nonlinear simulation is characterized by packets of synoptic-scale waves moving at about $17\,\mathrm{m\,s}^{-1}$ and have a planetary-scale wave envelope moving in the opposite direction at a slower speed of 5 to $6\,\mathrm{m\,s}^{-1}$, mimicking observed CCWs evolving within the MJO envelope. Consistent with the "self-similarity" of tropical convective systems across scales (Mapes et al. 2006; Majda 2007; Kiladis et al. 2009), the multicloud model exhibits a planetary scale envelope that has the same front-to-rear tilted structure as the synoptic-scale waves, though only the synoptic-scale waves are linearly unstable.

Moreover, when the beta effect is included the multicloud model exhibits instabilities corresponding to the full spectrum of convectively coupled waves seen in the observational records (Takayabu 1994; Wheeler and Kiladis 1999), with comparable length scales and phase speeds, namely, Kelvin, westward inertio-gravity waves, and $n = 0$ eastward mixed Rossby–gravity [MRG, also known as Yanai waves (Yanai and Maruyama 1966)] waves (Khouider and Majda 2008a; Han and Khouider 2010). Similarly, to the rotation-free case, all the simulated CCW solutions exhibit a front-to-rear tilted vertical structure as in nature. While westward MRG waves and Rossby waves are missing in the multicloud model linearized about a homogeneous RCE, they are recovered in the case of a meridional barotropic shear background, mimicking the climatological jet stream (Han and Khouider 2010). The combination of the results shown in Khouider and Majda (2008a) and Han and Khouider (2010), which are summarized in Fig. 9-5 demonstrate clearly that the multicloud model with rotation reproduces the full spectrum of convectively coupled waves that are reported in the observational literature (Takayabu 1994; Wheeler and Kiladis 1999).

To conclude this section, we note that the results in Figs. 9-3–9-5 are obtained with the typical synoptic-scale convective time scales, $\tau_{\mathrm{conv}} = 2\,\mathrm{h}$, $\tau_s = 3\,\mathrm{h}$, and $\tau_c = 1\,\mathrm{h}$, which are the estimated time scales of the deep convective, stratiform, and congestus clouds. Comparable values are also used in the stochastic version of the multicloud model (SMCM) (Khouider et al. 2010; Frenkel et al. 2012). In addition to the self-similar cloud morphology of multiscale tropical convective systems, the SMCM is introduced to capture the missing variability in GCMs due to unresolved convective processes. In a nutshell, it is based on the lattice system overlaid over each GCM grid box with an order parameter taking the values 1, 2, 3, or 0, on each lattice site, according to whether it is occupied by a cloud of a certain type (congestus, deep, or stratiform) or it is a clear-sky site. As time increases, the order parameter makes random transitions, with prescribed time scales, from one state to another according to whether the large-scale state and the microscopic configuration is favorable to that state or not.

4. The MJO analog wave

By exploiting the observed self-similarity of tropical convective systems (Majda 2007; Kiladis et al. 2009), the multicloud model can be configured toward a planetary and intraseasonal-scale instability of a wave that resembles the tropical intraseasonal oscillation, that is, the MJO, as regards flows along the equator. This was

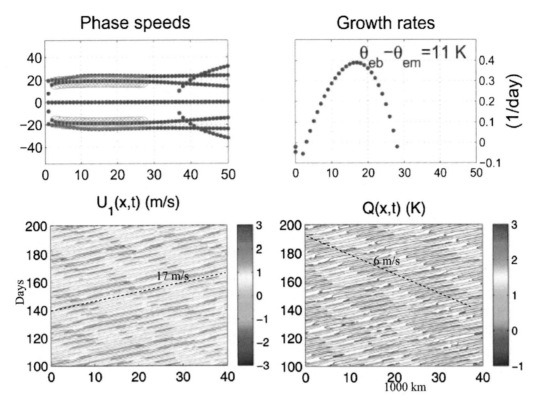

FIG. 9-3. (top left) Phase speeds and (top right) growth rates as functions of wavenumber depicting a scale-selective instability in the multicloud model with enhanced congestus closure. Contours of (x, t) of (bottom left) the first baroclinic zonal velocity and (bottom right) moisture anomaly showing streaks corresponding to synoptic-scale moist gravity waves moving to the right at $17\,\mathrm{m\,s^{-1}}$ and their planetary-scale wave envelopes moving in the opposite direction at $6\,\mathrm{m\,s^{-1}}$, respectively. [Figures 2 and 5 from Khouider and Majda (2008b). ©2008 American Meteorological Society. Reprinted with permission.]

accomplished in Majda et al. (2007) by setting the convective time scales to

$$\tau_{\mathrm{conv}} = 12 \text{ hours,} \quad \text{and} \quad \tau_c = \tau_s = 7 \text{ days.}$$

Recall that for the case of the synoptic-scale waves above, we have $\tau_{\mathrm{conv}} = 2\,\mathrm{h}$, $\tau_c = 1\,\mathrm{h}$, and $\tau_s = 3\,\mathrm{h}$. The main effect of this change in parameter value is to shift the instability band to larger scales in both time and space (see top panels of Fig. 9-3). Consistently, numerical simulations with these parameter values yield a solution with the following attractive features that essentially characterize the east–west flow of the MJO above the equator, as shown in Fig. 9-6 (from Majda et al. 2007):

(i) an actual propagation speed of roughly $5\,\mathrm{m\,s^{-1}}$ as predicted by linear theory;

(ii) a wavenumber 2 structure for the low-frequency planetary-scale envelope with distinct active and inactive phases of deep convection;

(iii) an intermittent turbulent chaotic multiscale structure within the wave envelope involving embedded westward- and eastward-propagating deep convection events; and

(iv) qualitative features of the low-frequency averaged planetary-scale envelope from the observational record in terms of, for example, vertical structure of heating and westerly wind burst.

5. GCM simulation of the MJO and convectively coupled waves

Here, we show an MJO solution produced by the multicloud model when implemented in the next generation climate model of the National Center for Atmospheric Research (NCAR), namely, the High-Order Method Modeling Environment (HOMME) dynamical core. HOMME is a highly scalable parallel code for the atmospheric general circulation based on the discretization of the primitive equations on the sphere using high-order spectral elements in the horizontal and finite differences in the vertical. The interested reader is invited to look into the code development papers (Dennis et al. 2005; Taylor et al. 2008) and the online documentation found on the NCAR website. Here, the discussion will be limited to a brief description of the strategy adopted to implement the multicloud

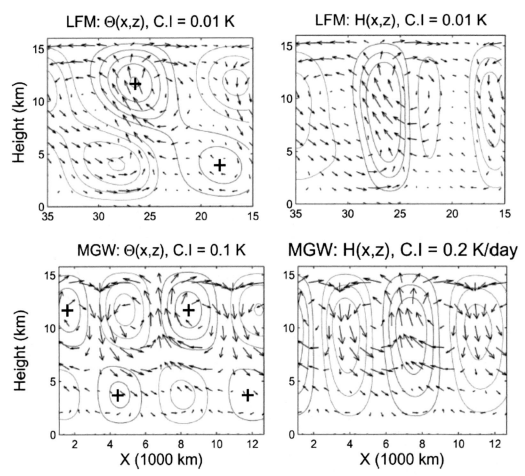

FIG. 9-4. (bottom) Filtered structure of the synoptic-scale moist gravity wave and (top) its low-frequency planetary-scale envelope. (left) Potential temperature contours, while (right) the heating anomalies are contoured with the corresponding u–w velocity arrows that are overlaid on top. The + signs on the temperature panels refer to positive anomalies. [Figure 7 from Khouider and Majda (2008b). ©2008 American Meteorological Society. Reprinted with permission.]

parameterization in HOMME, which provides the added condensational heating to the otherwise dry model to produce realistic MJO and convectively coupled waves' solutions as reported in Khouider et al. (2011).

The first step to incorporate the multicloud model in HOMME consists of designing the proper vertical profiles for the heating field associated with the three cloud types: congestus, deep, and stratiform. We use the vertical normal modes of Kasahara and Puri (1981) in place of the sine and cosine functions in (9-2). More precisely, the vertical normal modes result from a vertical mode expansion of the primitive equations by setting, for example, the horizontal velocity $\mathbf{v} = \sum_{j=0}^{\infty} \mathbf{v}_j \phi_1(z)$, where the ϕ_j's are recovered as the eigensolutions of a Sturm–Liouville problem. They are thus a generalization of the sine and cosine basis functions used above for the general case of a nonuniform stratification and without the rigid-lid assumption. The two eigenfunctions ϕ_1, ϕ_2,

corresponding to the first and second baroclinic modes, are plotted in Fig. 9-7a together with the associated heating profiles given by the temperature basis functions: $\psi_j = (p_B - p_T)^{-1} \int_{p_T}^{p} \phi_j(p')\,dp'$ and $j = 1, 2$, according to the hydrostatic balance equation. The function ψ_1 is used to fix the vertical profile for deep convective heating and ψ_2 is accordingly used for both stratiform and congestus clouds. The heating is set to zero above roughly 200 hPa to avoid spurious heating in the upper atmosphere.

With the heating rates H_d, H_s, and H_c parameterized as in the idealized case presented above and the vertical average moisture equation rederived according to the new basis functions ϕ_j, ψ_j (instead of the cosines and sines), the HOMME full primitive equations are forced by the total heating field $H(x, y, p) = H_d(x, y)\psi_1(p) + [H_c(x, y) - H_s(x, y)]\psi_2(p)$. The resulting coupled model is run on an aquaplanet with a uniform surface evaporation

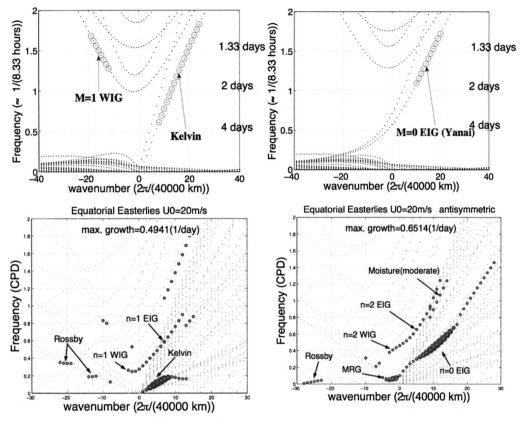

FIG. 9-5. Dispersion diagrams for the multicloud model with rotation. (top) With a homogeneous background. (bottom) With a barotropic meridional shear zonal wind mimicking the jet stream. (left) Symmetric waves and (right) antisymmetric waves are shown separately. [Figure 2 from Khouider and Majda (2008a) and Fig. 8 of Han and Khouider (2010). ©2008 and ©2010 American Meteorological Society. Reprinted with permission.]

and a standard value of radiative cooling: 1 K day^{-1}. The moisture and temperature anomalies are initialized to a tropical mean profile that is also used to fix the Brunt–Väisälä frequency profile in the Kasahara and Puri code

for the normal modes that is solved once for all at the beginning of the simulation.

In the appropriate parameter regime (Khouider et al. 2011), the coupled HOMME–multicloud model

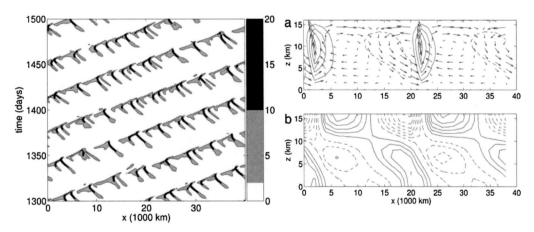

FIG. 9-6. MJO analog wave obtained by convective time scaling for the multicloud model. (left) x–t contours of precipitation showing slowing moving wave envelopes of mesoscale chaotic convective events that evolve within the active phase and propagate in the opposite direction. (right) Vertical structure of (a) the total heating, with the u–w velocity overlaid, and (b) the zonal velocity for the moving average of the planetary-scale envelope. [Figures 2 and 4 from Majda et al. (2007). ©2007 National Academy of Sciences, USA.]

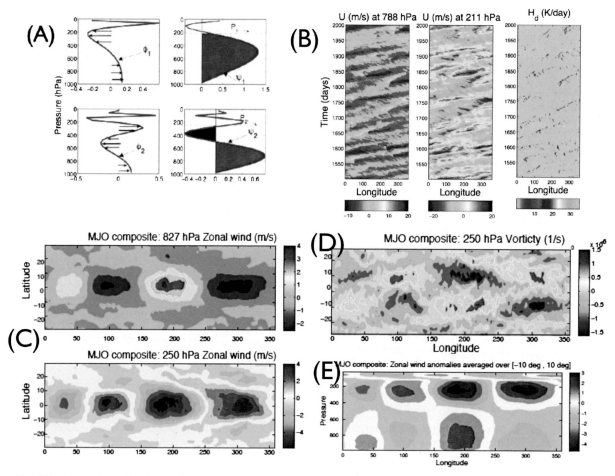

FIG. 9-7. Eigenmode and heating profiles used to implement the multicloud model in HOMME: (a) the resulting MJO solution; (b) the x–t contours of the zonal velocity and deep convection; (c) the zonal structure of the MJO filtered zonal velocity at two heights; (d) filtered vorticity; and (e) vertical structure of the filtered zonal velocity. [Figures 1, 3, 6, and 8 from Khouider et al. (2011). ©2010 American Meteorological Society. Reprinted with permission.]

yields a solution consisting of two MJO-like waves that move eastward at roughly $5 \, \text{m s}^{-1}$ somewhat similar to the two TOGA COARE MJO events reported in Yanai et al. (2000). This is illustrated by (Fig. 9-7b) the x–t contours of the zonal velocity and deep convection, (Fig. 9-7c) the zonal structure of the MJO filtered zonal velocity, (Fig. 9-7d) the filtered vorticity, and (Fig. 9-7e) the vertical structure of the filtered zonal velocity shown in Fig. 9-7. Note in particular that the MJO solution has the same tilted structure, intraseasonal period, and embedded mesoscale turbulent fluctuations in the deep convective heating as in the case of the MJO analog of Fig. 9-6, and, in addition, the vorticity field is characterized by a quadruple vortex surrounding the westerly wind burst. The interested reader is referred to the original paper (Khouider et al. 2011) for further discussion of this solution and others. In particular, the whole spectrum of synoptic-scale convectively coupled waves is reproduced in the appropriate parameter regime.

6. Conclusions

A simple model parameterization based on the systematic multiscale interactions of organized convective system is reviewed in this chapter. The main idea is to use the dynamics of mesoscale cloud systems as a building block to capture the multiscale interactions of tropical convection (Mapes et al. 2006). The model is based on the three cloud types that characterize organized tropical systems at the mesoscale, synoptic, and planetary scales (Mapes 1993).

As it is briefly illustrated here, despite its simplicity, the multicloud model captures the whole spectrum of convectively coupled equatorial waves from both the linear and nonlinear simulations stand points (Khouider and Majda 2006b,a, 2008b,a; Han and Khouider 2010; Khouider et al. 2012b). It is flexible enough to be tested as a cumulus parameterization in a GCM and as such it successfully captures the Madden–Julian oscillation and the whole spectrum of CCWs as observed in nature, while many GCM parameterizations failed badly to do so (Lin et al. 2006). Moreover, as already mentioned, a stochastic version (SMCM) has been invented and implemented in a GCM (HOMME) (Khouider et al. 2010; Qiang et al. 2014). In addition to recreating the physical features and morphology of organized convective systems, the SMCM captures well the intermittent variability of climate, CCWs, and the MJO when tested in both simplified models and in an aquaplanet GCM (Frenkel et al. 2012, 2013; Qiang et al. 2014). In particular, the coupled SMCM–HOMME model successfully simulates the MJO, in an aquaplanet setting, both in terms of the physical features and morphology listed in section 4 and in terms of its intermittent variability (Qiang et al. 2014).

Also the deterministic multicloud model is successfully used to study the parameterization of convective momentum transport (Majda and Stechmann 2008; Khouider et al. 2012a), the evolution of meso- and synoptic-scale convectively coupled waves in the MJO background (Han and Khouider 2010; Khouider et al. 2012b), and the diurnal cycle over land and over the ocean (Frenkel et al. 2011a,b). To make it adaptable to the land situation, the multicloud model is coupled to a dynamical boundary layer (Waite and Khouider 2009).

Acknowledgments. Many thanks to Dr. Wen wen Tung, who gave us this honorable opportunity to celebrate Professor's Yanai memory and achievements. Insightful comments by the anonymous referees helped improve the writing of this chapter.

REFERENCES

Arakawa, A., and W. H. Shubert, 1974: Interaction of a cumulus cloud ensemble with the large-scale environment, Part I. *J. Atmos. Sci.*, **31**, 674–701, doi:10.1175/1520-0469(1974)031<0674: IOACCE>2.0.CO;2.

Biello, J., and A. Majda, 2005: A multi-scale model for the Madden–Julian oscillation. *J. Atmos. Sci.*, **62**, 1694–1721, doi:10.1175/JAS3455.1.

Charney, J. G., and A. Eliassen, 1964: On the growth of the hurricane depression. *J. Atmos. Sci.*, **21**, 68–75, doi:10.1175/1520-0469(1964)021<0068:OTGOTH>2.0.CO;2.

Craig, G. C., and S. L. Gray, 1996: CISK or WISHE as the mechanism for tropical cyclone intensification. *J. Atmos. Sci.*, **53**, 3528–3540, doi:10.1175/1520-0469(1996)053<3528: COWATM>2.0.CO;2.

Dennis, J., A. Fournier, W. Spotz, A. St-Cyr, M. Taylor, S. J. Thomas, and H. Tufo, 2005: High-resolution mesh convergence properties and parallel efficiency of a spectral element atmospheric dynamical core. *Int. J. High Perform. Comput. Appl.*, **19**, 225–245, doi:10.1177/1094342005056108.

Dunkerton, T. J., and F. X. Crum, 1995: Eastward propagating ~2- to 15-day equatorial convection and its relation to the tropical

intraseasonal oscillation. *J. Geophys. Res.*, **100**, 25 781–25 790, doi:10.1029/95JD02678.

Emanuel, K. A., 1987: An air–sea interaction model of intraseasonal oscillations in the tropics. *J. Atmos. Sci.*, **44**, 2324–2340, doi:10.1175/1520-0469(1987)044<2324:AASIMO>2.0.CO;2.

——, J. D. Neelin, and C. S. Bretherton, 1994: On large-scale circulations in convecting atmosphere. *Quart. J. Roy. Meteor. Soc.*, **120**, 1111–1143, doi:10.1002/qj.49712051902.

Frenkel, Y., B. Khouider, and A. J. Majda, 2011a: Simple multicloud models for the diurnal cycle of tropical precipitation. Part I: Formulation and the case of the tropical oceans. *J. Atmos. Sci.*, **68**, 2169–2190, doi:10.1175/2011JAS3568.1.

——, ——, and ——, 2011b: Simple multicloud models for the diurnal cycle of tropical precipitation. Part II: The continental regime. *J. Atmos. Sci.*, **68**, 2192–2207, doi:10.1175/2011JAS3600.1.

——, A. J. Majda, and B. Khouider, 2012: Using the stochastic multicloud model to improve tropical convective parameterization: A paradigm example. *J. Atmos. Sci.*, **69**, 1080–1105, doi:10.1175/JAS-D-11-0148.1.

——, ——, and ——, 2013: Stochastic and deterministic multicloud parameterizations for tropical convection. *Climate Dyn.*, **41**, 1527–1551, doi:10.1007/s00382-013-1678-z.

Frierson, D., A. J. Majda, and O. Pauluis, 2004: Dynamics of precipitation fronts in the tropical atmosphere. *Commun. Math. Sci.*, **2**, 591–626, doi:10.4310/CMS.2004.v2.n4.a3.

Fuchs, Z., and D. Raymond, 2002: Large-scale modes of a nonrotating atmosphere with water vapor and cloud–radiation feedbacks. *J. Atmos. Sci.*, **59**, 1669–1679, doi:10.1175/1520-0469(2002)059<1669:LSMOAN>2.0.CO;2.

Haertel, P., and G. N. Kiladis, 2004: On the dynamics of two day equatorial disturbances. *J. Atmos. Sci.*, **61**, 2707–2721, doi:10.1175/JAS3352.1.

Han, Y., and B. Khouider, 2010: Convectively coupled waves in a sheared environment. *J. Atmos. Sci.*, **67**, 2913–2942, doi:10.1175/2010JAS3335.1.

Hayashi, Y., 1971: Instability of large-scale equatorial waves with a frequency-dependent CISK parameter. *J. Meteor. Soc. Japan*, **49**, 59–62.

Johnson, R. H., T. M. Rickenbach, S. A. Rutledge, P. E. Ciesielski, and W. H. Schubert, 1999: Trimodal characteristics of tropical convection. *J. Climate*, **12**, 2397–2418, doi:10.1175/1520-0442(1999)012<2397:TCOTC>2.0.CO;2.

Kasahara, A., and K. Puri, 1981: Spectral representation of three-dimensional global data by expansion in normal mode functions. *Mon. Wea. Rev.*, **109**, 37–51, doi:10.1175/1520-0493(1981)109<0037:SROTDG>2.0.CO;2.

Khouider, B., and A. J. Majda, 2006a: Multicloud convective parametrizations with crude vertical structure. *Theor. Comput. Fluid Dyn.*, **20**, 351–375, doi:10.1007/s00162-006-0013-2.

——, and ——, 2006b: A simple multicloud parametrization for convectively coupled tropical waves. Part I: Linear analysis. *J. Atmos. Sci.*, **63**, 1308–1323, doi:10.1175/JAS3677.1.

——, and ——, 2008a: Equatorial convectively coupled waves in a simple multicloud model. *J. Atmos. Sci.*, **65**, 3376–3397, doi:10.1175/2008JAS2752.1.

——, and ——, 2008b: Multicloud models for organized tropical convection: Enhanced congestus heating. *J. Atmos. Sci.*, **65**, 897–914, doi: 10.1175/2007JAS2408.1.

——, J. Biello, and A. J. Majda, 2010: A stochastic multicloud model for tropical convection. *Commun. Math. Sci.*, **8** (1), 187–216, doi:10.4310/CMS.2010.v8.n1.a10.

——, A. St-Cyr, A. J. Majda, and J. Tribbia, 2011: The MJO and convectively coupled waves in a coarse-resolution GCM with a simple multicloud parameterization. *J. Atmos. Sci.*, **68**, 240–264, doi:10.1175/2010JAS3443.1.

——, Y. Han, and J. Biello, 2012a: Convective momentum transport in a simple multicloud model. *J. Atmos. Sci.*, **69**, 281–302, doi:10.1175/JAS-D-11-042.1.

——, ——, A. J. Majda, and S. Stechmann, 2012b: Multiscale waves in an MJO background and CMT feedback. *J. Atmos. Sci.*, **69**, 915–933, doi:10.1175/JAS-D-11-0152.1.

Kiladis, G. N., K. H. Straub, and P. T. Haertel, 2005: Zonal and vertical structure of the Madden–Julian oscillation. *J. Atmos. Sci.*, **62**, 2790–2809, doi:10.1175/JAS3520.1.

——, M. C. Wheeler, P. T. Haertel, K. H. Straub, and P. E. Roundy, 2009: Convectively coupled equatorial waves. *Rev. Geophys.*, **47**, RG2003, doi:10.1029/2008RG000266.

Lin, J.-L., and Coauthors, 2006: Tropical intraseasonal variability in 14 IPCC AR4 climate models. Part I: Convective signals. *J. Climate*, **19**, 2665–2690, doi:10.1175/JCLI3735.1.

Lin, X., and R. H. Johnson, 1996: Heating, moistening, and rainfall over the western Pacific warm pool during TOGA COARE. *J. Atmos. Sci.*, **53**, 3367–3383, doi:10.1175/1520-0469(1996)053<3367:HMAROT>2.0.CO;2.

Lindzen, R. S., 1974: Wave-CISK in the tropics. *J. Atmos. Sci.*, **31**, 156–179, doi:10.1175/1520-0469(1974)031<0156:WCITT>2.0.CO;2.

Majda, A. J., 2007: New multiscale models and self-similarity in tropical convection. *J. Atmos. Sci.*, **64**, 1393–1404, doi:10.1175/JAS3880.1.

——, and M. Shefter, 2001a: Models for stratiform instability and convectively coupled waves. *J. Atmos. Sci.*, **58**, 1567–1584, doi:10.1175/1520-0469(2001)058<1567:MFSIAC>2.0.CO;2.

——, and ——, 2001b: Waves and instabilities for model tropical convective parametrizations. *J. Atmos. Sci.*, **58**, 896–914, doi:10.1175/1520-0469(2001)058<0896:WAIFMT>2.0.CO;2.

——, and B. Khouider, 2002: Stochastic and mesoscopic models for tropical convection. *Proc. Natl. Acad. Sci. USA*, **99**, 1123–1128, doi:10.1073/pnas.032663199.

——, and J. Biello, 2004: A multiscale model for the intraseasonal oscillation. *Proc. Natl. Acad. Sci. USA*, **101**, 4736–4741, doi:10.1073/pnas.0401034101.

——, and S. N. Stechmann, 2008: Stochastic models for convective momentum transport. *Proc. Natl. Acad. Sci. USA*, **105**, 17 614–17 619, doi:10.1073/pnas.0806838105.

——, and ——, 2016: Models for multiscale interactions. Part II: Madden–Julian oscillation, moisture, and convective momentum transport. *Multiscale Convection-Coupled Systems in the Tropics: A Tribute to Dr. Michio Yanai, Meteor. Monogr.*, No. 56, Amer. Meteor. Soc., doi:10.1175/AMSM-D-15-0005.1.

——, B. Khouider, G. Kiladis, K. H. Straub, and M. G. Shefter, 2004: A model for convectively coupled tropical waves: Nonlinearity, rotation, and comparison with observations. *J. Atmos. Sci.*, **61**, 2188–2205, doi:10.1175/1520-0469(2004)061<2188:AMFCCT>2.0.CO;2.

——, S. N. Stechmann, and B. Khouider, 2007: Madden–Julian oscillation analog and intraseasonal variability in a multicloud model above the equator. *Proc. Natl. Acad. Sci. USA*, **104**, 9919–9924, doi:10.1073/pnas.0703572104.

Mapes, B. E., 1993: Gregarious tropical convection. *J. Atmos. Sci.*, **50**, 2026–2037, doi:10.1175/1520-0469(1993)050<2026:GTC>2.0.CO;2.

——, 2000: Convective inhibition, subgridscale triggering energy, and "stratiform instability" in a toy tropical wave model. *J. Atmos. Sci.*, **57**, 1515–1535, doi:10.1175/1520-0469(2000)057<1515:CISSTE>2.0.CO;2.

——, S. Tulich, J. Lin, and P. Zuidema, 2006: The mesoscale convection life cycle: Building block or prototype for large-scale tropical waves? *Dyn. Atmos. Oceans*, **42**, 3–29, doi:10.1016/j.dynatmoce.2006.03.003.

Matsuno, T., 1966: Quasi-geostrophic motions in the equatorial area. *J. Meteor. Soc. Japan*, **44**, 25–41.

Nakazawa, T., 1974: Tropical super clusters within intraseasonal variation over the western Pacific. *J. Meteor. Soc. Japan*, **66**, 823–839.

Neelin, J. D., and J. Yu, 1994: Modes of tropical variability under convective adjustment and Madden–Julian oscillation. Part I: Analytical theory. *J. Atmos. Sci.*, **51**, 1876–1894, doi:10.1175/1520-0469(1994)051<1876:MOTVUC>2.0.CO;2.

Ooyama, K., 1964: A dynamical model for the study of tropical cyclone development. *Geofis. Int.*, **4**, 187–198.

Qiang, D., B. Khouider, and A. J. Majda, 2014: The MJO in a coarse-resolution GCM with a stochastic multicloud parameterization. *J. Atmos. Sci.*, **72**, 55–74, doi:10.1175/JAS-D-14-0120.1.

Straub, K. H., and G. N. Kiladis, 2002: Observations of a convectively coupled Kelvin wave in the eastern Pacific ITCZ. *J. Atmos. Sci.*, **59**, 30–53, doi:10.1175/1520-0469(2002)059<0030:OOACCK>2.0.CO;2.

——, and ——, 2003: The observed structure of convectively coupled Kelvin waves: Comparison with simple models of coupled wave instability. *J. Atmos. Sci.*, **60**, 1655–1668, doi:10.1175/1520-0469(2003)060<1655:TOSOCC>2.0.CO;2.

Takayabu, Y. N., 1994: Large-scale cloud disturbances associated with equatorial waves. Part I: Spectral features of the cloud disturbances. *J. Meteor. Soc. Japan*, **72**, 433–448.

Taylor, M., J. Edwards, and A. St-Cyr, 2008: Petascale atmospheric models for the Community Climate System Model: New developments and evaluation of scalable dynamical cores. *J. Phys. Conf. Ser.*, **125**, 012023, doi:10.1088/1742-6596/125/1/012023.

Waite, M. L., and B. Khouider, 2009: Boundary layer dynamics in a simple model for convectively coupled gravity waves. *J. Atmos. Sci.*, **66**, 2780–2795, doi:10.1175/2009JAS2871.1.

Wheeler, M., and G. N. Kiladis, 1999: Convectively coupled equatorial waves: Analysis of clouds and temperature in the wavenumber–frequency domain. *J. Atmos. Sci.*, **56**, 374–399, doi:10.1175/1520-0469(1999)056<0374:CCEWAO>2.0.CO;2.

——, ——, and P. J. Webster, 2000: Large-scale dynamical fields associated with convectively coupled equatorial waves. *J. Atmos. Sci.*, **57**, 613–640, doi:10.1175/1520-0469(2000)057<0613:LSDFAW>2.0.CO;2.

Yamasaki, M., 1969: Large-scale disturbances in a conditionally unstable atmosphere in low latitudes. *Pap. Meteor. Geophys.*, **20**, 289–336.

Yanai, M., and T. Maruyama, 1966: Stratospheric wave disturbances propagating over the equatorial Pacific. *J. Meteor. Soc. Japan*, **44**, 291–294.

——, B. Chen, and W. Tung, 2000: The Madden–Julian oscillation observed during the TOGA COARE IOP: Global view. *J. Atmos. Sci.*, **57**, 2374–2396, doi:10.1175/1520-0469(2000)057<2374:TMJOOD>2.0.CO;2.

Yano, J.-I., J. C. McWilliams, M. Moncrieff, and K. A. Emanuel, 1995: Hierarchical tropical cloud systems in an analog shallow-water model. *J. Atmos. Sci.*, **52**, 1723–1742, doi:10.1175/1520-0469(1995)052<1723:HTCSIA>2.0.CO;2.

——, M. Moncrieff, and J. C. McWilliams, 1998: Linear stability and single column analyses of several cumulus parametrization categories in a shallow-water model. *Quart. J. Roy. Meteor. Soc.*, **124**, 983–1005, doi:10.1002/qj.49712454715.

Zehnder, J. A., 2001: A comparison of convergence- and surface-flux-based convective parametrizations with applications to tropical cyclogenesis. *J. Atmos. Sci.*, **58**, 283–301, doi:10.1175/1520-0469(2001)058<0283:ACOCAS>2.0.CO;2.

Chapter 10

Models for Multiscale Interactions. Part II: Madden–Julian Oscillation, Moisture, and Convective Momentum Transport

ANDREW J. MAJDA

*Department of Mathematics, and Center for Atmosphere Ocean Science, Courant Institute of Mathematical Sciences,
New York University, New York, New York*

SAMUEL N. STECHMANN

*Department of Mathematics, and Department of Atmospheric and Oceanic Sciences, University of Wisconsin–Madison,
Madison, Wisconsin*

ABSTRACT

It is well known that the envelope of the Madden–Julian oscillation (MJO) consists of smaller-scale convective systems, including mesoscale convective systems (MCS), tropical cyclones, and synoptic-scale waves called "convectively coupled equatorial waves" (CCW). In fact, recent results suggest that the fundamental mechanisms of the MJO involve interactions between the synoptic-scale CCW and their larger-scale environment (Majda and Stechmann). In light of this, this chapter reviews recent and past work on two-way interactions between convective systems—both MCSs and CCW—and their larger-scale environment, with a particular focus given to recent work on MJO–CCW interactions.

1. Introduction

In this chapter, the multiscale hierarchy of organized convection will be divided into three broad categories: (i) the MJO on planetary spatial scales (roughly 20 000 km) and intraseasonal time scales (roughly 40 days) (Lau and Waliser 2005; Zhang 2005), (ii) convectively coupled equatorial waves (CCW) on equatorial synoptic scales (roughly 2000 km and 4 days) (Kiladis et al. 2009), and (iii) MCS on mesoscales (roughly 200 km and 0.4 days) (Houze 2004). This hierarchy has a remarkable multiscale structure: the MJO is an envelope of smaller-scale CCW, and, in turn, the CCW are envelopes of smaller-scale MCS. Figure 10-1 illustrates the structure of a generic large-scale envelope with smaller-scale fluctuations embedded within it. For such a multiscale envelope, one can imagine three scenarios for its physical mechanisms: (i) Does the envelope drive the fluctuations within it? (ii) Do the fluctuations drive the large-scale envelope? (iii) Or do the two evolve cooperatively? This situation is reminiscent of the classic conundrum: Which came first: the chicken or the egg? In atmospheric science, such questions are most familiar from contexts such as midlatitude eddies and jets. Here, instead, the focus is on the tropical case described above, and the evidence will suggest that, in fact, there are cooperative interactions between the envelopes and the fluctuations within them.

2. The chicken–egg question: Interactions between convective systems and their larger-scale environment

In this section, we review recent and past results on convection–environment interactions. MCS–environment interactions are reviewed first in order to set the stage for the relatively new topic of CCW–environment interactions and in order to highlight their similarities and differences.

It is well known that MCS can have important effects on the larger-scale atmospheric state in which they exist (Houze 2004). For example, precipitation and vertical transports of temperature and moisture can significantly

---

Corresponding author address: Samuel Stechmann, Department of Mathematics, University of Wisconsin–Madison, 480 Lincoln Drive, Madison, WI 53706.
E-mail: stechmann@wisc.edu

DOI: 10.1175/AMSMONOGRAPHS-D-15-0005.1

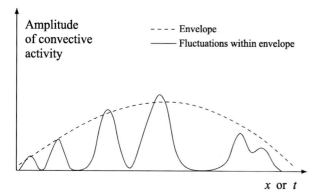

FIG. 10-1. A large-scale envelope with fluctuations embedded within it.

alter the larger-scale thermodynamic environment. A less-understood effect of MCS on their larger-scale environment is convective momentum transport (CMT). A pervasive aspect of this is the idea of "cumulus friction"; however, many studies have also shown that the CMT of some MCS can actually accelerate the background wind (Wu and Yanai 1994; LeMone and Moncrieff 1994; Tung and Yanai 2002a,b). On the other hand, these interactions also proceed in the opposite direction: the background state—including the wind shear and moist thermodynamic state—affects the MCS that form within it. The intensity of an MCS is related to the environmental thermodynamic state, and the vertical wind shear du/dz determines the propagation direction and morphology of MCS (Barnes and Sieckman 1984; LeMone et al. 1998; Liu and Moncrieff 2004).

On larger scales than MCS are the synoptic-scale phenomena known as CCW (Kiladis et al. 2009). With typical wavelengths of 2000–8000 km and periods of 2–10 days, CCW are the dominant synoptic-scale weather features in many regions in the tropics, but they are less understood than the smaller-scale MCS. Just as MCS interact with their larger-scale environment, so do CCW. However, CCW–environment interactions are in a relatively primitive stage of understanding.

On the one hand, one might suspect that CCW can affect their larger-scale environment as MCS do. Some results using diagnostic models for the CCW have demonstrated that this is true [see Haertel and Kiladis (2004), Biello et al. (2007), and references therein]. For example, CCW can act as heat sources and moisture sinks for their environment. Also, since they have tilts in the vertical/zonal direction (Takayabu et al. 1996; Straub and Kiladis 2003), CCW can transport momentum to larger scales. That is to say, eddy flux divergences can accelerate or decelerate the mean flow, where the CCW are the "eddies" in this context. This has important implications

for the MJO, as shown in the diagnostic multiscale models of Biello and Majda (2005) and Biello et al. (2007). They demonstrate that the westerly jet that appears during the westerly wind burst stage of the MJO can be driven by momentum transports from synoptic-scale CCW.

On the other hand, one might also suspect CCW–environment interactions to occur in the other direction; that is, one might suspect that the background state—including the background wind and moist thermodynamic states—can affect the CCW that form within it. Unfortunately, while several studies have documented the properties of CCW, few have documented them in different, distinct background environments [exceptions to this include Roundy and Frank (2004) and Yang et al. (2007)]. Besides observations, computer simulations of CCW have presently offered little insight into CCW–environment interactions, since general circulation models (GCMs) do not adequately capture CCW (Lin et al. 2006), and simulations of CCW with cloud-resolving models (CRMs) are challenging [but not impossible: see Grabowski and Moncrieff (2001), Tulich et al. (2007), and the discussion below]. One promising method for numerical simulations of CCW is the so-called multicloud model in the recent work of Khouider and Majda (2006, 2008), which has been successful in capturing observed features of CCW and in explaining their physical mechanisms.

Numerical simulations with the multicloud model offer some of the first glimpses of two-way CCW–environment interactions. Majda and Stechmann (2009a) show that the background vertical wind shear can favor either eastward- or westward-propagating CCW, which is analogous to the effect of wind shear on MCS as discussed above (although the details of the two cases are different). Also, because of their vertical/zonal tilts (Takayabu et al. 1996; Straub and Kiladis 2003), CCW can transport momentum to larger scales and alter the mean winds. Therefore, the combination of these effects allows for convectively coupled wave–mean flow interaction with dynamical two-way interactions between CCW and the background vertical shear. In this dynamical setting, Majda and Stechmann (2009a) showed that a low-level westerly jet, like that seen in the westerly wind burst stage of the MJO, can be driven by momentum transports due to convectively coupled Kelvin waves, which are often observed in the envelope of the MJO (Nakazawa 1988; Masunaga et al. 2006). These basic CMT features can also be illustrated using simple ordinary differential equations for wave amplitudes and the mean wind (Stechmann et al. 2013). Furthermore, Khouider et al. (2012b) investigated competing CMT effects due to multiscale waves. Using different MJO phases for the background environment, they compared the CMT due to a convective envelope (such as a convectively coupled wave) and the convective elements embedded within the envelope (such as squall lines).

Besides CCW–environment interactions involving momentum, one might also expect CCW to interact with their larger-scale moist thermodynamic environment. Indeed, the recent model of Majda and Stechmann (2009b) suggests that the essential physical mechanisms of the MJO involve such interactions.

3. The MJO's "skeleton" and "muscle"

The fundamental mechanism of the MJO proposed by Majda and Stechmann (2009b, 2011) involves neutrally stable interactions between (i) planetary-scale lower-tropospheric moisture anomalies and (ii) the planetary-scale envelope of smaller-scale convective activity, which can include both CCW and MCS. Using a simple dynamical model with these features, they captured all of the following fundamental features (i.e., the skeleton) of the MJO: (i) peculiar dispersion relation of roughly $dw/dk = 0$, (ii) slow eastward phase speed of roughly $5\,\mathrm{m\,s}^{-1}$, and (iii) horizontal quadrupole vortex structure (Hendon and Liebmann 1994; Wheeler and Kiladis 1999). The prediction of all of these features is strong evidence that CCW–moisture interactions are the essential physical mechanism of the MJO.

The basis for the mechanisms mentioned above comes from observations, theory, and modeling studies. On the one hand, CCW activity affects the heat and moisture budgets of the MJO through the planetary-scale envelope of its heating and drying. On the other hand, the moist thermodynamic environment regulates the growth/decay of CCW activity (Khouider and Majda 2008) in a way similar to the effects of a background wind shear on CCW (Majda and Stechmann 2009a). This is also suggested by observations that show the maximum of lower-tropospheric water vapor leading the maximum in convective activity (Kiladis et al. 2005), which indicates that the positive (negative) low-level water vapor anomalies have a tendency to increase (decrease) the envelope of CCW and MCS activity.

These CCW–moisture interactions are at the heart of the model of Majda and Stechmann (2009b, 2011) that produces the MJO's skeleton as described above. However, the MJO has other features beyond this basic skeleton, and we refer to these additional features as the MJO's muscle. One aspect of the MJO's muscle is its westerly wind burst, which can be driven by momentum transports of CCW as described in the previous section. Other asymmetries that fall within the realm of the MJO's muscle are potentially due to other upscale fluxes from synoptic scales and variations in surface heat fluxes (Sobel et al. 2010).

4. Implications for GCMs

Given the wide range of interactions involved in the hierarchy of organized tropical convection, it is maybe not surprising that GCMs, which must parameterize an important subset of the hierarchy, struggle with capturing CCW and the MJO (Moncrieff and Klinker 1997; Lin et al. 2006). With grid spacings of roughly 100 km, GCMs can at best hope to capture convection on scales of roughly 1000 km and larger (i.e., CCW and the MJO), assuming that about 10 grid points are needed to properly resolve a given feature. This also implies that features with scales of 100–1000 km and smaller must be parameterized; MCS fall within this category but are often ignored in parameterizations. These two problems—properly representing CCW and parameterizing MCS—are crucial to improving the MJO and the hierarchy of organized convection in GCMs.

A convective parameterization that captures realistic CCW is the multicloud parameterization of Khouider and Majda (2006, 2008), which includes several features that are not standard among contemporary convective parameterizations. The key aspect of this parameterization is the representation of three different cloud types: congestus, deep convection, and stratiform clouds (Houze 2004; Johnson et al. 1999). Each of these different phases of convection plays a different role in the heat and moisture budgets, and it is important to include all three phases rather than just deep convection (Haertel and Kiladis 2004; Khouider and Majda 2006, 2008). Another important component is convective triggering, for which the mid- and lower-tropospheric moisture plays a key role. While the multicloud parameterization has been developed using simplified dynamical models, it is currently being included in a GCM. Also, a stochastic version is being developed that includes stochastic transitions among different phases of convection (congestus, deep convection, and stratiform) (Khouider et al. 2010).

The effect of unresolved MCS on resolved GCM dynamics can occur in several ways, one of which is CMT. It has been recognized that CMT can either accelerate or decelerate the mean wind (Wu and Yanai 1994; LeMone and Moncrieff 1994), and representations of this feature have been included in parameterizations that have improved climatological-scale circulations (Wu et al. 2007; Richter and Rasch 2008). However, observations also show that the acceleration/deceleration due to CMT occurs in intense, intermittent bursts (Tung and Yanai 2002a,b), and that including this intermittency should be important for the variability of resolved convection. A promising method for including the intermittent aspects of unresolved CMT is to use a stochastic CMT parameterization (Majda and Stechmann 2008; Khouider et al. 2012a).

The general principles described above—different phases of convection and momentum transports—also appear to be important for the specific case of capturing the MJO in GCMs. In terms of different phases of convection, the congestus phase of convection appears to be important for moistening the lower troposphere during the MJO's suppressed stage (Kikuchi and Takayabu 2004; Tian et al. 2006). However, GCMs do not appear to properly represent congestus clouds and the accompanying moistening effects. It is possible that parameterizations with minimum entrainment parameters, which have improved the strength of intraseasonal variability in some GCMs, are employing a surrogate for congestus clouds and lower-tropospheric moistening (Tokioka et al. 1988; Sobel et al. 2010). Besides congestus clouds, another important physical effect in the multicloud model that is largely absent in contemporary GCMs is unsaturated downdrafts in the stratiform region, which are important for cooling and drying the lower troposphere. In terms of the second general principle—momentum transports—a stochastic parameterization of unresolved CMT, such as that of Majda and Stechmann (2008), should help increase the variability of CCW and the MJO, which tends to be significantly lower than in observations (Lin et al. 2006).

It is possible that lessons for GCMs can be learned from CRM simulations of multiscale convection, since CRMs are able to represent a CCW envelope and the MCS within it, whereas GCMs struggle to capture the MJO envelope and the CCW within it. For instance, the role of CMT and momentum damping can be seen in CRM simulations of CCW. The simulations of Grabowski and Moncrieff (2001) employ (parameterized) weak momentum damping with a time scale of 24 h, and the results show that resolved CMT from MCS plays an important role in driving the CCW envelope. On the other hand, the simulations of Tulich et al. (2007) employ (parameterized) stronger momentum damping with a time scale of 4 h, and they find little role of resolved CMT in driving the CCW envelope; instead, the key mechanism involves interactions between MCS and gravity waves generated by the MCS (Mapes 1993; Tulich and Mapes 2008; Stechmann and Majda 2009). Other CRM simulations of Held et al. (1993) show two extreme cases with (i) no CMT allowed and (ii) no momentum damping, and the results are completely different in the two cases. Given this variety of results, it is likely that parameterized momentum sources in GCMs strongly affect the resolved waves—that is, CCW and the MJO—and the simulated mechanisms behind them.

5. Summary

Recent research suggests that the fundamental mechanisms of the MJO involve interactions between CCW and their larger-scale environment (Majda and Stechmann 2009a,b, 2011), and this chapter is a review of these interactions (see also Khouider et al. 2013). Implications for GCMs were also discussed, both in terms of general principles for subgrid-scale parameterization and in terms of the MJO specifically. These ideas should be useful for improving our theoretical understanding and numerical simulations of the MJO.

Acknowledgments. The research of A.J.M. and S.N.S. is partially supported by ONR MURI Grant N00014-12-1-0912. The research of S.N.S. is also partially supported by ONR YIP Grant N00014-12-1-0744 and NSF Grant DMS–1209409.

REFERENCES

Barnes, G., and K. Sieckman, 1984: The environment of fast- and slow-moving tropical mesoscale convective cloud lines. *Mon. Wea. Rev.*, **112**, 1782–1794, doi:10.1175/1520-0493(1984)112<1782:TEOFAS>2.0.CO;2.

Biello, J. A., and A. J. Majda, 2005: A new multiscale model for the Madden–Julian oscillation. *J. Atmos. Sci.*, **62**, 1694–1721, doi:10.1175/JAS3455.1.

——, ——, and M. W. Moncrieff, 2007: Meridional momentum flux and superrotation in the multiscale IPESD MJO model. *J. Atmos. Sci.*, **64**, 1636–1651, doi:10.1175/JAS3908.1.

Grabowski, W. W., and M. W. Moncrieff, 2001: Large-scale organization of tropical convection in two-dimensional explicit numerical simulations. *Quart. J. Roy. Meteor. Soc.*, **127**, 445–468, doi:10.1002/qj.49712757211.

Haertel, P. T., and G. N. Kiladis, 2004: Dynamics of 2-day equatorial waves. *J. Atmos. Sci.*, **61**, 2707–2721, doi:10.1175/JAS3352.1.

Held, I., R. Hemler, and V. Ramaswamy, 1993: Radiative–convective equilibrium with explicit two-dimensional moist convection. *J. Atmos. Sci.*, **50**, 3909–3927, doi:10.1175/1520-0469(1993)050<3909:RCEWET>2.0.CO;2.

Hendon, H. H., and B. Liebmann, 1994: Organization of convection within the Madden–Julian oscillation. *J. Geophys. Res.*, **99**, 8073–8084, doi:10.1029/94JD00045.

Houze, R. A., 2004: Mesoscale convective systems. *Rev. Geophys.*, **42**, RG4003, doi:10.1029/2004RG000150.

Johnson, R. H., T. M. Rickenbach, S. A. Rutledge, P. E. Ciesielski, and W. H. Schubert, 1999: Trimodal characteristics of tropical convection. *J. Climate*, **12**, 2397–2418, doi:10.1175/1520-0442(1999)012<2397:TCOTC>2.0.CO;2.

Khouider, B., and A. J. Majda, 2006: A simple multicloud parameterization for convectively coupled tropical waves. Part I: Linear analysis. *J. Atmos. Sci.*, **63**, 1308–1323, doi:10.1175/JAS3677.1.

——, and ——, 2008: Equatorial convectively coupled waves in a simple multicloud model. *J. Atmos. Sci.*, **65**, 3376–3397, doi:10.1175/2008JAS2752.1.

——, J. Biello, and A. J. Majda, 2010: A stochastic multicloud model for tropical convection. *Commun. Math. Sci.*, **8**, 187–216.

——, Y. Han, and J. A. Biello, 2012a: Convective momentum transport in a simple multicloud model for organized convection. *J. Atmos. Sci.*, **69**, 281–302, doi:10.1175/JAS-D-11-042.1.

——, ——, A. J. Majda, and S. N. Stechmann, 2012b: Multiscale waves in an MJO background and convective momentum transport feedback. *J. Atmos. Sci.*, **69**, 915–933, doi:10.1175/JAS-D-11-0152.1.

——, A. J. Majda, and S. N. Stechmann, 2013: Climate science in the tropics: Waves, vortices, and PDEs. *Nonlinearity*, **26**, R1–R68, doi:10.1088/0951-7715/26/1/R1.

Kikuchi, K., and Y. N. Takayabu, 2004: The development of organized convection associated with the MJO during TOGA COARE IOP: Trimodal characteristics. *Geophys. Res. Lett.*, **31**, L10101, doi:10.1029/2004GL019601.

Kiladis, G. N., K. H. Straub, and P. T. Haertel, 2005: Zonal and vertical structure of the Madden–Julian oscillation. *J. Atmos. Sci.*, **62**, 2790–2809, doi:10.1175/JAS3520.1.

——, M. C. Wheeler, P. T. Haertel, K. H. Straub, and P. E. Roundy, 2009: Convectively coupled equatorial waves. *Rev. Geophys.*, **47**, RG2003, doi:10.1029/2008RG000266.

Lau, W. K.-M., and D. E. Waliser, Eds., 2005: *Intraseasonal Variability in the Atmosphere–Ocean Climate System.* Environmental Sciences Series, Springer, 437 pp.

LeMone, M., and M. Moncrieff, 1994: Momentum and mass transport by convective bands: Comparisons of highly idealized dynamical models to observations. *J. Atmos. Sci.*, **51**, 281–305, doi:10.1175/1520-0469(1994)051<0281:MAMTBC>2.0.CO;2.

——, E. Zipser, and S. Trier, 1998: The role of environmental shear and thermodynamic conditions in determining the structure and evolution of mesoscale convective systems during TOGA COARE. *J. Atmos. Sci.*, **55**, 3493–3518, doi:10.1175/1520-0469(1998)055<3493:TROESA>2.0.CO;2.

Lin, J.-L., and Coauthors, 2006: Tropical intraseasonal variability in 14 IPCC AR4 climate models. Part I: Convective signals. *J. Climate*, **19**, 2665–2690, doi:10.1175/JCLI3735.1.

Liu, C., and M. Moncrieff, 2004: Effects of convectively generated gravity waves and rotation on the organization of convection. *J. Atmos. Sci.*, **61**, 2218–2227, doi:10.1175/1520-0469(2004)061<2218:EOCGGW>2.0.CO;2.

Majda, A. J., and S. N. Stechmann, 2008: Stochastic models for convective momentum transport. *Proc. Natl. Acad. Sci. USA*, **105**, 17 614–17 619, doi:10.1073/pnas.0806838105.

——, and ——, 2009a: A simple dynamical model with features of convective momentum transport. *J. Atmos. Sci.*, **66**, 373–392, doi:10.1175/2008JAS2805.1.

——, and ——, 2009b: The skeleton of tropical intraseasonal oscillations. *Proc. Natl. Acad. Sci. USA*, **106**, 8417, doi:10.1073/pnas.0903367106.

——, and ——, 2011: Nonlinear dynamics and regional variations in the MJO skeleton. *J. Atmos. Sci.*, **68**, 3053–3071, doi:10.1175/JAS-D-11-053.1.

Mapes, B., 1993: Gregarious tropical convection. *J. Atmos. Sci.*, **50**, 2026–2037, doi:10.1175/1520-0469(1993)050<2026:GTC>2.0.CO;2.

Masunaga, H., T. L'Ecuyer, and C. Kummerow, 2006: The Madden–Julian oscillation recorded in early observations from the Tropical Rainfall Measuring Mission (TRMM). *J. Atmos. Sci.*, **63**, 2777–2794, doi:10.1175/JAS3783.1.

Moncrieff, M. W., and E. Klinker, 1997: Organized convective systems in the tropical western Pacific as a process in general circulation models: A TOGA COARE case-study. *Quart. J. Roy. Meteor. Soc.*, **123**, 805–827, doi:10.1002/qj.49712354002.

Nakazawa, T., 1988: Tropical super clusters within intraseasonal variations over the western Pacific. *J. Meteor. Soc. Japan*, **66**, 823–839.

Richter, J., and P. Rasch, 2008: Effects of convective momentum transport on the atmospheric circulation in the Community Atmosphere Model, version 3. *J. Climate*, **21**, 1487–1499, doi:10.1175/2007JCLI1789.1.

Roundy, P., and W. Frank, 2004: A climatology of waves in the equatorial region. *J. Atmos. Sci.*, **61**, 2105–2132, doi:10.1175/1520-0469(2004)061<2105:ACOWIT>2.0.CO;2.

Sobel, A. H., E. D. Maloney, G. Bellon, and D. M. Frierson, 2010: Surface fluxes and tropical intraseasonal variability: A reassessment. *J. Adv. Model. Earth Syst.*, **2** (2), doi:10.3894/JAMES.2010.2.2.

Stechmann, S. N., and A. J. Majda, 2009: Gravity waves in shear and implications for organized convection. *J. Atmos. Sci.*, **66**, 2579–2599, doi:10.1175/2009JAS2976.1.

——, ——, and D. Skjorshammer, 2013: Convectively coupled wave–environment interactions. *Theor. Comput. Fluid Dyn.*, **27**, 513–532, doi:10.1007/s00162-012-0268-8.

Straub, K. H., and G. N. Kiladis, 2003: The observed structure of convectively coupled kelvin waves: Comparison with simple models of coupled wave instability. *J. Atmos. Sci.*, **60**, 1655–1668, doi:10.1175/1520-0469(2003)060<1655:TOSOCC>2.0.CO;2.

Takayabu, Y. N., K. M. Lau, and C. H. Sui, 1996: Observation of a quasi-2-day wave during TOGA COARE. *Mon. Wea. Rev.*, **124**, 1892–1913, doi:10.1175/1520-0493(1996)124<1892:OOAQDW>2.0.CO;2.

Tian, B., D. Waliser, E. Fetzer, B. Lambrigtsen, Y. Yung, and B. Wang, 2006: Vertical moist thermodynamic structure and spatial–temporal evolution of the MJO in AIRS observations. *J. Atmos. Sci.*, **63**, 2462–2485, doi:10.1175/JAS3782.1.

Tokioka, T., K. Yamazaki, A. Kitoh, and T. Ose, 1988: The equatorial 30–60 day oscillation and the Arakawa–Schubert penetrative cumulus parameterization. *J. Meteor. Soc. Japan*, **66**, 883–901.

Tulich, S. N., and B. Mapes, 2008: Multiscale convective wave disturbances in the tropics: Insights from a two-dimensional cloud-resolving model. *J. Atmos. Sci.*, **65**, 140–155, doi:10.1175/2007JAS2353.1.

——, D. A. Randall, and B. E. Mapes, 2007: Vertical-mode and cloud decomposition of large-scale convectively coupled gravity waves in a two-dimensional cloud-resolving model. *J. Atmos. Sci.*, **64**, 1210–1229, doi:10.1175/JAS3884.1.

Tung, W., and M. Yanai, 2002a: Convective momentum transport observed during the TOGA COARE IOP. Part I: General features. *J. Atmos. Sci.*, **59**, 1857–1871, doi:10.1175/1520-0469(2002)059<1857:CMTODT>2.0.CO;2.

——, and ——, 2002b: Convective momentum transport observed during the TOGA COARE IOP. Part II: Case studies. *J. Atmos. Sci.*, **59**, 2535–2549, doi:10.1175/1520-0469(2002)059<2535:CMTODT>2.0.CO;2.

Wheeler, M., and G. N. Kiladis, 1999: Convectively coupled equatorial waves: Analysis of clouds and temperature in the wavenumber–frequency domain. *J. Atmos. Sci.*, **56**, 374–399, doi:10.1175/1520-0469(1999)056<0374:CCEWAO>2.0.CO;2.

Wu, X., and M. Yanai, 1994: Effects of vertical wind shear on the cumulus transport of momentum: Observations and parameterization. *J. Atmos. Sci.*, **51**, 1640–1660, doi:10.1175/1520-0469(1994)051<1640:EOVWSO>2.0.CO;2.

——, L. Deng, X. Song, and G. Zhang, 2007: Coupling of convective momentum transport with convective heating in global climate simulations. *J. Atmos. Sci.*, **64**, 1334–1349, doi:10.1175/JAS3894.1.

Yang, G., B. Hoskins, and J. Slingo, 2007: Convectively coupled equatorial waves. Part I: Horizontal and vertical structures. *J. Atmos. Sci.*, **64**, 3406–3423, doi:10.1175/JAS4017.1.

Zhang, C., 2005: Madden–Julian Oscillation. *Rev. Geophys.*, **43**, RG2003, doi:10.1029/2004RG000158.

Chapter 11

Influence of Cloud Microphysics and Radiation on Tropical Cyclone Structure and Motion

ROBERT G. FOVELL,* YIZHE PEGGY BU,* KRISTEN L. CORBOSIERO,[+] WEN-WEN TUNG,[#]
YANG CAO,* HUNG-CHI KUO,[@] LI-HUAN HSU,** AND HUI SU[++]

*Department of Atmospheric and Oceanic Sciences, University of California, Los Angeles, Los Angeles, California
[+]Department of Atmospheric and Environmental Sciences, University at Albany, State University of
New York, Albany, New York
[#]Department of Earth, Atmospheric and Planetary Sciences, Purdue University, West Lafayette, Indiana
[@]Department of Atmospheric Sciences, National Taiwan University, Taipei, Taiwan
**Taiwan Typhoon Flood and Research Institute, Taipei, Taiwan
[++]Jet Propulsion Laboratory, California Institute of Technology, Pasadena, California

ABSTRACT

The authors survey a series of modeling studies that have examined the influences that cloud microphysical processes can have on tropical cyclone (TC) motion, the strength and breadth of the wind field, inner-core diabatic heating asymmetries, outer-core convective activity, and the characteristics of the TC anvil cloud. These characteristics are sensitive to the microphysical parameterization (MP) in large part owing to the cloud-radiative forcing (CRF), the interaction of hydrometeors with radiation. The most influential component of CRF is that due to absorption and emission of longwave radiation in the anvil, which via gentle lifting directly encourages the more extensive convective activity that then leads to a radial expansion of the TC wind field. On a curved Earth, the magnitude of the outer winds helps determine the speed and direction of TC motion via the beta drift. CRF also influences TC motion by determining how convective asymmetries develop in the TC inner core. Further improvements in TC forecasting may require improved understanding and representation of cloud-radiative processes in operational models, and more comprehensive comparisons with observations are clearly needed.

1. Introduction

Professor Michio Yanai's life-long love of weather started in middle school, when he joined the school's meteorology club (*otenkikai*). He and his fellow "meteorology boys" were particularly interested in the tropical cyclones (TCs) that frequently visited Japan during the autumn months, going so far as issuing their own weather warnings, making their own measurements, and conducting their own damage surveys. After graduating from the University of Tokyo with a degree in geophysics, he stayed on for graduate studies

in meteorology. His master's thesis [published as Yanai (1958)] focused on a decaying typhoon, and this was soon followed by a series of seminal papers on TC genesis that appeared in rapid succession, including Yanai (1961a,b), Yanai (1963a,b), and Yanai (1964).

Subsequently, Professor Yanai shifted his research focus to different areas, which are represented by other papers in this volume, but he never relinquished a keen interest in TCs. He was concerned about their societal impacts, historical variations, and even their nomenclature. In his last years, he frequently returned to the subject of Cyclone Nargis (2008), which brought enormous devastation to Myanmar. Indeed, just a week before his untimely passing in October 2010, Professor Yanai was organizing yet another issue of his University of California, Los Angeles Tropical Meteorology Newsletter, started in 1996 and distributed via e-mail, dedicated to summarizing the

Corresponding author address: Prof. Robert Fovell, UCLA Atmospheric and Oceanic Sciences, 405 Hilgard Ave., Los Angeles, CA 90096-1565.
E-mail: rfovell@ucla.edu

DOI: 10.1175/AMSMONOGRAPHS-D-15-0006.1

TABLE 11-1. Models employed in the semi-idealized experiments.

Paper No.	Study	Model/version	No. of domains	Finest resolution (km)	Remarks
1	Fovell and Su (2007)	ARW 2.2	3	3	—
2	Fovell et al. (2009)	ARW 2.2.1	3	3	—
3–4	Fovell et al. (2010a,b)	ARW 2.2.1	1	4	—
5	Cao et al. (2011)	ARW 3.0	2	3	Moving nest
6	Hsu et al. (2013)	ARW 3.1.1	1	5	Water mountain
7	Bu et al. (2014)	HWRF pre-2013	3	3	Two moving nests
		CM1	1	5	Axisymmetric
—	This review	ARW 3.2	1	4	"Augmented P3"
—	This review	MPAS 2.0	1	25	Global, 92 to 25 km

ongoing Nargis research. In May 2005, Professor Yanai published a review of the origins of the words "typhoon," "tai-feng," and "tai-fu," which was cowritten with his last doctoral student, Professor Chih-wen Hung, and the first author, Robert G. Fovell.

The first author's interest in TCs commenced in the summer of 2004 with the release of version 2.0 of the Weather Research and Forecasting (WRF) Model's Advanced Research WRF (ARW) core. He decided to familiarize himself with this new, more powerful ARW system by simulating TCs in real time,[1] and the 2004 season proved compelling, with four major hurricanes striking in or very near Florida and a historically large number of typhoons making landfall at Japan. He soon noticed that the cloud microphysics parameterization (MP), which controls the evolution of condensed water, could exert a material impact on storm track. A literature survey revealed relatively little understanding of the role of cloud processes on TC motion, and Professor Yanai provided critical encouragement for this research with his knowledge, insight, and enthusiasm.

This review summarizes the findings of seven papers (see Table 11-1) produced by the first author and his collaborators concerning cloud microphysics and their direct and indirect influences on TC motion and structure. Fovell and Su [2007, hereinafter P1 (for paper 1)] conducted a physics ensemble (consisting of MPs and cumulus parameterizations) for Hurricane Rita (2005). They also introduced the "semi-idealized" model framework employed in our subsequent work, which utilizes the "real-data" versions of models such as WRF and configurations similar to those implemented in operations. However, these models are dramatically simplified with respect to initialization, with a guiding philosophy that can be summarized by a famous dictum

attributed to Albert Einstein: "make things as simple as possible, but not simpler." Fovell et al. (2009, hereinafter P2) demonstrated that varying microphysical assumptions resulted in different wind profiles in the outer core region (roughly 100–300 km from the eye), which contribute to distinct motions owing to the "beta drift" (see section 4a) that directly influences track. P2 further showed that track sensitivity was at least indirectly tied to particle terminal velocities.

Fovell et al. (2010b, hereinafter P3) introduced yet another simplification: they prevented hydrometeors from affecting longwave (LW) and shortwave (SW) radiation, effectively rendering clouds transparent. Track variation with respect to MP virtually disappeared, which demonstrated that the interplay of hydrometeors with radiation—which we term cloud-radiative forcing (CRF)—was a distinguishing factor among microphysics schemes. The interaction between condensed water and radiation is species dependent, and the MPs that generate more radiatively active particles also developed more radially extensive convective activity, different structural asymmetries with respect to diabatic forcing, and broader outer wind profiles. Fovell et al. (2010a, hereinafter P4) proposed that the cloud-top cooling and within-cloud warming resulting from CRF combined to help the anvil spread radially outward and thereby seeded the (normally dry) far environment, making it more favorable for the subsequent development of convection.

Naturally, microphysics alone cannot completely determine how a TC behaves. Cao et al. (2011, hereinafter P5) demonstrated that storm track and structure are both sensitive to the manner of TC initialization, and Hsu et al. (2013, hereinafter P6) showed how diabatic heating forced by flow over topography could explain speed variations of typhoons approaching and crossing an island like Taiwan. P4's explanation for outer region convective activity was finally assessed in Bu et al. (2014, hereinafter P7) and was determined to be insufficient to explain why CRF results in wider tropical cyclones. Their analysis identified LW absorption within the cloud anvil as the principal agent for

---

[1] Professor Fovell selected TCs because they seemed so much "simpler" than squall lines, which had been the principal focus of his research up to that time!

storm expansion, which potentially contributes to substantial track discrepancies.

The structure of this paper is as follows: The models employed in the experiments referenced herein are presented in section 2. Section 3 provides some cursory background information on cloud microphysics and radiative processes. The main findings are presented in section 4, and the final part of the paper summarizes this review.

2. Description of models and experiments

Studies P1–P7 have made use of different modeling systems in a variety of configurations and versions (see Table 11-1). This suite has included the WRF Model's ARW (Skamarock et al. 2007) and Nonhydrostatic Mesoscale Model (NMM) cores—the latter in its Hurricane WRF (HWRF) form (Gopalakrishnan et al. 2012)—and the Bryan Cloud Model 1 (CM1; Bryan and Fritsch 2002). For this review, additional experiments utilizing the above models as well as the global Model for Prediction across Scales (MPAS; Skamarock et al. 2012) have been made. All but CM1 started as "real-data" versions that were rendered semi-idealized by removing all land and setting the aquaplanet surface temperature at a uniform 29°C. These models are initialized with a horizontally homogeneous base state constructed from a single sounding that represents variants of Jordan's (1958) hurricane season composite. This approach facilitates analyses while retaining the dynamical frameworks that might be employed in operational settings.

The semi-idealized framework introduced in P1 used ARW v.2.2 with three telescoping and temporally fixed Mercator-projected domains with the highest horizontal resolution being 3 km. Model physics held fixed in those experiments included the Yonsei University (YSU; Hong et al. 2006) planetary boundary layer scheme, and the Dudhia (1989) SW and RRTM (Rapid Radiative Transfer Model; Mlawer et al. 1997) LW radiation parameterizations. Subsequent studies included alterations to the model version, domain width, depth, setup (including the incorporation of moving nests), and map projection (to Lambert conformal), as well as the horizontal and vertical resolutions. Adjustments were also made to the initial sounding and the model physics employed in the control configurations including, especially, new and improved radiation and microphysics schemes.

Many of our experiments have made use of a "bubble" initialization, in which a tropical cyclone is established over a 24-h period following the insertion of a synoptic-scale positive buoyancy perturbation at 20°N latitude into the otherwise horizontally homogeneous (and typically calm) base state. A convective parameterization is employed for the first 24 h at which time it is usually deactivated, depending on the domain setup and resolution. This technique can create a coherent cyclonic vortex of roughly tropical storm strength (with respect to the 10-m wind according to the Saffir–Simpson scale) in the first day. At that time, the microphysics parameterization is switched on (if it was not already active from the initial time). P5 contrasted this initialization, which was directly inspired by Hill and Lackmann (2009), with the customary technique of employing a "bogus" vortex of specified strength and structure.

The ARW simulations made expressly for this review used the configuration employed in P3, which consisted of a single 2700 km² domain with 4-km horizontal resolution but were made with version 3.2. Those experiments adopted the RRTM LW and Dudhia SW schemes. Along with P7's HWRF simulations, our MPAS runs employed RRTMG[2] for both longwave and shortwave.[3] Other radiation schemes used include Goddard (in CM1 for P7; Chou and Suarez 1994), HWRF's version of the GFDL parameterization (in P7; see also Gopalakrishnan et al. 2012), and CAM (Collins et al. 2006) and Fu–Liou–Gu (Fu and Liou 1993; Gu et al. 2010, 2011) in ARW (not shown). Finally, the CM1 model is employed solely in its axisymmetric configuration, with moist and dry simulations as described in P7. This model is used to test hypotheses relating to the influence of microphysics and cloud-radiative forcing on TCs.

This review combines results from these various experiments, using different models, configurations, and simulation strategies, because we are emphasizing the findings that are common to all studies. For example, we have found that, independent of the model or resolution employed, CRF invariably encourages the development of stronger winds in the TC's outer core region, as long as the MP schemes generate sufficient cloud ice and snow. Our first CRF experiment, made with ARW, suggested that TCs with transparent clouds were systematically more intense; however, this result was not found to be robust after simulations from other models such as HWRF, CM1, and MPAS were examined. As a consequence, TC intensity is largely ignored in this review, and remains an issue for further study.

---

[2] Rapid Radiative Transfer Model for general circulation models (Iacono et al. 2008).

[3] The P7 study used code provided by Greg Thompson to provide particle size information to the RRTMG radiation scheme. This improvement had little effect on the results, and is not employed herein. See P7 for further information.

3. Background

This section provides a brief background on some relevant aspects of cloud microphysics and cloud-radiative feedback. More comprehensive information is available in texts such as Liou (2002), Stensrud (2007), and Straka (2009), among others.

a. Cloud microphysics

Cloud microphysics comprises the processes that control the creation, evolution, and destruction of condensed water. The microphysical parameterizations that numerical models employ to handle these processes range from very simple to enormously complex. Condensed water particles come in a variety of basic species and sizes, and even in small volumes are far too numerous to individually track. Therefore, MPs have been developed that are either of the spectral or bulk varieties. In spectral (bin) microphysics the particle size distribution (PSD) is partitioned into discrete bins, and the evolution of particles through these partitions is modeled, which is usually a very expensive undertaking.

As a consequence, the vast majority of models employ bulk schemes in which the PSD for each species is specified separately. One of the earliest bulk MPs, the Kessler (1969) scheme, considered only "warm rain" processes involving cloud droplets and rainwater. Cloud droplets were presumed to be monodispersed (of constant size) and sufficiently small to be free-floating relative to still air. Raindrops were represented by an exponential size distribution (Fig. 11-1) characterized by an intercept (N_0) and a slope (λ), so that N_D, the number of drops of diameter D, is given by

$$N_D = N_0 e^{-\lambda D}. \tag{11-1}$$

This originated with the pioneering study of Marshall and Palmer (1948), who found that, except for the very smallest drop diameters, there was an exponential decrease in the number of rain drops collected as the diameter increased.

The total number of particles, N, is determined by integrating (11-1) over all possible drop sizes, which yields

$$N = \frac{N_0}{\lambda}.$$

If the drop is spherical, individual raindrops of diameter D have mass $M_D = \rho_l(\pi/6)D^3$, where ρ_l is the density of liquid water. The area under the line depicted in Fig. 11-1 is related to the total rainwater content in the model's grid volume, ρq_r, where ρ is the air density and q_r is the rainwater mixing ratio in kilograms of liquid per kilogram of air, which is presumed to be spread equally

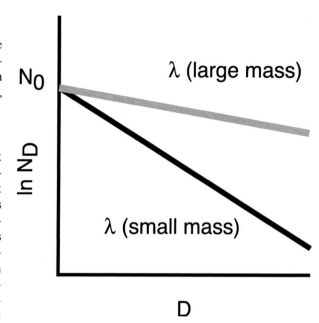

FIG. 11-1. Exponential (Marshall–Palmer) particle size distribution for number of particles (N_D) vs diameter (D), which assumes a fixed intercept N_0 with slope λ determined by particle mass content.

through the grid volume. Integrating over all drop sizes results in a relationship between the slope and intercept given by

$$\lambda = \left(\frac{\rho_l N_0 \pi}{\rho q_r}\right)^{1/4}. \tag{11-2}$$

The bulk terminal velocity applied to all raindrops in the volume is computed similarly, by taking the equation for a single drop of diameter D and integrating over all diameters.

"Single-moment" bulk MPs generally have fixed either N_0 or λ. While fixing the intercept is more common,[4] holding either constant is problematic. Constant N_0 means that the slope becomes more horizontal as the rainwater content increases, which implies that incremental increases in rain content come from assumed growth in the relatively larger drops within the PSD. "Double-moment" schemes relieve this constraint by prognosing the total mass and number of drops separately, which is more realistic but increases the computational complexity. More modern schemes have addressed some of the inherent limitations of particle size distributions based on (11-1) by adopting other functional shapes, such as the "gamma" distribution (cf. Willis 1984; Ziegler 1985; Seifert and Beheng 2006).

---

[4] Tripoli and Cotton (1980) is an example of the fixed λ approach.

Whatever assumptions are made, bulk MPs employ integrations of the continuous collection equation (CCE) over the presumed PSDs to handle processes such as the accretion of cloud droplets by more swiftly falling raindrops. The CCE is based on the idea that a particle sweeps out a cylindrical volume as it falls relative to still air and/or other particles (Kessler 1969; Lin et al. 1983). Kessler (1969) further approximated the CCE for this process and also handled the creation of new raindrops from cloud droplet aggregation (the autoconversion process) in a particularly simple way. Along with a saturation adjustment for transferring mass between vapor and cloud water, and an equation for rainwater evaporation (cloud droplets were assumed to evaporate instantly in subsaturated air), the Kessler scheme can be implemented in just a few lines of code.

By contrast, schemes that incorporate various species of ice (such as free-floating ice crystals, low-density snow, medium-density graupel, and/or high-density hail), more sophisticated PSDs and/or higher moments are much, much more complicated, and can claim a sizable fraction of the computing time required for a given simulation. These "ice-bearing schemes" embody many additional assumptions (including collision efficiencies for various two- and three-body collisions, ice multiplication, riming, melting, etc.) and require considerably more computational effort. Numerous MPs exist, which often differ solely with respect to ostensibly subtle factors such as the fall speed of graupel particles or how efficiently ice crystals evolve into snow. As discussed in section 4, research has shown these factors can have an outsized effect on TC structure, motion, and intensity.

b. Radiation and interaction with hydrometeors

Atmospheric gases such as water vapor, carbon dioxide, and ozone are selective absorbers of thermal radiation. These gases, other particles, and Earth's surface also scatter and emit radiation, nearly all of the latter occurring in the LW portion of the radiation spectrum. As radiative processes represent an important part of the energy balance of the atmosphere and Earth's surface, operational simulations of TCs typically make use of a radiation parameterization, of which several are available in the WRF platform (including RRTM, RRTMG, CAM, Goddard, Dudhia, Fu–Liou–Gu, and GFDL). These codes represent attempts to model very complex and time-consuming processes in a more efficient manner, but can still be quite computationally expensive.

When a multiday simulation is initialized with the Jordan (1958) sounding in which convection is not permitted (see P7), these radiation schemes tend to produce

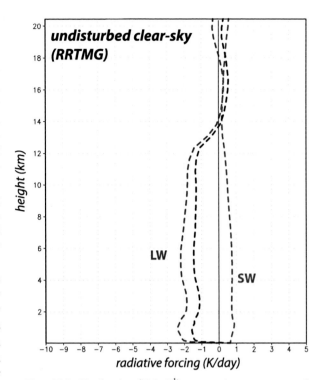

FIG. 11-2. Tendencies (K day^{-1}) averaged over space and through one diurnal cycle for LW (blue), SW (red), and net (black) radiation from an HWRF simulation undisturbed by convection using the RRTMG parameterization. From P7.

about 1 K day^{-1} of SW warming and about 2 K day^{-1} LW cooling in the lower-to-middle troposphere, for a net radiative forcing of about −1 K day^{-1} averaged through the diurnal cycle (Fig. 11-2). These profiles appear reasonable for a moist environment [cf. Figs. 3.18 and 4.15a in Liou (2002)] and variability among available schemes appears minor (not shown). By contrast, many idealized studies either neglect diabatic forcings associated with radiation or handle it in a very simple manner. As an example, Rotunno and Emanuel (1987) used Newtonian cooling for the express purpose of preventing the model atmosphere from straying too far from the initial hurricane environment.

The above describes background or "clear-sky" radiation, as it is independent of condensate content and convective activity. Hydrometeors also participate in the absorption of SW, and absorption and emission of LW radiation, representing the CRF. The various parameterizations handle resolved condensate, as well as subgrid-scale clouds, in different ways. Stephens (1978) related CRF to the cloud water path, which is the integral of cloud water content over depth. This concept was implemented in the original RRTM LW and Dudhia SW schemes via specified, species-dependent absorption and emission coefficients (ϵ). For LW (Table 11-2),

TABLE 11-2. LW absorption/emission coefficients (ϵ) used in the original RRTM scheme.

Species	Coefficient $(m^2\,g^{-1})$	Relative magnitude to cloud droplets for $1\,kg\,m^{-3}$ mass content
Cloud droplets q_c	0.144	—
Cloud ice q_i	0.0735	0.510
Snow q_s	0.00234	0.016
Rain drops q_r	0.00033	0.002
Graupel q_g	0.0	0.0

the coefficient assigned to free-floating cloud ice (ϵ_i) is one-half that used for liquid droplets, but is over 31 times that for snow (ϵ_s) and about 222 times the absorption coefficient applied for rainwater (for mass contents of $1\,kg\,m^{-3}$). In the RRTM LW scheme, graupel is completely ignored. In the P3 study, ϵ_i and ϵ_s were varied to crudely illustrate the effect of shifting condensate mass among various species, which in some cases had a nontrivial impact on TC track. RRTMG has introduced new ways of handling hydrometeor effects, which produce qualitatively similar results but are less straightforward to adjust.

4. Synopsis of CRF impacts

a. Microphysics influences on TCs

For some time, we have appreciated that cloud microphysical assumptions can materially influence TC intensity, but with considerable variability among the real and idealized cases examined, which suggests that microphysical processes are both important and an integral part of forecast uncertainty [e.g., see review by Tao et al. (2011)]. For example, Lord et al. (1984) concluded that including ice processes resulted in a significantly stronger storm, and McFarquhar et al. (2006) found intensity generally increased as graupel fall speeds were increased. In contrast, P2 found that faster tangential winds were simulated when graupel formation was suppressed in the Purdue–Lin (Lin et al. 1983; Chen and Sun 2002) ice MP scheme. Excluding graupel increased the storm intensity by about 10% relative to the original version of Purdue–Lin, which made it about 30% stronger than its warm rain (Kessler) counterpart (see P2's Fig. 14). In other studies, excluding ice produced TCs that intensified more rapidly and/or attained higher intensity at maturity (e.g., Hausman et al. 2006; Li and Pu 2008; Stern and Nolan 2012).

Thus, microphysical parameterizations may incorporate a variety of processes that individually may increase or suppress TC organization and/or intensity; however, the net result is sensitively dependent on precisely how the

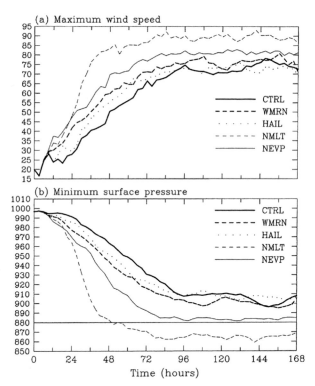

FIG. 11-3. Evolution of (a) maximum wind speed ($m\,s^{-1}$) at the lowest model level and (b) minimum SLP (hPa) for several experiments, including the control run (CTRL), a warm rain without ice (WMRN), a simulation including hail (HAIL), and runs in which melting (NMLT) or evaporation (NEVP) were neglected. From Wang (2002).

various processes combine. Many of these processes involve diabatic heating or cooling. Wang (2002) and Zhu and Zhang (2006) showed that excluding some sources of diabatic cooling encouraged more rapid intensification and lower central pressures during TC maturity. In particular, Wang's (2002) simulation called NMLT (for "no melting"), which neglected all melting of snow and graupel as well as rain evaporation, became organized much more rapidly and reached a substantially lower sea level pressure (SLP) than the other simulations (Fig. 11-3). Bu (2012) studied the organization of TCs in an axisymmetric version of the CM1 model, primarily using versions of the Kessler MP. She found that TC organization was most rapid and efficient when condensation was immediately removed upon creation, which excludes both diabatic cooling from evaporation and hydrometeor loading, but that simply preventing rainwater formation alone was not sufficient (Fig. 11-4) to hasten storm organization.

Prior studies have shown that TC behavior (e.g., Willoughby et al. 1984) and structure (e.g., P1–P7) are both sensitive to microphysics. The symmetric components of the 10-m wind speed from the 13 simulations conducted for the P3 study (Fig. 11-5) were obtained by

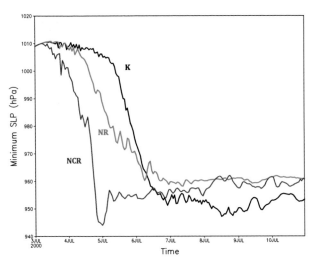

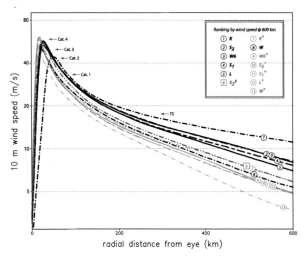

FIG. 11-4. Minimum SLP from axisymmetric CM1 model simulations made using microphysical variants of the warm rain Kessler (K) scheme. Simulations NR and NCR prevented the formation of raindrops and both cloud droplets and raindrops, respectively. From Bu (2012).

FIG. 11-5. Symmetric components of the 10-m wind speed (m s^{-1}), temporally averaged over a diurnal cycle in a vortex-following fashion, from ARW simulations conducted for the P3 study. Microphysical schemes considered include Kessler (K), two versions of Seifert–Beheng (S$_1$ and S$_2$), WSM3 (W), WSM6 (W6), and Purdue–Lin (L); see Table 11-3. Asterisks indicate CRF-off runs and S$_2^\#$ is the LW absorption sensitivity test from P3. Note the vertical axis is log scaled.

temporally averaging the field in a vortex-following fashion over the final day of a 4-day simulation, and then averaging again in the azimuthal direction; see P2 for more information. In addition to the warm rain Kessler (K) scheme, five ice-bearing schemes were considered: the single-moment WRF single moment three-class [WSM3 (W)] and six-class [WSM6 (W6)] and Purdue–Lin (L) MPs, as well as two versions of the Seifert and Beheng (2006) double-moment parameterization (S$_1$ and S$_2$) that differ with respect to ice-to-snow conversion (Table 11-3). Note that the warm rain scheme results in a storm with a uniquely wide eye and broad outer wind profile compared to the structure simulated with the ice-bearing schemes; this will be explored in more detail presently.

Even fairly subtle variations in outer wind strength are potentially very important as these winds influence TC motion owing to the "beta drift." Differential advection of planetary vorticity by the storm's cyclonic circulation creates gyres that combine to establish a "ventilation flow" across the vortex (e.g., Holland 1983; Fiorino and Elsberry 1989) that impels motion on a curved Earth, even in environments with no mean current. Using a barotropic model, Fiorino and Elsberry (1989) demonstrated that this ventilation flow influences the speed and direction of TC motion (Fig. 11-6). Even though the wind profiles (Fig. 11-6a) only varied beyond radius $r =$ 300 km from the center, distinctly different tracks (Fig. 11-6b) are predicted in the experiment, with the strongest outer winds resulting in both the most rapid and most relatively westerly motion.

A straightforward application of Fiorino and Elsberry's (1989) findings to the Fig. 11-5 profiles would predict that the relatively strong outer winds of the Kessler TC would lead to the fastest motion and the most northwestward track, while the simulated storms with weaker cyclonic flow at large radii would be expected to have slower and more northward motions. This is indeed what occurred in P3's experiment (Fig. 11-7a).[5] After an organizational stage, the warm rain storm moved northwest at over 9.7 km h^{-1} (2.7 m s^{-1}), while the other storm motions were much slower with direction during maturity that varied more than speed (4–6 km h^{-1}, 1–1.7 m s^{-1}). Such motions fall into the range of typical beta drift speeds of 1–4 m s^{-1} (cf. Holland 1983; Chan and Williams 1987). Figure 11-8 depicts motion during the last 12 h of the integration for a version of P3's experiment, augmented to include the Thompson and Ferrier (Thompson et al. 2008; Ferrier et al. 2002) MPs and additional radiative schemes (Table 11-1). The variation in motion directions implies a widening range of tracks with time.

In a warm-core vortex, the poleward-directed ventilation flow established by the beta gyres should weaken with height (e.g., Bender 1997), which then results in a northwesterly to northerly vertical shear across the TC that acts to enhance inner-core convective activity on its

---

[5] Keep in mind the model has no land, and a coastline was provided for scale only.

TABLE 11-3. Microphysics schemes referenced in this paper.

Symbol	Name	Reference(s)
K	Kessler (warm rain)	Kessler (1969)
L	Purdue–Lin	Lin et al. (1983), Chen and Sun (2002)
W	WSM3 (WRF single moment, 3-class)	Hong et al. (2004)
W6	WSM6 (WRF single moment, 6-class)	Hong et al. (2004), Dudhia et al. (2008)
T	Thompson	Thompson et al. (2008)
F	Ferrier	Ferrier et al. (2002)
S_1, S_2	Seifert–Beheng	Seifert and Beheng (2006), Fovell et al. (2010b)

downshear and downshear-left flanks (e.g., Frank and Ritchie 1999; Corbosiero and Molinari 2002). Consistent with this interpretation, a sample of cases using P5's experimental design reveals lower tropospheric average vertical motions that are generally concentrated on the storm's eastern and southeastern sides (Fig. 11-9), although notice that the patterns and degrees of asymmetry vary among the MP schemes. Note that the simulation with the warm rain MP generated the widest and most symmetric updraft pattern, which is consistent with its especially broad wind profile (Fig. 11-5).

This asymmetric diabatic heating also appears to modulate TC motion in ways that compete or cooperate with the ventilation flow, depending on the orientation of the asymmetry pattern that varies depending on microphysical assumptions. Wu and Wang (2000) employed a potential vorticity (PV) diagnostic for TC motion in which the relative contributions to the PV tendency (PVT) due to the advection (horizontal and vertical, herein labeled HA and VA), friction, and a term proportional to gradients of diabatic heating, Q (herein labeled DH). Specifically, the wavenumber-1 components of the PVT contributions were computed, using a least squares technique (see Wu and Wang 2000). The combination of the DH, VA, and the (typically small) residual terms will be called DH*. Since the contributions of all terms tended to shift with height (Fig. 11-10), however, they are not truly independent and strongly modulate each other (cf. P5). Papers P3 and P6 addressed the height variation by averaging the terms vertically through the lower troposphere above the boundary layer; P6 noted that the lack of independence encourages a fundamentally qualitative and comparative application.

The arrows on Fig. 11-9 represent storm motion (C) and the contributions to PVT from DH and DH*. Although not shown, the horizontal advection term can be inferred as the difference between C and DH*; as expected, HA is generally directed northwestward, since it is in large part the advection of PV by the ventilation flow. Since the Kessler TC (Fig. 11-9a) has the most symmetric vortex, the magnitude of DH is very small

and DH* is effectively zero, so the diabatic forcing in this simulated TC is not effective at opposing the motion due to the ventilation flow, which is substantial owing to the strength of this TC's outer wind profile.

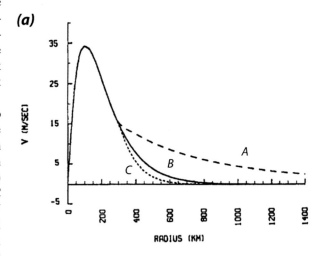

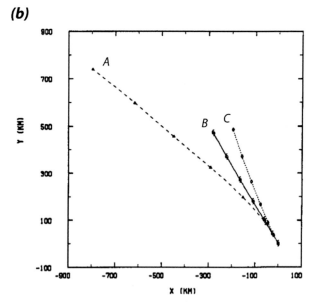

FIG. 11-6. (a) Initial tangential wind profiles and (b) corresponding tracks to 72 h from the Fiorino and Elsberry (1989) experiment. In (b), markers are separated by 12 h.

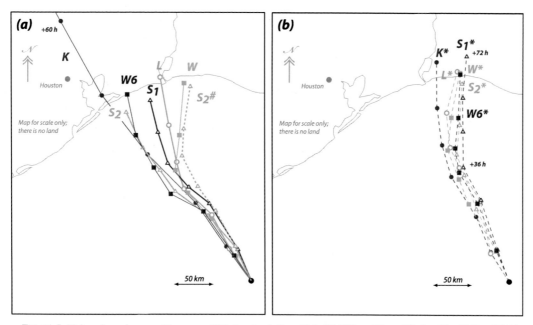

FIG. 11-7. 12-hourly cyclone positions over 72 h for simulations K, L, W, W6, and S_1 and S_2 (see Fig. 11-5 and Table 11-3) with CRF (a) on and (b) off. The 72-h K simulation position is beyond the subdomain depicted. The U.S. Gulf Coast segment is included only for scale as the model has no land. Simulations employed WRF v. 2.2.1, RRTM LW, and Dudhia SW. Adapted from P3.

Consequently, the Kessler TC has the fastest motion among the bubble-initialized storms.

In contrast, the simulations with ice-bearing MPs all have weaker outer wind profiles (Fig. 11-5) as well as substantially more asymmetric vertical velocity and heating patterns (Fig. 11-9). The S_2 (with RRTM/Dudhia), $T^@$ (Thompson, with RRTMG), and F (Ferrier, with GFDL radiation) cases (Figs. 11-9b–d) are examples of the range of TC structures and translations among these storms. Among them, the $T^@$ case has the fastest translation speed and moved north-northwestward at $5.9 \, \mathrm{km \, h^{-1}}$ ($1.6 \, \mathrm{m \, s^{-1}}$), which is substantially slower than the K storm but still exceeds the motions of S_2 ($3.6 \, \mathrm{km \, h^{-1}}$ or $1 \, \mathrm{m \, s^{-1}}$) and F/GFDL ($3.8 \, \mathrm{km \, h^{-1}}$ or $1.1 \, \mathrm{m \, s^{-1}}$). The somewhat faster storm motion in the $T^@$ simulation may have been due to the small contribution of the diabatic heating term in the direction of beta drift (Fig. 11-9b), even though the storm has weaker outer winds relative to the K case.[6] In the S_2 simulation (Fig. 11-9c), the diabatic heating term appears to be opposing the beta drift, while the concentrated asymmetric heating

on the F/GFDL storm's eastern flank (Fig. 11-9d) may explain why it developed the most eastward track (Fig. 11-8). Naturally, this particular aspect of storm motion could not be captured in Fiorino and Elsberry's

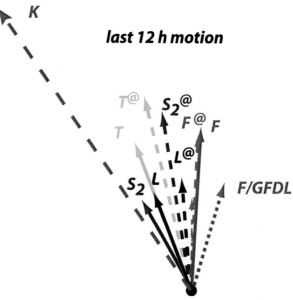

FIG. 11-8. Storm motion vectors for the 60–72-h period from simulations made for the P3 study (see Fig. 11-7 caption). The augmented P3 study adds simulations using WRF v. 3.2 employing Thompson (T) and Ferrier (F) microphysics, the RRTMG LW and SW (indicated by @ sign) schemes, and GFDL radiation. Adapted from P3.

_____

[6] It needs to be kept in mind that while microphysical diabatic heating, Q, logically tends to be well correlated with vertical velocity, the DH term actually consists of the gradients of Q multiplied by vorticity, along with other terms. This is why the DH vectors may not point toward where the air is ascending most strongly. See P6 for more information.

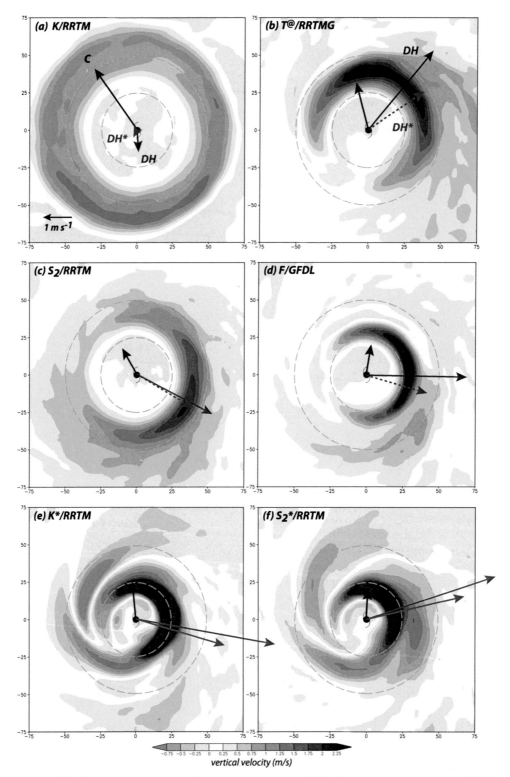

FIG. 11-9. Vertical velocity averaged from the surface to 500 hPa (mass-weighted) and over 150 km ×150 km portions of the model's single, 4-km resolution domain, from P3 and augmented P3 simulations (Table 11-1) using various MP and radiation schemes. Fields were averaged in a vortex-following manner over day 4 of the simulations. Superposed are vectors indicating storm motion (C) and contributions to motion from diabatic heating (DH) and a combination of DH and vertical advection (called DH*). The top of the figure represents north.

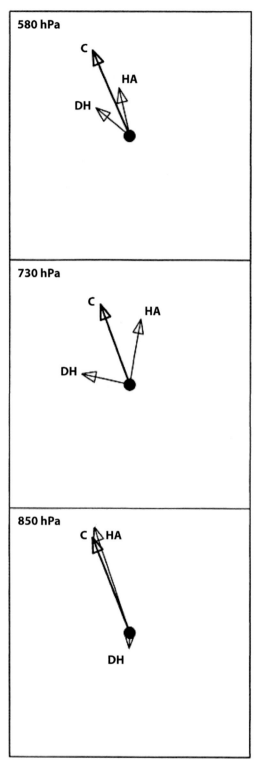

FIG. 11-10. Contributions of horizontal advection (HA), vertical advection (VA), and diabatic heating (DH) to storm translation (C) at 36 h at three levels from one of Wu and Wang's (2000) experiments. Maximum vector length is 2.9 m s^{-1}. Adapted from Wu and Wang (2000).

(1989) barotropic model, but it appears to make an important, and possibly crucial, contribution to the storm motion.

Sensitivity to microphysics has been most pronounced in our experiments using the bubble initialization. In P5's study, simulations commencing with artificially supplied initial outer wind profiles defined by a structural parameter, α (see section 4c), appeared to be quite "resilient" to the effects of the beta shear, and thus less likely than their bubble counterparts to develop asymmetric updraft (Fig. 11-11) and heating structures. As a consequence, these simulated TCs had significantly faster translation speeds than might have been expected from their symmetric outer wind profiles (Fig. 11-13). For example, note that the $\alpha = 0.75$ run had outer winds comparable to the bubble TC, but had a 2.6 times faster translation speed (2.43 vs 0.92 m s^{-2}; see P5's Table 2). The realism of the resilient bogussed vortices awaits closer examination.

PVT analysis was employed in P6 to examine how and why TCs tend to change direction and speed as they approach a mountainous island such as Taiwan, which contains a prominent central mountain range (CMR) roughly parallel to its east coast, by introducing an obstacle resembling Taiwan into the aquaplanet framework (Fig. 11-13). A novel element of the P6 study was that the Taiwan-like island was made of water, which removes potential complications such as changes in surface friction and fluxes after landfall. The bogus-initialized TC initially had little asymmetry in its heating and rainfall fields (reflecting vortex resilience) as it approached the obstacle island from the southeast. With the weak basic current included in this experiment, the TC had a relatively rapid translation during this period (Figs. 11-13a,b). Once the cyclonic storm circulation began being affected by the CMR, however, the TC direction and speed changed in a manner consistent with the influence of the DH* term (Figs. 11-13c,d). Speed and direction variations continued as the TC crossed the CMR, which affected the orientation of the topographically driven diabatic forcing (Figs. 11-13e,f). In summary, this study clearly reveals that terrain-induced asymmetries in diabatic heating suffice to profoundly impact the motion of a cyclonic vortex over an island barrier.

b. Cloud-radiative forcing

We have seen that microphysics clearly exerts an important influence on TC track and structure. The P3 experiment, however, revealed a surprising finding: these variations in track and structure largely disappeared when the radiative forcing owing to clouds was neglected, rendering clouds essentially transparent to radiation. The first indication that radiation was important came in P2, which showed that while the

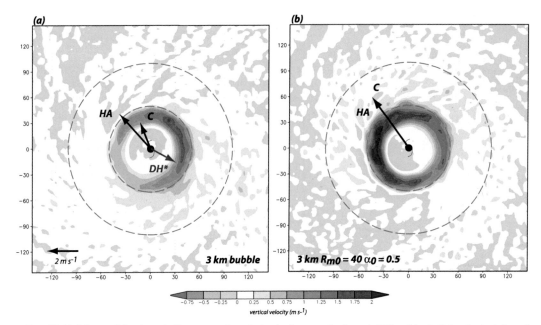

FIG. 11-11. Mass-weighted vertically averaged vertical velocity over the lowest 4.3 km (shaded) for the last day of 4-day simulations using (a) the bubble initialization and (b) a bogus vortex, from P5's high-resolution simulations (see Table 11-1); R_{m0} and α_0 are initial values used in the Rankine formula [(11-3)]. As in Fig. 11-9, vectors represent storm motion (C) and the HA and DH* motion contributions. For (b), the DH* term is essentially zero. Adapted from P5.

K storm's track was not sensitive to which radiation parameterization (RRTM or CAM) was selected, storm translation was dramatically slower when no radiative scheme was used (P2's Fig. 9). That was a crude version of P3's and P7's experiments in which only the specific influence of hydrometeors on radiation was deactivated, but the background (clear-sky) LW and SW forcings were retained. Figure 11-7b reveals the members of P3's microphysical ensemble had a similar speed and direction of motion evolution, including especially the Kessler version K*, which is clearly the most dissimilar from its "cloudy" counterpart.[7]

The shift to slower, more northward motions is consistent with the storms' weaker symmetric outer wind profiles (Fig. 11-5). Note further that deactivating CRF materially altered the storm structure (Figs. 11-9e,f; see also P3's Figs. 2e–h). The CRF-off cases tend to be narrower, even more asymmetric, and resemble each other far more than they do any of the CRF-on cases.[8] The shift with the warm rain MP (Figs. 11-9a,e) is especially dramatic, and the similarity between K* and S$_2^*$ is striking, especially with respect to the relatively strong and deep downdrafts that appear in the northwestern

quadrants of these storms that are absent when CRF is active. Note that the F/GFDL case (Fig. 11-9d) has intermediate characteristics between the CRF-on and CRF-off cases. P7 showed that this model physics

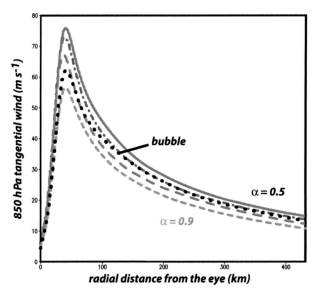

FIG. 11-12. Symmetric components of 850-hPa wind speed (m s^{-1}) for 3-km resolution experiments constructed from vortex-following composites averaged during the final 24-h period. The α and bubble designations refer to the initializations, and α values shown are 0.5 (solid), 0.625 (dashed-dotted), 0.75 (long dashed), and 0.9 (short dashed). Adapted from P5.

---

[7] Deactivation of cloud-radiative forcing is indicated with an asterisk (e.g., K*, S$_2^*$, etc.).

[8] As noted earlier, the intensity tendency noted in P3 is contradicted in other experiments, which is why it is not emphasized here.

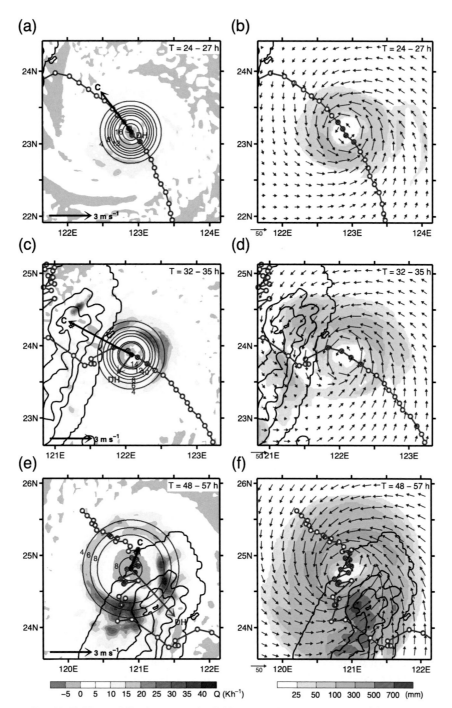

FIG. 11-13. Vortex-following composite fields averaged over three periods (T) from one of P6's experiments. (left) Vertically averaged diabatic heating Q (color shaded, $K\,s^{-1}$) and symmetric PV structure [blue contours, unit is PVU (1 PVU $= 10^{-6}\,m^2\,s^{-1}\,K\,kg^{-1}$)], along with motion C (black), and DH (red) terms. (right) Composite rainfall (shaded) and wind vectors at the lowest model level. Averaging periods are (a),(b) 24–27, (c),(d) 32–35, and (e),(f) 48–57 h, and indicated by filled circles on the superposed storm tracks. From P6.

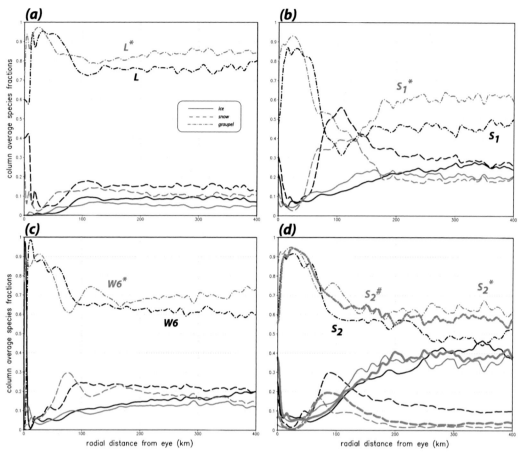

FIG. 11-14. Ice species fractions vs radial distance from the eye computed using symmetric components from temporally averaged vortex-following composites for simulations using versions of the (a) L, (b) S_1, (c) W6, and (d) S_2 MPs (see Table 11-3). Asterisks denote CRF is ignored; simulation $S_2^{\#}$ treats cloud ice as snow for radiative calculations. Augmented from P4.

combination, which closely resembles that employed in the operational HWRF, results in a substantially reduced cloud-radiative forcing owing to how that parameterization is implemented (see P7 for more information).

As indicated in section 3a, MP schemes may range from the simple to complex with respect to how many hydrometeor species are included and precisely how they are handled. However, it seems very clear that the major reason why microphysics can influence TC structure and motion is because condensate particles can influence radiative heating and cooling. CRF sensitivity therefore emerges precisely because MPs tend to produce different amounts of condensate species (Fig. 11-14) that have significantly disparate radiative impacts when CRF is active. A scheme that produces much more cloud ice than snow, such as S_2 (Fig. 11-14d), should have much larger CRF than ones with a swift evolutionary path to graupel such as the L (Fig. 11-14a) or W6 (Fig. 11-14b) MPs. Note that over 70% of the azimuthally and column-averaged ice mass in the L

storm is in the form of graupel (Fig. 11-14a), which observations (McFarquhar and Black 2004; McFarquhar et al. 2006) suggest is unreasonably large, and less than 10% is in cloud ice. Of course, the warm rain scheme generates copious amounts of cloud droplets, which are presumed to be more radiatively active than even cloud ice (not shown), which is why rendering them transparent to radiation had such a profound effect on the storm. It is probably easier to identify unrealistic condensate combinations than realistic ones, but these examples demonstrate that the consequences of these arguable microphysical assumptions are not small.

P3 further explored sensitivity to LW radiation via manipulation of the RRTM scheme's absorption coefficients for MP schemes S_2 and W6 by making snow either completely transparent to radiation (simulations $S_2\hat{}$ and W6$\hat{}$) or by treating cloud ice similarly to less radiatively active snow (simulations $S_2^{\#}$ and W6$^{\#}$). Rendering frozen condensate progressively more transparent resulted in eastward shifts of the simulated tracks

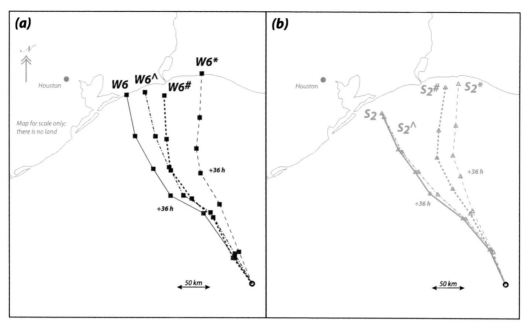

FIG. 11-15. As in Fig. 11-7, except for versions of W6 and S_2 that demonstrate LW absorption/emission coefficient sensitivity. Simulations designated with the hash symbol (#) treat cloud ice as snow for LW calculations, while those designated with the caret (^) ignore the contribution of snow. Adapted from P3.

(Fig. 11-15) as the magnitude of the storm outer winds decreased (Fig. 11-5), even though the relative distribution of frozen hydrometeors among the ice species was little affected (Figs. 11-14c,d). Among these three S_2 variants, we see that $S_2^{\#}$ falls between the other two with respect to the radial extent of its microphysical diabatic heating and tangential wind fields (Figs. 11-16a–c). Although the association between the $1\,K\,h^{-1}$ diabatic forcing and $20\,m\,s^{-1}$ wind contours with respect to position in these figures is very likely coincidental, it is clear that both extend radially farther outward as the radiative footprint of ice increases. These fields can be compared to the P7 study that used HWRF and Thompson microphysics (see their Figs. 4, 5, and 8).

The net cloud-radiative forcing from the combination of LW and SW averaged through the diurnal cycle consists of cooling along the top of the cloud anvil and warming within the cloudy region (Fig. 11-16d–f; see also P7's Figs. 5 and 6). This pattern depends on CRF (and thus is absent in S_2^{*}; Fig. 11-16e) and is far better developed when LW absorption and emission by ice is enhanced. Vertical profiles of temporally and spatially averaged net radiation (Fig. 11-17; see also P7's Fig. 7) reveal that only the CRF-on case possessed net warming in the troposphere within 350 km of the storm center.[9] P7 demonstrated that the sign reversal for net radiation relative to clear-sky

conditions in the CRF-active case was nearly all due to the hydrometeor effect on longwave absorption as SW radiation failed to penetrate the thick ice cloud (P7's Fig. 6). The CRF field is considerably more extensive radially for the CRF-active case (Fig. 11-16d), which reflects an expanded anvil (contoured field) in that case.

The S_2 variants examined herein represent the range of structural variations produced by other MP schemes in our experiment since altering the LW hydrometeor coefficients essentially mimics shifts in frozen water speciation. There is also some sensitivity to the radiation parameterization, especially with respect to the LW cooling at the cloud top. For example, employing RRTMG for LW and SW (in place of RRTM and Dudhia) with the S_2 scheme reduces the net radiative forcing by about one-half (simulation labeled $S_2^{@}$ in Fig. 11-17), which is a consequence of both reduced LW cooling and increased SW absorption (not shown). However, net radiative forcing is little affected at lower levels, and the role of cloud-top cooling in TC structure appears to be quite limited anyway (see next subsection).

c. How and why CRF influences TC structure

After documenting comparable differences between CRF-active and CRF-inactive TCs simulated using a semi-idealized version of the three-dimensional HWRF with Thompson MP and RRTMG radiation, P7 offered an explanation for how and why radiative forcing associated with hydrometeors causes radially expanded wind

---

[9] The clear-sky profiles shown for comparison were obtained in the manner described in P7.

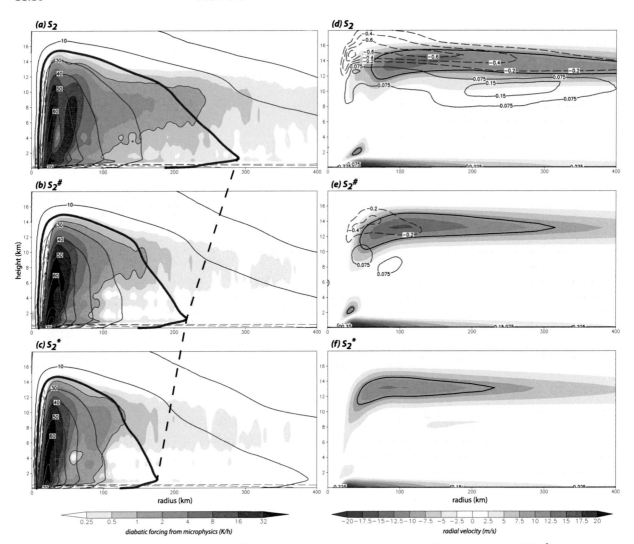

FIG. 11-16. Radius–height cross sections of (a)–(c) net diabatic heating from microphysics (shaded as shown, 1 K h^{-1} contour superposed) and tangential wind (20 m s^{-1} contour highlighted), and (d)–(f) radial wind (shaded as shown, 9 m s^{-1} contour superposed) and net radiation (contour intervals 0.2 and 0.075 K h^{-1} for negative and positive values, respectively) for three versions of the S$_2$ scheme. Each field represents the azimuthally symmetric components extracted from vortex-following composites constructed over the final 24 h.

and heating fields. An axisymmetric version of CM1 with 5-km radial resolution was used, and the simulations were integrated to maturity using a version of Thompson microphysics with the Goddard LW and SW schemes. Although some differences are apparent, which reflect alterations with respect to the model framework, initial sounding, and physical parameterizations, activating CRF is yet again found to widen the eye and enhance the secondary circulation, including the upper-level outflow (see P7's Fig. 9), and result in a substantially expanded wind field (Fig. 11-18). The net radiative forcing field is comparable in magnitude and spatial pattern to those from other models (Fig. 11-19a) despite the employment of a different radiation package. It is again seen that the condensation field is

expanded when CRF is active (see difference field in Fig. 11-19c), which reinforces the association between CRF and enhanced convective activity in the outer core region.

While it is logical that the cloud-radiative forcing field is only as wide as the cloudy area, this does not mean CRF actively helped expand the anvil or winds. To address this, P7 introduced the "CRF-fixed" experiment (Fig. 11-19d) in which the time-dependent CRF field was replaced with the CRF-on run's temporal average, which was computed during maturity and over multiple diurnal cycles. The field was then introduced at the initial time and held constant during the integration, thereby rendering it independent of convective activity. The CRF-fixed simulation produced a cloud shield

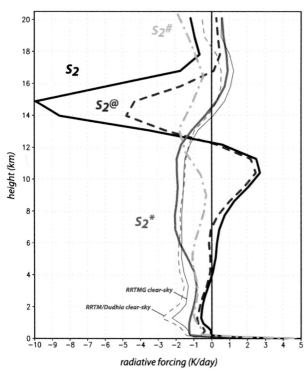

FIG. 11-17. Vertical profiles of net radiative forcing tendencies (K day^{-1}), averaged over a 350-km radius centered on the storm through a diurnal cycle, for four versions of S_2, along with the undisturbed clear-sky profiles for RRTMG and RRTM/Dudhia. In addition to CRF-on and CRF-off simulations with RRTM/Dudhia (S_2 and S_2^*), case $S_2^\#$ treats cloud ice as snow for LW calculations, and $S_2^@$ employed the RRTMG scheme.

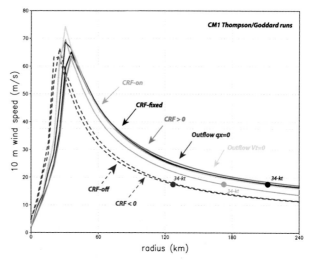

FIG. 11-18. Radial profiles of temporally averaged 10-m wind speed from CM1 experiments using Thompson MP and Goddard radiation and including CRF-on, CRF-off, and CRF-fixed cases. Simulations "CRF $\geq$ 0" and "CRF $\leq$ 0" are versions of CRF-fixed in which only the positive and negative CRF forcings were retained. Experiments "Outflow Vt = 0" and "Outflow qx = 0" are versions of CRF-fixed in which the terminal velocity and mixing ratio of hydrometeors in the outflow were set to zero, respectively. Radii of the 34-kt (17.5 m s^{-1}) wind are indicated. Averaged between days 9 and 12, inclusive. From P7.

(Fig. 11-19d) and tangential wind field (Fig. 11-18) that was comparable to the CRF-on case that supplied its radiative forcing (Fig. 11-19a).[10] Furthermore, the horizontal scale of the imposed CRF field can be altered arbitrarily, and the fact that the cloudy area directly responds to it (Fig. 11-20) clearly demonstrates that this relatively small diabatic forcing plays an integral role in determining anvil extent.

P7 also explored the direct and indirect impacts of CRF with a dry version of CM1 into which diabatic forcings from the moist experiments were inserted (Fig. 11-21). The dry model response to the full, temporally averaged CRF field was characterized by enhanced upper-tropospheric outflow (Fig. 11-21a), which may help explain why the CRF-active storms possessed stronger radial winds there. Note that the outflow also

transports the very hydrometeors that cause the radiative forcing in the first place, so strengthening the outflow should lead to a progressively wider anvil, at least in the upper troposphere (depicted in Fig. 11-22). This scenario represents an apparent positive feedback process between the CRF and the radial outflow, and a fundamentally similar idea was explored by Krueger and Zuluaf (2002) and Durran et al. (2009).

At first, it was believed that the primary agent of the outflow enhancement would be the LW cooling at cloud top, as this is relatively larger in magnitude and possesses sharper horizontal and vertical gradients. Yet, P7 showed that the net cooling at cloud top played almost no role in the radial enhancement. The "CRF < 0" version of the CRF-fixed experiment, which only retained the negative forcing, looks little different (Figs. 11-18 and 11-19e) from the case which neglected radiative forcing altogether (Fig. 11-19b). It is the subtle warming within the cloud shield, a consequence of the influence of hydrometeors on LW absorption and emission, that is relevant to the storm expansion, as demonstrated by the "CRF > 0" experiment (Fig. 11-19f).

The dry model experiments of P7 also suggested that the in-cloud LW warming is primarily responsible for the enhanced outflow in the upper troposphere (Figs. 11-21b,c). The direct result of the positive CRF forcing is to produce very weak but deep and persistent ascent

---

[10] The radial extent of the anvil and wind actually exceeded that of the CRF-on TC during the period shown because the cloud-radiative forcing was applied from the initial time, while the CRF-on simulation required several days to achieve forcing of comparable spatial extent and magnitude.

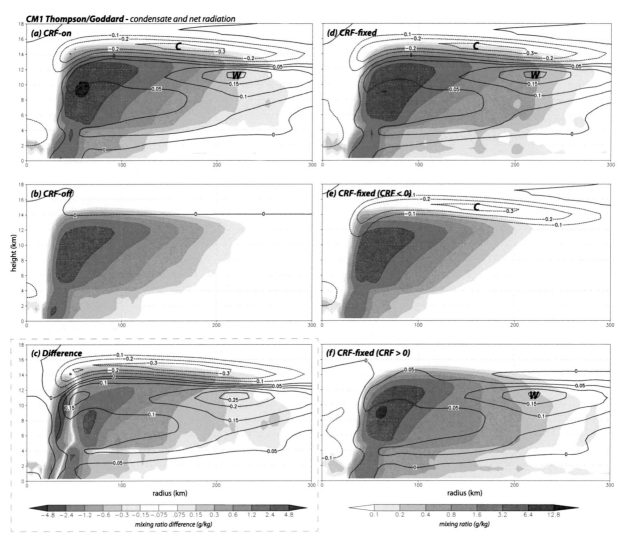

FIG. 11-19. Total condensate (shaded, note logarithmic scale) and net radiation [negative (dashed) contour interval 0.1 K h^{-1}, and positive (solid) interval 0.05 K h^{-1}] for CM1 Thompson/Goddard model storms, averaged as in Fig. 11-17: (a) CRF-on, (b) CRF-off, and (c) difference between CRF-on and CRF-off. At right, similar displays from members of the CRF-fixed experiment: (d) CRF-fixed, (e) CRF-fixed with only negative CRF forcing (CRF ≤ 0), and (f) CRF-fixed with only positive CRF forcing (CRF ≥ 0). Letters "C" and "W" highlight local maxima of diabatic cooling and warming, respectively. In (c), the color legend at bottom left is used; other panels use the bottom-right legend. From P7.

throughout the cloud shield (Fig. 11-21d). In a two-dimensional but slab-symmetric geometry, such concentrated rising motion could be expected to result in some amount of flow directed away from the heat source in both horizontal directions. In this axisymmetric framework, however, inertial stability strongly resists inward radial displacements (e.g., Eliassen 1951; Shapiro and Willoughby 1982; Holland and Merrill 1984), and thus the radial wind response is strongly biased toward outflow. The LW cooling at cloud top may be more extensive than the in-cloud LW warming, but it also occupies a much smaller volume. Vertically extensive, if gentle, ascent accomplishes the upward mass

transport necessary to enhance upper-tropospheric outflow of appreciable magnitude.

It was also initially believed that the enhanced upper-tropospheric radial outflow associated with CRF, whether responding to LW cooling or warming or some combination thereof, was directly responsible for enabling the enhanced outer region convection, with the concomitant radial broadening of the tangential wind field. We hypothesized (in P4) that outward transport of hydrometeors not only provided the aforementioned positive feedback but also, through fallout and subsequent reconversion to vapor, helped moisten the outer core, eventually rendering it more conducive to the convective

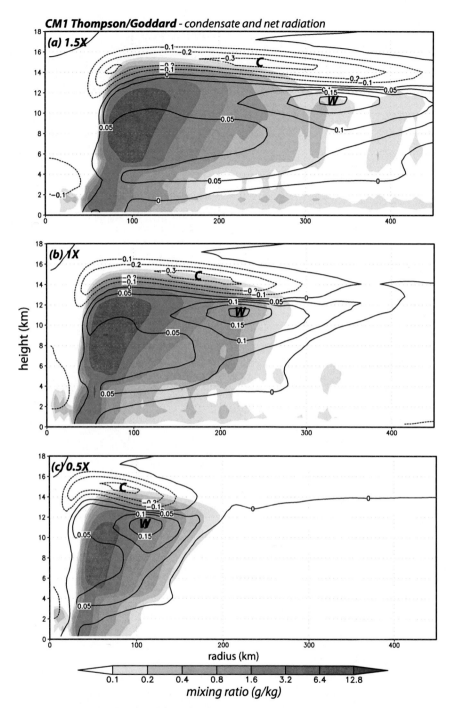

FIG. 11-20. As in Fig. 11-19d, but with the horizontal scale of the imposed CRF field varied from (a) 1.5 times, (b) 1.0 times, and (c) 0.5 times the original radial extent employed in the CRF-fixed simulation. Note the horizontal domain area shown has been increased to 450 km.

activity (see Fig. 11-22) that has been associated with broader wind fields. As the convective activity expanded outward, upward mass transport associated with it helped further augment and expand the upper-tropospheric outflow, and transport even more hydrometeors farther

outward. This is another positive feedback leading to storm expansion with respect to the anvil and winds.

This proposed explanation was tested (in P7) by artificially interfering with the radial transport of hydrometeors in the upper-tropospheric outflow. In the

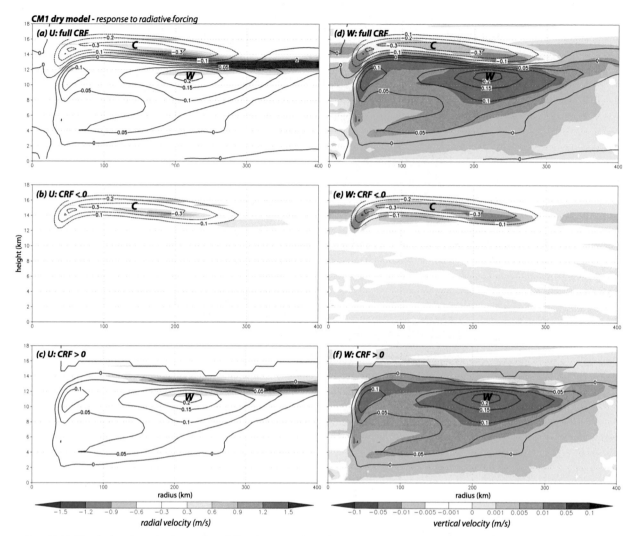

FIG. 11-21. Simulations from the dry version of CM1 forced by the difference between the CRF-on storm cloudy and clear-sky radiative tendencies (contoured as in Fig. 11-16d) averaged between days 9 to 12, inclusive. Radial velocity (shaded) response for the (a) full CRF forcing field, (b) CRF ≤ 0 component, and (c) CRF ≥ 0 component. Vertical velocity (shaded) response for the (d) full CRF forcing field, (e) CRF ≤ 0 component, and (f) CRF ≥ 0 component. Dry fields are averaged over the simulations first 4 days. From P7.

experiment "Outflow Vt = 0", condensation particles within the outflow beyond the radius of maximum wind (RMW) were given zero terminal velocity, which prevented them from easily settling into the dry midtropospheric region below the anvil shield. This restriction did not prevent convective activity (not shown) and the development of a wind field comparably broad to the CRF-on TC (Fig. 11-18). Forcing complete removal of hydrometeors within the radial outflow (experiment "Outflow qx = 0" in Fig. 11-18) also failed to prevent the development of a materially wider storm.

Thus, P7 concluded that the primary agent for inducing the convective activity was the very weak but deep and persistent ascent produced by the LW warming throughout the cloud shield (Fig. 11-21d). Its direct

effect is to lift, very gently, air parcels toward their saturation points. Once saturation is achieved in a particular area, the much larger diabatic forcings associated with vapor phase changes and other microphysical processes can establish and sustain the enhanced outercore convective activity that characterizes CRF-active simulations with enhanced diabatic heating and/or more extensive deep cloudiness (Figs. 11-16 and 11-19).

P7 completed this picture by linking the more extensive convective heating and the tangential wind field broadening. The extra diabatic heating generated in the outer region in the CRF-on TC was inserted in the dry model, which produced a circulation that directly enhanced the cyclonic winds beneath and radially outward from the heat source (see P7's Fig. 14). This response is

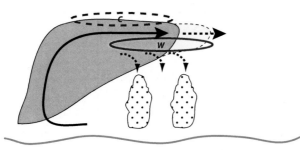

FIG. 11-22. Hypothesis for anvil self-spreading due to cloud-radiative processes, and presumed influence on outer convective activity advanced in P4 and largely refuted in P7. CRF associated with the cloudy area (depicted in orange) results in radial expansion (dashed black line) of the secondary circulation (solid black line), directly extending the cloudy area (depicted in yellow), which leads to expansion of the CRF field as well as hydrometeor fallout (blue dotted lines) that moistens the outer core, and thus was believed to eventually lead to enhanced convection (dotted area). Adapted from P4.

again due to the inertial stability distribution and could have been anticipated from prior work with Sawyer–Eliassen models (e.g., Shapiro and Willoughby 1982; Hack and Schubert 1986). Therefore, it is suggested that the in-cloud CRF warming directly results in lifting that, as a consequence of the enhanced convective activity it encourages, indirectly causes the anvil and wind field expansion in those simulated TCs that incorporate some measure of opaque clouds.

Finally, we note that although CRF provides one avenue for broadening the wind profile, it remains that even "cloudy" idealized and semi-idealized TCs often appear to be too radially compact relative to actual storms, especially in high-resolution simulations. This tendency becomes more obvious when the temporally and azimuthally averaged 700-hPa wind profiles are normalized with respect to the maximum tangential wind (V_m) and the RMW (R_m), as shown for simulations S_2, S_2^*, and $S_2^\#$ in Fig. 11-23a. Nondimensionalization essentially discounts the influence of eye size on outer wind strength. Also shown are modified Rankine (MR) wind profiles (cf. Depperman 1947; Anthes 1982), given by

$$\frac{V}{V_m} = \left(\frac{R}{R_m}\right)^\alpha, \qquad (11\text{-}3)$$

with decay (shape) parameters set to $\alpha = 0.5, 0.625$, and 0.75. This is one of a number of outer wind profile functions that have been proposed (e.g., Holland 1980; DeMaria 1987; Willoughby et al. 2006).

Mallen et al. (2005) analyzed flight-level (largely 700 hPa) data from 72 major hurricanes[11] and found

_____

[11] Category 3 or higher on the Saffir–Simpson scale, based on the 10-m wind speed.

values of $0.18 \le \alpha \le 0.67$, with an average of $\alpha = 0.48$, for the interval $1 \le R/R_m \le 3$. Over that range, the wind profiles of storms S* and $S_2^\#$ are very well described by MR profiles with $\alpha = 0.75$, which exceeds the largest decay parameters from the Mallen et al. (2005) survey. Even in the S_2 case, which has the broadest (non-dimensional) outer wind profile among the WRF-ARW TCs, the winds decayed with radius more rapidly than a substantial majority of the 72 cases. Similar compact wind profiles are found with HWRF simulations made for P7's study (Fig. 11-23b). In P5, we demonstrated that the outer wind structure also could also reflect model physics such as microphysics more so than the initial profile, even though bogussed vortices appeared to be resilient.

d. Global extension of the CRF experiment

Typically, several days are required for TCs to reach maturity in idealized and semi-idealized simulations, even when initialized with bogussed circulations. This prolonged evolution represents a significant shortcoming in the semi-idealized approach, which utilizes the real-data framework employed in operations. Real-data simulations require boundary tendencies from a "parent" model to guide how the regional-scale model's atmosphere evolves. In this semi-idealized situation, there is no parent—the bogus or bubble is placed in an otherwise horizontally homogeneous atmosphere—so those tendencies are zero, effectively sealing the model domain. This limits how long the simulations can be integrated as boundary influences eventually become important. Given the domain sizes employed, running for four days is acceptable, but longer simulations become problematic. The issue is the model storms generally have not finished intensifying prior to the end of these simulations.

A global model would permit longer integrations, but the required high resolution would come with a significant computational cost. As a pilot study, the global MPAS model is used with a variable-resolution, 163 842-cell mesh with coarse grid spacing of 92 km that is refined to 25 km over a circular area of roughly 30° latitude in radius, and with 41 vertical levels (see Table 11-1). A variety of bubble-initialized simulations were made using W6 microphysics with and without convective parameterization employed following the spinup period. For this example, the MP scheme was active from the initial time and the Kain–Fritsch convective scheme was switched off at the 48-h mark. Model runs extended over nine full days, during which time the TC cores remained comfortably within the refinement region. Other experiments yielded TCs that were quantitatively, but not qualitatively, different.

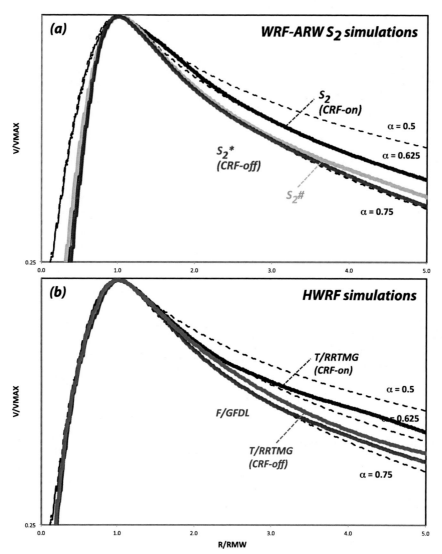

FIG. 11-23. Time-averaged symmetric component of 700-hPa tangential wind speed, non-dimensionalized with respect to maximum wind ($V_{\max}$) and radius of maximum wind (RMW or R_m), along with modified Rankine profiles using values of $\alpha = 0.5$, 0.625, and 0.75 in (11-3). Compare with P5's Fig. 9c.

Both the CRF-on and CRF-off storms appear to have reached maturity around 4–5 days, at least with respect to maximum 10-m wind speed and minimum SLP (Fig. 11-24). After a delay on the order of 12–24 h, the CRF-off storm attained roughly the same wind speeds seen in the "cloudy" counterpart, which corresponds to Saffir–Simpson category 1. Storm intensity is restrained by the coarse horizontal resolution and by the absence of the convective parameterization after spinup, although leaving the Kain–Fritsch scheme on only permits the storms to edge into category 2 (not shown). Averaged over the ninth day, the tangential wind field on the TCs' eastern flanks is clearly more radially extensive in the "cloudy" case and the MP forcing is also substantially

wider (Fig. 11-25), which is consistent with results from the regional model simulations.

In contrast with the regional simulations, the wind fields for these MPAS TCs are extremely broad by the end of the simulations, with 34-kt wind radii extending beyond 300 km (not shown, but it can be inferred from Fig. 11-25). This is in part a consequence of the coarse (25 km) resolution. Owing to the strength of these outer winds, the rapid northwestward motions of the MPAS TCs (compared to other bubble-initialized storms) are not surprising (Fig. 11-26). However, the relatively faster winds in the CRF-active case give this TC an even more substantial beta drift, and by the end of the ninth day this TC has translated over 330 km farther than the

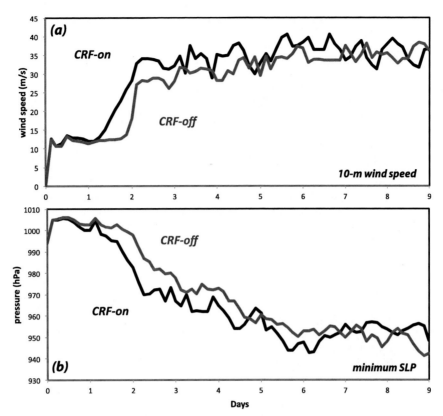

FIG. 11-24. Time series of domain (a) maximum 10-m wind speed and (b) minimum SLP from 9-day MPAS simulations with and without CRF.

transparent-cloud counterpart (Fig. 11-26). The translation speed difference between the two increases sharply after the fourth day of integration. While higher-resolution simulations should be attempted in the future, the principal results from the regional experiments have been verified in a global model with a less restrictive physical framework.

5. Summary

This paper surveyed a body of work (referred to as papers P1–P7; see Table 11-1) that focused on the influences that cloud microphysical processes can have on aspects of tropical cyclones (TCs) other than intensity. High-resolution simulations with semi-idealized and idealized numerical models were made, many of which employed a bubble initialization in which the cyclone was bred from a synoptic-scale buoyancy perturbation as an alternative to the imposition of artificially constructed bogus vortices. The TC characteristics of interest included the track, magnitude, and spatial extent of the radial and tangential winds, asymmetries with respect to the inner-core diabatic heating, convective activity through the outer rainband region, and the evolution and extent of the

TC anvil cloud. To a large degree these characteristics are sensitive to the microphysical parameterization (MP) owing to the interaction of hydrometeors with radiation, which we termed the cloud-radiative forcing (CRF).

In the absence of environmental steering, TC motion reflects a combination of beta drift and convective heating variations responding in part to beta-induced asymmetries (the beta shear). The beta drift is caused by a ventilation flow across the vortex that is generated as a consequence of differential advection of planetary vorticity on a curved Earth, and depends on the outer wind strength. Microphysical assumptions were shown to directly and indirectly modulate the strength of these winds, which result in motion variations with respect to both speed and direction (P1–P3). The beta shear is due to the warm core nature of the TC and encourages the development of asymmetric diabatic heating structures. These heating asymmetries, which are particularly important in the bubble-initialized model storms that incorporated ice microphysics, subtly but powerfully alter storm motion depending on the orientation of the asymmetry pattern, which varied from one MP to another.

In contrast, simulations with warm rain (Kessler) microphysics and/or bogussed vortices were more resilient,

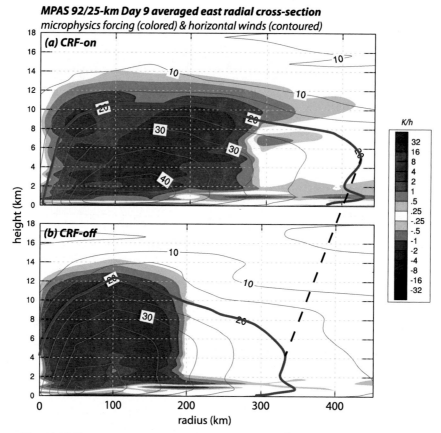

FIG. 11-25. Time-averaged fields of horizontal wind (5 m s^{-1} contours) and diabatic heating from microphysics (shaded) for (a) CRF-on and (b) CRF-off MPAS simulations. This cross section is computed following the vortex and averaged over the ninth day to represent structures that extend eastward from the storm center, and have not been azimuthally averaged.

successfully resisting the development of asymmetries (P5), at least until they were imposed on the storm by the environment, such as through the interaction of the TC with a topographic obstacle (P6). Without the development of asymmetric heating structures that tended to oppose the beta drift, relatively rapid translation was simulated and diagnosed. This was revealed through potential vorticity analyses applied to time-averaged, vortex-following composites (P3 and P5–P7). The realism of these symmetric and asymmetric structures deserves closer examination.

Variations among bubble-initialized model storms with respect to tracks, asymmetric convective patterns, and anvil extents largely vanished when the cloud-radiative forcing was removed in the aquaplanet model (P3). The various MPs produce different amounts, types, and distributions of particles, and ostensibly impact storm dynamics or thermodynamics through LW absorption and emission and SW absorption in a manner that depends largely on particle size (and thus species). Simulated storms with active CRF tended to have larger

eyes, more widespread convective activity in the outer region, and faster outer-core winds relative to their CRF-off counterparts (P3, P7). The CRF-on storms also possessed stronger secondary circulations with faster radial inflow at the lowest levels and stronger outflow aloft, and thicker and more radially extensive anvils. The bubble-initialized warm rain TC became quite symmetric because the effects of CRF were exaggerated by the huge radiative forcing associated with tiny cloud droplets.

Although the LW warming within the anvil is weak, it is the most significant component of CRF as it leads directly to stronger upper-tropospheric radial outflow as well as slow, yet sustained, ascent throughout the outer core (P7). This gentle ascent results in a moistening of the region outside the eyewall, enhances convective activity, elevates the equivalent potential temperature, and increases the radial extent of the TC, including its anvil and wind fields. These conclusions were reinforced with dry model experiments that examined in isolation the roles played by diabatic heat sources. The net heating associated with increased convection in the outer region directly acts to

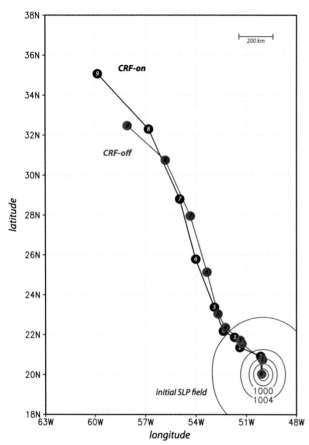

FIG. 11-26. Tracks from 9-day MPAS simulations for CRF-on (black) and CRF-off (red) storms. Day number indicated in position circles. Initial SLP field is shown in blue (4-hPa contours).

intensify both the cyclonic winds in the lower troposphere and the secondary circulation.

Finally, some semi-idealized simulations made with the global MPAS model were used to examine the influence of CRF on TC motion and structure within longer integrations without the limitations imposed by lateral boundary conditions required in regional models. Although the MPAS spatial resolution employed is rather coarse, these experiments confirmed that CRF acts to enhance convective activity in the outer region and broaden the wind profile, which lead to potentially different motions and hazards (e.g., as outer winds influence storm surges). In summary, the assumptions inherent in cloud microphysical parameterizations represent important contributions to forecast uncertainty, particularly because they are amplified by the role radiative processes can play in nontransparent clouds.

Acknowledgments. Professor Michio Yanai was a strong supporter and source of inspiration for this work. Drs. Kevin Hill and Gary Lackmann inspired the semi-idealized framework and the bubble initialization. Dr. Axel Seifert kindly lent us his two-moment microphysics scheme, and Dr. George Bryan permitted use of his CM1 model. WRF-ARW and HWRF analyses employed a customized version of the RIP package, created by Dr. Mark Stoelinga. We thank Dr. Bill Skamarock and Mr. Michael Duda for MPAS support, and Mr. Omar Nava for assisting with development of MPAS analysis tools. The authors also thank Drs. Kuo-nan Liou, Russ Elsberry, Gregory Thompson, Ligia Bernardet, Brad Ferrier, Mrinal Biswas, Yu Gu, Longtao Wu, and Mr. Donald Boucher for information, important interactions, and encouragement. Suggestions from two reviewers are gratefully acknowledged. Portions of this work were supported by the National Science Foundation (ATM-0554765), a Jet Propulsion Laboratory Strategic University Research Partnership grant, the Aerospace Corporation, the National Science Council and Central Weather Bureau of Taiwan, the National Atmospheric and Oceanic Administration's Hurricane Forecast Improvement Program (HFIP; NA12NWS4680009), and the National Aeronautics and Space Administration's Hurricane Science Research Program (HSRP; NNX12AJ83G).

REFERENCES

Anthes, R. A., 1982: *Tropical Cyclones: Their Evolution, Structure and Effects. Meteor. Monogr.*, No. 41, Amer. Meteor. Soc., 298 pp.

Bender, M., 1997: The effect of relative flow on the asymmetric structure in the interior of hurricanes. *J. Atmos. Sci.*, **54**, 703–724, doi:10.1175/1520-0469(1997)054<0703:TEORFO>2.0.CO;2.

Bryan, G. H., and M. Fritsch, 2002: A benchmark simulation for moist nonhydrostatic numerical models. *Mon. Wea. Rev.*, **130**, 2917–2928, doi:10.1175/1520-0493(2002)130<2917:ABSFMN>2.0.CO;2.

Bu, Y., 2012: Factors that can influence the onset time of rapid intensification of tropical cyclones. M.S. thesis, Atmospheric and Oceanic Sciences, University of California, Los Angeles, 75 pp.

——, R. G. Fovell, and K. L. Corbosiero, 2014: Influence of cloud-radiative forcing on tropical cyclone structure. *J. Atmos. Sci.*, **71**, 1644–1662, doi:10.1175/JAS-D-13-0265.1.

Cao, Y., R. G. Fovell, and K. L. Corbosiero, 2011: Tropical cyclone track and structure sensitivity to initialization in idealized simulations: A preliminary study. *Terr. Atmos. Ocean. Sci.*, **22**, 559–578, doi:10.3319/TAO.2011.05.12.01(TM).

Chan, J. C.-L., and R. T. Williams, 1987: Analytical and numerical studies of the beta-effect in tropical cyclone motion. *J. Atmos. Sci.*, **44**, 1257–1265, doi:10.1175/1520-0469(1987)044<1257:AANSOT>2.0.CO;2.

Chen, S.-H., and W.-Y. Sun, 2002: A one-dimensional time dependent cloud model. *J. Meteor. Soc. Japan*, **80**, 99–118, doi:10.2151/jmsj.80.99.

Chou, M. D., and M. J. Suarez, 1994: An efficient thermal infrared radiation parameterization for use in general circulation models. NASA Tech. Memo. 104606, 85 pp.

Collins, W. D., and Coauthors, 2006: The formulation and atmospheric simulation of the Community Atmosphere Model version 3 (CAM3). *J. Climate*, **19**, 2144–2161, doi:10.1175/JCLI3760.1.

Corbosiero, K. L., and J. Molinari, 2002: The effects of vertical wind shear on the distribution of convection in tropical cyclones. *Mon. Wea. Rev.*, **130**, 2110–2123, doi:10.1175/1520-0493(2002)130<2110:TEOVWS>2.0.CO;2.

DeMaria, M., 1987: Tropical cyclone track prediction with a barotropic spectral model. *Mon. Wea. Rev.*, **115**, 2346–2357, doi:10.1175/1520-0493(1987)115<2346:TCTPWA>2.0.CO;2.

Depperman, C. E., 1947: Notes on the origin and structure of Philippine typhoons. *Bull. Amer. Meteor. Soc.*, **28**, 399–404.

Dudhia, J., 1989: Numerical study of convection observed during the Winter Monsoon Experiment using a mesoscale two-dimensional model. *J. Atmos. Sci.*, **46**, 3077–3107, doi:10.1175/1520-0469(1989)046<3077:NSOCOD>2.0.CO;2.

——, S.-Y. Hong, and K.-S. Lim, 2008: A new method for representing mixed-phase particle fall speeds in bulk microphysics parameterizations. *J. Meteor. Soc. Japan*, **86A**, 33–44, doi:10.2151/jmsj.86A.33.

Durran, D. R., T. Dinh, M. Ammerman, and T. Ackerman, 2009: The mesoscale dynamics of thin tropical tropopause cirrus. *J. Atmos. Sci.*, **66**, 2859–2873, doi:10.1175/2009JAS3046.1.

Eliassen, A., 1951: Slow thermally or frictionally controlled meridional circulation in a circular vortex. *Astrophys. Norv.*, **5**, 19–60.

Ferrier, B. S., Y. Jin, T. Black, E. Rogers, and G. DiMego, 2002: Implementation of a new grid-scale cloud and precipitation scheme in NCEP Eta model. *15th Conf. on Numerical Weather Prediction*, San Antonio, TX, Amer. Meteor. Soc., 280–283. [Available online at https://ams.confex.com/ams/SLS_WAF_NWP/techprogram/paper_47241.htm.]

Fiorino, M. J., and R. L. Elsberry, 1989: Some aspects of vortex structure related to tropical cyclone motion. *J. Atmos. Sci.*, **46**, 975–990, doi:10.1175/1520-0469(1989)046<0975:SAOVSR>2.0.CO;2.

Fovell, R. G., and H. Su, 2007: Impact of cloud microphysics on hurricane track forecasts. *Geophys. Res. Lett.*, **34**, L24810, doi:10.1029/2007GL031723.

——, K. L. Corbosiero, and H.-C. Kuo, 2009: Cloud microphysics impact on hurricane track as revealed in idealized experiments. *J. Atmos. Sci.*, **66**, 1764–1778, doi:10.1175/2008JAS2874.1.

——, ——, and ——, 2010a: Influence of cloud-radiative feedback on tropical cyclone motion: Symmetric contributions. *29th Conf. on Hurricanes and Tropical Meteorology*, Tucson, AZ, Amer. Meteor. Soc., 13C.5. [Available online at https://ams.confex.com/ams/pdfpapers/168859.pdf.]

——, ——, A. Seifert, and K.-N. Liou, 2010b: Impact of cloud-radiative processes on hurricane track. *Geophys. Res. Lett.*, **37**, L07808, doi:10.1029/2010GL042691.

Frank, W. M., and E. A. Ritchie, 1999: Effects of environmental flow upon tropical cyclone structure. *Mon. Wea. Rev.*, **127**, 2044–2061, doi:10.1175/1520-0493(1999)127<2044:EOEFUT>2.0.CO;2.

Fu, Q., and K. N. Liou, 1993: Parameterization of the radiative properties of cirrus clouds. *J. Atmos. Sci.*, **50**, 2008–2025, doi:10.1175/1520-0469(1993)050<2008:POTRPO>2.0.CO;2.

Gopalakrishnan, S. G., S. Goldenberg, T. Quirino, X. Zhang, F. Marks Jr., K.-S. Yeh, R. Atlas, and V. Tallapragada, 2012: Toward improving high-resolution numerical hurricane forecasting: Influence of model horizontal grid resolution, initialization, and physics. *Wea. Forecasting*, **27**, 647–666, doi:10.1175/WAF-D-11-00055.1.

Gu, Y., K. N. Liou, W. Chen, and H. Liao, 2010: Direct climate effect of black carbon in China and its impact on dust storms. *J. Geophys. Res.*, **115**, D00K14, doi:10.1029/2009JD013427.

——, ——, S. C. Ou, and R. G. Fovell, 2011: Cirrus cloud simulations using WRF with improved radiation parameterization and increased vertical resolution. *J. Geophys. Res.*, **116**, D06119, doi:10.1029/2010JD014574.

Hack, J. J., and W. H. Schubert, 1986: Nonlinear response of atmospheric vortices to heating by organized cumulus convection. *J. Atmos. Sci.*, **43**, 1559–1573, doi:10.1175/1520-0469(1986)043<1559:NROAVT>2.0.CO;2.

Hausman, S. A., K. V. Ooyama, and W. H. Schubert, 2006: Potential vorticity structure of simulated hurricanes. *J. Atmos. Sci.*, **63**, 87–108, doi:10.1175/JAS3601.1.

Hill, K. A., and G. M. Lackmann, 2009: Influence of environmental humidity on tropical cyclone size. *Mon. Wea. Rev.*, **137**, 3294–3315, doi:10.1175/2009MWR2679.1.

Holland, G. J., 1980: An analytic model of the wind and pressure profiles in hurricanes. *Mon. Wea. Rev.*, **108**, 1212–1218, doi:10.1175/1520-0493(1980)108<1212:AAMOTW>2.0.CO;2.

——, 1983: Tropical cyclone motion: Environmental interaction plus a beta effect. *J. Atmos. Sci.*, **40**, 328–342, doi:10.1175/1520-0469(1983)040<0328:TCMEIP>2.0.CO;2.

——, and R. T. Merrill, 1984: On the dynamics of tropical cyclone structural changes. *Quart. J. Roy. Meteor. Soc.*, **110**, 723–745, doi:10.1002/qj.49711046510.

Hong, S.-Y., J. Dudhia, and S.-H. Chen, 2004: A revised approach to ice microphysical processes for the bulk parameterization of clouds and precipitation. *Mon. Wea. Rev.*, **132**, 103–120, doi:10.1175/1520-0493(2004)132<0103:ARATIM>2.0.CO;2.

——, Y. Noh, and J. Dudhia, 2006: A new vertical diffusion package with an explicit treatment of entrainment processes. *Mon. Wea. Rev.*, **134**, 2318–2341, doi:10.1175/MWR3199.1.

Hsu, L.-H., H.-C. Kuo, and R. G. Fovell, 2013: On the geographic asymmetry of typhoon translation speed across the mountainous island of Taiwan. *J. Atmos. Sci.*, **70**, 1006–1022, doi:10.1175/JAS-D-12-0173.1.

Iacono, M. J., J. S. Delamere, E. J. Mlawer, M. W. Shephard, S. A. Clough, and W. D. Collins, 2008: Radiative forcing by long-lived greenhouse gases: Calculations with the AER radiative transfer models. *J. Geophys. Res.*, **113**, D13103, doi:10.1029/2008JD009944.

Jordan, C. L., 1958: Mean soundings for the West Indies area. *J. Meteor.*, **15**, 91–97, doi:10.1175/1520-0469(1958)015<0091:MSFTWI>2.0.CO;2.

Kessler, E., 1969: *On the Distribution and Continuity of Water Substance in Atmospheric Circulations. Meteor. Monogr.*, Amer. Meteor. Soc., No. 32, 84 pp.

Krueger, S. K., and M. A. Zuluaf, 2002: Radiatively-induced anvil spreading. *Proc. 15th ARM Science Team Meeting*, Daytona Beach, FL, Department of Energy, 280–283.

Li, X., and Z. Pu, 2008: Sensitivity of numerical simulation of early rapid intensification of Hurricane Emily (2005) to cloud microphysical and planetary boundary layer parameterizations. *Mon. Wea. Rev.*, **136**, 4819–4838, doi:10.1175/2008MWR2366.1.

Lin, Y. L., R. D. Farley, and H. D. Orville, 1983: Bulk parameterization of the snow field in a cloud model. *J. Climate Appl. Meteor.*, **22**, 1065–1092, doi:10.1175/1520-0450(1983)022<1065:BPOTSF>2.0.CO;2.

Liou, K. N., 2002: *An Introduction to Atmospheric Radiation.* 2nd ed. Academic Press, 583 pp.

Lord, S. J., H. E. Willoughby, and J. M. Piotrowicz, 1984: Role of a parameterized ice-phase microphysics in an axisymmetric, nonhydrostatic tropical cyclone model. *J. Atmos. Sci.*, **41**, 2836–2848, doi:10.1175/1520-0469(1984)041<2836:ROAPIP>2.0.CO;2.

Mallen, K. J., M. T. Montgomery, and B. Wang, 2005: Re-examining the near-core radial structure of the tropical cyclone primary circulation: Implications for vortex resiliency. *J. Atmos. Sci.*, **62**, 408–425, doi:10.1175/JAS-3377.1.

Marshall, J. S., and W. M. Palmer, 1948: The distribution of raindrops with size. *J. Meteor.*, **5**, 165–166, doi:10.1175/1520-0469(1948)005<0165:TDORWS>2.0.CO;2.

McFarquhar, G. M., and R. A. Black, 2004: Observations of particle size and phase in tropical cyclones: Implications for mesoscale modeling of microphysical processes. *J. Atmos. Sci.*, **61**, 422–439, doi:10.1175/1520-0469(2004)061<0422:OOPSAP>2.0.CO;2.

——, H. Zhang, G. Heymsfield, R. Hood, J. Dudhia, J. B. Halverson, and F. Marks, 2006: Factors affecting the evolution of Hurricane Erin (2001) and the distributions of hydrometeors: Role of microphysical processes. *Mon. Wea. Rev.*, **136**, 127–150, doi:10.1175/JAS3590.1.

Mlawer, E. J., S. J. Taubman, P. D. Brown, M. J. Iacono, and S. A. Clough, 1997: Radiative transfer for inhomogeneous atmospheres: RRTM, a validated correlated-k model for the longwave. *J. Geophys. Res.*, **102** (D14), 16 663–16 682, doi:10.1029/97JD00237.

Rotunno, R., and K. A. Emanuel, 1987: An air–sea interaction theory for tropical cyclones. Part II: Evolutionary study using a nonhydrostatic axisymmetric numerical model. *J. Atmos. Sci.*, **44**, 542–561, doi:10.1175/1520-0469(1987)044<0542:AAITFT>2.0.CO;2.

Seifert, A., and K. D. Beheng, 2006: A two-moment cloud microphysics parameterization for mixed-phase clouds. Part I: Model description. *Meteor. Atmos. Phys.*, **92**, 45–66, doi:10.1007/s00703-005-0112-4.

Shapiro, L. J., and H. E. Willoughby, 1982: The response of balanced hurricanes to local sources of heat and momentum. *J. Atmos. Sci.*, **39**, 378–394, doi:10.1175/1520-0469(1982)039<0378:TROBHT>2.0.CO;2.

Skamarock, W. C., J. B. Klemp, J. Dudhia, D. O. Gill, D. M. Barker, and W. Wang, 2007: A description of the Advanced Research WRF version 2. NCAR Tech. Note NCAR/TN-468+STR, 88 pp, doi:10.5065/D6DZ069T.

——, ——, M. G. Duda, L. D. Fowler, S.-H. Park, and T. D. Ringler, 2012: A multiscale nonhydrostatic atmospheric model using centroidal Voronoi tesselations and C-grid staggering. *Mon. Wea. Rev.*, **140**, 3090–3105, doi:10.1175/MWR-D-11-00215.1.

Stensrud, D. J., 2007: *Parameterization Schemes: Keys to Understanding Numerical Weather Prediction Models.* Cambridge University Press, 488 pp.

Stephens, G., 1978: Radiation profiles in extended water clouds. II: Parameterization schemes. *J. Atmos. Sci.*, **35**, 2123–2132, doi:10.1175/1520-0469(1978)035<2123:RPIEWC>2.0.CO;2.

Stern, D. P., and D. S. Nolan, 2012: On the height of the warm core in tropical cyclones. *J. Atmos. Sci.*, **69**, 1657–1680, doi:10.1175/JAS-D-11-010.1.

Straka, J. M., 2009: *Cloud and Precipitation Microphysics: Principles and Parameterizations.* Cambridge University Press, 406 pp.

Tao, W.-K., J. J. Shi, S. S. Chen, S. Lang, P.-L. Lin, S.-Y. Hong, C. Peters-Lidard, and A. Hou, 2011: The impact of microphysical schemes on hurricane intensity and track. *Asia-Pac. J. Atmos. Sci.*, **47**, 1–16, doi:10.1007/s13143-011-1001-z.

Thompson, G., P. R. Field, R. M. Rasmussen, and W. D. Hall, 2008: Explicit forecasts of winter precipitation using an improved bulk microphysics scheme. Part II: Implementation of a new snow parameterization. *Mon. Wea. Rev.*, **136**, 5095–5115, doi:10.1175/2008MWR2387.1.

Tripoli, G. J., and W. R. Cotton, 1980: A numerical investigation of several factors contributing to the observed variable intensity of deep convection over south Florida. *J. Appl. Meteor.*, **19**, 1037–1063, doi:10.1175/1520-0450(1980)019<1037:ANIOSF>2.0.CO;2.

Wang, Y., 2002: An explicit simulation of tropical cyclones with a triply nested movable mesh primitive equation model: TCM3. Part II: Model refinements and sensitivity to cloud microphysics parameterization. *Mon. Wea. Rev.*, **130**, 3022–3036, doi:10.1175/1520-0493(2002)130<3022:AESOTC>2.0.CO;2.

Willis, P. T., 1984: Functional fits to some observed drop size distributions and parameterization of rain. *J. Atmos. Sci.*, **41**, 1648–1661, doi:10.1175/1520-0469(1984)041<1648:FFTSOD>2.0.CO;2.

Willoughby, H. E., F. D. Marks, and R. J. Feinberg, 1984: Stationary and moving convective bands in hurricanes. *J. Atmos. Sci.*, **41**, 3189–3211, doi:10.1175/1520-0469(1984)041<3189:SAMCBI>2.0.CO;2.

——, W. R. Darling, and M. E. Rahn, 2006: Parametric representation of the primary hurricane vortex. Part II: A new family of sectionally continuous profiles. *Mon. Wea. Rev.*, **134**, 1102–1120, doi:10.1175/MWR3106.1.

Wu, L., and B. Wang, 2000: A potential vorticity tendency diagnostic approach for tropical cyclone motion. *Mon. Wea. Rev.*, **128**, 1899–1911, doi:10.1175/1520-0493(2000)128<1899:APVTDA>2.0.CO;2.

Yanai, M., 1958: On the changes in thermal and wind structure in a decaying typhoon. *J. Meteor. Soc. Japan*, **36**, 141–155.

——, 1961a: A detailed analysis of typhoon formation. *J. Meteor. Soc. Japan*, **39**, 187–214.

——, 1961b: Dynamical aspects of typhoon formation. *J. Meteor. Soc. Japan*, **39**, 282–309.

——, 1963a: A comment on the creation of warm core in incipient tropical cyclone. *J. Meteor. Soc. Japan*, **41**, 183–187.

——, 1963b: A preliminary survey of large-scale disturbances over the tropical Pacific region. *Geofis. Int.*, **3**, 73–84.

——, 1964: Formation of tropical cyclones. *Rev. Geophys.*, **2**, 367–414, doi:10.1029/RG002i002p00367.

Zhu, T., and D.-L. Zhang, 2006: Numerical simulation of hurricane Bonnie (1998). Part II: Sensitivity to varying cloud microphysical processes. *J. Atmos. Sci.*, **63**, 109–126, doi:10.1175/JAS3599.1.

Ziegler, C. L., 1985: Retrieval of thermal and microphysical variables in observed convective storms. Part 1: Model development and preliminary testing. *J. Atmos. Sci.*, **42**, 1487–1509, doi:10.1175/1520-0469(1985)042<1487:ROTAMV>2.0.CO;2.

Chapter 12

Parameterization of Microphysical Processes in Convective Clouds in Global Climate Models

GUANG J. ZHANG AND XIAOLIANG SONG

Scripps Institution of Oceanography, La Jolla, California

ABSTRACT

The microphysical processes inside convective clouds play an important role in climate. They directly control the amount of detrainment of cloud hydrometeor and water vapor from updrafts. The detrained water substance in turn affects the anvil cloud formation, upper-tropospheric water vapor distribution, and thus the atmospheric radiation budget. In global climate models, convective parameterization schemes have not explicitly represented microphysics processes in updrafts until recently. In this paper, the authors provide a review of existing schemes for convective microphysics parameterization. These schemes are broadly divided into three groups: tuning-parameter-based schemes (simplest), single-moment schemes, and two-moment schemes (most comprehensive). Common weaknesses of the tuning-parameter-based and single-moment schemes are outlined. Examples are presented from one of the two-moment schemes to demonstrate the performance of the scheme in simulating the hydrometeor distribution in convection and its representation of the effect of aerosols on convection.

1. Introduction

The parameterization of atmospheric convection has been an active research area for over half a century (see reviews by Arakawa 2000, 2004). Convective transport of heat and moisture, and the latent heat release associated with it play a fundamental role in large-scale atmospheric circulation (Riehl and Malkus 1958). This role was quantitatively expressed in terms of apparent heat source and moisture sink by Yanai et al. (1973). Because of its small spatial and temporal scales compared to model grid spacing in numerical models for large-scale circulation and weather prediction, subgrid-scale convection has to be parameterized. Over the past 50 years, a range of parameterization schemes has been developed, varying from moist convective adjustment schemes (Manabe et al. 1965; Betts 1986) to more sophisticated, moisture-convergence-based (Kuo 1965, 1974; Tiedtke 1989) and convective-instability-based mass flux representations (Arakawa and Schubert

1974; Emanuel 1991; Donner 1993; Zhang and McFarlane 1995a; and many more). In all these parameterization studies, emphasis was placed on representing convective effects on temperature and moisture fields because of their obvious roles in atmospheric energetics and hydrological cycle. Later on, convective effects on the momentum field were also considered (Schneider and Lindzen 1976; Zhang and Cho 1991a,b; Wu and Yanai 1994; Zhang and McFarlane 1995b; Gregory et al. 1997). More recently, approaches to develop scale-aware (or unified) convective parameterization schemes, again largely for temperature and moisture fields, have been explored to meet the need of increasing global climate model (GCM) resolutions (Arakawa et al. 2011; Arakawa and Wu 2013).

On the other hand, the link between convection and stratiform anvil clouds has been very weak in large-scale models for historical reasons. In the early days of general circulation model development, large-scale cloud fraction needed for radiation calculation was specified (Manabe et al. 1965). The cloud liquid water path used for determining cloud optical properties was empirically related to the moisture field (e.g., Kiehl et al. 1998). Thus, convection in climate models affects large-scale

Corresponding author address: Guang J. Zhang, Scripps Institution of Oceanography, 9500 Gilman Dr., La Jolla, CA 92093.
E-mail: gzhang@ucsd.edu

DOI: 10.1175/AMSMONOGRAPHS-D-15-0015.1

clouds indirectly by modulating the moisture field. This is far from how nature works. In reality, deep convection generates massive amounts of anvil clouds by detraining cloud liquid water and ice from updrafts. The radiative effect of the anvil clouds in turn further affects convection (Fu et al. 1995; Stephens et al. 2008). Furthermore, these anvil clouds have a tremendous impact on the earth's radiative energy budget climatologically (Randall et al. 1989; Ramanathan and Collins 1991). The amount of detrained cloud liquid water and ice strongly depends on the strength of convective updrafts and the content of cloud water and ice within them. The latter is determined by convective microphysical processes. Part of the detrained condensate also moistens the upper-tropospheric ambient atmosphere, and can thus modify the upper-tropospheric water vapor distribution. This has important implications for water vapor feedbacks on climate change (Betts 1990; Lindzen 1990; Shine and Sinha 1991). Therefore, a proper treatment of convective microphysical processes in global climate models is crucial to reliable simulations of the present climate and future climate projection.

The impact of convective detrainment on climate feedback and climate change has been the center of a debate for over two decades. On one hand for example, Lindzen (1990) argues that convection has a negative water vapor feedback to climate warming. Under a warmer climate convection reaches higher altitudes, thus detrains less moisture from the saturated air at detrainment levels due to colder temperatures. The compensating subsidence of the drier air fills the upper troposphere, thus producing less greenhouse effect. On the other hand, Betts (1990) argues that this is an overly simplified view of convective effect on climate feedback. Convection, particularly in the tropics and summertime midlatitudes, often appears in the form of mesoscale convective systems with large anvils in the upper troposphere (Houze 1977; Zipser et al. 1981). The humidification of the upper troposphere from the decay of these anvils plays a much more important role than the detrainment of convective air into the ambient atmosphere there.

The generation of anvil and cirrus clouds strongly depends on the microphysics of precipitation formation within convective updrafts (Emanuel and Pierrehumbert 1996). Based on satellite data analysis, Lindzen et al. (2001) propose an "Iris hypothesis," arguing that anvil clouds associated with convection have a positive feedback on climate. They suggest that the area coverage of anvil clouds associated with tropical convection is less extensive when sea surface temperatures are higher, thus allowing for more

shortwave radiation to reach the sea surface and warm it. This claim was challenged by other studies using satellite data and simple model analysis (Lin et al. 2002; Hartmann and Michelsen 2002; Fu et al. 2002). Central to these debates are the effects of convection on upper-tropospheric moisture and cirrus/anvil clouds. Rennó et al. (1994) tested several convection parameterization schemes in a radiative–convective equilibrium model and found that the equilibrium climate was very sensitive to precipitation efficiency in convection. They found that clouds with high precipitation efficiency produced cold and dry climate and clouds with low precipitation efficiency produced moist and warm climate. They argue that convective parameterization schemes currently in use in GCMs bypass the microphysical processes by making arbitrary assumptions on convective precipitation and moistening, and thus are inadequate for climate change studies.

Early convection parameterization schemes treat cloud microphysics crudely, either by arbitrarily assigning a precipitation efficiency (Emanuel 1991) or by assuming that the conversion rate from cloud water to rainwater is proportional to the cloud water mixing ratio (Arakawa and Schubert 1974; Tiedtke 1989; Zhang and McFarlane 1995a). In recent years, more attention has been paid to representing microphysical processes in convection parameterization (Sud and Walker 1999; Zhang et al. 2005; Lohmann 2008; Song and Zhang 2011).

Convective microphysics is important not only to climate feedbacks and global climate change, but also to aerosol–convection interactions, an active research area lately (Koren et al. 2005; Khain et al. 2005; Tao et al. 2007; Rosenfeld et al. 2008; Li et al. 2011; Tao et al. 2012). Anthropogenic aerosols modify cloud microphysical properties by serving as cloud condensation nuclei (CCN) and ice nuclei (IN). More aerosols produce a larger number of smaller cloud droplets, which coalesce less efficiently into raindrops. The satellite observations demonstrated that aerosols suppress deep convective precipitation by reducing cloud droplet size (Rosenfeld 1999; Rosenfeld and Woodley 2000). On the other hand, the suppression of warm rain formation by aerosols in the lower part of the cloud allows a greater amount of cloud water to be lifted to above the freezing level by updrafts. This helps to release additional latent heat from freezing to invigorate convection (Koren et al. 2005; Khain et al. 2005; Rosenfeld et al. 2008; Li et al. 2011). Such microphysical processes must be incorporated into GCMs in order to have a better understanding of their climatic effects.

In this study, we review the development of the representation of convective microphysical processes associated with precipitation formation in GCMs. We will start with the simple treatment typical of that in the early years of convection parameterization development, and gradually move to more comprehensive treatments explored in recent years. Section 2 provides a description of tuning-parameter-based schemes. Section 3 outlines microphysical schemes that consider cloud liquid water and ice mass mixing ratios. Section 4 presents two-moment microphysical schemes for convective clouds that were explored in recent years. Section 5 presents a few examples of model simulation using two-moment convective microphysics schemes, and section 6 summarizes the paper.

2. Tuning-parameter-based schemes

The parameterization of convective microphysics arises from the general problem of convection parameterization. The effects of convection on temperature and moisture fields can be understood from the large-scale heat and moisture budget equations (Yanai et al. 1973):

$$\frac{\partial \bar{s}}{\partial t} + \mathbf{V} \cdot (\bar{\mathbf{v}}\bar{s}) + \frac{\partial(\bar{\rho}\,\bar{w}\,\bar{s})}{\bar{\rho}\partial z}$$
$$= \overline{Q}_R + L(c - e) - \frac{\partial}{\bar{\rho}\partial z}[M_u(s_u - \bar{s}) + M_d(s_d - \bar{s})],$$
$$(12\text{-}1)$$

and

$$\frac{\partial \bar{q}}{\partial t} + \mathbf{V} \cdot (\bar{\mathbf{v}}\bar{q}) + \frac{\partial(\bar{\rho}\,\bar{w}\,\bar{q})}{\bar{\rho}\partial z}$$
$$= -(c - e) - \frac{\partial}{\bar{\rho}\partial z}[M_u(q_u - \bar{q}) + M_d(q_d - \bar{q})],$$
$$(12\text{-}2)$$

where $s = c_p T + gz$ is the dry static energy, q is the specific humidity, $\mathbf{v}$ is the horizontal wind vector, w is the vertical velocity, and L is the latent heat of vaporization. The quantity ρ is air density, M is the cloud mass flux, and subscripts u and d denote updraft and downdraft properties, respectively. The overbar represents the average over the large-scale domain or a model grid box. The radiative heating rate is Q_R and $c - e$ represents the net condensation (condensation minus evaporation) within the GCM grid box. The latent heat release from condensation and the convective transport can be represented by a cloud model, which is used to determine such in-cloud properties as the mass flux, temperature, moisture, and the cloud hydrometeor.

The equations for these properties in convective updrafts are given by

$$\frac{\partial M_u}{\rho \partial z} = \varepsilon_u - \delta_u, \tag{12-3}$$

$$\frac{\partial M_u s_u}{\rho \partial z} = \varepsilon_u \bar{s} - \delta_u s_u + Lc_u, \quad \text{and} \tag{12-4}$$

$$\frac{\partial M_u q_u}{\rho \partial z} = \varepsilon_u \bar{q} - \delta_u q_u - c_u, \tag{12-5}$$

where c_u is the condensation rate, and ε_u and δ_u are the mass entrainment and detrainment. The equation for cloud water content is given by

$$\frac{\partial M_u q_l}{\rho \partial z} = -\delta_u q_l + c_u - R_r, \tag{12-6}$$

where q_l is the liquid water mixing ratio and R_r is the conversion rate from cloud water to rainwater. In early convection parameterization schemes, the conversion of cloud water to rainwater is parameterized based on simple intuitions. For example, in the Arakawa and Schubert scheme (1974), the conversion rate is assumed to be proportional to the cloud water mixing ratio (Lord 1982). In the diagnostic study of convective cloud properties, Yanai et al. (1973) did not use a constant proportionality factor but assumed it to be a function of height. In Emanuel's scheme (Emanuel 1991), instead of using Eq. (12-6) the cloud water mixing ratio is determined by

$$q_l = (1 - \varepsilon)q_{la}, \tag{12-7}$$

where q_{la} is the adiabatic liquid water mixing ratio. The conversion rate is represented through a precipitation efficiency parameter ε, which depends on cloud thickness. For updraft extent less than 150 hPa, there is no conversion from cloud water to rainwater, thus $\varepsilon = 0$. This is to mimic nonprecipitating shallow convection. For updraft extent greater than 500 hPa, the conversion rate is set to 1, meaning that all cloud water is converted to rainwater. For updraft extent between 150 and 500 hPa a linear interpolation is used. The Tiedtke (1989) scheme uses a similar expression to that by Arakawa and Schubert (1974), with the following equation for cloud water to rainwater conversion in Eq. (12-6):

$$R_r = \frac{M_u}{\rho} K(p) q_l, \tag{12-8}$$

where $K(p)$ is a height-dependent conversion coefficient given by

$$K(p) = \begin{cases} 0 & \text{for } p_B - p \le \Delta p_{\text{crit}}, \\ 6 \times 10^{-4}\,\text{m}^{-1} & \text{for } p_B - p > \Delta p_{\text{crit}} \end{cases} \tag{12-9}$$

where p_B is the cloud-base pressure, and Δp_{crit} is the critical pressure thickness for cloud layer and is set to 150 mb over ocean and 300 mb over land. Zhang and McFarlane (1995a) set K to a constant following Lord (1982).

In short, all these schemes invoke crude representation for cloud water to rainwater conversion, with no explicit consideration of cloud microphysical processes. Whereas this may be adequate for convection parameterization in GCMs in the past when diagnostic cloud parameterizations were common (e.g., Xu and Krueger 1991; Kiehl et al. 1998), it is far from satisfactory in today's state-of-the-art GCMs, where sophisticated cloud parameterizations are used to represent stratiform cloud processes in addition to assumed distribution of subgrid-scale moisture/heat variations (e.g., Morrison and Gettelman 2008).

3. Single-moment convective microphysics schemes

In more recent studies, better treatment of cloud water to rainwater conversion was incorporated. Sud and Walker (1999) introduced an autoconversion from cloud water to rainwater following Sundqvist (1978, 1988), who pioneered the prognostic cloud parameterization for the state-of-the-art GCMs. The conversion term in Eq. (12-6) is expressed as

$$R_r = \frac{M_u}{\rho w_u} R_p q_l, \qquad (12\text{-}10)$$

where w_u is the vertical velocity of the updraft and $1/w$ reflects the time it takes for the air parcel to travel through a model layer. The longer it takes, the more cloud water is converted to rainwater. The parameter R_p represents precipitation formation rate (kg kg^{-1} s^{-1}) through autoconversion following Sundqvist (1988):

$$R_p = \frac{f_c}{C_0}\left\{1 - \exp\left[-\left(\frac{q_l}{q_{lcrit}}\right)^2\right]\right\}, \qquad (12\text{-}11)$$

where q_{lcrit} is the critical value of cloud water for autoconversion, f_c is the cloud fraction determined from updraft mass flux and vertical velocity, and C_0 is an autoconversion coefficient, which is related to factors such as precipitation intensity to mimic the coalescence process and subfreezing temperatures to mimic the Bergeron–Findeisen process following Sundqvist et al. (1989). Note that in Eq. (12-11) cloud water to rainwater conversion is independent of the rainwater content. Thus, accretion of cloud water by rainwater is not considered explicitly. The vertical velocity of updrafts determines how long the parcel stays in the layer, thus the amount of conversion. Sud and Walker (1999)

used a cloud-scale vertical momentum equation to estimate w:

$$w\frac{\partial w}{\partial z} = \frac{gB}{1+\gamma} - \frac{1}{M_u}\frac{\partial M_u}{\partial z}\frac{w^2}{2} - gq_l - gq_r - K_d f(w), \qquad (12\text{-}12)$$

where γ is a coefficient that accounts for the pressure gradient effect; the q_l and q_r are terms that represent the loading from cloud water and rainwater, respectively; and the last term represents the frictional drag associated with the turbulent flow in updrafts.

An effort to develop a comprehensive convective microphysics parameterization was made by Zhang et al. (2005). Instead of using a height-dependent K as in Eq. (12-9) in the Tiedtke (1989) scheme, they included more sophisticated cloud microphysical processes for both cloud water and ice mixing ratios. The following two diagnostic equations for cloud water and ice in convective updrafts are used:

$$\frac{\partial M_u q_1}{\rho \partial z} = -\delta_u q_1 + c_{ul} - G_{pl}, \quad \text{and} \qquad (12\text{-}13)$$

$$\frac{\partial M_u q_i}{\rho \partial z} = -\delta_u q_i + c_{ui} - G_{pi}, \qquad (12\text{-}14)$$

where q_l and q_i are cloud liquid and ice mixing ratio, respectively. The conversion terms G_{pl} and G_{pi}, which represent sinks to cloud water and ice due to conversion to rain and snow, are parameterized through more sophisticated cloud microphysical processes than in Sud and Walker (1999):

$$G_{pl} = f_c(Q_{aut} + Q_{racl} + Q_{sacl} + Q_{fho} + Q_{fhe} - Q_{mlt}), \quad \text{and} \qquad (12\text{-}15)$$

$$G_{pi} = f_c(Q_{agg} + Q_{saci} - Q_{fho} - Q_{fhe} + Q_{mlt}), \qquad (12\text{-}16)$$

where f_c is cloud fraction, related to convective cloud mass flux empirically based on Xu and Krueger (1991); subscript aut is for autoconversion, racl and sacl are for accretion of cloud droplets by raindrops and snow, respectively; fho and fhe are for homogeneous and heterogeneous freezing, respectively; mlt is for melting of ice crystals; agg is for aggregation of ice crystals; and saci for accretion of ice crystals by snow. Thus, in Zhang et al. (2005) the cloud water is depleted by autoconversion to rain, accretion by rain and snow, and (homogeneous and heterogeneous) freezing, but increased by the melting of cloud ice. The cloud ice is decreased by aggregation to snow, accretion by snow and melting, and increased by freezing.

The condensation c_{ul} and deposition c_{ui} in Eqs. (12-13) and (12-14) are calculated based on supersaturation:

$$c_{ul} = f_c(q - q_l^*) \quad \text{if} \quad T > 0°C$$
$$c_{ui} = f_c(q - q_i^*) \quad \text{if} \quad T < -35°C,$$

where q_l^* and q_i^* are saturation water vapor mixing ratio with respect to water and ice, respectively. For temperatures in between, condensation and deposition are treated in a more complicated way, depending on the ice water content. Condensation is allowed to occur first and q^* is calculated with respect to water. The scheme then assumes that cloud ice forms through heterogeneous freezing. If the cloud ice water content from heterogeneous freezing (see below) is less than a threshold value of $0.5 \, \text{mg kg}^{-1}$, condensation will continue to take place. On the other hand, if the ice water content exceeds the threshold value, Bergeron–Findeisen process sets in, allowing ice crystal to grow at the expense of cloud water. Further condensation is suppressed and only deposition is allowed. From this point on, q^* is calculated with respect to ice.

Each of the microphysical terms in Eqs. (12-15) and (12-16) is parameterized. Homogeneous freezing Q_{fho} takes place at temperatures below $-35°C$ and all cloud water freezes instantaneously. Between $-35°$ and $0°C$, heterogeneous freezing of cloud droplets Q_{fhe} takes place, and is parameterized following Bigg (1953):

$$Q_{\text{fhe}} = a_1\{\exp[b_1(T_0 - T)] - 1\}\frac{\rho q_l^2}{\rho_l N_l}, \quad (12\text{-}17)$$

where parameters a_1 and b_1 are specified constants. Here $T_0 = 273.16 \, \text{K}$; N_l is the number concentration of cloud droplets, which is prescribed; and ρ_l is water density. The heterogeneous freezing increases exponentially with decreasing temperature and increases with the square of cloud liquid water mixing ratio. Autoconversion is parameterized in terms of cloud water content and droplet number concentration:

$$Q_{\text{aut}} = [6\gamma_1 10^{28} n^{-1.7}(10^{-6}N_l)^{-3.3}(10^{-3}\rho q_l)^{4.7}]/\rho, \quad (12\text{-}18)$$

where γ_1 is a tunable constant, which determines the efficiency of rain formation and is set to 15, and $n = 10$ is the width parameter of the cloud droplet spectrum, assumed to be a gamma function. The functional form of dependence on cloud water content and droplet number follows that in Beheng (1994). High cloud water content and low droplet number concentration favor autoconversion from cloud water to rainwater. The aggregation of cloud ice to snow is parameterized as

$$Q_{\text{agg}} = \gamma_2 q_i/\Delta t, \quad (12\text{-}19)$$

where γ_2 is another tunable parameter determining the efficiency of snow formation. The term Δt is the time

needed for the ice crystal number concentration to decrease by a certain factor, and inversely depends on cloud ice content and the collection efficiency between ice crystals.

The parameterization of accretion of cloud water by raindrops in Zhang et al. (2005) also follows the work of Beheng (1994) and the accretion rate is proportional to the product of cloud water and rainwater contents:

$$Q_{\text{racl}} = a_3\rho q_l q_r, \quad (12\text{-}20)$$

where $a_3 = 6 \, \text{s}^{-1}$ and q_r is the rainwater mixing ratio. The accretion of cloud water by snow follows the work of Levkov et al. (1992) and Lin et al. (1983). It is based on the concept of the geometric sweep-out of cloud droplets by snow particles integrated over all snow sizes, which is assumed to have the exponential distribution, and is given by

$$Q_{\text{sacl}} = \gamma_3 \frac{\pi E_{sl} n_{0s} a_4 q_l \Gamma(3 + b_4)}{4\lambda_s^{3+b_4}}\left(\frac{\rho_0}{\rho}\right)^{1/2}, \quad (12\text{-}21)$$

where $a_4 = 4.83$, $b_4 = 0.25$, and $\gamma_3 = 0.1$. Here $E_{sl} = 1$ is the collection efficiency of cloud droplets by snow, Γ is the gamma function, $n_{0s} = 3 \times 10^6 \, \text{m}^{-4}$ is the intercept parameter, and λ_s is the slope of the snow particle size distribution. It is related to snow mass mixing ratio:

$$\lambda_s = \left(\frac{\pi \rho_s n_{0s}}{\rho q_s}\right)^{1/4},$$

where q_s is the snow mixing ratio and $\rho_s = 100 \, \text{kg m}^{-3}$ is the snow density. The accretion of ice crystals by snow takes a similar form to Eq. (12-21) except with the collection efficiency depending on temperature, $E_{si} = \exp[0.025(T - T_0)]$. For melting ($Q_{\text{mlt}}$), all ice crystals melt to become cloud water when temperature is above $0°C$. The rainwater and snow form as a result of autoconversion, accretion, and aggregation.

Zhang et al. (2005) tested this convective microphysics scheme in a single-column model of ECHAM5 (the GCM developed by the Max Planck Institute for Meteorology in Hamburg, Germany, based on the ECMWF model) using the U.S. Department of Energy Atmospheric Radiation Measurement (ARM) program observations in the Southern Great Plains. They found that the in-cloud convective water content (liquid plus ice) is very sensitive to the introduction of convective microphysics and to assumptions on precipitation formation in the lower levels of convection (Fig. 12-1).

Although single-moment schemes are a significant step forward in representing convective microphysical processes in GCMs, their lack of information on droplet

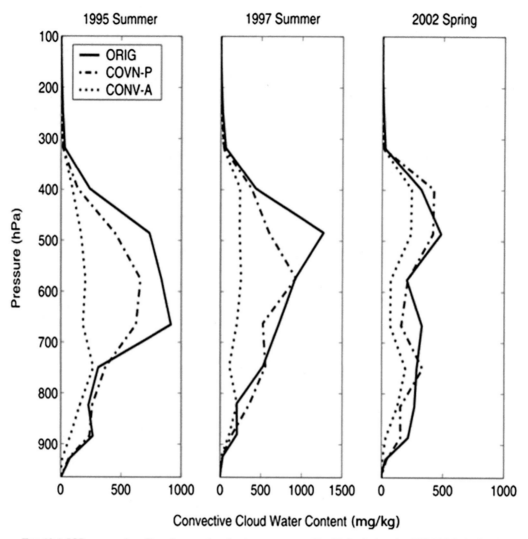

FIG. 12-1. IOP-averaged profiles of convective cloud water content (liquid plus ice) at the ARM SGP site for three IOPs (summers of 1995 and 1997, and spring of 2002). In CONV-A, precipitation is allowed to form at all levels of convective clouds, whereas in CONV-P, precipitation formation is suppressed in the lowest 300 mb above the cloud base. ORIG is the original Tiedtke (1989) scheme without detailed microphysics. Adapted from Zhang et al. (2005).

and ice crystal sizes and number concentration is a clear weakness. In particular, they cannot be used to understand aerosol–convection interaction. To overcome this shortcoming, two-moment schemes have recently been developed for convective clouds.

4. Two-moment convective microphysics schemes

Recognizing the need for considering cloud droplet and ice crystal number concentrations for understanding aerosol–convection interaction in climate models, Lohmann (2008) extended the single-moment convective microphysics scheme of Zhang et al. (2005) using a double-moment cloud microphysics scheme (Lohmann et al. 2007). In particular, droplet activation by aerosols,

which is linked to updraft vertical velocity, was included. The updraft velocity used in aerosol activation was formulated as the sum of grid mean vertical velocity, contributions of turbulent kinetic energy (TKE), and convective available potential energy (CAPE):[1]

$$w = \overline{w} + \sqrt{2\text{CAPE}} + 1.33\sqrt{\text{TKE}}. \quad (12\text{-}22)$$

The droplet activation was related to vertical velocity and internally mixed aerosol number concentration. She

---

[1] In Lohmann (2008) the term $\sqrt{2\text{CAPE}}$ was written as $2\sqrt{\text{CAPE}}$, which we believe is a typo, as CAPE is an estimate of updraft kinetic energy.

Two-moment microphysics scheme for convective clouds

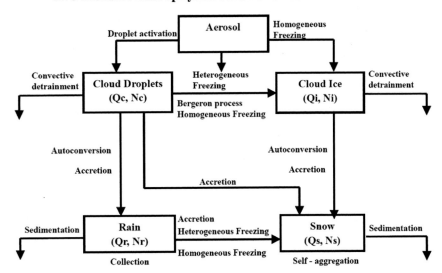

FIG. 12-2. Schematic diagram of the two-moment microphysics scheme for convection. Adapted from Song et al. (2012).

found that by including droplet activation by aerosols in convective cloud microphysics parameterization the ECHAM5 model is able to simulate the aerosol effect on convection invigoration by comparing results from the twentieth-century simulations.

Recently Song and Zhang (2011) and Song et al. (2012) developed a two-moment microphysics parameterization scheme for convective clouds based on the work of Morrison and Gettelman (2008). This diagnostic microphysics scheme explicitly treats mass mixing ratio and number concentration of four hydrometeor species (cloud water, cloud ice, rain, and snow) in convective updrafts. Microphysical processes in downdrafts are not considered. The budget equations for mass mixing ratios q_x and number concentrations N_x of hydrometeor species x (subscript l for cloud water, i for cloud ice, r for rain, and s for snow) in saturated updrafts are given by

$$\frac{\partial M_u q_x}{\rho \partial z} = -\delta_u q_x + \frac{M_u}{\rho w_u} S_x^q, \quad \text{and} \quad (12\text{-}23)$$

$$\frac{\partial M_u N_x}{\rho \partial z} = -\delta_u N_x + \frac{M_u}{\rho w_u} S_x^N. \quad (12\text{-}24)$$

The quantities S_x^q and S_x^N are the in-cloud source/sink terms for q_x and N_x, respectively, from microphysical processes. Figure 12-2 shows the schematic of microphysical processes considered in the scheme (hereafter the SZ scheme). They include droplet activation and ice nucleation by aerosols; autoconversion of cloud water/ice to rain/snow; accretion of cloud water by rain; accretion of cloud water,

cloud ice, and rain by snow; homogeneous and heterogeneous freezing of rain to form snow; Bergeron–Findeisen process; fallout of rain and snow; condensation/deposition; self-collection of rain drops; and self-aggregation of snow. Compared to the single-moment scheme of Zhang et al. (2005), there are several differences in the consideration of the microphysical processes besides including the number concentrations of the cloud hydrometeors. First, the droplet activation and ice nucleation by aerosols are included. Second, autoconversion instead of aggregation of ice to snow is considered. Third, sedimentation of rainwater and snow is considered, which affects the accretion of cloud water/ice by rain and snow. The explicit treatment of cloud particle number concentration enables the SZ microphysics scheme to account for the impact of aerosols on convection. The main features of the scheme are described below.

a. Representation of convective vertical velocity

Different from the approach of Lohmann (2008), which uses a simple formulation of vertical velocity for aerosol activation, the SZ microphysics scheme estimates the updraft velocity of convective clouds using the vertical momentum (or kinetic energy) equation:

$$\frac{\partial K_u}{\partial z} = -\frac{\nu_u}{M_u}(1 + \beta C_d)K_u + \frac{1}{f(1+\gamma)} g \frac{T_{vu} - T_{ve}}{T_{ve}}, \quad (12\text{-}25)$$

where $K_u = w_u^2/2$, ν_u is a mixing coefficient equal to the larger of entrainment or detrainment of mass flux, and g

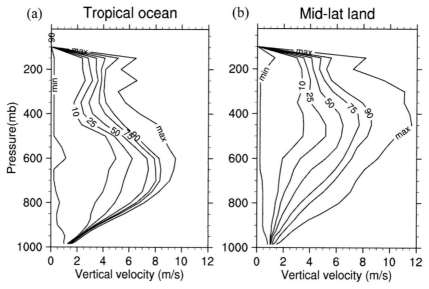

FIG. 12-3. The maximum, minimum, 10th, 25th, 50th, 75th, and 90th percentiles of vertical velocity distribution in convective clouds averaged over (a) the tropical (10°S–10°N) ocean and (b) the Northern Hemisphere middle latitude (20°–40°N) land for June–August (JJA) from CAM5 with the SZ convective microphysics scheme. Adapted from Song et al. (2012).

is gravitational acceleration. The value for β is 1.875. Here $C_d = 0.506$ is the drag coefficient and $\gamma = 0.5$ is the virtual mass coefficient. The factor $f = 2$ is used to account for highly turbulent flows in updrafts. The quantities T_{vu} and T_{ve} are the virtual temperature of the updraft and environment, respectively. This updraft velocity is also used for deriving the cloud fraction and in droplet activation and ice nucleation parameterizations in the SZ microphysics scheme.

Figure 12-3 shows the probability distribution of vertical velocity in deep convection produced by the NCAR Community Atmosphere Model, version 5 (CAM5) with the SZ microphysics scheme (Song et al. 2012). The convective vertical velocities over tropical (10°S–10°N) oceans range from 0.5 to 9 m s^{-1}, peaking at about 3–4 km with maximum and median velocity of 9 and 7 m s^{-1}, respectively. The altitude of maximum vertical velocity is in good agreement with observations for oceanic convection by LeMone and Zipser (1980) and Lucas et al. (1994), but lower than that in a cloud-resolving model simulation (Xu and Randall 2001). Xu and Randall (2001) suggested that this may be due to the low vertical resolution in the aircraft observations. Lucas et al. (1994) also compared observed vertical velocity of oceanic convection with that of land convection from the Thunderstorm Project (Byers and Braham 1949) and found a higher peak with a greater intensity (15 m s^{-1} at 8 km) in land convection. A recent study using airborne Doppler radar found strong updraft vertical velocities exceeding 15 m s^{-1} in both oceanic and land convection, with peak

altitudes above the 10-km level (Heymsfield et al. 2010). Compared to oceanic convection, the vertical velocities in midlatitude (20°–40°N) land convection estimated by the SZ scheme peak at a higher altitude (400 mb) with a greater maximum value ($\sim$12 m s^{-1}), are in good agreement with the top 2% of updrafts simulated in a cloud-resolving model for the ARM Southern Great Plains (SGP) region (Xu and Randall 2001).

It should be noted that the probability distribution of vertical velocities from CAM5 are sampled from broad latitude bands stratified by ocean versus land, whereas observations or cloud-resolving model simulations are from limited geographical locations. The observations have a broad range of vertical velocities even under a limited range of large-scale meteorological conditions. On the other hand, the parameterized vertical velocities are from a bulk plume approach, and are thus not expected to capture the large variability under a given meteorological condition. The broad range seen in Fig. 12-3 mostly reflects the large variability of meteorological conditions in the latitude band. Since microphysical processes and aerosol activation of cloud particles are highly nonlinear, use of a spectral plume model for convective updrafts is more desirable.

b. Autoconversion processes

The autoconversion of cloud water to rainwater in general depends on cloud water content, from linear dependence in the original work of Kessler (1969) to an exponential relationship [Sundqvist 1988; and Eq.

(12-11)] and a power relationship [Eq. (12-18)]. In Song and Zhang (2011), following Khairoutdinov and Kogan (2000), it is parameterized to be proportional to $q_l^{2.47}N_l^{-1.79}$. Compared to that in Zhang et al. (2005), although both are proportional nonlinearly to some power of cloud water mixing ratio and the inverse of droplet number concentration, the degree of nonlinearity is much higher in Zhang et al. (2005). The parameterization of autoconversion of cloud ice to snow follows Ferrier (1994) by integrating the cloud ice size distribution over the range greater than the specified threshold value, which is set to $200\,\mu m$ to separate cloud ice from snow. The ice mixing ratio and number concentrations are converted to snow over a specified time scale (3 min).

c. Accretion processes

The accretion processes include accretion of cloud water by rain and snow, accretion of cloud ice by snow, and accretion of rain by snow. The accretion of cloud water by rainwater follows Khairoutdinov and Kogan (2000), who found it to be proportional to $(q_lq_r)^{1.15}$ based on high-resolution cloud simulations. The accretion for droplet numbers is simply that for cloud water content divided by the average mass of a droplet (q_l/N_l). The accretion of cloud water by snow follows Thompson et al. (2004), which is based on Lin et al. (1983). It takes a similar form to Eq. (12-21) except with different values for tunable parameters. Likewise, the accretion of cloud ice by snow also follows Thompson et al. (2004) and is expressed in the same functional form of gamma function dependence. Finally, the accretion of rain by snow in subfreezing conditions is parameterized according to Ikawa and Saito (1990), in which the accretion rate is linked to the difference of the fall speed between rain and snow and the slopes of the spectral size distributions of rain and snow.

In updrafts of convective clouds, the hydrometeor budget equations in the SZ microphysics scheme are integrated from cloud base to cloud top. A well-known problem with integrating the rainwater/snow equation from the bottom up is that precipitation falling from above into the layer cannot be accounted for (Zhang et al. 2005). This may result in an underestimation of rain/snow in the layer and thus a lower efficiency of the accretion process. Song et al. (2012) considered the effect of accretion by falling precipitation in the SZ scheme by integrating the hydrometeor equations twice. The first integration provides the provisional values of rain/snow. The iteration takes into account the accretion effect of precipitation falling from above using the provisional values.

d. Self-collection and aggregation

Self-collection of rain does not change the rain mass, but it does change the number concentration.

Therefore, it is considered in the two-moment convective microphysics scheme. In Song and Zhang (2011) it is assumed to be proportional to the product of rainwater content and the number concentration following Beheng (1994). Similarly, self-aggregation of snow is also included for snow number concentration following Reisner et al. (1998). It is approximately proportional to $q_s^{0.8}N_s^{1.2}$.

e. Sedimentation

Rainwater and snow fall out of updrafts at certain terminal fall speeds. In Zhang et al. (2005), rain and snow are assumed to fall out instantly. On the other hand, Song and Zhang (2011) consider the sink of rain and snow and their number concentration due to sedimentation following Kuo and Raymond (1980):

$$P_{\text{fallout}}^{q_x} = \frac{V_{q_x}}{\Delta z}q_x, \quad \text{and} \tag{12-26}$$

$$P_{\text{fallout}}^{N_x} = \frac{V_{N_x}}{\Delta z}N_x. \tag{12-27}$$

The mass- and number-weighted terminal fall speeds for all precipitation species are obtained by integrating over the particle size distributions with the appropriate weight of mixing ratio or number concentration, and are dependent on the gamma function and the intercepts of the raindrop and snow size distributions. The V_{N_x} and V_{q_x} are limited to maximum values of $10\,\text{m s}^{-1}$ for rain and $3.6\,\text{m s}^{-1}$ for snow. Once the precipitation particles fall out of the updrafts, a Sundqvist (1988) style evaporation of the convective precipitation is employed in the subsaturated model layer. Note that updrafts are assumed upright. Cheng and Arakawa (1997) showed that considering rainwater and vertical momentum budgets in a combined updraft–downdraft model would lead to tilted updrafts, allowing precipitation to fall out of the updrafts and into the downdrafts. Such an updraft–downdraft configuration is yet to be incorporated in convective microphysics schemes.

f. Cloud droplet activation

The primary activation of aerosols occurs at the cloud base because of high supersaturation and relatively low condensate there. Above the cloud base, the depletion of excess water vapor by condensation on previously activated particles reduces the supersaturation in a constant updraft. Therefore, most of the cloud microphysics studies consider the droplet activation at the cloud base only. However, in real-world deep convective clouds, the increasing updraft strength (e.g., Warner 1969; Pinsky and Khain 2002), the inevitable depletion of droplets formed at cloud base by accretion (Lamb and Verlinde 2011; Phillips et al. 2005), and entrainment (Brenguier and Grabowski 1993; Su et al. 1998; Lasher-Trapp et al.

2005) may also produce supersaturation conditions and thus droplet activation in cloud updrafts. Previous model simulations (Slawinska et al. 2012; Phillips et al. 2005; Morrison and Grabowski 2008) and observations (Prabha et al. 2011) have confirmed the importance of droplet activation in cumulus updrafts. For this reason, the aerosol activation parameterization of Abdul-Razzak and Ghan (2000) is modified and implemented both at and above the cloud base in the SZ microphysics scheme.

g. Ice nucleation

The ice nucleation includes both homogeneous freezing and heterogeneous freezing. Homogeneous freezing occurs through spontaneous freezing of cloud droplets or aerosols at temperatures colder than $-38°$ to about $-40°C$ without the action of ice nuclei, whereas heterogeneous freezing involves the action of ice nuclei and occurs at warmer temperatures. There are four modes of heterogeneous freezing: deposition nucleation, condensation freezing, immersion freezing, and contact freezing (Pruppacher and Klett 1997). The SZ convective microphysics scheme parameterizes all these ice nucleation processes. Homogeneous droplet freezing is performed by instantaneous conversion of the super-cooled cloud liquid water to cloud ice at temperatures below $-40°C$. Between $-5°$ and $-35°C$, the immersion freezing of black carbon and dust is parameterized after Diehl and Wurzler (2004) and the contact freezing of dust follows Liu et al. (2007). Below $-35°C$, ice nucleation is based on Liu et al. (2007), which includes heterogeneous immersion freezing of dust competing with the homogeneous freezing of sulfate, and depends on updraft velocity, air temperature, and aerosol properties. The deposition/condensation nucleation on mineral dust between $-37°$ and $0°C$ is represented by Meyers et al. (1992) and secondary ice production between $-3°$ and $-8°C$ (i.e., the Hallett–Mossop process; Hallett and Mossop 1974) is also included based on Cotton et al. (1986). In addition, the SZ microphysics scheme also considers the competition between homogeneous aerosol freezing and homogeneous droplet freezing. A recent study assessed the relative roles of nucleation processes in deep convection (Phillips et al. 2007). It shows that homogeneous aerosol freezing occurs only in regions of weak ascent, while heterogeneous droplet freezing is dominant in stronger updrafts. Thus, the homogeneous freezing of sulfate is suppressed when updraft vertical velocity is greater than $4\,m\,s^{-1}$.

5. Evaluation in a GCM

Song and Zhang (2011) incorporated the SZ microphysics scheme in the Zhang and McFarlane (1995a,

hereafter ZM95a) convection scheme of the single-column version of NCAR CAM version 3.5 (SCAM3.5) and evaluated its performance with the Tropical Warm Pool–International Cloud Experiment (TWP-ICE) observations. Compared to satellite and C-POL radar retrievals, the standard SCAM3.5 underestimates the cloud ice water and liquid water contents by approximately 75% during the active monsoon period. In contrast, when the SZ microphysics scheme is used, the cloud ice and liquid water contents are increased by more than a factor of 3, making them agree well with observations. With the detrainment of more realistic convective cloud ice and liquid water contents as sources for stratiform clouds, the surface stratiform precipitation, which is seriously underestimated in the model, is increased by a factor of 2.5, and therefore is closer to the observations (Schumacher and Houze 2003).

Song et al. (2012) further implemented the convective microphysics scheme in the ZM95a convection scheme of the NCAR CAM version 5 (CAM5) to evaluate its performance in the global climate model. The simulated cloud ice water content (IWC) is compared to month-long pixel-scale ice water content in convective clouds from *CloudSat* observations. Figure 12-4 shows the boreal summer (July) IWC profiles averaged in three major convective regimes—the tropical (20°S–20°N) ocean, tropical land, and midlatitude [20°–40°N(S)] land—from *CloudSat* retrievals and model simulations. The largest difference in IWC is found between the standard CAM5 simulation and *CloudSat* over the tropical ocean. In the standard CAM5 run (CTL), the maximum IWC is less than 20% of the observations. With the SZ microphysics parameterization, the peak of IWC is in much better agreement with the *CloudSat* observations.

Because the cloud liquid water content (LWC) retrievals in intense convection from satellites are either missing or unreliable due to attenuation of radar signal by precipitation, Song et al. (2012) compared the simulated LWC to aircraft observations. The statistics of aircraft observations by Borovikov et al. (1963) show that characteristic values for dense cumulus congestus and cumulonimbus are $0.5–3\,g\,m^{-3}$. Recent aircraft observations of the LWC for Indian monsoon clouds (Prabha et al. 2012) show that the LWC above cloud base at 2 km is in the range of $0.5–2.5\,g\,m^{-3}$. Figure 12-5 shows the cloud LWC as functions of height above the cloud base from 802 cumulus clouds (Wallace and Hobbs 2006) and CAM5 simulations. The blue dots are averaged LWC, and the squares are the largest values measured (Fig. 12-5a). The observed cloud LWC within 2 km above the cloud base is in the range of $0.5–2\,g\,m^{-3}$, with typical values between 1 and $1.3\,g\,m^{-3}$. The maximum of LWC simulated by the standard CAM5 in

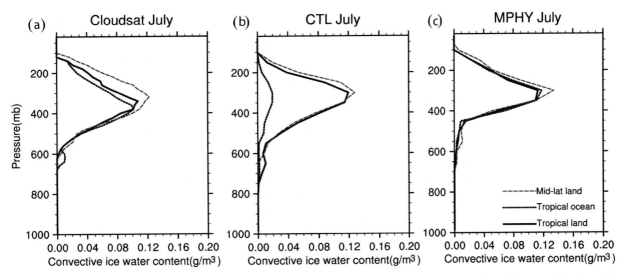

FIG. 12-4. Cloud ice water content (g m^{-3}) profiles in convective clouds averaged over the tropical (20°S–20°N) ocean (blue dotted), tropical land (black solid), and middle latitude [20°–40°N(S)] land (red dashed) for July from (a) *CloudSat* (2007), (b) standard CAM5, and (c) CAM5 with the SZ convective microphysics scheme for convection. Adapted from Song et al. (2012).

July is only about 0.15 g m^{-3} over the tropical ocean, indicating a serious underestimation of LWC. With the SZ microphysics scheme, the LWC reaches a maximum of 1.3 g m^{-3} at 800 hPa. Over tropical land, the simulated LWC peaks near 700 hPa, with a magnitude of 0.5 g m^{-3} for the standard CAM5 simulation and 1.0 g m^{-3} for the simulation with the microphysics scheme. The LWC distribution over midlatitude land produced by the SZ microphysics scheme peaks at a higher altitude near

600 hPa with a maximum of 1.2 g m^{-3}, twice as large as that in the standard CAM5.

Observations show that cloud droplet number concentration (CDNC) in active maritime cumulus is typically 20–60 cm^{-3}, while it is much higher in continental cumulus, in the range of 50–300 cm^{-3} or higher (Squires 1958; Pruppacher and Klett 1997; Wood et al. 2011; Wallace and Hobbs 2006). This contrast in CDNC between maritime and continental convection is easily

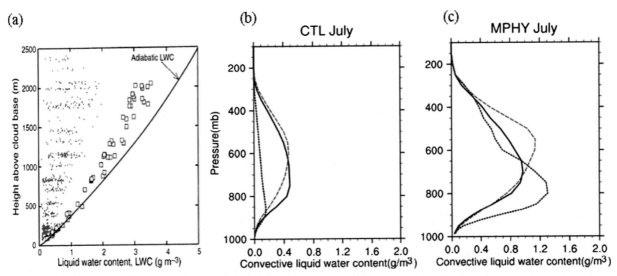

FIG. 12-5. (a) Variation of cloud liquid water content (LWC) with height above the cloud base measured in traverses of 802 cumulus clouds. The blue dots are average LWC, while the squares are the largest values measured (adapted from Wallace and Hobbs 2006); (b),(c) cloud LWC (g m^{-3}) profiles in convective clouds averaged over the tropical (20°S–20°N) ocean (blue dotted), tropical land (black solid), and middle latitude [20°–40°N(S)] land (red dashed) for July from (b) standard CAM5 and (c) CAM5 with the SZ convective microphysics scheme.

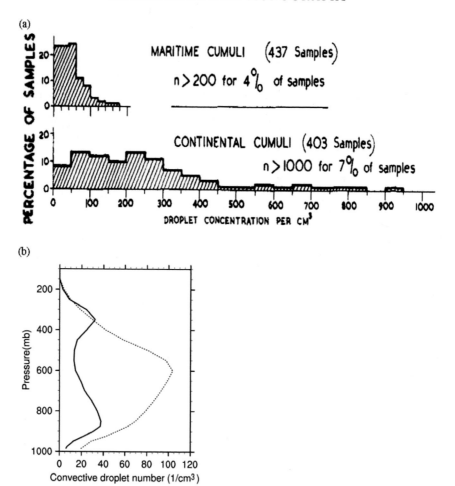

FIG. 12-6. (a) Cloud droplet number concentration (CDNC) in active maritime and continental cumuli (adapted from Squires 1958), and (b) JJA mean cloud droplet in convective clouds averaged over the tropical (20°S–20°N) ocean (solid) and Northern Hemisphere middle latitude (20°–40°N) land (dotted) from CAM5 with the SZ convective microphysics scheme.

seen in Figs. 12-6a and 12-6b. There is a large contrast in CDNC between maritime (tropical ocean) and continental (midlatitude land) convection in the CAM5 simulation with the SZ scheme. The maximum CDNC is about $40\,cm^{-3}$ in maritime convection, and $100\,cm^{-3}$ in continental convection, in qualitative agreement with observations. This contrast in CDNC between maritime and continental convection directly reflects the aerosol effect on cloud microphysics in convection: higher aerosol amount in continental convection produces more cloud droplets and cleaner air over the ocean produces fewer cloud droplets in convective updrafts.

To date, the observation of ice particle number concentration in convective clouds remains a challenging problem because of measurement uncertainties of complicated crystal properties (e.g., shapes, size) and safety concerns arising from aircraft penetrating convective cores above the freezing level. Song et al. (2012) compared the model simulations with microphysical

characteristics of convectively generated ice clouds from two field campaign observations, the Central Equatorial Pacific Experiment (CEPEX; McFarquhar and Heymsfield 1996) and the Kwajalein Experiment (KWAJEX; Heymsfield et al. 2002), and those from cloud system resolving model simulations (Phillips et al. 2007). As shown in Fig. 12-7a, the observed values for ice crystal number concentration (ICNC) vary from $1\,cm^{-3}$ above 9 km in CEPEX to $0.02-0.07\,cm^{-3}$ from 5 to 11 km in KWAJEX, while the WRF simulation of TOGA COARE convection gives ICNC in the range of $0.02-2.5\,cm^{-3}$ (Phillips et al. 2007). The CAM5 simulation with SZ microphysics scheme shows that the ICNC ranges from 0.02 to $1.4\,cm^{-3}$, roughly in line with available observational and cloud resolving model results.

One of the major benefits a detailed microphysics parameterization for convective clouds can bring to GCMs is that it makes investigation of aerosol effects on convection in GCMs possible. To check whether the SZ

(a)

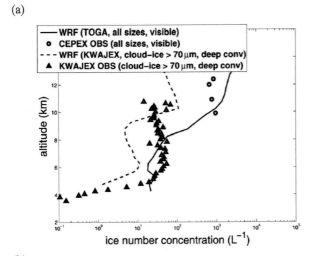

(b)

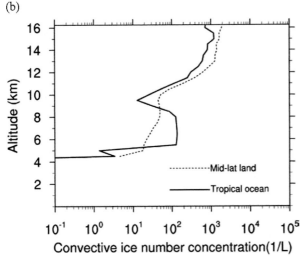

FIG. 12-7. (a) Profiles of ice crystal number concentration obtained in CEPEX and KWAJEX experiments and WRF simulations (adapted from Phillips et al. 2007), and (b) JJA mean cloud ice crystal number concentration (cm^{-3}) profiles in convective clouds averaged over the tropical (20°S–20°N) ocean (solid) and Northern Hemisphere middle latitude (20°–40°N) land (dotted) from CAM5 with the SZ convective microphysics scheme.

convective microphysics scheme can realistically represent the possible impact of aerosols on convective clouds, Song et al. (2012) conducted a sensitivity simulation, in which the aerosol concentration in convection is reduced by a factor of 10. As a result, both the CDNC and ICNC are reduced by more than a factor of 5, indicating that droplet activation and ice nucleation are weakened due to reduced aerosol concentration. They further found that when the aerosol loading was decreased there was more autoconversion from cloud droplets to raindrops, and less ice production in convection averaged over the major convective regime (20°S–40°N) (Fig. 12-8). This is in agreement with the

argument that aerosols suppress warm rain formation and enhance freezing, thereby invigorating convection (Khain et al. 2005; Rosenfeld et al. 2008).

One of the advantages of using convective microphysics scheme is physically based partitioning of convective condensate between precipitation particles and detrained condensate. As an important water source, convective detrainment of cloud liquid/ice water can affect the large-scale clouds and precipitation. Song et al. (2012) found that including the SZ convective microphysics scheme in the CAM5 indeed increased the detrainment of cloud water from convective updrafts, which in turn leads to increased cloud amount and large-scale precipitation.

6. Summary

This paper reviews the schemes for parameterizing the microphysical processes in convective updrafts. The tuning-parameter-based schemes, as presented in section 2, are largely based on intuitive arguments and some of them are still used today in convective parameterization schemes in GCMs. In general, in these schemes the conversion rate of cloud water to rainwater is related to the amount of cloud water itself in some simple forms. This is the simplest way to mimic the microphysical processes for rain production. Although careful parameter tuning can produce reasonable cloud water in convection, the lack of detailed representation of microphysical processes hampers further progress in physically based parameterization of convective processes. The single-moment microphysics schemes for convective clouds presented in section 3 allows for some interaction of convection, clouds, and climate, and these schemes incorporate the details of microphysical processes for cloud water and ice to varying degrees. However, although they consider autoconversion and accretion processes for mass mixing ratios in sufficient details (Sud and Walker 1999; Zhang et al. 2005), the number concentrations of the hydrometeor species are specified. The role of aerosols serving as condensation and ice nuclei for droplet activation and ice nucleation is not considered. Thus, these schemes cannot meet the need of current GCMs to provide accurate estimates of aerosol indirect effects not only for stratiform clouds, but also for convective clouds. The two-moment convective microphysics schemes developed recently (Lohmann 2008; Song and Zhang 2011) fill this void. Cloud droplet and ice crystal numbers are calculated with sources and sinks from different microphysical processes including droplet activation and ice nucleation by aerosols. This is particularly important, as recent observational and cloud model studies suggest that aerosols can modify the microphysical properties of

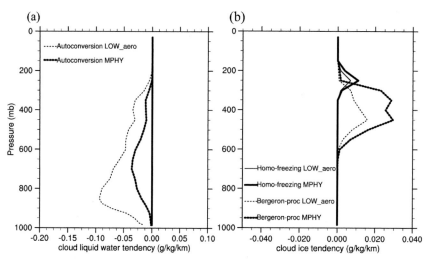

FIG. 12-8. (a) The autoconversion rate of cloud liquid water to rain ($g\,kg^{-1}\,km^{-1}$) and (b) cloud ice production ($g\,kg^{-1}\,km^{-1}$) through the Bergeron process (dotted lines) and homogeneous droplet freezing (dashed lines) in convective clouds averaged over 20°S–40°N for JJA from simulation with the present aerosol loading (MPHY, thick lines) and that with lower aerosol loading (LOW_aero, thin lines). Adapted from Song et al. (2012).

convection and affect the intensity and time evolution of convection and associated precipitation (Fan et al. 2007; Khain et al. 2008; Li et al. 2011; Tao et al. 2012).

As an example to demonstrate the performance of the two-moment scheme of Song and Zhang (2011), results of the cloud ice and water mass and number concentrations are compared with available observations in section 5. With the scheme included, the CAM5 model simulates the vertical distribution of these fields well, in broad agreement with both observations and simulations from cloud-resolving models (Xu 1995; Phillips et al. 2007). In addition, it realistically simulates the effect of aerosols on cloud microphysics in convection. When aerosol loading is increased, autoconversion from cloud water to rainwater is suppressed; however, ice production is enhanced, which can lead to more freezing heating in convection.

The GCM results are encouraging. Nevertheless, further improvements on the parameterization are needed. First, vertical velocity in updrafts strongly affects the microphysical processes. Thus, it should be calculated as accurately as possible. At present, either a simple relationship linking it to CAPE [Eq. (12-22)] or a bulk calculation [Eq. (12-25)] is used. For different updraft plumes, the vertical velocity should be different. Thus, a spectral plume model for updrafts and their vertical velocities is desirable. Second, the two-moment convective microphysics schemes consider updrafts only. It is well known that downdrafts play important roles in convection (Cheng and Arakawa 1997) and planetary boundary layer via cold pools and gust fronts

(Johnson and Nicholls 1983). In deep convection, they are mostly produced by evaporation of precipitation and condensates (Knupp and Cotton 1985) through microphysical processes. Cheng and Arakawa (1997) showed that downdrafts could affect the tilting of updrafts, which in turn impacts the sedimentation of hydrometeors in updrafts and evaporation in downdrafts. Therefore, microphysics parameterization should also be applied to downdrafts and incorporated in determining the intensity of downdrafts.

Although comprehensive cloud microphysics schemes have been used for many years to represent large-scale clouds in regional and global atmospheric and climate models, the use of schemes of similar complexity for convective clouds is relatively new. Since convective detrainment of cloud water and ice plays an important role in anvil cloud formation and evolution, parameterizing convective cloud microphysical processes opens the door for new lines of research in climate change, including aerosol effect on convection invigoration and aerosol indirect effect. The aerosol effects on convection have been explored extensively in recent years, both observationally and numerically using cloud-resolving models (Li et al. 2011; Fan et al. 2012; also see Tao et al. 2012 for a review). With the development of two-moment convective microphysics schemes, it is expected that more GCM investigations of aerosol effects on convection will emerge.

Acknowledgments. This work was supported by the Office of Science (BER), U.S. Department of Energy under

Grant DE-SC0008880, the U.S. National Oceanic and Atmospheric Administration Grants NA08OAR4320894 and NA11OAR431098, and the National Science Foundation Grant AGS-1015964. The authors thank two anonymous reviewers and Dr. Kuan-Man Xu for their valuable comments that helped improve the presentation of the paper.

REFERENCES

Abdul-Razzak, H., and S. J. Ghan, 2000: A parameterization of aerosol activation: 2. Multiple aerosol types. *J. Geophys. Res.*, **105**, 6837–6844, doi:10.1029/1999JD901161.

Arakawa, A., 2000: Future development of general circulation models. *General Circulation Model Development: Past, Present, and Future*, D. A. Randall, Ed., Academic Press, 721–780.

——, 2004: The cumulus parameterization problem: Past, present, and future. *J. Climate*, **17**, 2493–2525, doi:10.1175/1520-0442(2004)017<2493:RATCPP>2.0.CO;2.

——, and W. H. Schubert, 1974: The interactions of a cumulus cloud ensemble with the large-scale environment. Part I. *J. Atmos. Sci.*, **31**, 674–701, doi:10.1175/1520-0469(1974)031<0674:IOACCE>2.0.CO;2.

——, and C.-M. Wu, 2013: A unified representation of deep moist convection in numerical modeling of the atmosphere. Part I. *J. Atmos. Sci.*, **70**, 1977–1992, doi:10.1175/JAS-D-12-0330.1.

——, J.-H. Jung, and C.-M. Wu, 2011: Toward unification of the multiscale modeling of the atmosphere. *Atmos. Chem. Phys.*, **11**, 3731–3742, doi:10.5194/acp-11-3731-2011.

Beheng, K. D., 1994: A parameterization of warm cloud microphysical conversion processes. *Atmos. Res.*, **33**, 193–206, doi:10.1016/0169-8095(94)90020-5.

Betts, A. K., 1986: A new convective adjustment scheme. Part I: Observational and theoretical basis. *Quart. J. Roy. Meteor. Soc.*, **112**, 677–691, doi:10.1002/qj.49711247307.

——, 1990: Greenhouse warming and the tropical water vapor budget. *Bull. Amer. Meteor. Soc.*, **71**, 1465–1467.

Bigg, E. K., 1953: The supercooling of water. *Proc. Phys. Soc. London*, **66B**, 688–694, doi:10.1088/0370-1301/66/8/309.

Borovikov, A. M., I. I. Gaivoronskii, V. V. Kostarev, I. P. Mazin, V. E. Minervin, A. K. Khrgian, and S. M. Shmeter, 1963: *Cloud Physics*. Israel Program for Scientific Translations, U.S. Department of Commerce, 392 pp.

Brenguier, J. L., and W. Grabowski, 1993: Cumulus entrainment and cloud droplet spectra: A numerical model within a two-dimensional dynamical framework. *J. Atmos. Sci.*, **50**, 120–136, doi:10.1175/1520-0469(1993)050<0120:CEACDS>2.0.CO;2.

Byers, H. R., and R. R. Braham, 1949: The Thunderstorm Project. U.S. Weather Bureau, U.S. Department of Commerce Tech. Rep., 287 pp. [NTIS PB-234515.]

Cheng, M.-D., and A. Arakawa, 1997: Inclusion of rainwater budget and convective downdrafts in the Arakawa–Schubert cumulus parameterization. *J. Atmos. Sci.*, **54**, 1359–1378, doi:10.1175/1520-0469(1997)054<1359:IORBAC>2.0.CO;2.

Cotton, W. R., G. J. Tripoli, R. M. Rauber, and E. A. Mulvihill, 1986: Numerical simulation of the effects of varying ice crystal nucleation rates and aggregation processes on orographic snowfall. *J. Climate Appl. Meteor.*, **25**, 1658–1680, doi:10.1175/1520-0450(1986)025<1658:NSOTEO>2.0.CO;2.

Diehl, K., and S. Wurzler, 2004: Heterogeneous drop freezing in the immersion mode: Model calculations considering soluble and insoluble particles in the drops. *J. Atmos. Sci.*, **61**, 2063–2072, doi:10.1175/1520-0469(2004)061<2063:HDFITI>2.0.CO;2.

Donner, L. J., 1993: A cumulus parameterization including mass fluxes, vertical momentum dynamics, and mesoscale effects. *J. Atmos. Sci.*, **50**, 889–906, doi:10.1175/1520-0469(1993)050<0889:ACPIMF>2.0.CO;2.

Emanuel, K. A., 1991: A scheme for representing cumulus convection in large-scale models. *J. Atmos. Sci.*, **48**, 2313–2335, doi:10.1175/1520-0469(1991)048<2313:ASFRCC>2.0.CO;2.

——, and R. T. Pierrehumbert, 1996: Microphysical and dynamical control of tropospheric water vapor. *Clouds, Chemistry and Climate*, P. J. Crutzen and V. Ramanathan, Eds., NATO ASI Series, Vol. 135, Springer-Verlag, 17–28.

Fan, J., R. Zhang, G. Li, and W.-K. Tao, 2007: Effects of aerosols and relative humidity on cumulus clouds. *J. Geophys. Res.*, **112**, D14204, doi:10.1029/2006JD008136.

——, L. R. Leung, Z. Li, H. Morrison, H. Chen, Y. Zhou, Y. Qian, and Y. Wang, 2012: Aerosol impacts on clouds and precipitation in eastern China: Results from bin and bulk microphysics. *J. Geophys. Res.*, **117**, D00K36, doi:10.1029/2011JD016537.

Ferrier, B. S., 1994: A double-moment multiple-phase four-class bulk ice scheme. Part I: Description. *J. Atmos. Sci.*, **51**, 249–280, doi:10.1175/1520-0469(1994)051<0249:ADMMPF>2.0.CO;2.

Fu, Q., S. K. Krueger, and K. N. Liou, 1995: Interactions of radiation and convection in simulated tropical cloud clusters. *J. Atmos. Sci.*, **52**, 1310–1328, doi:10.1175/1520-0469(1995)052<1310:IORACI>2.0.CO;2.

——, M. Baker, and D. L. Hartmann, 2002: Tropical cirrus and water vapor: An effective Earth infrared iris feedback? *Atmos. Chem. Phys.*, **2**, 31–37, doi:10.5194/acp-2-31-2002.

Gregory, D., R. Kershaw, and P. M. Innes, 1997: Parameterization of momentum transport by convection. II: Tests in single column and general circulation models. *Quart. J. Roy. Meteor. Soc.*, **123**, 1153–1183, doi:10.1002/qj.49712354103.

Hallett, J., and S. C. Mossop, 1974: Production of secondary ice particles during the riming process. *Nature*, **249**, 26–28, doi:10.1038/249026a0.

Hartmann, D. L., and M. L. Michelsen, 2002: No evidence for Iris. *Bull. Amer. Meteor. Soc.*, **83**, 249–254, doi:10.1175/1520-0477(2002)083<0249:NEFI>2.3.CO;2.

Heymsfield, A. J., A. Bansemer, P. R. Field, S. L. Durden, J. L. Stith, J. E. Dye, W. Hall, and C. Grainger, 2002: Observations and parameterizations of particle size distributions in deep tropical cirrus and stratiform precipitating clouds: Results from in situ observations in TRMM field campaigns. *J. Atmos. Sci.*, **59**, 3457–3491, doi:10.1175/1520-0469(2002)059<3457:OAPOPS>2.0.CO;2.

Heymsfield, J. M., L. Tian, A. J. Heymsfield, L. Li, and S. Guimond, 2010: Characteristics of deep tropical and subtropical convection from nadir-viewing high-altitude airborne Doppler radar. *J. Atmos. Sci.*, **67**, 285–308, doi:10.1175/2009JAS3132.1.

Houze, R. A., Jr., 1977: Structure and dynamics of a tropical squall-line system. *Mon. Wea. Rev.*, **105**, 1540–1567, doi:10.1175/1520-0493(1977)105<1540:SADOAT>2.0.CO;2.

Ikawa, M., and K. Saito, 1990: Description of the nonhydrostatic model developed at the Forecast Research Department of the MRI. Meteorological Research Institute MRI Tech. Rep. 28, 238 pp., doi:10.11483/mritechrepo.28. [Available online at http://www.mri-jma.go.jp/Publish/Technical/DATA/VOL_28/28_en.html.]

Johnson, R. H., and M. E. Nicholls, 1983: A composite analysis of the boundary layer accompanying a tropical squall line. *Mon. Wea. Rev.*, **111**, 308–319, doi:10.1175/1520-0493(1983)111<0308:ACAOTB>2.0.CO;2.

Kessler, E., 1969: *On the Distribution and Continuity of Water Substance in Atmospheric Circulations. Meteor. Monogr.*, No. 32, Amer. Meteor. Soc., 84 pp.

Khain, A., D. Rosenfeld, and A. Pokrovsky, 2005: Aerosol impact on the dynamics and microphysics of deep convective clouds. *Quart. J. Roy. Meteor. Soc.*, **131**, 2639–2663, doi:10.1256/qj.04.62.

——, N. BenMoshe, and A. Pokrovsky, 2008: Factors determining the impact of aerosols on surface precipitation from clouds: An attempt at classification. *J. Atmos. Sci.*, **65**, 1721–1748, doi:10.1175/2007JAS2515.1.

Khairoutdinov, M. F., and Y. Kogan, 2000: A new cloud physics parameterization in a large-eddy simulation model of marine stratocumulus. *Mon. Wea. Rev.*, **128**, 229–243, doi:10.1175/1520-0493(2000)128<0229:ANCPPI>2.0.CO;2.

Kiehl, J. T., J. J. Hack, G. B. Bonan, B. B. Boville, D. L. Williamson, and P. J. Rasch, 1998: The National Center for Atmospheric Research Community Climate Model: CCM3. *J. Climate*, **11**, 1131–1150, doi:10.1175/1520-0442(1998)011<1131:TNCFAR>2.0.CO;2.

Knupp, K. R., and W. R. Cotton, 1985: Convective cloud downdraft structure: An interpretive survey. *Rev. Geophys.*, **23**, 183–215, doi:10.1029/RG023i002p00183.

Koren, I., Y. J. Kaufman, D. Rosenfeld, L. A. Remer, and Y. Rudich, 2005: Aerosol invigoration and restructuring of Atlantic convective clouds. *Geophys. Res. Lett.*, **32**, L14828, doi:10.1029/2005GL023187.

Kuo, H. L., 1965: On formation and intensification of tropical cyclones through latent heat release by cumulus convection. *J. Atmos. Sci.*, **22**, 40–63, doi:10.1175/1520-0469(1965)022<0040:OFAIOT>2.0.CO;2.

——, 1974: Further studies of the parameterization of the influence of cumulus convection on large-scale flow. *J. Atmos. Sci.*, **31**, 1232–1240, doi:10.1175/1520-0469(1974)031<1232:FSOTPO>2.0.CO;2.

——, and W. H. Raymond, 1980: A quasi-one-dimensional cumulus cloud model and parametrization of cumulus heating and mixing effects. *Mon. Wea. Rev.*, **108**, 991–1009, doi:10.1175/1520-0493(1980)108<0991:AQODCC>2.0.CO;2.

Lamb, D., and J. Verlinde, 2011: *Physics and Chemistry of Clouds.* Cambridge University Press, 584 pp.

Lasher-Trapp, S. G., W. A. Cooper, and A. M. Blyth, 2005: Broadening of droplet size distributions from entrainment and mixing in a cumulus cloud. *Quart. J. Roy. Meteor. Soc.*, **131**, 195–220, doi:10.1256/qj.03.199.

LeMone, M. A., and E. J. Zipser, 1980: Cumulonimbus vertical velocity events in GATE. Part I: Diameter, intensity, and mass flux. *J. Atmos. Sci.*, **37**, 2444–2457, doi:10.1175/1520-0469(1980)037<2444:CVVEIG>2.0.CO;2.

Levkov, L., B. Rockel, H. Kapitza, and E. Raschke, 1992: 3D mesoscale numerical studies of cirrus and stratus clouds by their time and space evolution. *Beitr. Phys. Atmos.*, **65**, 35–58.

Li, Z., F. Niu, J. Fan, Y. Liu, D. Rosenfeld, and Y. Ding, 2011: Long-term impacts of aerosols on the vertical development of clouds and precipitation. *Nat. Geosci.*, **4**, 888–894, doi:10.1038/ngeo1313.

Lin, B., B. A. Wielicki, L. H. Chambers, Y. Hu, and K.-M. Xu, 2002: The Iris hypothesis: A negative or positive cloud feedback? *J. Climate*, **15**, 3–7, doi:10.1175/1520-0442(2002)015<0003:TIHANO>2.0.CO;2.

Lin, Y. L., R. D. Farley, and H. D. Orville, 1983: Bulk parameterization of the snow field in a cloud model. *J. Climate Appl. Meteor.*, **22**, 1065–1092, doi:10.1175/1520-0450(1983)022<1065:BPOTSF>2.0.CO;2.

Lindzen, R. S., 1990: Some coolness concerning global warming. *Bull. Amer. Meteor. Soc.*, **71**, 288–299, doi:10.1175/1520-0477(1990)071<0288:SCCGW>2.0.CO;2.

——, M.-D. Chou, and A. Y. Hou, 2001: Does the Earth have an adaptive infrared iris? *Bull. Amer. Meteor. Soc.*, **82**, 417–432, doi:10.1175/1520-0477(2001)082<0417:DTEHAA>2.3.CO;2.

Liu, X., J. E. Penner, S. Ghan, and M. Wang, 2007: Inclusion of ice microphysics in the NCAR Community Atmospheric Model version 3 (CAM3). *J. Climate*, **20**, 4526–4547, doi:10.1175/JCLI4264.1.

Lohmann, U., 2008: Global anthropogenic aerosol effects on convective clouds in ECHAM5-HAM. *Atmos. Chem. Phys.*, **8**, 2115–2131, doi:10.5194/acp-8-2115-2008.

——, P. Stier, C. Hoose, S. Ferrachat, S. Kloster, E. Roechner, and J. Zhang, 2007: Cloud microphysics and aerosol indirect effects in the global climate model ECHAM5-HAM. *Atmos. Chem. Phys.*, **7**, 3425–3446, doi:10.5194/acp-7-3425-2007.

Lord, S. J., 1982: Interaction of a cumulus cloud ensemble with large-scale environment. Part III: Semi-prognostic test of the Arakawa–Schubert cumulus parameterization. *J. Atmos. Sci.*, **39**, 88–103, doi:10.1175/1520-0469(1982)039<0088:IOACCE>2.0.CO;2.

Lucas, C., E. Zipser, and M. LeMone, 1994: Vertical velocity in oceanic convection off tropical Australia. *J. Atmos. Sci.*, **51**, 3183–3193, doi:10.1175/1520-0469(1994)051<3183:VVIOCO>2.0.CO;2.

Manabe, S., J. Smagorinsky, and R. F. Strickler, 1965: Simulated climatology of a general circulation model with a hydrological cycle. *Mon. Wea. Rev.*, **93**, 769–798, doi:10.1175/1520-0493(1965)093<0769:SCOAGC>2.3.CO;2.

McFarquhar, G. M., and A. J. Heymsfield, 1996: Microphysical characteristics of three cirrus anvils sampled during the Central Equatorial Pacific Experiment. *J. Atmos. Sci.*, **53**, 2401–2423, doi:10.1175/1520-0469(1996)053<2401:MCOTAS>2.0.CO;2.

Meyers, M. P., P. J. DeMott, and W. R. Cotton, 1992: New primary ice nucleation parameterizations in an explicit cloud model. *J. Appl. Meteor.*, **31**, 708–721, doi:10.1175/1520-0450(1992)031<0708:NPINPI>2.0.CO;2.

Morrison, H., and A. Gettelman, 2008: A new two-moment bulk stratiform cloud microphysics scheme in the Community Atmosphere Model, version 3 (CAM3). Part I: Description and numerical tests. *J. Climate*, **21**, 3642–3659, doi:10.1175/2008JCLI2105.1.

——, and W. W. Grabowski, 2008: Modeling supersaturation and subgrid-scale mixing with two-moment bulk warm microphysics. *J. Atmos. Sci.*, **65**, 792–812, doi:10.1175/2007JAS2374.1.

Phillips, V. T. J., and Coauthors, 2005: Anvil glaciation in a deep cumulus updraught over Florida simulated with an explicit microphysics model. I: Impact of various nucleation processes. *Quart. J. Roy. Meteor. Soc.*, **131**, 2019–2046, doi:10.1256/qj.04.85.

——, L. J. Donner, and S. T. Garner, 2007: Nucleation processes in deep convection simulated by a cloud-system-resolving model with double-moment bulk microphysics. *J. Atmos. Sci.*, **64**, 738–761, doi:10.1175/JAS3869.1.

Pinsky, M., and A. P. Khain, 2002: Effects of in-cloud nucleation and turbulence on droplet spectrum formation in cumulus clouds. *Quart. J. Roy. Meteor. Soc.*, **128**, 501–533, doi:10.1256/003590002321042072.

Prabha, T. V., A. Khain, R. S. Maheshkumar, G. Pandithurai, J. R. Kulkarni, M. Konwar, and B. N. Goswami, 2011: Microphysics of premonsoon and monsoon clouds as seen from in situ measurements during the Cloud Aerosol Interaction and Precipitation Enhancement Experiment (CAIPEEX). *J. Atmos. Sci.*, **68**, 1882–1901, doi:10.1175/2011JAS3707.1.

——, and Coauthors, 2012: Spectral width of premonsoon and monsoon clouds over Indo-Gangetic valley. *J. Geophys. Res.*, **117**, D20205, doi:10.1029/2011JD016837.

Pruppacher, H. R., and J. D. Klett, 1997: *Microphysics of Clouds and Precipitation.* 2nd ed. Kluwer, 997 pp.

Ramanathan, V., and W. Collins, 1991: Thermodynamic regulation of ocean warming by cirrus clouds deduced from observations of the 1987 El Niño. *Nature*, **351**, 27–32, doi:10.1038/351027a0.

Randall, D. A., Harshvardhan, D. A. Dazlich, and T. G. Corsetti, 1989: Interactions among radiation, convection, and large-scale dynamics in a general circulation model. *J. Atmos. Sci.*, **46**, 1943–1970, doi:10.1175/1520-0469(1989)046<1943:IARCAL>2.0.CO;2.

Reisner, J., R. M. Rasmussen, and R. T. Bruintjes, 1998: Explicit forecasting of supercooled liquid water in winter storms using the MM5 forecast model. *Quart. J. Roy. Meteor. Soc.*, **124**, 1071–1107, doi:10.1002/qj.49712454804.

Rennó, N. O., K. A. Emanuel, and P. H. Stone, 1994: Radiative-convective model with an explicit hydrologic cycle: 1. Formulation and sensitivity to model parameters. *J. Geophys. Res.*, **99**, 14 429–14 441, doi:10.1029/94JD00020.

Riehl, H., and J. S. Malkus, 1958: On the heat balance in the equatorial trough zone. *Geophysica*, **6**, 503–538.

Rosenfeld, D., 1999: TRMM observed first direct evidence of smoke from forest fires inhibiting rainfall. *Geophys. Res. Lett.*, **26**, 3105–3108, doi:10.1029/1999GL006066.

——, and W. L. Woodley, 2000: Deep convective clouds with sustained supercooled liquid water down to −37.5°C. *Nature*, **405**, 440–442, doi:10.1038/35013030.

——, U. Lohmann, G. B. Raga, C. D. O'Dowd, M. Kulmala, S. Fuzzi, A. Reissell, and M. O. Andreae, 2008: Flood or drought: How do aerosols affect precipitation? *Science*, **321**, 1309–1313, doi:10.1126/science.1160606.

Schneider, E. K., and R. S. Lindzen, 1976: A discussion of the parameterization of momentum exchange by cumulus convection. *J. Geophys. Res.*, **81**, 3158–3161, doi:10.1029/JC081i018p03158.

Schumacher, C., and R. A. Houze Jr., 2003: Stratiform rain in the tropics as seen by the TRMM Precipitation Radar. *J. Climate*, **16**, 1739–1756, doi:10.1175/1520-0442(2003)016<1739:SRITTA>2.0.CO;2.

Shine, K. P., and A. Sinha, 1991: Sensitivity of the Earth's climate to height-dependent changes in the water vapour mixing ratio. *Nature*, **354**, 382–384, doi:10.1038/354382a0.

Slawinska, J., W. W. Grabowski, H. Pawlowska, and H. Morrison, 2012: Droplet activation and mixing in large-eddy simulation of a shallow cumulus field. *J. Atmos. Sci.*, **69**, 444–462, doi:10.1175/JAS-D-11-054.1.

Song, X., and G. J. Zhang, 2011: Microphysics parameterization for convective clouds in a global climate model: Description and single-column model tests. *J. Geophys. Res.*, **116**, D02201, doi:10.1029/2010JD014833.

——, ——, and J.-L. F. Li, 2012: Evaluation of microphysics parameterization for convective clouds in the NCAR

Community Atmosphere Model CAM5. *J. Climate*, **25**, 8568–8590, doi:10.1175/JCLI-D-11-00563.1.

Squires, P., 1958: The microstructure and colloidal stability of warm clouds. Part I—The relation between structure and stability. *Tellus*, **10A**, 256–261, doi:10.1111/j.2153-3490.1958.tb02011.x.

Stephens, G. L., S. van den Heever, and L. Pakula, 2008: Radiative–convective feedbacks in idealized states of radiative–convective equilibrium. *J. Atmos. Sci.*, **65**, 3899–3916, doi:10.1175/2008JAS2524.1.

Su, C.-W., S. K. Krueger, P. A. McMurtry, and P. H. Austin, 1998: Linear eddy modeling of droplet spectral evolution during entrainment and mixing in cumulus clouds. *Atmos. Res.*, **47–48**, 41–58, doi:10.1016/S0169-8095(98)00039-8.

Sud, Y. C., and G. K. Walker, 1999: Microphysics of clouds with the Relaxed Arakawa–Schubert Scheme (McRAS). Part I: Design and evaluation with GATE Phase III data. *J. Atmos. Sci.*, **56**, 3196–3220, doi:10.1175/1520-0469(1999)056<3196:MOCWTR>2.0.CO;2.

Sundqvist, H., 1978: A parameterization of non-convective condensation including prediction of cloud water content. *Quart. J. Roy. Meteor. Soc.*, **104**, 677–690, doi:10.1002/qj.49710444110.

——, 1988: Parameterization of condensation and associated clouds in models for weather prediction and general circulation simulation. *Physically Based Modelling and Simulation of Climate and Climatic Change*, M. E. Schlesinger, Ed., Reidel, 433–461.

——, E. Berge, and J. E. Kristjansson, 1989: Condensation and cloud parameterization studies with a mesoscale numerical weather prediction model. *Mon. Wea. Rev.*, **117**, 1641–1657, doi:10.1175/1520-0493(1989)117<1641:CACPSW>2.0.CO;2.

Tao, W.-K., X. Li, A. Khain, T. Matsui, S. Lang, and J. Simpson, 2007: Role of atmospheric aerosol concentration on deep convective precipitation: Cloud-resolving model simulations. *J. Geophys. Res.*, **112**, D24S18, doi:10.1029/2007JD008728.

——, J.-P. Chen, Z. Li, C. Wang, and C. Zhang, 2012: Impact of aerosols on convective clouds and precipitation. *Rev. Geophys.*, **50**, RG2001, doi:10.1029/2011RG000369.

Thompson, G., R. M. Rasmussen, and K. Manning, 2004: Explicit forecasts of winter precipitation using an improved bulk microphysics scheme. Part I: Description and sensitivity analysis. *Mon. Wea. Rev.*, **132**, 519–542, doi:10.1175/1520-0493(2004)132<0519:EFOWPU>2.0.CO;2.

Tiedtke, M., 1989: A comprehensive mass flux scheme for cumulus parameterization in large-scale models. *Mon. Wea. Rev.*, **117**, 1779–1800, doi:10.1175/1520-0493(1989)117<1779:ACMFSF>2.0.CO;2.

Wallace, J. M., and P. V. Hobbs, 2006: *Atmospheric Science: An Introductory Survey.* 2nd ed. Academic Press, 504 pp.

Warner, J., 1969: The microstructure of cumulus cloud. Part II: The effect on droplet size distribution of the cloud nucleus spectrum and updraft velocity. *J. Atmos. Sci.*, **26**, 1272–1282, doi:10.1175/1520-0469(1969)026<1272:TMOCCP>2.0.CO;2.

Wood, R., C. S. Bretherton, D. Leon, A. D. Clarke, P. Zuidema, G. Allen, and H. Coe, 2011: An aircraft case study of the spatial transition from closed to open mesoscale cellular convection. *Atmos. Chem. Phys.*, **11**, 2341–2370, doi:10.5194/acp-11-2341-2011.

Wu, X., and M. Yanai, 1994: Effects of vertical wind shear on the cumulus transport of momentum: Observations and parameterization. *J. Atmos. Sci.*, **51**, 1640–1660, doi:10.1175/1520-0469(1994)051<1640:EOVWSO>2.0.CO;2.

Xu, K.-M., 1995: Partitioning mass, heat, and moisture budgets of explicitly simulated cumulus ensembles into convective and stratiform components. *J. Atmos. Sci.*, **52**, 551–573, doi:10.1175/1520-0469(1995)052<0551:PMHAMB>2.0.CO;2.

——, and S. K. Krueger, 1991: Evaluation of cloudiness parameterizations using a cumulus ensemble model. *Mon. Wea. Rev.*, **119**, 342–367, doi:10.1175/1520-0493(1991)119<0342:EOCPUA>2.0.CO;2.

——, and D. A. Randall, 2001: Updraft and downdraft statistics of simulated tropical and midlatitude cumulus convection. *J. Atmos. Sci.*, **58**, 1630–1649, doi:10.1175/1520-0469(2001)058<1630:UADSOS>2.0.CO;2.

Yanai, M., S. Esbensen, and J.-H. Chu, 1973: Determination of bulk properties of tropical cloud clusters from large-scale heat and moisture budgets. *J. Atmos. Sci.*, **30**, 611–627, doi:10.1175/1520-0469(1973)030<0611:DOBPOT>2.0.CO;2.

Zhang, G. J., and H. R. Cho, 1991a: Parameterization of the vertical transport of momentum by cumulus clouds. Part I: Theory. *J. Atmos. Sci.*, **48**, 1483–1492, doi:10.1175/1520-0469(1991)048<1483:POTVTO>2.0.CO;2.

——, and ——, 1991b: Parameterization of the vertical transport of momentum by cumulus clouds. Part II: Application. *J. Atmos. Sci.*, **48**, 2448–2457, doi:10.1175/1520-0469(1991)048<2448:POTVTO>2.0.CO;2.

——, and N. A. McFarlane, 1995a: Sensitivity of climate simulations to the parameterization of cumulus convection in the Canadian Climate Centre general circulation model. *Atmos.–Ocean*, **33**, 407–446, doi:10.1080/07055900.1995.9649539.

——, and ——, 1995b: Role of convective-scale momentum transport in climate simulation. *J. Geophys. Res.*, **100**, 1417–1426, doi:10.1029/94JD02519.

Zhang, J. H., U. Lohmann, and P. Stier, 2005: A microphysical parameterization for convective clouds in the ECHAM5 climate model: Single-column model results evaluated at the Oklahoma Atmospheric Radiation Measurement Program site. *J. Geophys. Res.*, **110**, D15S07, doi:10.1029/2004JD005128.

Zipser, E. J., M. J. Meitin, and M. A. LeMone, 1981: Mesoscale motion fields associated with a slow moving GATE convective band. *J. Atmos. Sci.*, **38**, 1725–1750, doi:10.1175/1520-0469(1981)038<1725:MMFAWA>2.0.CO;2.

Chapter 13

Characterizing and Understanding Cloud Ice and Radiation Budget Biases in Global Climate Models and Reanalysis

J.-L. F. LI, D. E. WALISER, G. STEPHENS, AND SEUNGWON LEE

Jet Propulsion Laboratory, California Institute of Technology, Pasadena, California

ABSTRACT

The authors present an observationally based evaluation of the vertically resolved cloud ice water content (CIWC) and vertically integrated cloud ice water path (CIWP) as well as radiative shortwave flux downward at the surface (RSDS), reflected shortwave (RSUT), and radiative longwave flux upward at top of atmosphere (RLUT) of present-day global climate models (GCMs), notably twentieth-century simulations from the fifth phase of the Coupled Model Intercomparison Project (CMIP5), and compare these results to those of the third phase of the Coupled Model Intercomparison Project (CMIP3) and two recent reanalyses. Three different *CloudSat* and/or *Cloud–Aerosol Lidar and Infrared Pathfinder Satellite Observations* (*CALIPSO*) combined ice water products and two methods are used to remove the contribution from the convective core ice mass and/or precipitating cloud hydrometeors with variable sizes and falling speeds so that a robust observational estimate can be obtained for model evaluations.

The results show that, for annual mean CIWC and CIWP, there are factors of 2–10 (either over- or underestimate) in the differences between observations and models for a majority of the GCMs and for a number of regions. Most of the GCMs in CMIP3 and CMIP5 significantly underestimate the total ice water mass because models only consider suspended cloud mass, ignoring falling and convective core cloud mass. For the annual means of RSDS, RLUT, and RSUT, a majority of the models have significant regional biases ranging from -30 to 30 W m^{-2}. Based on these biases in the annual means, there is virtually no progress in the simulation fidelity of RSDS, RLUT, and RSUT fluxes from CMIP3 to CMIP5, even though there is about a 50% bias reduction improvement of global annual mean CIWP from CMIP3 to CMIP5. It is concluded that at least a part of these persistent biases stem from the common GCM practice of ignoring the effects of precipitating and/or convective core ice and liquid in their radiation calculations.

1. Introduction

Representing atmospheric convection, precipitating/ nonprecipitating clouds, and their multiscale organization as well as their radiation interaction in GCMs remains a pressing challenge to reduce and quantify uncertainties associated with climate change projections (Randall et al. 2007; Stephens 2005). Atmospheric radiative structures, such as fluxes and the vertical/ horizontal distributions of heating, are one of the most important factors determining global weather and climate. In particular, clouds can exert a strong influence on these radiative structures in regional radiative balance by reflecting shortwave (SW) radiation back to space and trapping longwave (LW) radiation and radiating it back to the surface, providing one of the strongest feedbacks in the climate system. The balance of these fluxes is essential for understanding Earth's climate system and constraining the energy balance for climate models (Stephens 2005).

Global constraints and information for developing and evaluating clouds and radiation in GCM simulations were typically derived from cloud cover observations from the International Satellite Cloud Climatology Project (ISCCP) and related products (e.g., Han et al. 1999; Rossow and Zhang 1995; Rossow and Schiffer 1999) and from radiation budget observations from the Earth Radiation Budget Experiment/Clouds and the Earth's Radiant Energy System (ERBE/CERES)

Corresponding author address: J.-L. F. Li, Jet Propulsion Laboratory, California Institute of Technology, 4800 Oak Grove Dr., Pasadena, CA 91109.
E-mail: juilin.f.li@jpl.nasa.gov

DOI: 10.1175/AMSMONOGRAPHS-D-15-0007.1

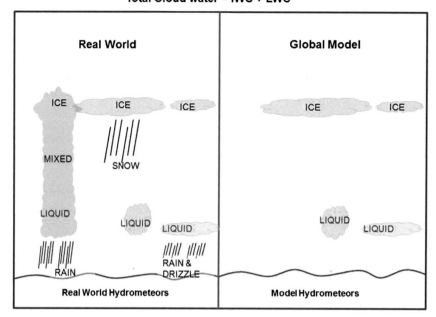

FIG. 13-1. (left) A schematic depiction of real-world cloud hydrometeors and (right) an incomplete representation of cloud hydrometeors in most GCMs. Note that, in some cases, observations are not a perfect representation of these components and cautious/mindful considerations have to be made to use them.

(Wielicki et al. 1996). In the last decade, the first satellite simulator was available for the ISCCP (Rossow and Schiffer 1999) to serve for evaluation and intercomparison of climate model clouds (e.g., Norris and Weaver 2001; Lin and Zhang 2004; Zhang et al. 2005; Schmidt et al. 2006; Cole et al. 2011; Kay et al. 2012). More recently, the Cloud Feedback Model Intercomparison Project (CFMIP) (e.g., Bony et al. 2011) has been coordinating development of the CFMIP Observation Simulator Package (COSP) and includes a number of new satellite observations from the Multiangle Imaging SpectroRadiometer (MISR), Moderate Resolution Imaging Spectroradiometer (MODIS), *CALIPSO*, and *Polarization and Anisotropy of Reflectances for Atmospheric Sciences coupled with Observations from a Lidar* (*PARASOL*). COSP has been used widely to understand and quantify climate model cloud biases (e.g., Chepfer et al. 2008; Bodas-Salcedo et al. 2008, 2011; Zhang et al. 2010; Kay et al. 2012; Kodama et al. 2012; Nam and Quaas 2012).

A key step of obtaining an accurate top of atmosphere (TOA) and surface radiation budget is the representation of clouds; for GCMs, TOA balance is often gotten by tuning models' TOA radiative fluxes toward observations through quantities such as cloud cover, cloud particle effective radius, and cloud mass, which have been largely unconstrained because of the lack of observations for cloud water mass and particle size. This is

especially the case for the vertical structure information of cloud water mass leaving too many degrees of freedom unconstrained. The recent availability of the first tropospheric vertically resolved cloud radar reflectivity and derived ice/liquid profiles from *CloudSat* (Austin et al. 2009), combined with *CALIPSO* (Deng et al. 2010, 2013; Delanoë and Hogan 2008, 2010), provide new means for global cloud mass evaluation (e.g., Chepfer et al. 2008; Li et al. 2012, 2013; Waliser et al. 2009; Chen et al. 2011; Jiang et al. 2012; Gettelman et al. 2010; Klein and Jakob 1999; Webb et al. 2001; Delanoë and Hogan 2008, 2010; Delanoë et al. 2011; Bodas-Salcedo et al. 2008, 2011; Zhang et al. 2010; Kay et al. 2012; Kodama et al. 2012). Among those exploring this issue, Li et al. (2011, 2012, 2013) and Waliser et al. (2009, 2011) strive to point out that considerable care and caution are required in order to make judicious comparisons/ interpretations regarding atmospheric liquid/ice and its associated interactions with radiation. This is because most GCMs typically only represent the "suspended" hydrometeors associated with some/most clouds, while satellite observations include both clouds and falling hydrometeors (e.g., rain or snow) as well as convective core cloud mass, as illustrated in Fig. 13-1. Note that the observations from sensors such as the *CloudSat* Radar and the CERES instruments are sensitive to a broader range of particles for ice/liquid water mass, including clouds, falling snow/rain, and convective core water mass. In contrast, most GCMs, including all CMIP3 models and most of the models in the CMIP5, only model the radiation impacts from the cloud-related hydrometeors and, in some cases, not even all the clouds (e.g., deep convection). Given that most models from CMIP3 and CMIP5, for example, significantly underestimate (or do not explicitly model all) the total water mass, this may result in possible biases in the radiation fields. An observation-based modeling study by Waliser et al. (2011) led to the hypothesis that the typical practice of ignoring the impacts of precipitating hydrometeors would account for at least a portion of this systematic bias. As the persistence of this practice continues, it is worth examining if the same systematic bias might be evident in CMIP5.

In this chapter, we highlight the recent evaluations of the model representations of cloud ice water content (CIWC) and cloud ice water path (CIWP) from a number of studies performed in recent years on cloud ice (e.g., Li et al. 2005, 2007, 2008, 2012; Waliser et al. 2009; Chen et al. 2011) and the radiation budget (e.g., Li et al. 2013). This includes a measure of observational uncertainty, and the illustrative and quantitative set of evaluation diagnostics for the fidelity of the models may have changed between CMIP3 and CMIP5. We then discuss systematic radiation budget biases in CMIP3 and CMIP5, in particular, by evaluating the radiative shortwave flux downward at the surface (RSDS) at the surface and reflected shortwave (RSUT) and radiative longwave flux upward at TOA (RLUT) in examining their biases in light of the ice water biases.

The model simulations considered in this study are from twentieth-century CMIP3 and CMIP5 simulations as well as the NASA Goddard Earth Observing System version 5 (GEOS5) AGCM with prescribed sea surface temperatures (SSTs) and Modern-Era Retrospective Analysis for Research and Applications (MERRA) data if available. Observation-based reference data for the TOA fluxes is derived from contemporary satellite radiation measurements, while surface fluxes are derived from satellite-constrained model calculations using a radiative transfer model.

In section 2, we describe the observational resources for cloud water (IWC/IWP), including the way the different retrievals and other methodologies are combined to form a robust observational estimate with some quantitative information on uncertainty as well as the observed and derived radiative fluxes at TOA and at surface, respectively. In section 3, we briefly describe the models and reanalyses datasets utilized in this evaluation study. In section 4, we illustrate and discuss the results of our model evaluation. Section 5 summarizes and draws conclusions.

2. Observed cloud water and radiation

a. Observed estimates of IWC and IWP

The A-Train constellation of satellites, which includes *CloudSat* and *CALIPSO* flying only 15 s apart, provides a global view of the vertical structure of clouds, including cloud condensate, such as IWC. *CloudSat* provides vertical profiles of radar reflectivity measured by a 94-GHz cloud profiling radar (CPR) with a minimum sensitivity of $\sim -30\,$dBZ. The profiles extend between the surface and 30-km altitude with a vertical resolution of 240 m and have a footprint of about 2.5 km along track and 1.4 km across track. The *CALIPSO* lidar measures parallel and perpendicular backscattered laser energy at 532 nm and total backscattering at 1064 nm at altitude-dependent vertical resolutions and footprints (75 m vertically with about a 0.3-km along-track footprint above 8.2 km and 30 m vertically with about a 1.0-km along-track footprint below 8.2 km). To date, a series of retrieval algorithms using either *CloudSat* radar or *CALIPSO* lidar or both provide global retrievals of IWC, effective radius (Re), and the extinction coefficient from the thinnest cirrus (seen only by the lidar) to the thickest ice cloud (Austin and Stephens 2001; Hogan 2006; Delanoë and Hogan 2008, 2010; Mace et al. 2009; Young and Vaughan 2009; Sassen et al. 2009; Deng et al. 2013; Stein et al. 2011).

There are three IWC and IWP products retrieved from three different algorithms available from *CloudSat* CPR data combined with other satellites' data that can be used to help account for observational uncertainty. They are as follows:

(i) 2B-CWC-RO4 (Austin et al. 2009): The *CloudSat* Science Team level-2B radar-only cloud water content product (2B-CWC-RO4) provides estimates of IWC and Re using measured radar reflectivity from *CloudSat* 2B-GEOPROF to constrain the retrieved IWC. The retrieved IWC profiles are obtained by assuming constant ice particle density with a spherical shape and a lognormal particle-size distribution (PSD). An a priori PSD is specified based on its temperature dependencies obtained from European Centre for Medium-Range Weather Forecasts (ECMWF) operational analyses. The cloud water contents for both liquid and ice phases are retrieved for all heights using separate assumptions. Then a composite profile is created by using the retrieved ice properties at temperatures colder than $-20°$C, the retrieved liquid water content at temperatures warmer than $0°$C, and a linear combination of the two in intermediate temperatures. This reduces the total

IWC as the temperature approaches 0°C. The sensitivity and uncertainty of this retrieval algorithm are discussed in Austin et al. (2009). The time period of this dataset is from January 2007 to December 2010. The vertical and horizontal resolutions are the same as the *CloudSat* instrument discussed above.

(ii) DARDAR (Hogan 2006; Delanoë and Hogan 2008, 2010): DARDAR is a synergistic ice cloud retrieval product derived from the combination of the *CloudSat* radar and *CALIPSO* lidar using a variational method for retrieving profiles of the extinction coefficient, IWC, and Re of the ice cloud. DARDAR assumes a unified PSD given by Field et al. (2005). The mass-size and area-size relations of nonspherical particles are considered using in situ measurements (Brown and Francis 1995; Francis et al. 1998; Delanoë et al. 2011; Stein et al. 2011). For DARDAR, *CALIPSO* backscatter and temperature were used to find supercooled water in the 0° to −40°C range, while the depolarization is too noisy to use at the *CALIPSO* resolution (Delanoë and Hogan 2010). The time period of this dataset is from July 2006 to June 2009.

(iii) 2C-ICE (Deng et al. 2013): Similar to DARDAR, the *CloudSat* level-2C ice cloud property product (2C-ICE) is a synergistic ice cloud retrieval derived from the combination of the *CloudSat* radar and *CALIPSO* lidar using a variational method for retrieving profiles of the extinction coefficient, IWC, and Re in ice clouds. The *CALIPSO* attenuated backscattering coefficients are collocated to the *CloudSat* vertical and horizontal resolutions. The ice cloud microphysical model assumes a first-order Gamma particle-size distribution of idealized nonspherical ice crystals (Yang et al. 2000). The Mie scattering of radar reflectivity is calculated in a forward model lookup table according to a discrete dipole approximation calculation (Hong 2007). The 2C-ICE cloud identification is provided by the *CloudSat* cloud classification (CLDCLASS)-lidar product, which takes advantage of *CALIPSO* lidar backscatter (sensitive to water clouds), lidar depolarization (sensitive to nonspherical ice particles), and *CloudSat* radar (sensitive to large ice particles). Readers desiring a more in-depth description of the 2C-ICE algorithm should refer to Deng et al. (2010) for details. The time period of this dataset is from January 2007 to December 2008.

There is one important aspect to keep in mind regarding model and observation compatibility. All three products, to first order, represent total tropospheric ice, including "floating" ice and the precipitating cloud hydrometeors with variable size and falling speed, as the measurements are sensitive to a wide range of particle sizes, including small (quasi-suspended/cloud) particles and large (falling/precipitating) particles. The latter, including those particles associated with convective clouds, are generally not included as prognostic variables (see Fig. 13-1) in most GCMs (e.g., Li et al. 2008, 2011, 2012, 2013; Waliser et al. 2009). It is generally assumed that convective core areas in a GCM grid box are small for a GCM with a gridbox size that is commonly larger than 100 km². Thus, its contribution to total water content is not very large. Even though they are prognostically determined, the relative contribution does not change. However, the gridbox resolution in most current state-of-the-art GCMs is much higher, with gridbox sizes smaller than 100 km² or less so that their IWC contribution from the convective core should be considered. Thus, for a meaningful model–observation comparison between the satellite-estimated and model-simulated IWC, an estimate of the convective/precipitating ice mass needs to be removed from the satellite-derived IWC/IWP values.

Two independent approaches to distinguish ice mass associated with clouds from ice mass associated with precipitation and convection follow:

1) FLAG method (Li et al. 2008, 2012; Waliser et al. 2009): All the retrievals in any profile that are flagged as precipitating at the surface and any retrieval within the profile whose cloud type is classified as deep convection or cumulus (from *CloudSat* 2B-CLDCLASS data) are excluded. By excluding these portions of the ice mass, an estimate of the cloud-only portion of the IWP/IWC (CIWP/CIWC) is obtained. This methodology of estimating CIWP/CIWC was used in model–data comparisons in many studies (e.g., Li et al. 2008, 2012; Waliser et al. 2009; Chen et al. 2011; Gettelman et al. 2010; Song et al. 2012; Donner et al. 2011; Ma et al. 2012).

2) PSD method (Chen et al. 2011): The ice PSD parameters associated with each *CloudSat* retrieval to separate the total IWC into mass with particle sizes smaller and larger than a selected particle-size threshold are also used. Based on the analysis in Chen et al. (2011) and references therein, the size separation of cloud ice and precipitating ice on a global mean basis likely falls between 100 and 200 μm in diameter. A threshold of 150 μm is chosen for the present study, and the integrated mass of particles with diameter smaller than this size is considered representative of the CIWC/CIWP. In

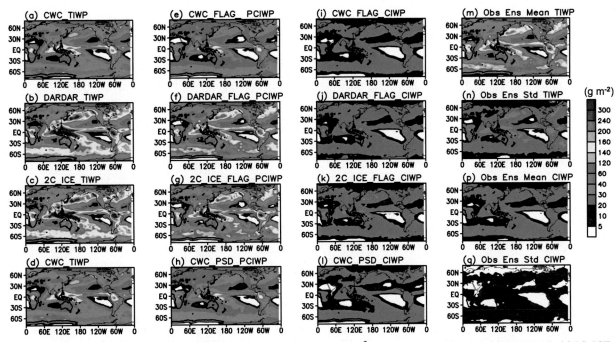

FIG. 13-2. (a)–(d) Annual mean maps of TIWP (cloud + precipitating) (g m^{-2}) estimated from (a) CWC, (b) DARDAR, (c) 2C_ICE, and (d) CWC [repeated from (a)] satellite products. (e)–(h) As in (a)–(d), but for estimates of the precipitating ice water path (PIWP). (e)–(g) Maps of this estimate calculated based on surface precipitation and/or convective cloud flags and (h) using a method based on the PSD and a 150-mm cutoff. Further details and references for these methods are given in section 2. (i)–(l) As in (e)–(h), but for estimates of the CIWP. (m) Ensemble mean and (n) standard deviation of (a)–(d); (o) ensemble mean and (p) standard deviation of (i)–(l). (m),(o) Observed estimates of TIWP and CIWP used in this study.

this case, such estimates are based on a quantitative, microphysical characterization (i.e., PSD) regardless of the presence of surface precipitation or cloud type; thus, the vertical distributions of cloud ice versus precipitating ice mass can be derived from each *CloudSat* profile. The CIWC derived by this method has been shown to agree well with estimates based on the FLAG method (Li et al. 2012), and these CIWC have been applied to evaluate the atmospheric ice in the ECMWF IFS and the NASA Goddard Multiscale Modeling Framework (fvMMF) GCM (Chen et al. 2011).

It should be underscored that, with present satellite/retrieval technology, it is not possible to absolutely separate floating/cloudy forms of ice from falling/precipitating forms, yet models often try to make this distinction. Specific retrievals of this sort will require collocated vertical velocity information, such as from a Doppler radar capability, and/or multiple frequency radar to better characterize particle size.

To account for observational uncertainty Li et al. (2012) produce four different estimates of cloud ice (i.e., CIWP/CIWC) from the three retrieval products and two precipitation/convection filtering methods described above. These include the FLAG method applied to all

three of the retrieval products as well as the PSD method applied to the 2B-CWC-RO4 product. The ensemble mean of these four estimates as the "observed" or "reference" values (herein referred to as such) and the spread between the four estimates can be used as a measure of observational uncertainty.

Figure 13-2 shows multiple annual mean maps of IWP quantities associated with our observational estimates. The four columns represent estimates of total ice (TIWP; Figs. 13-2a–d), precipitating and convective ice (PCIWP: Figs 13-2e–h), CIWP (Figs. 13-2i–l), and ensemble information (Figs. 13-2m–q). Overall, it is evident that the cloud ice (Figs. 13-2i–l) represents a smaller contribution to the total ice mass (Figs. 13-2a–d) than the precipitating/convective contribution (Figs 13-2e–h), ranging from 10%–30% depending on the product and location. It accounts for a smaller contribution in the two radar + lidar products (Figs. 13-2b,f,j,n and Figs. 13-2c,g,k,p) and in the tropics and storm-track regions in all products. In general, the CIWP estimates typically agree relatively well, typically within a factor of two, and most of the differences can be explained by the different microphysical assumptions (Deng et al. 2013; Delanoë et al. 2011). The difference in mass-size–area-size relations and the cloud occurrence identified

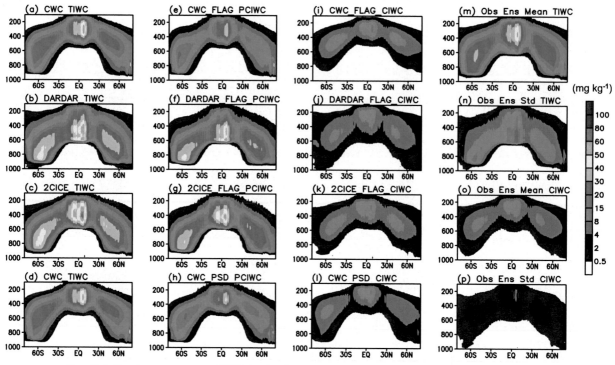

FIG. 13-3. Zonal mean averages of IWC (mg kg^{-1}) for each of the datasets and methods shown in Fig. 13-2.

criteria between *CloudSat* 2C-ICE and DARDAR contributes to some subtle differences between those two datasets (Deng et al. 2013). Figures 13-2m and 13-2n show the ensemble mean and standard deviation of the three observational estimates of TIWP (Figs. 13-2a–c), while the same is shown in Figs. 13-2o and 13-2p for the four estimates of CIWP. It is the latter two maps, which are based on the four individual estimates in the third column of Fig. 13-2, that represent the observational basis for CIWP and the GCM evaluations in this study.

Figure 13-3 is similar to Fig. 13-2, but displays the data as zonally averaged annual mean IWC as a function of height, rather than as vertically integrated IWP. The general commonalities and differences between the products and different filtering methods are the same as described above for CIWP and Fig. 13-2. Apart from that, the overall vertical structure of IWC in each exhibits three local maxima; one is in the tropics near 300 hPa and is associated with deep convection, and the other two are at 800 and 700 hPa in the midlatitude of the Southern and Northern Hemisphere, respectively, and correspond to the storm tracks. In general, these maxima tend to be lower in the two radar + lidar products (Figs. 13-3b,f,j,n and Figs. 13-3c,g,k,o) and higher in the radar-only products (Figs. 13-3a,e,i,m and Figs. 13-3d,h,l,p). Moreover, there is a tendency for the maxima to be higher in the CIWC profiles (Figs. 13-3i–l) compared to the PCIWC profiles (Figs. 13-3e–h),

particularly in the midlatitudes, which has some intuitive merit as the larger, precipitating particles with larger falling velocity present in the cloud(s) present at lower altitudes. As with Fig. 13-2, the four estimates of CIWC (Figs. 13-3i–l) and their ensemble information (Figs. 13-3o,p) are used to evaluate the model/reanalysis representations of CIWC. The information in Figs. 13-3m and 13-3n can be used to compare with two GCMs [i.e., the GFDL atmospheric general circulation model (AM3) and CM3] examined in this study that provide outputs of TIWC/TIWP (details see section 3).

b. Radiation data sources

The most up-to-date source of the RLUT and RSUT fluxes is the CERES Energy Balanced and Filled (EBAF) product (CERES_EBAF-TOA_Ed2.6r) (Loeb et al. 2012, 2009). The CERES EBAF product includes the latest instrument calibration improvements, algorithm enhancements, and other updates. CERES TOA SW and LW fluxes in the EBAF product are adjusted within their range of uncertainty to remove the inconsistency between average global net TOA flux and heat storage in the earth–atmosphere system, as determined primarily from ocean heat content anomaly (OHCA) data [see supplementary information in Loeb et al. (2012) for more details]. Note, the RLUT and RSUT fluxes used for direct model evaluation are CERES-EBAF, which is directly measured and adjusted/balanced with global energy.

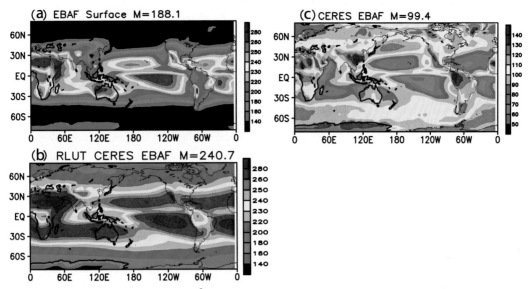

FIG. 13-4. Annual mean maps ($W\,m^{-2}$) of RSDS estimated from satellite estimates of (a) EBAF-Surface; (b) RLUT; (c) as in (b), but for RSUT from satellite measurement of CERES EBAF, which are used for the model evaluations presented in this chapter. Variable M represents the global area mean value.

The sources of RSDS are available from the EBAF-Surface product. This surface flux radiation product is constrained by TOA CERES–derived flux with EBAF adjustments (Kato et al. 2010, 2011, 2013, 2012) and is based on two CERES data products. Edition 3-lite SYN1deg-Month provides computed irradiances to be adjusted, and EBAF Ed2.6r (Loeb et al. 2009, 2012) provides the constraint. In addition, temperature and humidity profiles used in the computations are from the GEOS Data Assimilation System reanalysis (GEOS-4 and 5). MODIS-derived cloud properties (Minnis et al. 2011) are combined with geostationary satellite (GEO)-derived cloud properties (Minnis et al. 1994) to resolve the diurnal cycle used in the SYN1deg-Month flux computations. Note that, unlike TOA irradiances, the global estimate of irradiance at the surface is only possible by using a radiative transfer model. The errors in cloud and atmospheric properties used as inputs, therefore, directly affect the accuracy and stability of modeled surface irradiances (Kato et al. 2013). In addition, model-computed TOA irradiances do not necessarily agree with CERES-derived observed TOA irradiances. Therefore, to mitigate these problems, computed fluxes in the EBAF-Surface product are constrained to be consistent with CERES-derived observed TOA fluxes to within their uncertainties. In the constraining process, *Cloudsat* radar- and *CALIPSO* lidar-derived cloud vertical profiles (Kato et al. 2010), as well as Atmospheric Infrared Sounder (AIRS)-derived temperature and humidity profiles, are used to determine the uncertainty in cloud and atmospheric properties.

Figure 13-4a shows the annual mean maps of RSDS estimated from EBAF-Surface, while Fig. 13-4b shows annual mean maps of RLUT ($W\,m^{-2}$) from CERES EBAF. Shown in Fig. 13-4c is an annual mean map of RSUT from CERES EBAF. The data is the monthly mean, collected from January 2000 to December 2010. For more details on the observed radiation reference datasets and their uncertainties, readers are referred to Li et al. (2013) and Kato et al. (2013, 2012) for RSDS and Loeb et al. (2012) for RSUT and RLUT.

3. Modeled values of cloud water and radiative fluxes

Using the observations described in the previous section, we present results from the evaluation of the CIWP/CIWC and radiative fluxes in Li et al. (2013) in reanalysis datasets, including ECMWF [ERA-Interim; Dee et al. (2011)] and NASA MERRA, coupled atmosphere–ocean GCMs (CGCMs) from the CMIP3 (for CIWP only), CGCMs from CMIP5, and two additional state-of-the-art GCMs, including the University of California, Los Angeles, (UCLA) CGCM (Ma et al. 2012) and the NASA GEOS5 GCM. The CMIP3 simulations are the same as those described in Waliser et al. (2009). Note that CMIP3 model output did not include CIWC. The models used for CMIP5 simulations are listed in Table 13-1. The outline of cloud microphysics parameterizations and SW/LW radiation parameterizations used in the selected CMIP5 models, as well as in the UCLA CGCM, GEOS5, MERRA, and

TABLE 13-1. Model label, resolution, institution, and model name for the CMIP5 GCMs examined in this study. See section 3 for more details.

Model Label	Model resolution	Institution/model name
BCC_CSM1	64 × 128 × 26	Beijing Climate Center, China/BCC_CSM1.1
BCC	64 × 128 × 26	Beijing Climate Center, China/ BCC_CSM1.1 ESM
CanESM2	64 × 128 × 35	Canadian Centre for Climate Modeling and Analysis, Canada/CanESM2
CCSM4	288 × 192 × 26	National Center for Atmospheric Research, United States of America/CCSM4
CNRM-CM5	128 × 256 × 17	Centre National de Recherches Météorologiques, France/CNRM-CM5
CSIRO	96 × 192 × 18	Australian Commonwealth Scientific and Industrial Research Organization, Australia/CSIRO Mk3.6.0
GISS-E2-H	90 × 144 × 29	NASA/Goddard Institute for Space Studies, United States of America/GISS-E2-H
GISS-E2-R	90 × 144 × 29	NASA/Goddard Institute for Space Studies, United States of America/GISS-E2-R
HadGEM2-ES	145 × 192 × 38	Met Office Hadley Centre for Climate Change, United Kingdom/HadGEM2-ES
INM-CM4.0	120 × 180 × 21	Institute of Numerical Mathematics, Russia/INM-CM4.0
INM-CM4.0 ESM	120 × 180 × 21	Institute of Numerical Mathematics, Russia/INM-CM4.0 ESM
IPSL	96 × 96 × 39	L'Institute Pierre-Simon Laplace, France/ IPSL-CM5A-LR
MIROC	64 × 128 × 80	National Institute for Environmental Studies, University of Tokyo, and JAMSTEC, Japan/MIROC-ESM-CHEM
MIROC4h	320 × 640 × 56	National Institute for Environmental Studies, University of Tokyo, and JAMSTEC, Japan/MIROC4h
MIROC5	128 × 256 × 40	National Institute for Environmental Studies, University of Tokyo, and JAMSTEC, Japan/MIROC5
MRI-CGCM3	160 × 320 × 35	Meteorological Research Institute, Japan/MRI-CGCM3
NorESM	96 × 144 × 26	Norwegian Climate Centre, Norway/NorESM1-M
MPI	96 × 192 × 25	Max Planck Institute for Meteorology/MPI-ESM-LR
GFDL CM3	90 × 144 × 40	NOAA GFDL/GFDL CM3
CESM1	192 × 288 × 30	DOE–National Science Foundation–NCAR/CESM1(CAM5)

ERA-Interim models, are available in Li et al. (2012) and Li et al. (2013).

Unlike all other models examined, which do not include ice mass from convective-type clouds in their CIWC, the two GFDL models include grid means over shallow cumulus, deep cumulus cells, and convective mesoscale clouds, weighted by their respective area fractions. In the GFDL CM3, precipitating ice, however, that has fallen out of large-scale stratiform clouds and into clear areas is not included. Thus, the GFDL models should be considered somewhat carefully with respect to the others, as they are including cloud mass from clouds whose contributions have been typically ignored, and their IWC/IWP fields would be more commensurate with TIWC/TIWP. In the CSIRO, diagnostic falling precipitation is considered while the convective-type clouds of cloud hydrometeors are not included. Thus, the CSIRO model should somewhat be considered between the cloud-only and total ice water content/path. For both the GCM and observational datasets, all fields have been regridded to 40 levels (with a constant pressure interval of 25 hPa) and mapped onto common 8° × 4° longitude by latitude grids.

The specific experimental scenario in CMIP5 is the historical twentieth-century simulation, which used observed twentieth-century greenhouse gas, ozone, aerosol, and solar forcing. The time period used for the long-term mean is 1970–2005, and if a model provided an ensemble

of simulations, only one of them was chosen for this evaluation. For both the GCM and observational datasets, all fields have been regridded and mapped onto common 2° × 2° latitude by longitude grids for radiative fluxes and mapped onto common 8° × 4° latitude by longitude grids for IWC/IWP because of the narrow track of *CloudSat–CALIPSO*.

4. Characterizing and understanding cloud ice and radiation budget biases

a. Biases of modeled IWC and IWP

Figure 13-5 shows the long-term annual mean spatial distributions of simulated values of CIWP from the 15 CMIP5 CGCMs (see Table 13-1), the multimodel ensemble mean from the 15 CMIP5 models (Fig. 13-5p), GEOS5 (Fig. 13-5s), UCLA CGCM (Fig. 13-5t), and the two analyses ERA-Interim (Fig. 13-5u) and MERRA (Fig. 13-5v), as well as the ensemble mean (Fig. 13-5y) and standard deviation (Fig. 13-5z) of the four observed estimates of CIWP discussed above. Overall, the multimodel mean CMIP5 CIWP values are spatially similar to observations but nonetheless are biased high. Individually, most models tend to qualitatively capture the global and regional CIWP patterns. This includes the relatively high values of CIWP in the intertropical convergence zone (ITCZ), South Pacific convergence zone (SPCZ), warm pool, and storm tracks from the

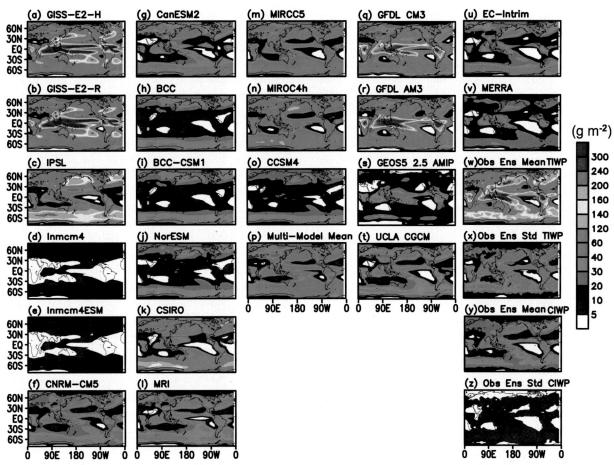

FIG. 13-5. (a)–(o) Annual mean maps of IWP (g m^{-2}) from the twentieth-century GCM simulations that contributed to the CMIP5; (p) their multimodel mean; (q) GFDL CM3 TIWP; (r) GFDL AM3 TIWP; (s) NASA GEOS5 2.5 AMIP; (t) UCLA CGCM; (u) ERA-Interim; and (v) MERRA, as well as the (w) ensemble-mean; (x) standard deviation of three TIWP observed estimates; (y) ensemble-mean and observed estimate; and (z) standard deviation of four CIWP-observed estimates (see section 2).

subtropics to high latitudes and over convectively active continental areas over central Africa and South America. Note that the relative magnitudes between tropical and midlatitude values can be quite different across models (this will be more evident when discussing Fig. 13-8 below). About three of the CMIP5 models do a good job at representing both the observed patterns and magnitudes of CIWP (i.e., CNRM-CM5, CanESM2, MRI). A number of models, however, significantly (~factor of 2) underestimate tropical CIWP (i.e., NorESM, BCC, BCC_CSM1, CCSM4) and two severely (~factor of 10) underestimate CIWP (i.e., INM-CM4.0 and INM-CM4.0-ESM). The two GISS GCMs greatly overestimate (~factor of 5) tropical CIWP. The IPSL, CSIRO, MIROC5, MIROC4h, and the two GISS GCMs moderately overestimate CIWP in the extratropics. For the non-CMIP5 GCMs, the GEOS5 AGCM significantly underestimates (~factor of 3) CIWP in the storm tracks, while the UCLA CGCM does remarkably well over most

of the globe. The two reanalyses, ECMWF and MERRA, show relatively good CIWP patterns and magnitudes, with MERRA being biased a bit low in midlatitudes which is not surprising given that the base model (GEOS5) exhibits such a strong negative bias. While the above model–observation differences are still substantial in many regards, it is worth noting that the ensemble of CMIP5 CIWP values examined here appears to exhibit improvement compared to the ensemble CMIP3 models evaluated in our previous study (Li et al. 2012, appendix Fig. A1); this will be discussed and quantified further below. The two GFDL models (Figs. 13-5q,r) that simulate and provide output on TIWC each exhibit fairly good TIWP in the tropical ITCZ, warm pool, and convectively active continental regions but significantly underestimate TIWP in the extratropics storm-track regions compared to ensemble mean TIWP shown in Fig. 13-5w.

To further quantify and synthesize the comparative information discussed above, we can use a Taylor diagram (Taylor 2001) to summarize both the degree of agreement in overall CIWP spatial pattern correlations along with the standard deviation among the CMIP5 CGCMs, including their multimodel mean, two analyses, three other GCMs, and four observed CWIP estimates. The ensemble mean of the latter is used as the reference dataset and their spread to help quantify observational uncertainty. The Taylor diagram relates three statistical measures of model fidelity: the centered root-mean-square error, the spatial correlation, and the spatial standard deviations. These statistics are calculated for the long-term time mean and over the global domain (area weighted). The reference dataset is plotted along the x axis at the value 1.0.

Figure 13-6 shows Taylor diagrams for CMIP3 (Fig. 13-6a) and CMIP5 (Fig. 13-6b). The observed estimates are plotted in blue, the CMIP GCMs in red, their ensemble means in green, and the reanalyses and non-CMIP GCMs are in black. The red rectangular-like region illustrates a measure of observational uncertainty developed and shown in conjunction with Fig. 13-2. Not surprisingly, the four individual observed estimates, reanalyses, and AGCM simulations (i.e., specified SST; GEOS5 version 2.5) perform as a group considerably better than the CMIP coupled GCMs. The former all tend to have correlations at around 0.9 or better and standard deviation ratios of between about 0.8 and 1.5. For the CMIP values (red), most of them have correlations between about 0.4 and 0.7 with standard deviation ratios well above 1, with some well above 3 and even up to 5. The CMIP3 and CMIP5 multimodel means do not exhibit the best overall performance relative to the individual models because of the few strong outliers in the ensembles. Noteworthy, however is that the CMIP3 and CMIP5 multimodel means (green) have correlations of 0.54 and 0.76, and standard deviation ratios of about 3.1 and 1.4, respectively, indicating a rather considerable performance improvement from CMIP3 to CMIP5 for representing CWIP. While this progress is encouraging, keep in mind also that all models shown still exhibit a very poor correlation against the reference dataset, with values less than 0.8, and none of the CMIP GCMs fall within the (red box) range of observational uncertainty. In regards to details of specific models' performance, readers are referred to Li et al. (2012).

Next, we present the fidelity of the models' CIWC vertical structure. A comparison is given in Fig. 13-7 showing the CIWC zonal and annual mean values from the 13 CMIP5 CGCMs (note that the CNRM-CM5 CGCM CIWC is not available from the CMIP5 data portal at this time), the GEOS5 AGCM (Fig. 13-7q) and the UCLA CGCM (Fig. 13-7p), as well as the ERA-Interim (Fig. 13-7r) and MERRA (Fig. 13-7s). These

models provide output specifically on cloud ice. The two GFDL GCMs, on the other hand, are shown in Figs. 13-7n and 13-7o and provide output for TIWC. Overall, there are significant disparities between the CMIP5 CGCMs against the observed ensemble mean (Fig. 13-7v), with overall discrepancies ranging from multiplicative factors of about 0.25 of the observations (i.e., INM-CM4.0) to factors of 10 (i.e., GISS GCMs). Moreover, the general character of their vertical distributions with respect to pressure levels is considerably different. For example, the IPSL exhibits significant overestimates of CIWC over the storm-track regions. About five of the CMIP5 models do a fair job at representing the vertical structure and magnitude of IWC [i.e., CanESM2 (Fig. 13-7f), BCC_CSM1.1 ESM (Fig. 13-7g), NorESM1 (Fig. 13-7h), MIROC5 (Fig. 13-7k), MRI (Fig. 13-7l), and CCSM4 (Fig. 13-7m)]. The rest of the models [CSIRO (Fig. 13-7i), MIROC4h (Fig. 13-7j), and ERA-Interim (Fig. 13-7o)] generally tend to qualitatively capture the patterns but overestimate CIWC over midlatitudes and below 700 hPa. The GEOS5 model, on the other hand, tends to slightly overestimate CIWC in the tropics but significantly underestimates CIWC in the mid-to-high latitudes by about a factor of 2–3. The analyses from MERRA as well as the simulation from the UCLA CGCM show realistic CIWC vertically with values close to the observed ensemble mean, albeit not extending as close to the surface when compared to the observed estimate. However, it is reasonable to exercise caution when considering the robustness of the observed values in these lower-tropospheric regions or anywhere below the freezing level, as there are artificial limitations applied to the retrievals that involve separating ice from liquid contributions. Compared to the observed TIWC (Fig. 13-7t), the two GFDL models all capture the ITCZ in tropical regions pretty well but significantly underestimate TWIC in the extratropical storm track and polar regions. A realistic ITCZ is found in the GFDL uncoupled AGCM (Fig. 13-7o), while a more notable double ITCZ is evident in the GFDL CGCM (Fig. 13-7n). However, the higher values in midtropospheric tropics in these models relative to the observed value might be due to an unrealistic amount of cloud ice for temperatures above freezing in the tropics.

Apart from gross qualitative agreement with observations among many CMIP5 models, it is still apparent from Figs. 13-5, 13-6, and 13-7 that significant disparities exist not only horizontally but also in the vertical structure.

b. Bias of modeled radiative fluxes

While most GCMs (including all CMIP3 and most of CMIP5) typically only represent the suspended hydrometeors associated with some/most clouds, satellite observations include both clouds and falling hydrometeors

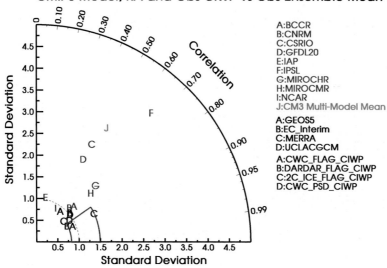

CMIP3 Model, RA and Obs CIWP vs Obs Ensemble Mean CIWP

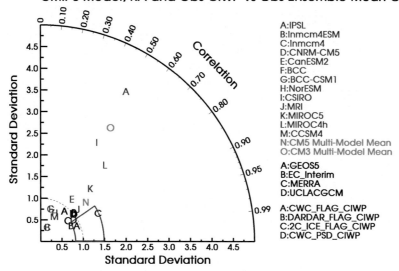

CMIP5 Model, RA and Obs CIWP vs Obs Ensemble Mean CIWP

FIG. 13-6. (top) Taylor diagram for GCM annual mean (80°N–80°S) CIWP (g m^{-2}), where the reference value is the observed estimate shown in Fig. 13-1o. Modeled values include the CMIP3 GCMs (red) shown in Fig. A1 in Li et al. (2012) and their multimodel mean (green), as well as the GEOS5 GCM, UCLA CGCM, and the MERRA and ERA-Interim (black). Models with standard ratios that exceed 5 (GISS-E2-R and GISS-E2-H) are not shown in the Taylor diagram. The red box is defined in Fig. 13-4 and discussed in section 2. (bottom) As in (top), but for the CMIP5 GCMs instead of CMIP3, although the CMIP3 multimodel mean value is also replicated here. Note that the correlation and the standard deviation ratio of the CMIP5 multimodel mean when excluding the two GISS GCMs are 0.614 and 1.236, respectively. When excluding the two GISS and two INM-CM4.0 models, they are 0.645 and 1.09, respectively. For reference, when for including all models, the values are 0.765 and 1.4, respectively.

(e.g., snow) as well as convective core cloud mass (see section 1 and Fig. 13-1). These GCMs, however, have typically tuned their radiation and cloud fields to the observations, which naturally are sensitive to all/most hydrometeors in the atmosphere despite the fact that most of the models typically only represent the suspended

hydrometeors associated with clouds, and usually this does not include ice and liquid in convective cores. Figure 13-8 presents CMIP3 (Fig. 13-8a) and CMIP5 (Fig. 13-8b) multimodel mean biases of TIWP (cloud + convective core + precipitating), where the observed cloud ice estimate is from the ensemble mean of three

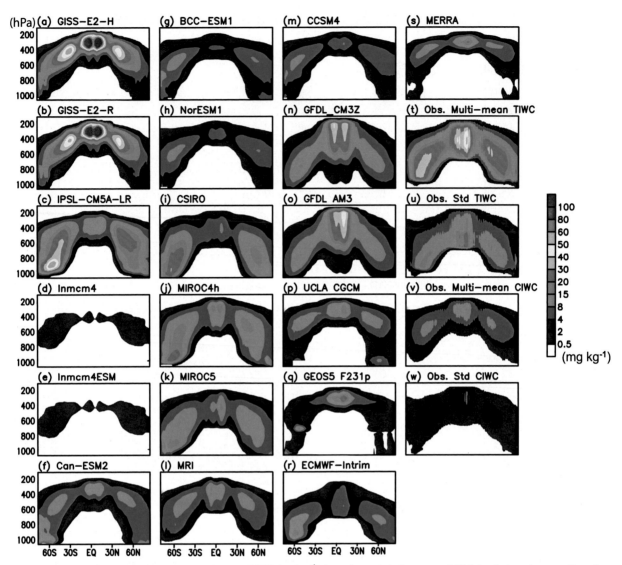

FIG. 13-7. (a)–(m) Zonally averaged annual mean CIWC (mg kg^{-1}) from the twentieth-century GCM simulations that contributed to the CMIP5; TIWC from (n) GFDL CM3 and (o) AM3; CIWC from (p) UCLA CGCM, (q) NASA GEOS5, (r) ERA-Interim, and (s) MERRA, as well as the (t) ensemble-mean and (u) standard deviation of three TIWC-observed estimates and the (v) ensemble-mean and observed estimate (w) standard deviation of four CIWC observed estimates (see section 2). Note that the CNRM-CM5 CGCM CIWC is not available from the CMIP5 data portal at the time.

total ice water path observed estimates from the standard *CloudSat*, a version of DARDAR (Delanoë and Hogan 2008, 2010), and a version of 2C-ICE (Deng et al. 2013) satellite products. The modeled IWP values include only the contributions of suspended clouds, as that is all they typically represent, with no contribution from convective cores or precipitation. The observed IWP values are based on an ensemble of *CloudSat* + *CALIPSO* estimates (Li et al. 2012), which do include contributions from precipitation and all clouds, including convective cores. It shows that both CMIP3 and CMIP5 models significantly underestimate ice mass over the ITCZ, the SPCZ, a part of Southern Ocean

and tropical continents, and the Indian monsoon regions. Thus, contributions from falling/precipitating hydrometeors are unaccounted for and/or erroneously accounted for by other processes, such as interaction with radiation calculation and hydrological cycle, etc. The models are trying to achieve a radiative balance at TOA without representing all the ice mass in the atmosphere (e.g., Waliser et al. 2011; Li et al. 2013). However, the observations from sensors such as the *CloudSat* radar and the CERES instruments are sensitive to a broader range of particles for ice/liquid water mass. It is not surprising, yet it is important to highlight that this missing water mass will have some interaction with

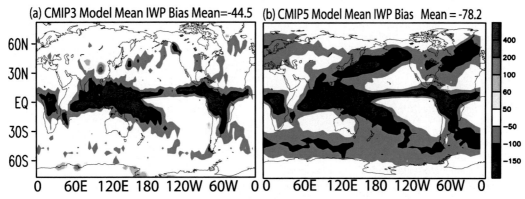

FIG. 13-8. (a) Multimodel mean bias of CIWP (g m^{-2}) from the CMIP3 GCMs against observed estimates of TIWP (cloud + convective core + precipitating IWP); (b) as in (a), but for the multimodel mean bias of CIWP in the CMIP5 GCMs. The observed estimate is from the ensemble mean of the TIWP using the standard *CloudSat* product CWC, a version of DARDAR, and a version of 2C-ICE satellite products. For details, readers are referred to Li et al. (2012, 2013).

radiation. Here, we highlight the radiative biases for CMIP3/CMIP5 in terms of only presenting the multimodel mean biases for each radiative flux. For each individual model's performance in radiative fluxes at TOA and at surface, readers are referred to Li et al. (2013).

Figure 13-9 illustrates the multimodel mean biases (Figs. 13-9a,b) and the multimodel mean standard deviation of the error (SDE), which is defined as a root-mean-square error with the mean bias removed (Figs. 13-9c,d), against the observed estimate, which is calculated

across the models for each of the ensembles. Overall, CMIP3 RSDS (Fig. 13-9a) shows low negative biases globally and more uniformly, except over the ITCZ, off the coast of the Peru/California regions, and over Indian monsoon regions. While the CMIP5 global area average bias (Fig. 13-9b) is reduced (~30%) to the value of 2.5 W m^{-2} from the value of -6.9 W m^{-2} in CMIP3, it exhibits more distinct spatial gradients with greater local extreme biases and a higher bias over most land regions. While the bias figure emphasizes

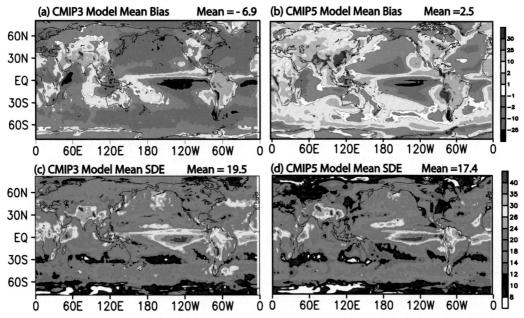

FIG. 13-9. (a) Multimodel mean bias of RSDS (W m^{-2}) for the CMIP3 GCMs shown in Fig. A1 in Li et al. (2012); (b) multimodel mean bias for the CMIP5 GCMs in Fig. 4 in Li et al. (2013); (c) as in (a), but for root-mean-square error; (d) as in (b), but for SDE for CMIP5. In all cases, the observed reference is shown in Fig. 13-4a. Note the mean bias has been removed in the calculation of the SDE.

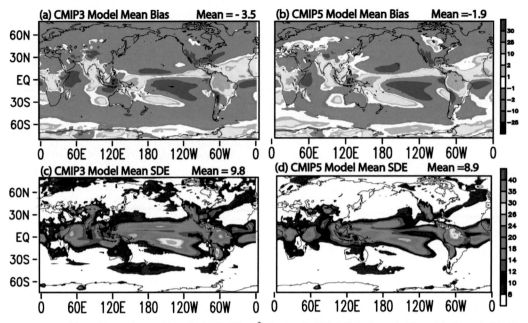

FIG. 13-10. (a) Multimodel mean bias of RLUT (W m^{-2}) for the CMIP3 GCMs shown in Fig. A2 in Li et al. (2013); (b) multimodel mean bias for the CMIP5 GCMs in Fig. 7 in Li et al. (2013); (c) as in (a), but for standard deviation of the error (SDE); (d) as in (b), but for SDE. In all cases, the observed reference is shown in Fig. 13-4b. Note the mean bias has been removed in the calculation of the SDE.

the sign and magnitude of systematic biases across the two model archives, the SDE figure emphasizes errors irrespective of the sign. The similar pattern of the SDE between the two model ensembles indicates that CMIP3 and CMIP5 share many of the same systematic errors in simulating radiation fluxes. The fact that the magnitude of the SDE is about the same between CMIP3 and CMIP5 indicates that little improvement from CMIP3 has been made. The high SDE values in the equatorial regions of the Pacific and Atlantic (Figs. 13-9c,d), combined with the low bias in the same regions (Figs. 13-9a,b), indicates significant disparity in the manner in which models represent surface radiation in these regions. This is similarly true over the mountainous regions of Asia and South America and, to a lesser extent, over the storm tracks.

The multimodel performance of CMIP3 and CMIP5 in representing the time-mean pattern of RLUT shown in Fig. 13-10 illustrates the multimodel mean bias (Figs. 13-10a,b) and the multimodel mean SDE (Figs. 13-10c,d) against the observed estimate. Interestingly, both CMIP3 (Fig. 13-10a) and CMIP5 (Fig. 13-10b) exhibit very similar RLUT bias patterns and magnitudes. Notable is the consistency of having a high bias over the ITCZ, SPCZ, a part of Southern Ocean and tropical continents, and the Indian monsoon regions. The CMIP5 global area average RLUT bias value of -1.9 W m^{-2} is about a factor of 2 smaller compared to the CMIP3 value of -3.5 W m^{-2}. The SDE figure indicates a similar pattern of systematic

errors in the tropics, with no substantial change in the global mean SDE in CMIP3 (9.8 W m^{-2}) and CMIP5 (8.9 W m^{-2}). The bias and SDE together indicate that the models are performing relatively well in the mid and high latitudes but exhibit significant shortcomings in the tropics. Given that all but one of these models is coupled, it is possible that some part of the SDE errors could result from spatial variations in the location of the ITCZ and associated cloud structure that have a substantive impact on the RLUT field.

Figure 13-11 illustrates the multimodel mean biases (Figs. 13-11a,b) and the SDE (Figs. 13-11c,d) of RSUT against the observed estimate. Similar to RSDS and RLUT, the RSUT SDE and bias in both CMIP3 and CMIP5 exhibit very similar patterns, and there is a clear systematic underestimation over ITCZ, SPCZ, and parts of the Southern Ocean, tropical continents, and the Indian monsoon regions. While the CMIP5 global area average RSUT bias is about a factor of 2 smaller (2.5 W m^{-2}) compared to the CMIP3 value of 4.5 W m^{-2}, they share very similar pattern distributions in bias and SDE. The SDE figure indicates a slight improvement of CMIP5 (14.1 W m^{-2}) over CMIP3 (14.7 W m^{-2}).

To illustrate the effect of the underestimated cloud mass due to excluding the cloud mass from precipitation and convective core in the conventional GCMs, we draw a conceptual sketch of cloud–precipitation–radiation

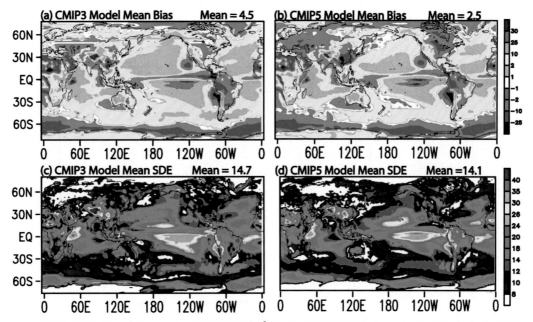

FIG. 13-11. (a) Multimodel mean bias of RSUT (W m^{-2}) for the CMIP3 GCMs shown in Fig. A3 in Li et al. (2013); (b) multimodel mean bias for the CMIP5 GCMs in Fig. 10 in Li et al. (2013); (c) as in (a), but for SDE shown in Fig. 10 in Li et al. (2013); (d) as in (b), but for SDE shown in Fig. 16 in Li et al. (2013). In all cases, the observed reference is shown in Fig. 13-4c. Note the mean bias has been removed in the calculation of the SDE.

interactions for the real world versus those for conventional GCMs in Fig. 13-12. The figure shows the underestimated values of cloud mass in the CMIP3 and CMIP5 multimodel means might directly, and in part, lead to the overestimations of RSDS (Fig. 13-9) and RLUT (Fig. 13-10) and the underestimation RSUT (Fig. 13-11) across the models and multimodel mean in the heavily precipitating regions. This conjecture is supported by Fig. 13-13, which shows that the latitude of the maximum zonal mean precipitation (180°–360°; 0°–15°N) is strongly correlated with the latitude of the maximum/minimum multimodel mean bias of RSDS, RLUT, and RSUT of the CMIP5 models. In other word, the RSDS, RLUT, and RSUT biases in CMIP3 and CMIP5 are attributed to ignoring the radiative impacts from the precipitating clouds and convective core water mass. As cloud–climate feedback will undoubtedly represent a key uncertainty in the next Intergovernmental Panel on Climate Change (IPCC) assessment report, it is essential that cloud and radiation observations be utilized to their full extent and in concert to provide more complete constraints and that clouds, convection, precipitation, and radiation be treated in a consistent manner, as shown in the left panel of Fig. 13-12.

5. Summary and discussion

In this study, with the missing falling particles/convective mass interactions with radiation in mind, we evaluate the representation of surface and TOA fluxes of atmospheric radiation in GCMs, namely CMIP GCMs, with the focus on the fluxes most strongly influenced by clouds (i.e., RSDS, RSUT, and RLUT) associated with heavily precipitating regions. This means that, apart from a general assessment of the fidelity of the models' representation of radiation, we seek to relate the impacts of ignoring the interaction of radiation with precipitating and convective clouds, a common practice in most CGCMs contributing to CMIP3 and CMIP5 [see cloud liquid and ice evaluations in Li et al. (2012)].

We first presented an evaluation of the representation of atmospheric ice and radiation in model simulations of present-day climatology, including those of CMIP5 and their comparison to those of CMIP3. Observational reference values and their uncertainty were addressed by using four different estimates of CIWC/CIWP that accounted for different approaches to the retrieval and to the methods of filtering out the contribution to the ice mass in the retrievals because of large-particle/precipitating components, which is a contribution to the mass that is typically not represented in GCMs as a prognostic or column-resolved quantity (see section 2). The models evaluated included 15 simulations of present-day climate available to date in CMIP5 and two other GCMs of interest (GEOS-5 AGCM and UCLA CGCM). The evaluation also included two modern reanalyses (MERRA and ERA-Interim).

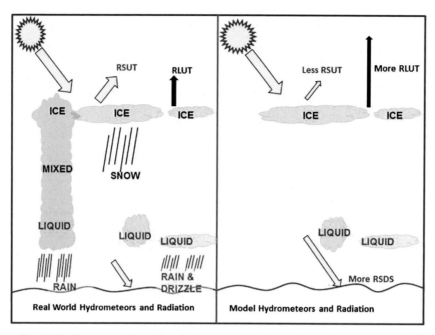

FIG. 13-12. Conceptual illustration of cloud–precipitation–radiation interactions for (left) the real world and for (right) conventional GCMs. The light gray clouds represent floating ice clouds and floating liquid clouds in the real world and the model. The darker gray cloud represents convective cloud core in the real world. The blue (gray) lines represent falling snow (rain) in the real world.

Overall, based on a number of diagnostics, there is a fairly wide disparity in the fidelity of CIWP/CIWC representations in the models examined. Even for the annual mean maps considered, there are easily factors of 2 and nearly up to 10 for the differences between observations and modeled values for most of the GCMs for a number of regions (Figs. 13-5–13-8). As expected the two reanalyses examined performed relatively well compared to the group as a whole because of their use of observed SSTs and the incorporation of a wide array of constraining observations: a result that is still notable though since they do not assimilate cloud ice observations and thus rely on (parameterized) model physics to represent this quantity. However, even with the assimilation of many other/related quantities and the benefit of observed SSTs, neither MERRA's nor ERA-Interim's performance was within the level of uncertainty of the observations for both the standard deviation ratio and pattern correlation. Considering even these results alone and the remaining disparities between the observations and modeled values of CIWP, it is evident that while the models may be providing roughly the correct radiative energy budget at TOA, many are accomplishing it by means of unrealistic cloud characteristics of cloud ice mass at a minimum, which, in turn, likely indicates unrealistic cloud particle sizes and cloud cover.

Examining the overall performance between CMIP3 and CMIP5, based on a number of diagnostics, we find that there has been significant and quantitative improvement in the representation of CIWP between CMIP3 and CMIP5 by the reduction (by about 50% or more) in the multimodel mean bias of the annual mean maps of CIWP from CMIP3 to CMIP5 (Fig. 13-6). Note that the overall assessment of the improvement from CMIP3 to CMIP5 is done with different sets of models because the participating centers/models in CMIP3/CMIP5 are very different. Only six modeling centers have participated in both CMIP3 and CMIP5, and the models from the same modeling centers might be very different from CMIP3 to CMIP5.

The shortcomings of GCMs ignoring the cloud mass associated with precipitating hydrometeors (e.g., snow and rain) and convective core ice and liquid mass, are clearly presented in the CMIP3 (Fig. 13-8a) and CMIP5 (Fig. 13-8b) multimodel mean biases of total ice water path (TIWP = cloud + convective core + precipitating). It is clearly shown in Fig. 13-8 that both CMIP3 and CMIP5 models significantly underestimate ice mass over the ITCZ, the SPCZ, a part of the Southern Ocean and tropical continents, and the Indian monsoon regions, which also tend to all be high-precipitation regions. In addition, Fig. 13-8 suggests the IWP biases against observed total IWP are significantly worse in CMIP5

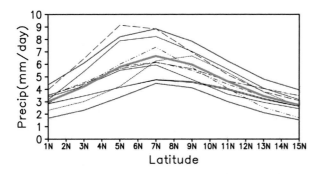

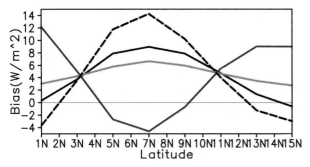

FIG. 13-13. (top) Models and multimodel mean (thick green) values of precipitation (mm day^{-1}) for the CMIP5 GCMs. The red lines represent one standard deviation from the models' population. (bottom) Multimodel mean bias (W m^{-2}) of RLUT (blue), RSDS (dashed black), and RSUT (red) for the CMIP5. In all cases, the observed reference is from CERES EBAF for RLUT and RSUT, while RSDS is from SRB EBAF. Zonal mean is averaged over the longitude section from 180° to 360° and latitude from 0° to 15°N for the ITCZ section.

models compared to CMIP3 models (in the multimodel means), yet the radiative fields are slightly improved overall in terms of global area average shown in Figs. 13-9–13-11.

Observational reference values and their uncertainties for RSDS are addressed by using the EBAF-Surface and CERES TOA radiative fluxes of RLUT and RSUT. Overall, there is a fairly wide disparity in the fidelity of RSDS, RLUT, and RSUT representations in the models examined. Even for the annual mean bias maps considered, there are local biases easily as high (low) as 30 (-30) W m^{-2}, respectively. Based on a number of diagnostics, there has been a small degree of improvement in the representation of RSDS, RLUT, and RSDS from CMIP3 to CMIP5. This is demonstrated, in terms of global mean, by the reduction of RSDS (by about 30%), RLUT (about 50%), and RSUT (about 40%) in the multimodel mean bias. In particular, the multimodel mean bias of RSDS has been reduced from CMIP3 (-6.9 W m^{-2}) to CMIP5 ($\sim$2.5 W m^{-2}). However, this is mostly because of the positive biases over land that became larger while the negative biases over the ocean remained about the same. In addition, an indication of overall improvement in the representation of the

quantities studied from CMIP3 to CMIP5 is not evident when considering the SDE computed across the models (i.e., Figs. 13-9–13-11).

Persistent and systematic spatial pattern biases across most of the models with the multimodel ensemble means values are underestimated in RSUT and overestimated in RSDS and in RLUT in the convectively active regions of the tropics (i.e., ITCZ/SPCZ, warm pool, Indian monsoon, South America, and central Africa) (i.e., Figs. 13-9–13-11). Given that a number of these RSDS, RLUT, and RSUT biases occur in conjunction with heavy precipitation and with biases in cloud liquid and ice (Li et al. 2012), we hypothesize that at least a part of these persistent radiation biases stems from GCMs ignoring the effects of precipitating and/or convective core ice and liquid in their radiation calculations illustrated in Fig. 13-12 (e.g., Waliser et al. 2011; Li et al. 2013).

The fact that viable observed estimates of TOA radiation fields and observation-driven modeled values at the surface have been available for many years and yet the biases are still sizeable suggests challenges to utilizing the observations by the modeling groups or that there are still too many degrees of freedom unconstrained (e.g., cloud cover, cloud mass, particle size, vertical structure, and particle shape). There is certainly evidence of this in regards to cloud liquid and ice content (e.g., Waliser et al. 2009; Li et al. 2012). In addition, GCMs have typically tuned their radiation and cloud fields to the observations, which naturally are sensitive to all/most hydrometeors in the atmosphere despite the fact that most of the models typically only represent the suspended hydrometeors associated with clouds and usually do not include ice and liquid in convective cores. Thus, contributions from falling/precipitating hydrometeors are unaccounted for and/or erroneously accounted for by other processes, such as the calculation of radiation and the hydrological cycle.

While it is beyond the scope of this study to probe the causes of the model-to-model differences and model-to-observation biases in cloud water and radiation, based on the results, we hypothesize that the lack of an explicit representation of the cloudy, precipitating, and convective core components of the ice (and liquid) mass might play an important role for the biases in RSDS, RLUT, and RSUT. Our recent study (Waliser et al. 2011) showed that ignoring radiative effects of the precipitating components of the ice mass can result in nontrivial biases in the shortwave and longwave radiation budgets at the surface and top of atmosphere and even more significant impacts in the vertical radiative heating profile. While more work needs to be pursued in this area, there is a strong suggestion from these studies that GCMs should strive to explicitly represent a broader range of ice and liquid hydrometeors—namely, the larger falling hydrometeors (rain and snow)—as well as convective core mass and include their effects in the radiative heating calculations, which, for the moment, are largely ignored. Moreover, the evaluation results of this study show that the radiation balance in the CMIP class of GCMs is still underconstrained and, in many cases, is likely to have been achieved in unrealistic ways.

Taken together, these points indicate the need for additional observational resources to adequately characterize and constrain cloud–precipitation–radiation interactions. Some potentially useful observational resources are a multichannel radar/lidar measurement to characterize the profile and spectrum of cloud and precipitation particle sizes, as well as a Doppler radar capability to provide information on cloud and precipitation dynamics. In addition, satellite observations are affected by spatiotemporal sampling, instrument sensitivity, and retrieval assumptions. Simulators are one method available to emulate these idiosyncrasies within a climate model and thus can be an invaluable tool for robust evaluation of model-simulated clouds. In the future, we plan to integrate these methodologies into our evaluation studies. The use of these additional observational resources, in conjunction with systematic model experimentation practices, will likely be a constructive strategy for improving the cloud–precipitation–radiation interactions alluded to above.

Acknowledgments. We thank Dr. Bo-Wen Shen for useful comments. This work has been supported in part by AIST-11 JPL Advanced Information Systems Technology. The contributions by DEW, SWL, and JLL to this study were carried out on behalf of the Jet Propulsion Laboratory, California Institute of Technology, under a contract with the National Aeronautics and Space Administration.

REFERENCES

Austin, R. T., and G. L. Stephens, 2001: Retrieval of stratus cloud microphysical parameters using millimeter-wave radar and visible optical depth in preparation for Cloudsat: 1. Algorithm formulation. *J. Geophys. Res.*, **106**, 28 233–28 242, doi:10.1029/2000JD000293.

——, A. J. Heymsfield, and G. L. Stephens, 2009: Retrieval of ice cloud microphysical parameters using the CloudSat millimeter-wave radar and temperature. *J. Geophys. Res.*, **114**, D00A23, doi:10.1029/2008JD010049.

Bodas-Salcedo, A., M. J. Webb, M. E. Brooks, M. A. Ringer, K. D. Williams, S. F. Milton, and D. R. Wilson, 2008: Evaluating cloud systems in the Met Office global forecast model using simulated CloudSat radar reflectivities. *J. Geophys. Res.*, **113**, D00A13, doi:10.1029/2007JD009620.

——, and Coauthors, 2011: COSP: Satellite simulation software for model assessment. *Bull. Amer. Meteor. Soc.*, **92**, 1023–1043, doi:10.1175/2011BAMS2856.1.

Bony, S., M. Webb, C. Bretherton, S. Klein, P. Siebesma, G. Tselioudis, and M. Zhang, 2011: CFMIP: Towards a better evaluation and understanding of clouds and cloud feedbacks in CMIP5 models. *CLIVAR Exchanges*, No. 56, International CLIVAR Project Office, Southampton, United Kingdom, 20–22.

Brown, P. R. A., and P. N. Francis, 1995: Improved measurements of the ice water content in cirrus using a total-water probe. *J. Atmos. Oceanic Technol.*, **12**, 410–414, doi:10.1175/1520-0426(1995)012<0410:IMOTIW>2.0.CO;2.

Chen, W.-T., C. P. Woods, J.-L. F. Li, D. E. Waliser, J.-D. Chern, W.-K. Tao, J. H. Jiang, and A. M. Tompkins, 2011: Partitioning CloudSat ice water content for comparison with upper-tropospheric ice in global atmospheric models. *J. Geophys. Res.*, **116**, D19206, doi:10.1029/2010JD015179.

Chepfer, H., S. Bony, D. Winker, M. Chiriaco, J.-L. Dufresne, and G. Sèze, 2008: Use of CALIPSO lidar observations to evaluate the cloudiness simulated by a climate model. *Geophys. Res. Lett.*, **35**, L15704, doi:10.1029/2008GL034207.

Cole, J., H. W. Barker, N. G. Loeb, and K. von Salzen, 2011: Assessing simulated clouds and radiative fluxes using properties of clouds whose tops are exposed to space. *J. Climate*, **24**, 2715–2727, doi:10.1175/2011JCLI3652.1.

Dee, D. P., and Coauthors, 2011: The ERA-Interim reanalysis: Configuration and performance of the data assimilation system. *Quart. J. Roy. Meteor. Soc.*, **137**, 553–597, doi:10.1002/qj.828.

Delanoë, J., and R. J. Hogan, 2008: A variational scheme for retrieving ice cloud properties from combined radar, lidar, and infrared radiometer. *J. Geophys. Res.*, **113**, D07204, doi:10.1029/2007JD009000.

——, and ——, 2010: Combined CloudSat–CALIPSO–MODIS retrievals of the properties of ice clouds. *J. Geophys. Res.*, **115**, D00H29, doi:10.1029/2009JD012346.

——, ——, R. M. Forbes, A. Bodas-Salcedo, and T. H. M. Stein, 2011: Evaluation of ice cloud representation in the ECMWF and UK Met Office models using CloudSat and CALIPSO data. *Quart. J. Roy. Meteor. Soc.*, **137**, 2064–2078, doi:10.1002/qj.882.

Deng, M., G. G. Mace, Z. Wang, and H. Okamoto, 2010: Tropical Composition, Cloud and Climate Coupling Experiment validation for cirrus cloud profiling retrieval using CloudSat radar and CALIPSO lidar. *J. Geophys. Res.*, **115**, D00J15, doi:10.1029/2009JD013104.

——, ——, ——, and R. P. Lawson, 2013: Evaluation of several A-Train ice cloud retrieval products with in situ measurements collected during the SPARTICUS campaign. *J. Appl. Meteor. Climatol.*, **52**, 1014–1030, doi:10.1175/JAMC-D-12-054.1.

Donner, L. J., and Coauthors, 2011: The dynamical core, physical parameterizations, and basic simulation characteristics of the atmospheric component of the GFDL coupled model CM3. *J. Climate*, **24**, 3484–3519, doi:10.1175/2011JCLI3955.1.

Field, P. R., R. J. Hogan, P. R. A. Brown, A. J. Illingworth, T. W. Choularton, and R. J. Cotton, 2005: Parameterization of ice-particle size distributions for mid-latitude stratiform cloud. *Quart. J. Roy. Meteor. Soc.*, **131**, 1997–2017, doi:10.1256/qj.04.134.

Francis, P. N., P. Hignett, and A. Macke, 1998: The retrieval of cirrus cloud properties from aircraft multi-spectral reflectance measurements during EUCREX'93. *Quart. J. Roy. Meteor. Soc.*, **124**, 1273–1291, doi:10.1002/qj.49712454812.

Gettelman, A., and Coauthors, 2010: Global simulations of ice nucleation and ice supersaturation with an improved cloud scheme in the Community Atmosphere Model. *J. Geophys. Res.*, **115**, D18216, doi:10.1029/2009JD013797.

Han, Q. Y., W. B. Rossow, J. Chou, K. S. Kuo, and R. M. Welch, 1999: The effects of aspect ratio and surface roughness on satellite retrievals of ice-cloud properties. *J. Quant. Spectrosc. Radiat. Transfer*, **63**, 559–583, doi:10.1016/S0022-4073(99)00039-4.

Hogan, R. J., 2006: Fast approximate calculation of multiply scattered lidar returns. *Appl. Opt.*, **45**, 5984–5992, doi:10.1364/AO.45.005984.

Hong, G., 2007: Radar backscattering properties of non-spherical ice crystals at 94 GHz. *J. Geophys. Res.*, **112**, D22203, doi:10.1029/2007JD008839.

Jiang, J. H., and Coauthors, 2012: Evaluation of cloud and water vapor simulations in CMIP5 climate models using NASA A-Train satellite observations. *J. Geophys. Res.*, **117**, D14105, doi:10.1029/2011JD017237.

Kato, S., S. Sun-Mack, W. F. Miller, F. G. Rose, Y. Chen, P. Minnis, and B. A. Wielicki, 2010: Relationships among cloud occurrence frequency, overlap, and effective thickness derived from CALIPSO and CloudSat merged cloud vertical profiles. *J. Geophys. Res.*, **115**, D00H28, doi:10.1029/2009JD012277.

——, and Coauthors, 2011: Computation of top-of-atmosphere and surface irradiance computations with with CALIPSO-, CloudSat-, and MODIS-derived cloud and aerosol properties. *J. Geophys. Res.*, **116**, D19209, doi:10.1029/2011JD016050.

——, N. G. Loeb, D. A. Rutan, F. G. Rose, S. Sun-Mack, W. F. Miller, and Y. Chen, 2012: Uncertainty estimate of surface irradiances computed with MODIS-, CALIPSO-, and CloudSat-derived cloud and aerosol properties. *Surv. Geophys.*, **33**, 395–412, doi:10.1007/s10712-012-9179-x.

——, ——, F. G. Rose, D. R. Doelling, D. A. Rutan, T. E. Caldwell, L. Yu, and R. A. Weller, 2013: Surface irradiances consistent with CERES-derived top-of-atmosphere shortwave and longwave irradiances. *J. Climate*, **26**, 2719–2740, doi:10.1175/JCLI-D-12-00436.1.

Kay, J. E., and Coauthors, 2012: Exposing global cloud biases in the Community Atmosphere Model (CAM) using satellite observations and their corresponding instrument simulators. *J. Climate*, **25**, 5190–5207, doi:10.1175/JCLI-D-11-00469.1.

Klein, S. A., and C. Jakob, 1999: Validation and sensitivities of frontal clouds simulated by the ECMWF model. *Mon. Wea. Rev.*, **127**, 2514–2531, doi:10.1175/1520-0493(1999)127<2514:VASOFC>2.0.CO;2.

Kodama, C., A. T. Noda, and M. Satoh, 2012: An assessment of the cloud signals simulated by NICAM using ISCCP, CALIPSO, and CloudSat satellite simulators. *J. Geophys. Res.*, **117**, D12210, doi:10.1029/2011JD017317.

Li, J.-L. F., and Coauthors, 2005: Comparisons of EOS MLS cloud ice measurements with ECMWF analyses and GCM simulations: Initial results. *Geophys. Res. Lett.*, **32**, L18710, doi:10.1029/2005GL023788.

——, J. H. Jiang, D. E. Waliser, and A. M. Tompkins, 2007: Assessing consistency between EOS MLS and ECMWF analyzed and forecast estimates of cloud ice. *Geophys. Res. Lett.*, **34**, L08701, doi:10.1029/2006GL029022.

——, and Coauthors, 2008: Comparisons of satellites liquid water estimates to ECMWF and GMAO analyses, 20th century IPCC AR4 climate simulations, and GCM simulations. *Geophys. Res. Lett.*, **35**, L19710, doi:10.1029/2008GL035427.

——, D. E. Waliser, and J. H. Jiang, 2011: Correction to "Comparisons of satellites liquid water estimates to ECMWF and GMAO analyses, 20th century IPCC AR4 climate simulations, and GCM simulations." *Geophys. Res. Lett.*, **38**, L24807, doi:10.1029/2011GL049956.

——, and Coauthors, 2012: An observationally based evaluation of cloud ice water in CMIP3 and CMIP5 GCMs and contemporary reanalyses using contemporary satellite data. *J. Geophys. Res.*, **117**, D16105, doi:10.1029/2012JD017640.

——, D. E. Waliser, G. Stephens, S. Lee, T. L'Ecuyer, S. Kato, N. Loeb, and H.-Y. Ma, 2013: Characterizing and understanding radiation budget biases in CMIP3/CMIP5 GCMs, contemporary GCM, and reanalysis. *J. Geophys. Res.*, **118**, 8166–8184, doi:10.1002/jgrd.50378.

Lin, W. Y., and M. H. Zhang, 2004: Evaluation of clouds and their radiative effects simulated by the NCAR Community Atmospheric Model against satellite observations. *J. Climate*, **17**, 3302–3318, doi:10.1175/1520-0442(2004)017<3302:EOCATR>2.0.CO;2.

Loeb, N. G., B. A. Wielicki, D. R. Doelling, G. L. Smith, D. F. Keyes, S. Kato, N. Manalo-Smith, and T. Wong, 2009: Toward optimal closure of the earth's top-of-atmosphere radiation budget. *J. Climate*, **22**, 748–766, doi:10.1175/2008JCLI2637.1.

——, J. M. Lyman, G. C. Johnson, R. P. Allan, D. R. Doelling, T. Wong, B. J. Soden, and G. L. Stephens, 2012: Observed changes in top-of-the-atmosphere radiation and upper-ocean heating consistent within uncertainty. *Nat. Geosci.*, **5**, 110–113, doi:10.1038/ngeo1375.

Ma, H.-Y., M. Köhler, J.-L. F. Li, J. D. Farrara, C. R. Mechoso, R. M. Forbes, and D. E. Waliser, 2012: Evaluation of an ice cloud parameterization based on a dynamical–microphysical lifetime concept using CloudSat observations and the ERA-Interim reanalysis. *J. Geophys. Res.*, **117**, D05210, doi:10.1029/2011JD016275.

Mace, G. G., Q. Zhang, M. Vaughan, R. Marchand, G. Stephens, C. Trepte, and D. Winker, 2009: A description of hydrometeor layer occurrence statistics derived from the first year of merged CloudSat and CALIPSO data. *J. Geophys. Res.*, **114**, D00A26, doi:10.1029/2007JD009755.

Minnis, P., W. L. Smith Jr, D. P. Garber, J. K. Ayers, and D. R. Doelling, 1994: Cloud properties derived from GOES-7 for spring 1994 ARM intensive observing period using version 1.0.0 of ARM satellite data analysis program. NASA Reference Publication 1366, 59 pp. [Available online at http://citeseerx.ist.psu.edu/viewdoc/download?doi=10.1.1.45.3023&rep=rep1&type=pdf.]

——, and Coauthors, 2011: CERES Edition-2 cloud property retrievals using TRMM VIRS and Terra and Aqua MODIS data—Part I: Algorithms. *IEEE Trans. Geosci. Remote Sens.*, **49**, 4374–4400, doi:10.1109/TGRS.2011.2144601.

Nam, C. C. W., and J. Quaas, 2012: Evaluation of clouds and precipitation in the ECHAM5 general circulation model using *CALIPSO* and *CloudSat* satellite data. *J. Climate*, **25**, 4975–4992, doi:10.1175/JCLI-D-11-00347.1.

Norris, J. R., and C. P. Weaver, 2001: Improved techniques for evaluating GCM cloudiness applied to the NCAR CCM3. *J. Climate*, **14**, 2540–2550, doi:10.1175/1520-0442(2001)014<2540:ITFEGC>2.0.CO;2.

Randall, D. A., and Coauthors, 2007: Climate models and their evaluation. *Climate Change 2007: The Physical Science Basis*, S. Solomon et al., Eds., Cambridge University Press, 589–662. [Available online at https://www.ipcc.ch/pdf/assessment-report/ar4/wg1/ar4-wg1-chapter8.pdf.]

Rossow, W. B., and Y. C. Zhang, 1995: Calculation of surface and top of atmosphere radiative fluxes from physical quantities based on ISCCP data sets: 2. Validation and first results. *J. Geophys. Res.*, **100**, 1167–1197, doi:10.1029/94JD02746.

——, and R. A. Schiffer, 1999: Advances in understanding clouds from ISCCP. *Bull. Amer. Meteor. Soc.*, **80**, 2261–2287, doi:10.1175/1520-0477(1999)080<2261:AIUCFI>2.0.CO;2.

Sassen, K., Z. Wang, and D. Liu, 2009: Cirrus clouds and deep convection in the tropics: Insights from CALIPSO and CloudSat. *J. Geophys. Res.*, **114**, D00H06, doi:10.1029/2009JD011916.

Schmidt, G. A., and Coauthors, 2006: Present-day atmospheric simulations using GISS ModelE: Comparison to in-situ, satellite, and reanalysis data. *J. Climate*, **19**, 153–192, doi:10.1175/JCLI3612.1.

Song, X., G. J. Zhang, and J.-L. F. Li, 2012: Evaluation of microphysics parameterization for convective clouds in the NCAR Community Atmosphere Model CAM5. *J. Climate*, **25**, 8568–8590, doi:10.1175/JCLI-D-11-00563.1.

Stein, T. H. M., J. Delanoë, and R. J. Hogan, 2011: A comparison among four different retrieval methods for ice–cloud properties using data from *CloudSat*, *CALIPSO*, and MODIS. *J. Appl. Meteor. Climatol.*, **50**, 1952–1969, doi:10.1175/2011JAMC2646.1.

Stephens, G. L., 2005: Cloud feedbacks in the climate system: A critical review. *J. Climate*, **18**, 237–273, doi:10.1175/JCLI-3243.1.

Taylor, K. E., 2001: Summarizing multiple aspects of model performance in a single diagram. *J. Geophys. Res.*, **106**, 7183–7192, doi:10.1029/2000JD900719.

Waliser, D. E., and Coauthors, 2009: Cloud ice: A climate model challenge with signs and expectations of progress. *J. Geophys. Res.*, **114**, D00A21, doi:10.1029/2008JD010015.

——, J.-L. F. Li, T. S. L'Ecuyer, and W.-T. Chen, 2011: The impact of precipitating ice and snow on the radiation balance in global climate models. *Geophys. Res. Lett.*, **38**, L06802, doi:10.1029/2010GL046478.

Webb, M., C. Senior, S. Bony, and J. J. Morcrette, 2001: Combining ERBE and ISCCP data to assess clouds in the Hadley Centre, ECMWF and LMD atmospheric climate models. *Climate Dyn.*, **17**, 905–922, doi:10.1007/s003820100157.

Wielicki, B. A., B. R. Barkstrom, E. F. Harrison, R. B. Lee, G. L. Smith, and J. E. Cooper, 1996: Clouds and the Earth's Radiant Energy System (CERES): An Earth observing system experiment. *Bull. Amer. Meteor. Soc.*, **77**, 853–868, doi:10.1175/1520-0477(1996)077<0853:CATERE>2.0.CO;2.

Yang, P., K. Liou, K. Wyser, and D. Mitchell, 2000: Parameterization of the scattering and absorption properties of individual ice crystals. *J. Geophys. Res.*, **105**, 4699–4718, doi:10.1029/1999JD900755.

Young, S. A., and M. A. Vaughan, 2009: The retrieval of profiles of particulate extinction from Cloud–Aerosol Lidar Infrared Pathfinder Satellite Observations (CALIPSO) data: Algorithm description. *J. Atmos. Oceanic Technol.*, **26**, 1105–1119, doi:10.1175/2008JTECHA1221.1.

Zhang, M. H., and Coauthors, 2005: Comparing clouds and their seasonal variations in 10 atmospheric general circulation models with satellite measurements. *J. Geophys. Res.*, **110**, D15S02, doi:10.1029/2004JD005021.

Zhang, Y., S. A. Klein, J. Boyle, and G. G. Mace, 2010: Evaluation of tropical cloud and precipitation statistics of Community Atmosphere Model version 3 using CloudSat and CALIPSO data. *J. Geophys. Res.*, **115**, D12205, doi:10.1029/2009JD012006.

Chapter 14

A Synoptic-Scale Cold-Reservoir Hypothesis on the Origin of the Mature-Stage Super Cloud Cluster: A Case Study with a Global Nonhydrostatic Model

KAZUYOSHI OOUCHI

Research Institute for Global Change, Japan Agency for Marine-Earth Science and Technology, Yokohama, Kanagawa, Japan

MASAKI SATOH

Research Institute for Global Change, Japan Agency for Marine-Earth Science and Technology, Yokohama, Kanagawa, and Atmosphere and Ocean Research Institute, University of Tokyo, Tokyo, Japan

ABSTRACT

This chapter proposes a working assumption as a way of conceptual simplification of the origin of Madden–Julian oscillation (MJO)-associated convection, or super cloud cluster (SCC). To develop the simplification, the importance of the synoptic-scale cold reservoir underlying the convection and its interaction with the accompanying zonal–vertical circulation is highlighted. The position of the convection with respect to that of climatological warm pool is postulated to determine the effectiveness of this framework. The authors introduce a prototype hypothesis to illustrate the usefulness of the above assumption based on a numerical simulation experiment with a global nonhydrostatic model for the boreal summer season.

Premises for the hypothesis include 1) that the cloud cluster (CC) is a basic building block of tropical convection accompanying the precipitation-generated cold reservoir in its subcloud layer and 2) that a warm-pool-induced quasi-persistent zonal circulation is key for the upscale organization of CCs. The theory of squall-line structure by Rotunno, Klemp, and Weisman (hereafter RKW) is employed for the interpretation. No account is taken regarding the influences of equatorial waves as a first-order approximation. Given the premises, an SCC of $O(1000)$ km scale is interpretable as a gigantic analog of a multicellular squall line embedded in the quasi-stationary westerly shear branch of the zonal circulation east of the warm water pool. A CC corresponds to the "cell," and its successive formation to the east and westward movement represents an upshear-tilting core of intense updraft. The upshear-tilted SCC is favorably maintained with the precipitating area being separated from the gust front boundary between the cold reservoir and a low-level easterly, which is supported in the realm of the RKW theory where two horizontal vortices associated with the cold reservoir and vertical shear are opposite in sign but cold reservoir's vorticity can be inferred to be larger, leading to upshear-tilted and multicellular behavior. As a counterexample, CCs to the west of the warm pool (Indian Ocean and Arabian Sea) are embedded in the easterly shear and organized into a less coherent cloud cluster complex (CCC) given the situation of RKW where two horizontal vortices associated with the cold reservoir and vertical shear are still opposite in sign, but the smaller vertical shear west of the warm pool causes even more suboptimal vorticity imbalance in the western flank of cold reservoir, leading to larger tilt with height and intermittent, less viable storm situations.

A cold pool or cold reservoir, having been prevalent in mesoscale convection research, is argued to be important for the MJO as pointed out by the emerging evidence in the international field campaign for the MJO called Cooperative Indian Ocean Experiment on Intraseasonal Variability (CINDY)/DYNAMO. The simplified and idealistic hypothesis proposed here does not cover all aspects of MJO and its validation awaits further modeling and observational studies, but it can offer a framework for characterizing a fundamental aspect of the origin of MJO-associated convection.

Corresponding author address: Dr. Kazuyoshi Oouchi, Research Institute for Global Change, Japan Agency for Marine-Earth Science and Technology, 3173-25 Showamachi, Kanazawa-ku, Yokohama, Kanagawa 236-0001, Japan.
E-mail: k-ouchi@jamstec.go.jp

DOI: 10.1175/AMSMONOGRAPHS-D-15-0008.1

1. Introduction

The aim of this chapter is to provide a framework that we hope will be of value for understanding the origin of tropical large-scale organized convection over the tropical warm water pool. A super cloud cluster (SCC; Hayashi and Sumi 1986; Nakazawa 1988), a typical example of this type of convection, is a loosely defined, albeit salient, *ensemble of tropical convection of 3000–5000-km scale that propagates eastward over the high sea surface temperature waters in the equatorial open oceans.* Its definition in this chapter is as described in italics. An SCC is frequently observed in the convectively active phase of the Madden–Julian oscillation (MJO), and it is hardly observed outside of the Pacific warm water pool (Madden and Julian 1971, 1972; Hayashi and Nakazawa 1989). It forms somewhere over or to the west of the Maritime Continent; develops into the active, well-identifiable form over the western Pacific; and decays as it approaches the central Pacific (e.g., Lau et al. 1991).

Over the Indian Ocean, on the other hand, there are relatively fewer observed cases of the purely eastward-propagating SCC sometimes associated with the propagation of the MJO (Julian and Madden 1981; Wang and Rui 1990; Mapes and Houze 1993; Chen and Houze 1997). The original schematic of Madden and Julian (1972) illustrating a prototype of the large-scale convection associated with the MJO perceptively indicates such a different nature—almost stationary or a slow-moving buildup—of convection over the Indian Ocean; the inference was supported by a cloudiness data analysis in almost the same target period (Julian and Madden 1981). The convection over the Indian Ocean illustrated in the schematic is decoupled from the underlying surface pressure anomaly, and the surface pressure and the low-level zonal flow are negatively correlated with each other. The negative correlation, being at odds with the eastward-propagating gravity wave dynamics, implies that some other mechanisms can be at work. Such a geographical preference (e.g., over the Pacific warm water pool) and the origin of the existence of SCC from a more general standpoint remain big unanswered mysteries; this chapter attempts to provide a simple hypothesis on this.

One of the long-standing interests in SCCs is that they involves features in common with the equatorial Kelvin wave (Hayashi and Sumi 1986). The previous modeling and theoretical studies have provided substantial progress in clarifying the association of SCCs with the equatorial (Kelvin) wave dynamics under the influence of tropical convection, and the importance of the Kelvin wave in SCC dynamics is well established by now. The major viewpoints include an instability of or interaction with the equatorial wave using a mechanistic model and a general circulation model with the parameterized effects of cumulus convection. There is extensive literature on this issue (e.g., Lau and Peng 1987; Takahashi 1987;

Chang and Lim 1988; Numaguti and Hayashi 1991; Hendon and Salby 1994; Oouchi and Yamasaki 1997), and the viewpoints remain challenging research topics.

What makes the SCC even more unique is that its inner structure consists of a cloud cluster (CC) of $O(100)$ km scale; the life cycle of the CC ultimately determines the pattern and behavior of the SCC. A typical picture indicates that a new CC develops successively to the east of the existing one, thereby creating an eastward-propagating SCC (Nakazawa 1988). Since the CC is ubiquitous in the tropical atmosphere, the mechanism of its upscale organization is at the heart of the investigation of the SCC (Chao and Lin 1994). From this standpoint, the resolution of the hierarchical convective feature consisting of the SCC and the CC is a prerequisite for successfully understanding the SCC; a cloud-resolving model (CRM) has come into wide use as a unique, fascinating tool for this purpose. Major CRM research on SCCs has used 2D zonal–vertical (Oouchi 1999; Grabowski and Moncrieff 2001; Tulich and Mapes 2008) and 3D global cloud-resolving frameworks (Nasuno et al. 2007). These studies have suggested that mesoscale processes play a critical role in the maintenance of SCCs by pointing out the importance of convectively induced gravity waves (Oouchi 1999; Mapes 2000), or gust-front waves (Tulich and Mapes 2008), and the precipitation-generated cold pool (Oouchi and Yamasaki 2001; Nasuno et al. 2007), as in the case of typical mesoscale tropical convection (Yamasaki 1988). A key finding is that such mesoscale processes are equally important for explaining the well-organized nature of the SCC, in addition to explaining the contributions of the equatorial waves and MJO (Oouchi 1999; Oouchi and Yamasaki 2001).

To look at the hierarchy of the SCC further down the scale, each CC consists of mesoscale convection of $O(10)$ km (Chen and Houze 1997; Oouchi 1999). The class of convection is emphasized as an important ingredient for suitably representing any tropical disturbance in a coarser grid model (Yamasaki 1984), and there is no exception for the SCC (Oouchi 1999). Although a suitable examination of mesoscale convection in the SCC is very significant from a different viewpoint (Oouchi 1999), it is not directly relevant in the context of the present chapter, and therefore discussion is omitted.

Acknowledging the fact that all these hierarchical features are important for understanding the origin of the SCC, in this chapter, we take another simple view on the importance of the SCC–CC hierarchy. We focus on a two-way interaction of the SCC and the CC to understand the geographical preference of the SCC as a first-order approximation. The rationale behind this is the fact that the CC is the most fundamental building block of tropical convection on a synoptic scale in the tropics and that the SCC has a distinctive and quite systematic characteristic involving the life cycle of a CC. The latter is, in a sense, reminiscent of a multicellular squall line. This type of squall line is a coherently organized ensemble of convective cells at the mesobeta scale (Houze 1977). It moves forward against the environmental flow, which is essentially a manifestation of the life cycle of the inner convective cell with meso-gamma scale that occurs by a convergence at the leading edge of the system-scale cold pool; the flow feeds moisture into the inner cells and interacts with the cold pool to sustain the cells in the entire system and, as a result, the mature stage shifts from one cell to another, thereby yielding an entire squall-line system. Apparently, this feature is analogous to the SCC, although the scale of the SCC is one order larger.

In light of this analogy, we provide a bold, yet intuitively simple, assumption that the SCC can be viewed as a gigantic multicell squall line, being independent of, if not at the expense of, the widely held discussion of its association with equatorial waves. It is well known that the evolution of squall line is controlled significantly by the effect of the orientation of line-normal vertical wind shear (e.g., Yamasaki 1984; Rotunno et al. 1988, hereafter RKW), and the analogy is applicable to the SCC. On this assumption, we employ a working hypothesis that the SCC is maintained or suppressed under climatologically different vertical shear regimes of the warm water pool origin. Whether or not CCs are organized into SCCs depends on the location of the CCs with respect to the tropical warm water pool and the associated vertical shear regimes. The MJO can modulate the vertical shear and, therefore, indirectly control the organization of the SCC. We explore this hypothesis and derive a new interpretation of the origin of the SCC by focusing on some simulated cases of SCCs and other ensemble CCs. The results of this chapter are based solely on a simulation using the global cloud-resolving Nonhydrostatic Icosahedral Atmospheric Model (NICAM; Satoh et al. 2008), with less emphasis being placed on the comparison with observations. A reason for this is that the aim of this chapter is to propose a concept as a step before discussing the details of observational verification. This is an important step in pointing out the possible essential physics of the tropical convection even in an idealized framework. In the future, by pursuing its applicability through observational verifications, the interpretation is expected to help answer the question of why the SCC preferentially occurs in the western Pacific but not in the Indian Ocean and Arabian Sea, thereby providing new insight into the origin of the SCC.

Section 2 describes briefly the method of the experiments. Section 3 explains the background atmosphere condition and the convective regimes of interest for introducing the key factors for interpreting SCCs. Section 4 documents the typical cases of convective organization and flow fields for the proposed regimes with a comparison with observations. Sections 5 and 6 provide an interpretation and conceptual model for it, which is followed by discussion and summary in section 7.

2. Experimental design

This study used NICAM (Satoh et al. 2008; Tomita and Satoh 2004). The horizontal grid spacing was 14 km, which was fine enough to resolve CCs abundant in the tropics and the gross features of mesoscale convection (Yamasaki 1984). The horizontal domain covered Earth, and the vertical domain, consisting of 40 layers, extended up into the upper stratosphere with its top at 40 km. We used the cloud microphysics scheme of Grabowski (1998), which includes two classes of solid phases of ice and snow. To formulate turbulent boundary layer, an improved version (Noda et al. 2010) of the scheme of Nakanishi and Niino (2006) was employed.

The time integration was performed from 1 June to 10 November 2004, using the initial atmospheric conditions interpolated from the National Centers for Environmental Prediction (NCEP) Global Tropospheric Analyses at 0000 UTC 1 June 2004. The experiment included the sea surface temperature (SST) dataset from the weekly National Oceanic and Atmospheric Administration (NOAA) Optimum Interpolation SST. No nudging techniques were applied during the course of the time integration; therefore, the genesis of clouds and convection and their interactions with atmospheric disturbances were expressed as an internally driven spontaneous process in the global cloud-resolving model. The time integration period included the active period of the Asian monsoon season, which was long enough to discuss a couple of MJO cycles and the associated tropical cyclogenesis. For other details of the experimental design, refer to Oouchi et al. (2009).

A major aim of the experiment was to investigate the performance of NICAM in representing the boreal summer seasonal variation of tropical convective disturbances including the tropical cyclone and MJO, the SCC, and the associated seasonal change in Asian monsoons and relevant phenomena. It is regarded as a companion experiment of the boreal winter experiment that demonstrated a good performance of the MJO-related convection and the accompanying genesis of tropical cyclones (Miura et al. 2007). Boreal summer is a challenging season under the control of Asian summer monsoons. The insight obtained from the present experiment will help improve seasonal forecasts of the tropical phenomena in that season. Highlights of the

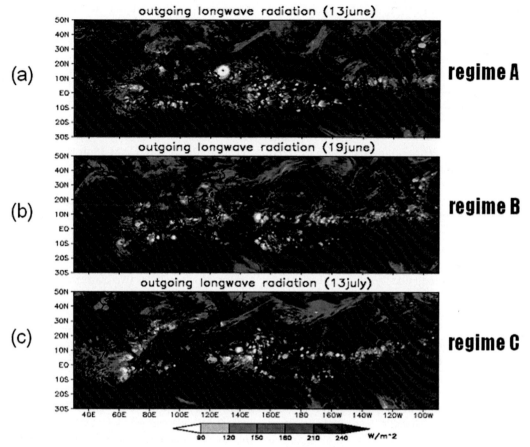

FIG. 14-1. Snapshots of outgoing longwave radiation (W m^{-2}) in typical convective regimes over the Indian Ocean and the Pacific region (30°S–50°N, 30°E–90°W): (a) regime A on 13 June, (b) regime B on 19 June, and (c) regime C on 13 July.

results and details of the performance of the NICAM revealed in the experiment are reported elsewhere, including a well-simulated relationship between the MJO and tropical cyclogenesis (Oouchi et al. 2009) and an improved representation of low-level/boundary layer clouds (Noda et al. 2010).

3. Convective regime and flow field diagnosis

Figure 14-1 plots snapshots of the outgoing longwave radiation as a proxy for convective activity in the Indian Ocean and western Pacific region. The three panels are selected to facilitate the following discussion about the three regimes—A, B, and C—of convective organization in the region. Each regime roughly corresponds to a different phase of an MJO event, as will be explained with Fig. 14-3. More systematic categorization with respect to various phases of MJO's propagation or other atmospheric conditions is under development. The selection of the regimes here serves as our understanding of the typical patterns of

organized convection in the region as seen in the global cloud-resolving model.

In all panels, we can see that the tropical atmosphere over the Indian Ocean and the western Pacific is characterized by an occurrence of active convection in a variety of forms. Typically, various types of convection are present in the tropical region. Some types are organized into a tropical cyclone with a clear eye near the Philippine Islands, as in Fig. 14-1a, and others are transient, creating ensembles in vague forms. What is immediately evident from these panels is the fact that the convection is identified as a patch of CCs of $O(100)$ km scale. Cloud clusters would be accompanied by a mesoscale circulation and interact with various disturbances in the tropics, including equatorial waves. Cloud clusters are known to be a basic, important building block of tropical convection (Nakajima and Matsuno 1988; Satoh et al. 2008), and their own origins are an important research topic. Here, we regard CCs just as a basic element of tropical convection, and we focus on the issue of further upscale organization of CCs. A

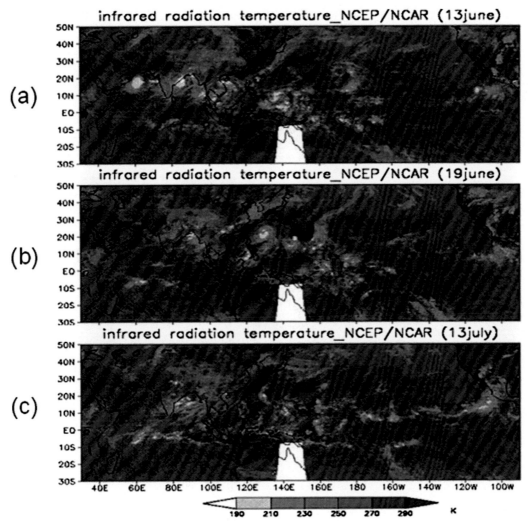

FIG. 14-2. Snapshots of infrared radiation temperature (K) from NCEP–NCAR over the Indian Ocean and the Pacific region (30°S–50°N, 30°E–90°W) for (a) 13 June, (b) 19 June, and (c) 13 July.

well-known upscale organization of CCs is an SCC that is associated with the propagation or existence of the MJO. In Fig. 14-1a, an SCC is identified around 140°–180°E as an ensemble of CCs with more than a 4000-km scale flanked by a tropical cyclone. The coexistence of SCCs and tropical cyclones is a typically observed situation in the active phase of the MJO (Nakazawa 1986; Sui and Lau 1992; McBride et al. 1995).

In other periods and regions such as the Indian Ocean, we can see some transient, loosely detectable, upscale organization of CCs other than SCCs; we call this a cloud cluster complex (CCC). The behaviors of CCs in the real atmosphere and in the NICAM experiment are not so straightforward that one can objectively categorize them into a simple picture. The argument in this chapter is just one way of ad hoc categorization to highlight the background physics that might work behind the CCC and SCC.

The aim of this chapter is not to pursue the similarity between the observation and simulation, but to propose a conceptual model as a first step toward improving our understanding of SCCs. However, it is important to check whether the simulated convective fields deviate significantly from observations. Figure 14-2 plots the infrared radiation (IR) temperature compiled by NCEP, which serves as a proxy of convective activity—like outgoing longwave radiation. Note that the IR temperature is a variable that represents the physical quantity differently from the outgoing longwave radiation in Fig. 14-1, so the legends are different between the two figures. In Fig. 14-2a, we can see that an ensemble of convection is present near the equator, at 140°–160°E. The ensemble we called "SCC" is a focus of this chapter; it is reasonably simulated in the in regime A in Fig. 14-1. The SCC became less clear 6 days later (Fig. 14-2b), and

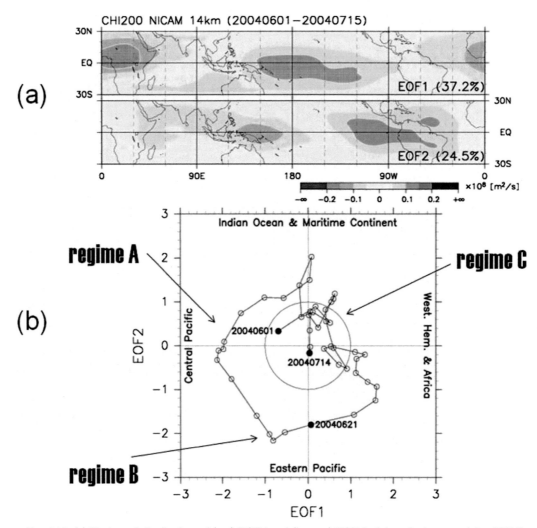

FIG. 14-3. (a) Horizontal distributions of (top) EOF 1 and (bottom) EOF 2 of the velocity potential at 200 hPa calculated for the period from 1 June to 15 July, with divergence in red and convergence in blue. (b) Phase space points for PC 1 and PC 2, which were plotted as daily estimates as in Taniguchi et al. (2010). Also labeled are the approximate locations of the enhanced convective signal of the MJO for that location of the phase space. Convective regimes of interest are marked on the plot (regimes A, B, and C in Fig. 14-1).

the weakening is also simulated to some degree in the model as regime B in Fig. 14-1. After approximately 20 days (Fig. 14-2c), the convective activity associated with the SCC looked further weakened, and a less intense cloud cluster occupied the western Pacific, a situation resembling simulated regime C in Fig. 14-1. Another important feature is that in Figs. 14-1 and 14-2, convection is less coherently organized in the Indian Ocean than in the western Pacific in all regimes. In this chapter, we call the less coherent convection "CCC." A mechanism for contrasting convective features of the western Pacific and Indian Oceans is the main subject of this chapter.

The association of the convective regimes in Fig. 14-1 with the simulated MJO event is illustrated based on the MJO index of Taniguchi et al. (2010) and Wheeler and

Hendon (2004). The index was shown to serve as a good quantitative diagnosis of the MJO event simulated in the boreal winter experiment with NICAM (Taniguchi et al. 2010; Miura et al. 2007). It is derived from the leading two modes of empirical orthogonal function (EOF) analysis of the velocity potential at 200 hPa. Figure 14-3a plots the horizontal distributions of EOF 1 and EOF 2, and Fig. 14-3b plots principal component (PC) 1 and 2 on the phase space diagram. The MJO is identified by a combination of the leading two EOFs that represent, as a pair, a large-scale eastward propagation of the signals. EOF 1 represents a situation in which the MJO-associated convection is enhanced (suppressed) and yields divergence (convergence) in the upper troposphere (200 hPa) at the longitudes of the African continent, and convergence

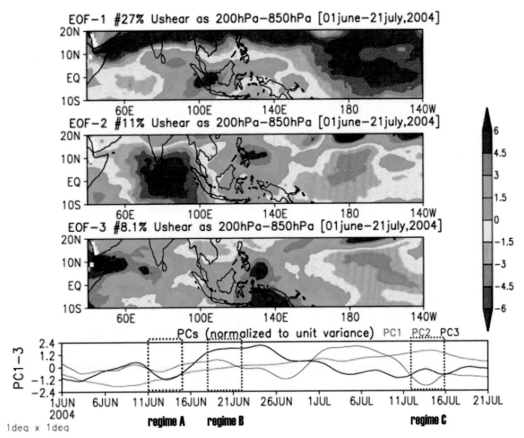

FIG. 14-4. (top) Horizontal distributions of EOF 1, EOF 2, and EOF 3 of the vertical shear represented as U200 hPa minus U850 hPa, and (bottom) the corresponding PCs for the region 10°S–20°N, 40°E–140°W. Each of the dotted rectangles in red marks a consecutive 4-day selected in the analysis for the regimes A, B, and C in Fig. 14-1.

(divergence) in the mid-Pacific, while EOF 2 exhibits a pattern nearly in quadrature to that of EOF 1. A combination of Figs. 14-3a and 14-3b shows the behavior of the simulated MJO-associated large-scale circulation that propagates eastward over the central Pacific (regime A), while showing intensified divergence over the western Pacific (regime B) and reentering of the divergence signal into the Western Hemisphere and African continent (regime C). The regimes thus cover a cycle of the MJO that travels over the globe in the tropical regions; the regimes explain reasonably the enhanced (suppressed) activity of the MJO in the Eastern (Western) Hemisphere. The associated flow patterns are more clearly identified in the following arguments.

The key factor in this study is a vertical shear of the horizontal zonal velocity that changes systematically in association with the propagation of the MJO. This factor significantly controls the large-scale wind fields responsible not only for the favorable occurrence of tropical cyclogenesis (Gray 1968; Oouchi et al. 2009), but also for the manner of convective organization in the Indian Ocean and western Pacific as suggested in this

chapter. To define the typical spatial patterns of the simulated vertical shear regimes, EOF analysis of the vertical shear of zonal winds (U200 minus U850) is performed for the region of interest covering the Indian Ocean and western Pacific (10°S–20°N, 40°E–140°W). Figure 14-4 shows the leading three EOFs and a set of corresponding principal components representing a normalized time series.

EOF 1 explains 27% of the total velocity variance and represents an elongated easterly shear (low-level westerly) in the subtropical regions and a westerly shear (low-level easterly) that is flanked on the southern side up to the equatorial region in and around the Maritime Continent (80°–150°E). The pattern in this region therefore represents a stronger meridional gradient of the horizontal shear and tends to generate an anticyclonic vortex in the lower troposphere. A comparison of the spatial pattern with PC 1 reveals that in the second half of June, the lower troposphere is favorable for cyclogenesis in view of the negative polarity of PC 1 (cyclone-generating sense). It is interesting that the polarity of EOF 1 becomes reversed on a time scale of more than

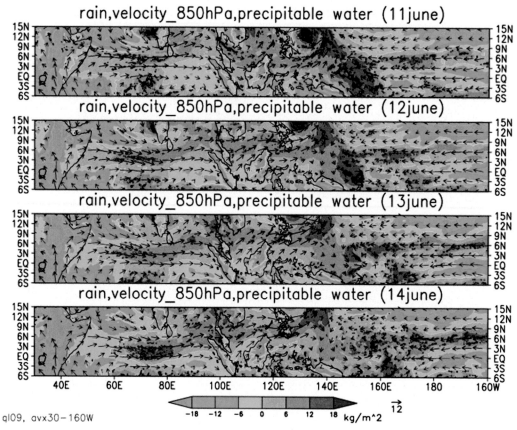

FIG. 14-5. Daily snapshots of the horizontal velocity at 850 hPa (arrows, m s^{-1}), precipitable water (shading, kg m^{-2}), and precipitation location (blue shading) during the regime A period for 11–14 June.

30 or 40 days, which is closely associated with the behavior of the MJO simulated in this period (Oouchi et al. 2009). Given its substantial amplitude in the off-equatorial region, the regime can also be controlled by monsoon-associated signals, which is beyond the scope of this study.

On the other hand, EOF 2, which explains 11% of the total variance, exhibits a temporal variation on the shorter time scale (approximately submonthly). Its spatial pattern shows a zonally aligned westerly–easterly shear region with its amplitude maximized around the equator. The strongest easterly shear is located in the Indian Ocean, that is, to the west of the Maritime Continent or the warm-pool region, while the westerly shear counterpart is in the equatorial western Pacific; its amplitude is smaller than the easterly shear in the Indian Ocean. As discussed in the following, this regime can be interpreted as resulting from a different manner of coupling between the vertical shear and CCC or SCC in different oceanic basins (the Indian Ocean and the western Pacific).

EOF 3 appears to reflect the regime with a time scale similar to that of EOF 2 and shows a reversed polarity, particularly during the period from mid-June to early

July. However, the scattered nature of the spatial pattern in the maximum amplitude makes its interpretation difficult. The leading two EOF patterns serve as the key references of the environment by which the following interpretation of the organized convective regime becomes more easily adopted.

4. Overview of the regimes

a. Simulation

The evolution of the convective ensemble in each regime in Fig. 14-1 is displayed in Fig. 14-5 (regime A), Fig. 14-6 (regime B), and Fig. 14-7 (regime C) as the daily snapshots of four consecutive days for precipitable water (shades), horizontal flow at 850 hPa (arrows) and precipitation (blue contours). In Fig. 14-5, it is evident that an envelope of convection in the tropical western Pacific (around 140°–150°E on 11 June) propagates eastward, which we define here as SCC. The SCC propagates in the low-level easterly duct that stretches across the western and central Pacific and is nearly collocated with the region with higher precipitable water. On 11 and 12 June, an off-equatorial precipitation

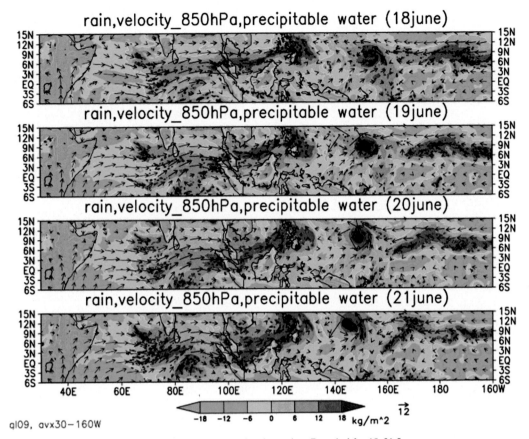

FIG. 14-6. As in Fig. 14-5, but for the regime B period for 18–21 June.

pattern accompanied by a tropical cyclone is flanked by the SCC. As argued in Oouchi et al. (2009), SCCs and tropical cyclones are frequently observed in the convectively active period of the MJO in this experiment, and this is also the case here. As to the convection in the Indian Ocean, that is, to the west of the Maritime Continent or the western Pacific warm pool, it exhibited a pattern of organization different from the SCC in the western Pacific; it had a scale of about 2000 km (e.g., 60°–80°E on 12 and 14 June) and was therefore regarded as an ensemble of CCs (defined as CCC). However, the ensemble tended to stay in the low-level westerly region, and it decayed after 14 June.

Such a simultaneous occurrence of the low-level westerly and precipitation and the tendency for precipitation to stay within the Indian Ocean (or within the oceanic basin of the Arabian Sea or the Bay of Bengal) can be confirmed also in Figs. 14-6 and 14-7. Note that the convective pattern and the moisture and flow fields in the western Pacific in Figs. 14-6 and 14-7 are different from those in Fig. 14-5. We see neither conspicuous eastward movement corresponding to SCC nor extended easterly with little sign of meridional flow in the equatorial western and central Pacific; instead, the

precipitation pattern takes the form of vortices, some of which can be identified as tropical cyclones.

On the other hand, the flow and precipitation patterns in the Indian Ocean and the Maritime Continent are different between Fig. 14-6 and Fig. 14-7. In Fig. 14-6, a persistent westerly stretches from off the eastern coast of Africa (as a Somali jet) to around the Maritime Continent (100°–120°E). The signature of the tendency of such an eastward extension of the westerly is obvious if one traces the snapshots from 11 June (Fig. 14-5) to 21 July (Fig. 14-6). The extension is obviously caused by the propagation of the MJO as suggested by Fig. 14-3b, which indicates the eastward propagation of the MJO from around the Indian Ocean to around the Maritime Continent, and the accompanying change in precipitation pattern, such as enhanced precipitation (another form of CCC) over 100°–120°E on 18–21 June in Fig. 14-7 (regime A to B). In Fig. 14-7, such widespread features of precipitation patterns and flow fields become weakened, and the coupling among the precipitation, flow, and moisture fields seems to be only locally activated in the Indian Ocean and the western Pacific. This is the regime, according to the MJO index in Fig. 14-3b, when the center of the MJO action propagates away from the

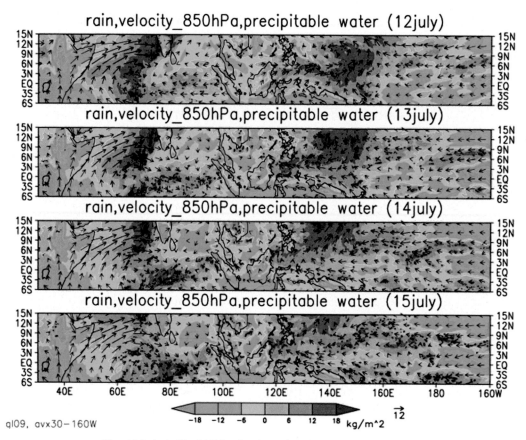

FIG. 14-7. As in Fig. 14-5, but for the regime C period for 12–15 July.

Maritime Continent and the western Pacific region and approaches, or passes over, the Western Hemisphere. This is also suggested by the lower precipitable water in 80°–120°E on 12–15 July, which is likely to be caused by the subsidence associated with the MJO.

These results suggest that the SCC is seen favorably in the western Pacific, in association with the presence of low-level zonal (small meridional component) easterlies ahead and rotational flows behind, and that this is closely associated with the propagation of the MJO. This is consistent with Oouchi et al. (2009) and the widely held scenario (Hayashi and Sumi 1986; Hayashi and Nakazawa 1989, and many others). This is, therefore, just a confirmation of previous studies. However, we stress here that the different convective regimes (SCC and CCC) as shown in Figs. 14-5–14-7 can be controlled by the warm-pool-induced zonal circulation, which is explored in the following.

The specific question here is what is the difference between the SCC and other convective ensembles (CCC) regarding the relative location of the evaporatively driven cold pool, active cloud condensates, and environmental velocity fields, and how can one explain the favorable or unfavorable conditions for the occurrence of SCCs in light of this. To explore the

problem, the vertical–zonal cross section along the active convection area in each regime in Fig. 14-1 is shown in Figs. 14-8–14-10, and for the SCC over the western Pacific in Fig. 14-8, CCC in the Indian Ocean in Figs. 14-9 and 14-10. In Fig. 14-8, the SCC is identified as an ensemble of the cloud cluster as is well known (Nakazawa 1988). We can see that the lower troposphere in the SCC region is occupied by a negative temperature anomaly (purple boxes) that might result from an evaporative cooling of rainwater, called "cold reservoir" here. The cold reservoir is defined as the entity of the negative temperature anomaly floating in the low- to midlevel air (up to about 700 hPa) of the troposphere plus the conventional cold pool, which refers to the negative temperature anomaly on Earth's surface. The original cold pool is an important feature in mesoscale phenomena such as squall lines, and also in the SCC (Oouchi 1999; Nasuno et al. 2007). The bottom panel in Fig. 14-8 plots the temperature anomaly averaged in the vertical layer spanning 925–850 hPa (black line) and surface precipitation (blue line). We can see that with respect to the precipitation center (142°–155°E), the cold anomaly spreads to the west. As to the velocity fields shown in the middle panel of Fig. 14-8, a subsidence of

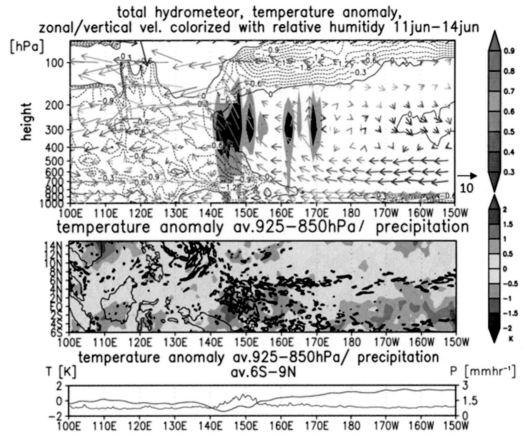

FIG. 14-8. (top) Vertical cross section (average over 6°S–9°N) of the total hydrometeor (color shading), zonal and vertical velocity (vectors) that are colorized with relative humidity (see the color bar for values), and "negative" temperature anomaly (dotted contours) for 100°E–150°W. These are averaged over the consecutive 4-day periods during regime A for 11–14 June. (middle) Horizontal distribution of the temperature anomaly for the layers average (925–850 hPa, shading) and precipitation (contours) in the corresponding area and period. (bottom) Temperature anomaly for the layers (925–850 hPa) from the area average (black), and the area-averaged precipitation (blue) in the corresponding area and period. Note that y-axis labels are on the left for the temperature and the right for the precipitation.

the dry air in the mid- to upper troposphere is evident to the east of the SCC (east of 170°E). The low-level easterly over the central Pacific (160°E–150°W) flows into the cold pool beneath the active convection area (140°–160°E), and this should maintain the entire convective system of the SCC. As indicated by the flow field, in an averaged sense, over the active convection area of 140°–170°E, the axis of the updraft tilts westward with height. We propose that this is an important feature for maintaining SCCs; the precipitation from the active convective area of the SCC falls to the west of the area for interaction between the cold pool and the easterly inflow and therefore does not prevent the interaction from feeding moisture to the SCC. This situation highlighting the importance of the interaction between the cold pool and environmental flow resembles the maintenance of tropical mesoscale convection proposed by Yamasaki (1984) and is also in consistent with Oouchi (1999) and Nasuno et al. (2007) for SCCs. However, it is

important to note that they argued the importance of the cold pool only for maintaining the inner convective element of the SCC, not for the SCC as a whole.

On the other hand, for the synoptic-scale CCC in the Indian Ocean (Figs. 14-9, 14-10), the location of the cold pool and the flow fields surrounding the CCC are not favorable for maintaining the CCC. For example, we can see in the bottom and middle panels of Figs. 14-9 and 14-10 that the cold pool and the low-level inflow take a pattern unfavorable for long-lived convective activity. Particularly in Fig. 14-10, we can see that the conspicuous low-level westerly enters into the system around 40°–60°E, and the location of the interaction between the low-level flow and the cold pool is to the west of the main convective region. In addition, the location is almost collocated with the area of the falling precipitation, and therefore the situation is unfavorable for maintaining the convective system. The situation is more clearly interpreted by using the theory of RKW; over the Indian Ocean, two horizontal vortices

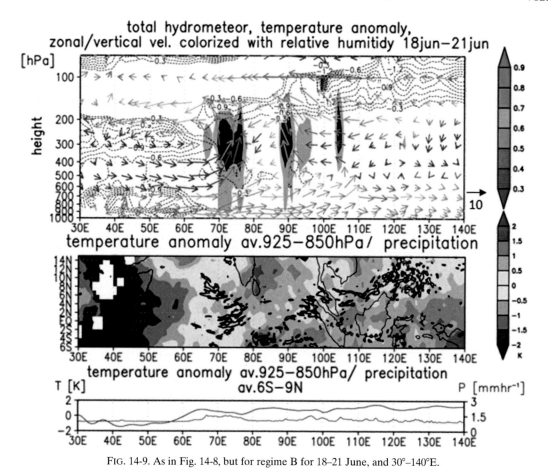

FIG. 14-9. As in Fig. 14-8, but for regime B for 18–21 June, and 30°–140°E.

associated with the cold reservoir and vertical shear are still opposite in sign, but the smaller vertical shear west of the warm pool causes even more suboptimal vorticity imbalance in the western flank of cold reservoir, leading to larger tilt with height and intermittent, less viable storm situation. This is the situation frequently seen in the other CCC cases in the Indian Ocean. We can therefore infer that this might be responsible for the absence of well-organized SCC in the Indian Ocean (to the west of the warm water pool), if one views the SCC as a convective ensemble involving the intense, westward-tilting updraft core.

b. Comparison with the observations

The aim of this chapter is to propose and characterize the convective regimes associated with SCCs based on numerical simulations. At a next step, it should be instructive to verify their existence in the real atmosphere. As a first step, our aim is to investigate the morphological similarities between the simulated regimes (Figs. 14-8–14-10) and their observational counterparts. Figures 14-11, 14-12, and 14-13 indicate the plots of TRMM 3B42 (precipitation) and NCEP reanalysis (other variables) as the observational counterparts of each of the regimes—A, B, and C,

respectively. Note that even though the period and region displayed happened to be the same as those in the numerical simulation, we cannot expect rigorous correspondence of the simulated regime-specific features with the observed counterparts because of the intrinsic nature of the initial value problem and the transience of cloud systems.

1) REGIME A

A comparison of the simulation (Fig. 14-8) and the observation (Fig. 14-11) indicates a common feature of regime A in terms of the active convective area—represented as total hydrometeors (top panel) and precipitation (middle panel)—that is mostly collocated with the negative temperature anomaly in the lower atmosphere, that is, the cold reservoir. This reinforces the key premise of the hypothesis. The relationship between the convection and the zonal–vertical flow fields (top panel) is also similar in that the updraft core tilts westward with height with respect to the active convection cell(s). To describe it another way, the low-level warm and moist air advances into the cold reservoir from the east, ascends to the middle troposphere, and constitutes a significant

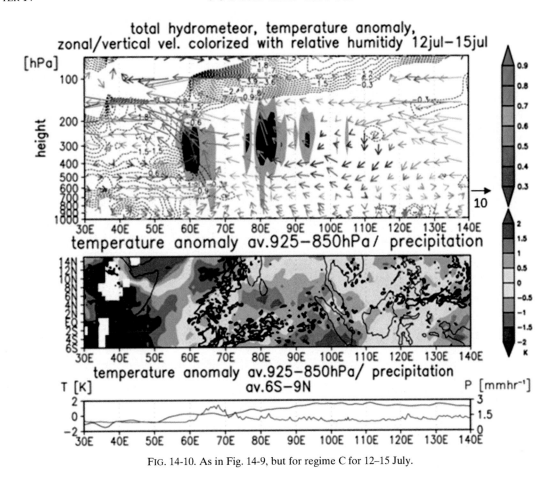

FIG. 14-10. As in Fig. 14-9, but for regime C for 12–15 July.

easterly outflow core at the upper level to the west/rear of the entire convective system. This picture again reminds us of a typical feature of a squall line and suggests that the conceptual framework for regime A is plausible.

2) REGIME B

On the other hand, the vertical cross section (top panel) does not show close similarity between the simulation and the observation (Fig. 14-12) in terms of the structures of convection, but the low-level zonal wind is characterized by the common features of the predominant westerly. As a key feature of regime B, there is evidence of correspondence between the relatively less organized convection and the underlying positive temperature anomaly (i.e., a situation contrary to a cold reservoir), both in the simulation (Fig. 14-9, middle) and observation (Fig. 14-12, middle), which may be a result of inactive or decayed convective activity as compared to regime A.

3) REGIME C

Both in the simulation (Fig. 14-10) and the observation (Fig. 14-13), the flow fields (top panel) are characterized by a less-clear signature typical of serene conditions

when convection is relatively inactive as compared with regimes A and B. The situation is consistent with this particular period when the MJO center propagates eastward away from the warm-pool region and reaches the Western Hemisphere (Fig. 14-3).

In summary, the comparison between the simulation and observation reveals a reasonable correspondence for regime A and, to a lesser degree, regimes B and C, in terms of the qualitative relationships among the key physical parameters constituting the proposed hypothesis. Given the nature of the systematic influences of the MJO on the intensification of synoptic-scale tropical convective activity for this particular period (Oouchi et al. 2009), the regime-dependent correspondences seem to be acceptable as the absence or the weaker phases of the MJO is likely to diminish the structured pattern of tropical convection and the related circulation.

5. Interpretation

a. Three-dimensional conceptual schematic

Figures 14-14 and 14-15 illustrate the schematics summarizing the results and associated inferences derived

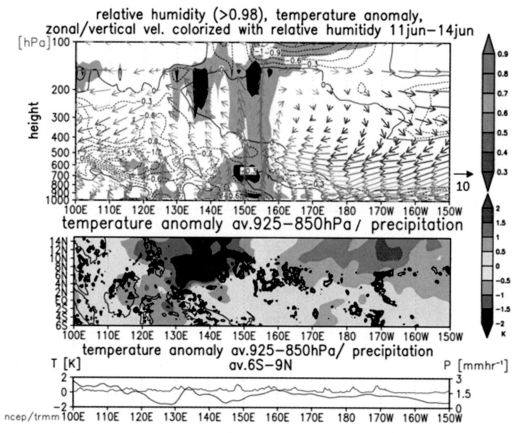

FIG. 14-11. (top) Vertical cross section (average over 6°S–9°N) of high (>0.98) relative humidity areas (color shading), zonal and vertical velocity (vectors) that are colorized with relative humidity (see the color bar for values), and "negative" temperature anomaly (dotted contours) for 100°E–150°W. These are averaged over the consecutive 4-day periods during regime A for 11–14 June. (middle) Horizontal distribution of the temperature anomaly for the layer (925–850 hPa, shading) and precipitation (contours) in the corresponding area and period. (bottom) Temperature anomaly for the layers (925–850 hPa) from the area average (black), and the area-averaged precipitation (blue) in the corresponding area and period. Note that y-axis labels are on the left for the temperature and the right for the precipitation. The datasets are from TRMM 3B42 for precipitation and NCEP analyses for the others.

from the above findings. First, we stand on the premise that contributions of wave-associated characteristics of the MJO and the equatorial waves including moist Kelvin waves are undoubtedly important in explaining some aspects of the SCC, including its organization and behavior (e.g., eastward movement). Details of the comprehensive analysis of the waves will be reported elsewhere. Independent of these factors, the discussion in this chapter focuses rather on the interaction between the ensemble of CCs and the warm-pool-induced climatological vertical circulation as well as on a possible background modulation of CCs by the MJO. This facilitates clarifying why SCCs are preferably organized in the western Pacific region and how the climatological vertical circulation contributes to maintaining the SCC.

Figure 14-14 depicts a large-scale overview of the boreal summer atmosphere of the Asian monsoon–susceptible region. As a basic environmental factor, the existence of the warm water pool in the western Pacific is very important, with the ascending motions staying over the warmest sea surface temperatures. Highlighted here are the MJO and the organized convection in the Indian Ocean (to the west of the warm water pool) and in the western-to-central Pacific (to the east of the warm water pool). Since a large part of the tropical convection is known to exist as a form of CC (Nakajima and Matsuno 1988), we call the convection ensemble exceeding the scale of CC a cloud cluster complex (CCC). An SCC can be interpreted as a highly organized special form of CCC in the eastern branch of the quasi-stationary climatological circulation under some influence of the MJO. These atmospheric components are classified into three regimes depending on the possible background influence of the MJO (Fig. 14-3; Taniguchi et al. 2010; Oouchi et al. 2009).

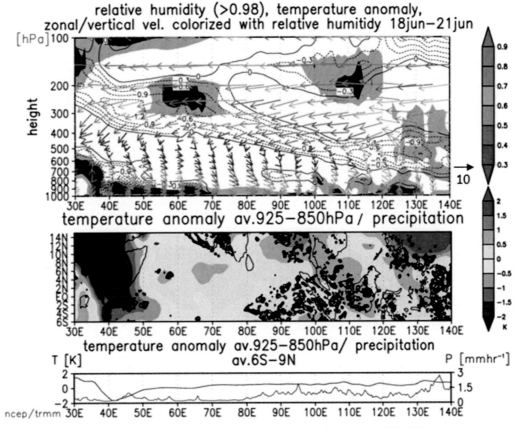

relative humidity (>0.98), temperature anomaly,
zonal/vertical vel. colorized with relative humitidy 18jun–21jun

temperature anomaly av.925–850hPa / precipitation

temperature anomaly av.925–850hPa/ precipitation
av.6S–9N

FIG. 14-12. As in Fig. 14-11, but for regime B for 18–21 June, and 30°–140°E.

The first regime (regime A) is the active MJO period, in which the center of the MJO propagates over the east of the Maritime Continent, and the quasi-stationary warm pool circulation and convective mode undergo significant modulation. The regime is characterized by a preponderance of SCCs in the western Pacific and the genesis of a tropical cyclone (TC). Both are important components of the MJO-related signal that appear in a combined form of Kelvin and Rossby waves (e.g., Matsuno 1966; Gill 1980; Lau and Peng 1987; Miyahara 1987; Oouchi et al. 2009). On the other hand, in the Indian Ocean, CCs stay transient, or the SCC is not coherently maintained as the region is under the influence of a strong low-level westerly jet associated with the Somali jet.

In the second regime (regime B) called the "decay of the MJO period," the MJO (strictly a Kelvin wave signature) propagates away from the central Pacific, and low-level westerly anomalies overtake the western Pacific region. The low-level westerly jet also becomes enhanced and stretches up to the southwest of the Maritime Continent (see the change of polarity of PC 1 and PC 2 late in June in Fig. 14-4). The intensified jet is obviously influenced by the effect of the

MJO-associated anomalous low-level westerly; we call this westerly anomaly of sizable extent the Somali–MJO westerly burst (SMWB). The SMWB is sometimes observed in the boreal summer monsoon season (Lestari and Iwasaki 2006). The Somali–MJO combined flow may originate essentially from a westerly jet in response to a heat source imposed somewhere north of the equator in the typical monsoon season. As the center of the action of the MJO shifts northward with respect to the equator in this season, the low-level anomalous westerly of the MJO can act to enhance and enlarge the original westerly jet. The realistic setup (topography and sea surface temperature) may modify this simple picture by strengthening the linkage between the SMWB and the land–ocean coupled system, and as a result, the region of the westerly jet stretches across a wide-spanning area from the Arabian Sea to the southeastern region of Asia. Under this regime, the SCC decays as the control by MJO's control subsides, and some of the vortical disturbances in the wake of the MJO's westerlies grow into a tropical cyclone (Miura et al. 2007; Oouchi et al. 2009).

The third regime (regime C), the inactive MJO period, is characterized by a revisit or reintensification of

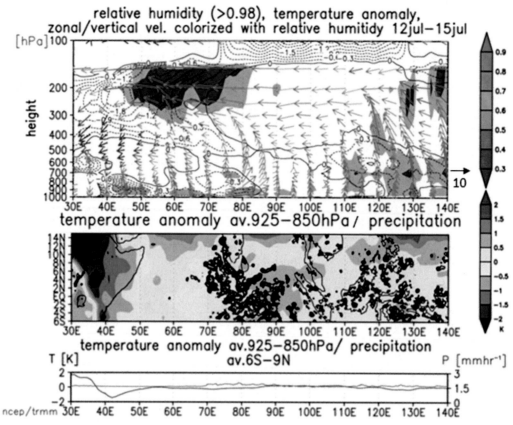

FIG. 14-13. As in Fig. 14-12, but for regime C for 12–15 July.

the MJO-associated circulation in the Indian Ocean. It is a convectively quiescent period both in the Indian Ocean and the western Pacific, with some transient CCs developing and sometimes loosely organizing into CCCs (see the negative polarity of PC 2 after 10 July in Fig. 14-4).

The hypothesized control (propagation) of the MJO over the three regimes above is a possible enhancement of the moisture convergence to the east of the Maritime Continent, its interaction with CCs in that area, and the intensification of the SMWB, in combination with typical monsoon atmospheric conditions. Most of the features may have already been discussed in previous studies, and the schematic is just to reorganize the simulated situations in and around the warm water pool region. Investigation of these factors is beyond the scope of this study and will be an important future work.

b. Two-dimensional (vertical–zonal) conceptual schematic

To focus again on the organization of a CC into an SCC and the mechanism of its maintenance, the schematic of the SCC in association with the vertical circulation is shown in Fig. 14-15. The question here is what

vertical shear regime is favorable for maintaining the SCC? Here, we propose a simple interpretation in terms of its analogy to a multicellular squall line. This should facilitate conceptual understanding. We assume that the convection under consideration is in the near-equatorial open ocean in the tropics and under a weaker rotational constraint. The schematic illustrates two types of convection, SCC and suboptimal SCC, as a counter example of SCC. The latter is suboptimal in the sense that there would be little chance such a type of organization would develop under the environmental factors in that region. The SCC, as was frequently observed in the western Pacific, is under the quasi-stationary climatological westerly shear regime as a result of the existence of the warm water pool. Given its eastward propagation and its inclusion of CCs exhibiting the successive formation of the new CC to the east while the matured one is accompanied by a trailing upper anvil to the west, the SCC generates a cold pool and heavier rainfall to its more western part (Fig. 14-8). From another viewpoint, the active core of the SCC, or the active CC inside the SCC, tilts westward with height as suggested by the westward tilt of the updraft core in SCC (Fig. 14-8). This is

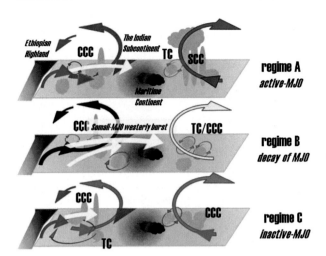

FIG. 14-14. Conceptual schematic of the quasi-3D, simplified convective features in the Indian Ocean and the western Pacific for regimes (top) A, (middle) B, and (bottom) C.

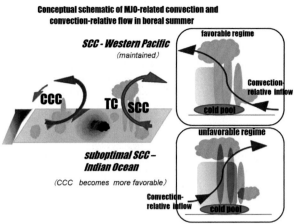

FIG. 14-15. (left) Conceptual schematic of the vertical structure and associated zonal–vertical circulation of the (top right) SCC and (bottom right) suboptimal SCC. Details in text.

consistent with an analysis of the convective momentum transport associated with SCCs that indicated an upward transport of easterly momentum (Moncrieff and Klinker 1997). Now that moist low-level flows move into the SCC from the east, the resultant interaction between the easterly and the cold pool is realized in such a way that heavier precipitation does not prevent the formation of a new CC on the eastern side of the SCC. The SCC is thus favorably maintained under the vertical shear regime.

On the other hand, if the suboptimal SCC in the Indian Ocean was in place, it would be under the easterly shear regime, as the region is to the west of the warm water pool. Compared to the western Pacific SCC, the reversed line of reasoning explains the less favorable condition for SCCs there. That is, moist low-level westerly flows move into the SCC from the west, and the resultant interaction between the westerly and the cold pool (Fig. 14-10) would occur in the region of heavier precipitation. This would be unfavorable for maintaining the ensemble of CCs, and therefore, the SCC would decay. Instead of an SCC, such a vertical shear regime might favor less organized, transient forms of convection (i.e., CCCs). The CCC involves no systematically aligned multicellular CCs. What is argued here is, in essence, analogous to that for a tropical squall line and the convection associated with an easterly wave (Yamasaki 1984), albeit in the larger (synoptic scale) context here.

The theory of squall line structure by RKW is helpful for interpreting and concisely defining the two situations. An SCC of $O(1000)$ km scale is interpretable as a gigantic analog of a multicellular squall line embedded in the quasi-stationary westerly shear branch of the zonal circulation east of the warm water pool. A CC corresponds to the "cell," and its successive formation to the east and westward movement represents an upshear-tilting core of intense updraft. The upshear-tilted SCC is favorably maintained with the precipitating area being separated from the gust-front boundary between the cold reservoir and a low-level easterly (the situation of RKW, where two horizontal vortices associated with the cold reservoir and vertical shear are opposite in sign, but the cold reservoir's vorticity can be inferred to be larger, leading to upshear-tilted and multicellular behavior). As a counterexample, a CC to the west of the warm pool (Indian Ocean and Arabian Sea) is embedded in the easterly shear and, from the reversed line of reasoning, organized into a less coherent CCC given the SCC's geometry as defined above (the situation of RKW where two horizontal vortices associated with the cold reservoir and vertical shear are still opposite in sign, but the smaller vertical shear west of the warm pool causes even more suboptimal vorticity imbalance in the western flank of cold reservoir, leading to larger tilt with height and intermittent less viable storm situations). Note that the above interpretation in no way precludes the widely held viewpoint of interpreting the CC as the inner element of the SCC and in terms of most equatorial waves (e.g., Takayabu 1994; Wheeler and Kiladis 1999). Instead, we offer a simple, yet intuitively plausible, alternative interpretation of the origin of the SCC.

6. Some implications to test the hypothesis

To organize the new features and related inferences from this study, additional notes are presented in this

TABLE 14-1. Comparison of some features and related proposed mechanisms between SCC and MCSL in the tropical atmosphere. The items proposed in this chapter are in bold. Note that the list is not all inclusive.

	SCC	MCSL
Horizontal scale and its origin	3000–5000 km (Nakazawa 1988) • **Area affected by a combination of MJO circulation and warm-pool-induced easterly shear** • The equatorial radius of deformation (Hayashi and Sumi 1986)	$O(100)$ km • Nearly homogeneous (quasi-equilibrium, Fovell and Ogura 1988) condition where convective motion is sustained • Latent instability of the tropical atmosphere
Major building block	• Cloud cluster (CC) • Mesoconvection (MC)[a] embedded in CC (Oouchi 1999)	• Convective cell (thunder storm) or cumulus convection (Zipser 1977)
Mechanism for the building block	• Cold pool/mesoscale gravity wave (Oouchi 1999) • 2-day inertia–gravity wave plus 1-day convective regime (Chen and Houze 1997)	• CISK (cumulus cloud) under conditionally unstable atmosphere
Mechanism for the entire system	• **Warm-pool-induced circulation/vertical shear and cold reservoir** The other proposed mechanisms include • unstable Kelvin wave (Hayashi and Sumi 1986) • diabatic heating of westward-propagating CC (Takayabu and Murakami 1991) • MC-related mechanism (cold pool/mesoscale gravity wave) plus Kelvin-wave instability (Oouchi and Yamasaki 2001)	• Significant low- to midlevel vertical shear • Cold pool • Conditionally unstable atmosphere

[a] More generally, MC systems are preferred forms under latent instability condition, as argued in Matsuno et al. (2011).

section. These instructive notes can serve as a guide to motivate future studies on the origin of the SCC. This chapter highlights the role of the vertical shear and the cold reservoir inherent in the convective activity associated with SCCs, employing an analogy to multicellular squall lines (MCSLs).

Table 14-1 summarizes a comparison of some notable characteristics between SCCs and MCSLs. Each of the identified or inferred characteristics represents a possible pedagogic implication that will be further explored and validated in a future study. General comparisons can be made involving, horizontal size, possible origin, major building blocks of convection, and mechanisms of maintenance, as well as the entire system. Following are notes for some related points included or not covered in the table.

a. Horizontal scale

One might ask about the origin of the scales of the SCC (3000–5000 km) and the MCSL [$O(100)$ km] as a premise for comparison. An inference we could make is, as indicated in Table 14-1, SCCs may be related to the equatorial radius of deformation that determines the maximum limit of the convection scale in a certain state of equilibrium. The association of the equatorial radius of deformation with the size of a diabatically induced tropical large-scale convection around the equator was also proposed in the aquaplanet experiments by Hayashi and Sumi (1986). They inferred that the scale of $O(1000)$ km is the largest stable state of convection and

one plausible manifestation of organized tropical convection in the tropics, although its internal dynamics had been left for future investigation.

On the other hand, a typical scale of MCSL measures $O(100)$ km, which is interpreted as a preferred scale with the specific form of organized convection under preferred atmospheric conditions, including vertical shear profile and strength.

The other important feature of the SCC and the MCSL that is associated with the scale consideration and suggested in this chapter includes the morphological similarity, in that successive generation and development of convective cells or CCs at different stages coexist in a leading-edge, normal-direction line under a specific vertical shear environment (i.e., easterly shear for SCC). This feature may also be important to determine the scale, because a typical configuration for organizing (enlarging) the entire system size consists of individual convective elements (Takeda 1971); in other words, the size is no smaller than the distance propagated by the mesoscale convection (CC) evolving in the single rearward direction, at a certain velocity range from the leading edge at a nearly periodic genesis rate [shorter than half a day (a few days)] for the MCSL (SCC). This point will be investigated in a future study.

b. Vertical shear and the organization of CCs (regimes A, B, and C)

Vertical shear is known to control large-scale organization of tropical convective systems (Yamasaki 1988), and this chapter explores the further implications associated with SCCs, as mentioned above in the "horizontal size" issue. In MCSLs, vertical shear is necessary[1] to cause successive generations of new convective cells, thereby creating longer-lived (longer than that of individual cumulus clouds), organized convective systems at $O(100)$ km scale. Likewise, in SCCs, the vertical shear is helpful for successive generations of new cloud clusters, thereby creating and maintaining its ensemble at $O(1000)$ km scale.

An underlying emphasis of this chapter is the role of the cold pool (or cold reservoir) that can act to cooperate with the vertical shear for the organization and longevity of the entire convective system. This chapter focuses on this aspect rather than on other possible effects (e.g., equatorial waves) for the organization of SCCs. The idea is simple enough. As long as convective activity persists, evaporation from the rain maintains downdrafts and cold air. The initial cooling creates a cold pool that spreads upstream at low levels, thus producing convergence with the ambient warmer air/flows. The location of the convergence and its effects on the longevity of the entire system are controlled by the tilt of the convection. It is likely to dissipate if a downshear updraft slope releases rain into the buoyant inflow air; the upshear slope of the updraft is therefore typical of tropical squall lines (Thorpe et al. 1982).

In terms of the SCC, similar reasoning should be applicable when cold pool is replaced with a cold reservoir and the ambient vertical shear is assumed to be affected by warm-pool-induced circulation. The preference of SCCs or other regimes of organization also can be determined by how long and how much the vertical shear works to sustain the CC's organization. This is, in a sense, analogous to the longevity and maintenance of a given squall line in a nondestructive (homogeneous) environment, as suggested by Fovell and Ogura (1988). They indicated that a state of quasi-equilibrium can be attained and that squall lines do not decay as long as environmental conditions ahead of the storms are favorable and essentially unchanged during the course of the simulations. It can be postulated in this chapter that the environmental homogeneity in SCCs is controlled by the location of the SCC with respect to the tropical warm pool that creates the background shear for it.

c. Possible association of the Kelvin wave

One of the important questions related to the SCC is what mechanism sustains its eastward movement? If regarded as an analog of the MCSL, as assumed in this chapter, it is interesting to explore how the arguments here can or cannot be reconciled to the existing theoretical framework on equatorial Kelvin waves (Hayashi and Sumi 1986).

A key viewpoint is if the coexistence of the Kelvin wave and the MCSL-based mechanism above can be realized, and which contribution is more dominant in SCC. A hint as to the answer comes from Oouchi (1999), who noted coexistence of the synoptic-scale extension of the cold pool and an eastward-propagating gravity wave convective instability of the second kind (CISK) with a 2D cloud-resolving model for the realization of the SCC. As an unstable mode, the CISK worked to maintain a positive correlation between temperature anomaly and vertical motion (generating kinetic energy), in addition to a collocation of convective heating and synoptic-scale upward motion (generating available potential energy). One may find it interesting to relate the low- to midtropospheric negative temperature signal as a part of a cold reservoir behind the main convective region of the SCC (regime A, Fig. 14-8) to a part of the typical temperature signal associated with eastward-propagating gravity (Kelvin) waves. A close inspection of the temperature anomaly would confirm that the temperature is generally lower to the west of the active precipitation center than to the east. As the convection-associated low-level flow is westerly to the west, and easterly to the east of the precipitation center, this indicates that zonal velocity and temperature fields are negatively correlated with each other. Therefore, SCCs and CCCs can be somehow associated with eastward-propagating gravity waves, whatever the degree of coupling that might turn out to be significant. This is an important feature associated with the mechanism of generation of the eastward-propagating equatorial wave and its interaction with convections, which is a subject for investigation at the next step.

Note that the CISK defined in Oouchi (1999) is more comprehensive than what is generally thought in that it includes the effect of the cold pool. In the study, a cold pool (reservoir) spanning the synoptic SCC scale can create an ensemble of mesohighs and therefore horizontal pressure gradients and induced synoptic-scale circulation—a kind of rear-to-front/front-to-rear flow as seen in a squall line. The argument forms a basis for the proposed hypothesis of this chapter. We need further studies and observations of SCCs to assess the MCSL-related hypothesis and its relation to Kelvin-wave control on SCCs. It is hoped that the arguments above will solicit further discussion to better understand this aspect.

---

[1] Given RKW, vertical shear that can at least partially oppose the cold pool circulation is needed for storms to be viable enough to be unsteady.

7. Discussion and remarks

This chapter provides a simplified interpretation of the origin of large-scale convection at its mature stage as associated with the MJO [or super cloud cluster (SCC)] by focusing on the existence of the warm water pool in the western Pacific and its accompanying climatological zonal circulation. The interpretation is proposed as a hypothesis that needs to be tested in future investigations using observational and various modeling datasets of MJO-related convection. A drastic simplification of the hypothesis is that the SCC is independent of the widely held viewpoint of its association with equatorial wave dynamics. Even though we do recognize that the equatorial waves play an important role in producing some aspects of the SCC, they alone cannot explain its origin and geographical preference (e.g., its dominance over the western Pacific as opposed to the global propagating feature of convection in the aquaplanet experiments). The hypothesis in this chapter proposes an interpretation in a broader working framework in which details of the existing MJO convection-associated theories could be based or complemented.

a. Vertical tilt of convection and convective regime selection

Focus is placed on the presence (absence) of SCCs in the western Pacific (Indian Ocean and Arabian Sea). We propose that the different situation in the oceanic basins is interpretable, by employing an analogy of a multicellular squall line, in terms of the preferred regimes of CCs (SCCs or CCCs) under different regimes of the warm-pool-induced climatological vertical circulation with some modification by the MJO. Given the well-known nature of SCCs in which CCs develop from the eastern portion of the entire system (as shallow convection) and move westward, growing and accompanied by anvils aloft, the center of the action in the entire ensemble of CCs tilts westward with height (upshear tilt) in the warm-pool-induced eastern branch of the climatological zonal circulation. This is a situation in which the interaction between the cold reservoir and low-level easterly inflow to the system stays intact, being separated from the heavier precipitation on the western portion of the system, and therefore SCC is maintained (the situation of RKW where two horizontal vortices associated with the cold reservoir and vertical shear are opposite in sign but cold reservoir's vorticity can be inferred to be larger, leading to upshear-tilted and multicellular behavior). On the other hand, CCCs tend to be short lived or less systematically organized as compared to SCCs, making it harder to observe (the situation of RKW where two horizontal vortices associated with the cold reservoir and vertical shear are still opposite in sign, but the smaller vertical shear west of the warm pool causes even more suboptimal vorticity imbalance in the western flank of cold reservoir, leading to larger tilt with height and intermittent less viable storm situations). Favorable occurrences of SCC in the active phase of the MJO can be explained in terms of this interpretation.

The core concept described above is associated with Yamasaki (1984), who reported a significant relationship between a preferred regime of organized cumulus convection (ensemble of mesoscale convection) that is interactive with tropical large-scale disturbances and the environmental vertical shear and argued that the convection of the upshear-tilting type tends to be long lasting, while that of downshear-tilting type does not. It is surprising that the essence of this picture is also logically consistent with that proposed for a single long-lasting or short-lived cloud regarding the relationship between a cloud's lifetime and a vertical shear regime (Takeda 1971). A recent modeling study has also indicated that the westward tilt is an important feature of MJO-associated convection in which convective momentum transport can support its coherent organization and slow phase speed (Miyakawa et al. 2012). It remains controversial if upshear-titled or downshear-tilted convection is an optimal or suboptimal state for the storm's strength even when applied to multicellular storms. At least we can note that convection under an upshear-tilted regime is likely to be more multicellular (Fovell and Ogura 1988), which is relevant to the types of convection both in the western Pacific and the Indian Ocean in this chapter.

An important factor for controlling the longevity of each convective regime is the cold reservoir. Its realization in NICAM as associated with the diurnal cycle of tropical precipitation was demonstrated by Sato et al. (2009). Although its importance in SCCs and the MJO was proposed by a cloud-resolving model (Oouchi 1999; Oouchi and Yamasaki 2001), the existence of those specific phenomena had not been observationally verified. Evidence is emerging as to the existence of the cold reservoir and its possible association with the development of the MJO over the Indian Ocean during the Cooperative Indian Ocean Experiment on Intraseasonal Variability (CINDY)/DYNAMO period (Yoneyama et al. 2013). Further analysis of the dataset and comparison with model results will serve to assess the importance of the cold reservoir and the hypothesis proposed in this chapter.

b. Eastward-propagating phase velocity of the SCC

The observed SCC propagates eastward against the low-level easterly at the ground-relative phase velocity of less than $10\,\mathrm{m\,s}^{-1}$. Likewise, a multicellular squall line propagates against the surrounding low- to midlevel airflow at a similar velocity. In both systems, system-relative flows contribute to feeding moisture into the inner elements and maintaining the systematic life cycle of inner convective structures. As a manifestation of this, whereas a squall line includes successive generations of new cells in the upstream side of the existing cells, the SCC follows a similar cycle for the CC. Determining the mechanism for creating the larger-scale moisture field by which the SCC's eastward propagation is prompted is an important problem from a different perspective, and we may need to clarify the interaction between the SCC and equatorial waves as a possible control for this aspect.

To further pursue the characteristic of the eastward propagation of the SCC, it is tempting to note that there is a significant difference in the feature of propagation of SCCs between the aquaplanet setup and the realistic setup of topography and SST (existence of the warm water pool). The typical SCC in the aquaplanet setup propagates over the equatorial circumference at a phase velocity of $15–25\,\mathrm{m\,s}^{-1}$, both in the conventional GCMs (Hayashi and Sumi 1986; Numaguti and Hayashi 1991) and global cloud-resolving models (GCRMs; Satoh et al. 2005; Tomita et al. 2005; Nasuno et al. 2007). This is faster than the observed velocity ranging typically from less than $5\,\mathrm{m\,s}^{-1}$ (Nakazawa 1995) to $10°–12°\,\mathrm{day}^{-1}$ (less than $15\,\mathrm{m\,s}^{-1}$; Takayabu and Murakami 1991). On the other hand, the typical SCC in the realistic setup propagates at a phase velocity similar to the observed one over the limited area of, in, and around the warm water pool, and specifically over the western Pacific. The reason the simulated phase velocity of the SCCs in the realistic setup is slower than that in the aquaplanet setup and is comparable to that of the observed SCC has to do with the existence of the warm water pool or the existence of zonal nonuniformity of the moisture field in the realistic setup, the latter being suggested by Miura et al. (2007) as a key controlling factor of the MJO's propagation.

Here, we take the argument one step further and suggest that the existence of the quasi-stationary westerly shear (low-level easterly) to the east of the warm-pool center and its interaction with the CC of inherently nonpropagating nature are important factors that retard the eastward propagation of the SCC. From the interpretation discussed in this chapter, the westerly shear regime tends to favor the occurrence of SCCs, in which successive generation of new CCs to the east of the existing ones leads to the SCC's propagation. In this situation, the new CC does not propagate so rapidly eastward as does the low-level easterly, and the vertical

shear is capable of putting the center of the action of the CC more westward than otherwise. This speculation is supported when we consider the effects of the convective momentum transport associated with a simulated MJO case (Miyakawa et al. 2012). On the other hand, the opposite tendency—to drive the eastward propagation of the SCC as a whole—can be prompted by an equatorial (Kelvin) wave, although we excluded any discussion on the possible contribution of equatorial waves in this chapter. If the coupling between the SCC or, equivalently, the envelope of the CCs, and the wave is tight enough, eastward propagation would be more favorably prompted under the sufficiently conditionally unstable condition. The observed SCCs, however, suggest that this is not the case. The unrealistically clear propagation of SCCs in the aquaplanet setup is hypothesized to result from either or both of the factors: the absence of the warm-pool-induced vertical circulations and too strong a coupling between the CC and the low-level convergence being forced by a formulation of cumulus parameterization. The speculation needs to be checked against the observed SCCs and the climatology of flow fields.

c. Further remarks

The arguments in this chapter are limited to a highly simplified framework to focus on the basic aspects of SCCs. They excluded investigation into their relationship with environmental disturbances, including equatorial waves and the MJO. More satisfactory explanations as to the origin of SCCs should require considering these waves. The necessary factors still unclear include the Kelvin-wave instability viewpoint, teleconnection via mesoscale waves (Chao and Lin 1994; Oouchi 1999; Mapes 2000), and the buildup and advection of moisture associated with the topography around the Maritime Continent and westward-propagating disturbances (Miura et al. 2007). Additionally, a significant theme that has not been handled appropriately in the conventional GCMs is how the dynamics and thermodynamics of the mesoscale convections work in developing and maintaining the hierarchy of SCCs. Such an investigation was attempted by Oouchi (1999), Grabowski and Moncrieff (2001), and Oouchi and Yamasaki (2001), and more recently by Tulich and Mapes (2008). As suggested by these studies, the origin of mesoscale convection, or cloud clusters in broader meaning, is the key to understanding the mechanism of tropical convection. The classification of convective regimes presented in this chapter should be based on more rigorous criteria by taking into account more quantitative metrics of the organization regimes of mesoscale convection, as well as the activities of equatorial waves. The suggested regimes here are classified solely by the phases of a single MJO and introduced for easier understanding of the hypothesis.

In terms of the multicellular feature of the SCC as an analogy to a squall line, it is interesting to note the view that the multicell feature of a squall line is a manifestation of vertically trapped gravity waves (Yang and Houze 1995). The multi-CC nature of the SCC is likely to be a result of its close linkage with 2-day inertia–gravity waves (Takayabu et al. 1996), or a result of the combined effect of the mesoscale gravity wave dynamics at the initial stage and the mesoscale cold reservoir at the mature stage (Oouchi 1999). A gravity wave control may also play a role in making the upper troposphere cloud last longer (Mapes et al. 2006). Investigating from this viewpoint will further our understanding of the origin of the SCC, and GCRM is a fascinating framework for this purpose.

The interpretation proposed in this chapter may be highly idealized and its elaboration is necessary, as is the test of its applicability to existing observational datasets. Moncrieff (2004) proposed a nonlinear dynamical model of the tropical organized convection associated with the MJO and discussed that the SCC-like organized structure is a self-maintaining system that contains westward-tilted structure with countergradient momentum transport in the equatorial vertical cross section. The present study supports this view and proposed that an additional factor—warm-pool-induced circulation—explains the observed preference or absence of SCCs. It is important to note that, in reality, there may be no clear-cut distinction between SCCs and CCCs and the underlying respective vertical-shear regime. The characteristic climatological environment of the SCC varies on more than a seasonal time scale (Wang and Rui 1990), and this should provide additional important modification to the climatological vertical shear and the manner of organizing the SCC. The influence of the boreal summer Asian monsoon should be enormous over the vertical shear and the MJO (Yasunari 1979) on a seasonal and longer time scale. Among the interactions that could work within the complex monsoon–MJO–SCC network, the simple interpretation proposed in this chapter highlights an aspect of the link between the warm-pool-induced circulation and SCCs; this is expected to offer a new insight into the mechanism of upscale organization of CCs around the tropical warm water pool.

Acknowledgments. The authors appreciate the late Professor Yanai for the encouragement and fruitful discussions on the research in the NICAM group. The insightful and constructive comments by reviewers helped improve the manuscript. The numerical experiments were performed on the Earth Simulator/JAMSTEC, under the framework of the Innovative Program of Climate Change Projection for the 21st century (KAKUSHIN) project "Global Cloud Resolving Simulations toward More Accurate and Sophisticated Climate Prediction of Cloud/Precipitation Systems" funded by Ministry of Education, Culture, Sports, Science and Technology (MEXT). We thank NICAM team members for discussion and support for the simulation and analysis. The original idea of this paper was motivated from the discussions with Drs. R. A. Madden, W. W. Grabowski, and A. Kasahara while the first author stayed at the National Center for Atmospheric Research in 1999 under the JSPS fellowship. The author is grateful to them for the thought-provoking inputs on the tropical large-scale convections and Madden–Julian oscillation. The discussions with Drs. M. Yamasaki, B. E. Mapes, T. Nakazawa, Y. N. Takayabu, and the late Drs. Y. Hayashi and A. Numaguti were also enlightening.

REFERENCES

Chang, C. P., and H. Lim, 1988: Kelvin-wave CISK: A possible mechanism for the 30-50 day oscillations. *J. Atmos. Sci.*, **45**, 1709–1720, doi:10.1175/1520-0469(1988)045<1709:KWCAPM>2.0.CO;2.

Chao, W. C., and S. J. Lin, 1994: Tropical intraseasonal oscillation, super cloud clusters, and cumulus convection schemes. *J. Atmos. Sci.*, **51**, 1282–1297, doi:10.1175/1520-0469(1994)051<1282:TIOSCC>2.0.CO;2.

Chen, S. S., and R. A. Houze, 1997: Interannual variability of deep convection over the tropical warm pool. *J. Geophys. Res.*, **102**, 25 783–25 795, doi:10.1029/97JD02238.

Fovell, R. G., and Y. Ogura, 1988: Numerical simulation of a midlatitude squall line in two dimensions. *J. Atmos. Sci.*, **45**, 3846–3879, doi:10.1175/1520-0469(1988)045<3846:NSOAMS>2.0.CO;2.

Gill, A. E., 1980: Some simple solutions for heat-induced tropical circulation. *Quart. J. Roy. Meteor. Soc.*, **106**, 447–462, doi:10.1002/qj.49710644905.

Grabowski, W. W., 1998: Toward cloud resolving modeling of large-scale tropical circulation: A simple cloud microphysics parameterization. *J. Atmos. Sci.*, **55**, 3283–3298, doi:10.1175/1520-0469(1998)055<3283:TCRMOL>2.0.CO;2.

——, and M. W. Moncrieff, 2001: Large-scale organization of tropical convection in two-dimensional explicit numerical simulations. *Quart. J. Roy. Meteor. Soc.*, **127**, 445–468, doi:10.1002/qj.49712757211.

Gray, W., 1968: Global view of the origin of tropical disturbances and storms. *Mon. Wea. Rev.*, **96**, 669–700, doi:10.1175/1520-0493(1968)096<0669:GVOTOO>2.0.CO;2.

Hayashi, Y.-Y., and A. Sumi, 1986: The 30–40 day oscillations simulated in an "aqua planet" model. *J. Meteor. Soc. Japan*, **64** (4), 451–467.

——, and T. Nakazawa, 1989: Evidence of the existence and eastward motion of superclusters at the equator. *Mon. Wea. Rev.*, **117**, 236–243, doi:10.1175/1520-0493(1989)117<0236:EOTEAE>2.0.CO;2.

Hendon, H., and M. Salby, 1994: The life cycle of the Madden–Julian oscillation. *J. Atmos. Sci.*, **51**, 2225–2237, doi:10.1175/1520-0469(1994)051<2225:TLCOTM>2.0.CO;2.

Houze, R. A., Jr., 1977: Structure and dynamics of a tropical squall-line system. *Mon. Wea. Rev.*, **105**, 1540–1567, doi:10.1175/1520-0493(1977)105<1540:SADOAT>2.0.CO;2.

Julian, P. R., and R. A. Madden, 1981: Comments on a paper by T. Yasunari, A quasi-stationary appearance of 30 to 40-day period in the cloudiness fluctuations during the summer monsoon over India. *J. Meteor. Soc. Japan*, **59**, 435–437.

Lau, K.-M., and L. Peng, 1987: Origin of low-frequency (intra-seasonal) oscillations in the tropical atmosphere. Part I: Basic theory. *J. Atmos. Sci.*, **44**, 950–972, doi:10.1175/1520-0469(1987)044<0950:OOLFOI>2.0.CO;2.

——, T. Nakazawa, and C. H. Sui, 1991: Observations of cloud cluster hierarchies over the tropical western Pacific. *J. Geophys. Res.*, **96**, 3197–3208, doi:10.1029/90JD01830.

Lestari, R. K., and T. Iwasaki, 2006: A GCM study on the roles of the seasonal marches of the SST and land-sea thermal contrast in the onset of the Asian summer monsoon. *J. Meteor. Soc. Japan*, **84**, 69–83, doi:10.2151/jmsj.84.69.

Madden, R. A., and P. R. Julian, 1971: Detection of a 40–50 day oscillation in the zonal wind in the tropical Pacific. *J. Atmos. Sci.*, **28**, 702–708, doi:10.1175/1520-0469(1971)028<0702:DOADOI>2.0.CO;2.

——, and ——, 1972: Description of global-scale circulation cells in the tropics with a 40-50 day period. *J. Atmos. Sci.*, **29**, 1109–1123, doi:10.1175/1520-0469(1972)029<1109:DOGSCC>2.0.CO;2.

Mapes, B. E., 2000: Convective inhibition, subgrid-scale triggering energy, and stratiform instability in a toy tropical wave model. *J. Atmos. Sci.*, **57**, 1515–1535, doi:10.1175/1520-0469(2000)057<1515:CISSTE>2.0.CO;2.

——, and R. A. Houze, 1993: Cloud clusters and superclusters over the oceanic warm pool. *Mon. Wea. Rev.*, **121**, 1398–1415, doi:10.1175/1520-0493(1993)121<1398:CCASOT>2.0.CO;2.

——, S. Tulich, J. Lin, and P. Zuidema, 2006: The mesoscale convection life cycle: Building block or prototype for large-scale tropical waves? *Dyn. Atmos. Oceans*, **42**, 3–29, doi:10.1016/j.dynatmoce.2006.03.003.

Matsuno, T., 1966: Quasi-geostrophic motions in the equatorial area. *J. Meteor. Soc. Japan*, **44**, 24–42.

——, M. Satoh, H. Tomita, T. Nasuno, S. I. Iga, A. T. Noda, and K. Oouchi, 2011: Cloud-cluster-resolving global atmosphere modelling—A challenge for the new age of tropical meteorology. *The Global Monsoon System: Research and Forecast*, 2nd ed., C.-P. Chang et al., Eds., 455–473.

McBride, J. L., N. E. Davidson, K. Puri, and G. C. Tyrell, 1995: The flow during TOGA COARE as diagnosed by the BMRC tropical analysis and prediction system. *Mon. Wea. Rev.*, **123**, 717–736, doi:10.1175/1520-0493(1995)123<0717:TFDTCA>2.0.CO;2.

Miura, H., M. Satoh, T. Nasuno, A. T. Noda, and K. Oouchi, 2007: A Madden-Julian oscillation event simulated using a global cloud-resolving model. *Science*, **318**, 1763–1765, doi:10.1126/science.1148443.

Miyahara, S., 1987: A simple model of the tropical intraseasonal oscillation. *J. Meteor. Soc. Japan*, **65**, 341–351.

Miyakawa, T., Y. N. Takayabu, T. Nasuno, H. Miura, M. Satoh, and M. W. Moncrieff, 2012: Convective momentum transport by rainbands within a Madden–Julian oscillation in a global nonhydrostatic model with explicit deep convective processes. Part I: Methodology and general results. *J. Atmos. Sci.*, **69**, 1317–1338, doi:10.1175/JAS-D-11-024.1.

Moncrieff, M. W., 2004: Analytic representation of the large-scale organization of tropical convection. *J. Atmos. Sci.*, **61**, 1521–1538, doi:10.1175/1520-0469(2004)061<1521:AROTLO>2.0.CO;2.

——, and E. Klinker, 1997: Organized convective systems in the tropical western Pacific as a process in general circulation models: A TOGA COARE case-study. *Quart. J. Roy. Meteor. Soc.*, **123**, 805–827, doi:10.1002/qj.49712354002.

Nakajima, K., and T. Matsuno, 1988: Numerical experiments concerning the origin of cloud clusters in the tropical atmosphere. *J. Meteor. Soc. Japan*, **66**, 309–329.

Nakanishi, M., and H. Niino, 2006: An improved Mellor-Yamada level-3 model: Its numerical stability and application to a regional prediction of advection fog. *Bound.-Layer Meteor.*, **119**, 397–407, doi:10.1007/s10546-005-9030-8.

Nakazawa, T., 1986: Mean features of 30–60 day variations as inferred from 8-year OLR data. *J. Meteor. Soc. Japan*, **64**, 777–786.

——, 1988: Tropical super clusters within intraseasonal variations over the western Pacific. *J. Meteor. Soc. Japan*, **66**, 823–839.

——, 1995: Intraseasonal oscillation during TOGA-COARE IOP. *J. Meteor. Soc. Japan*, **73**, 305–319.

Nasuno, T., H. Tomita, S. Iga, H. Miura, and M. Satoh, 2007: Multiscale organization of convection simulated with explicit cloud processes on an aquaplanet. *J. Atmos. Sci.*, **64**, 1902–1921, doi:10.1175/JAS3948.1.

Noda, A. T., K. Oouchi, M. Satoh, H. Tomita, S. Iga, and Y. Tsushima, 2010: Importance of the subgrid-scale turbulent moist process: Cloud distribution in global cloud-resolving simulations. *Atmos. Res.*, **96**, 208–217, doi:10.1016/j.atmosres.2009.05.007.

Numaguti, A., and Y.-Y. Hayashi, 1991: Behavior of cumulus activity and the structures of circulations in an "aqua planet" model. Part I: The structure of the super clusters. *J. Meteor. Soc. Japan*, **69**, 541–561.

Oouchi, K., 1999: Hierarchical organization of super cloud cluster caused by WISHE, convectively induced gravity waves and cold pool. *J. Meteor. Soc. Japan*, **77**, 907–927.

——, and M. Yamasaki, 1997: Kelvin wave-CISK controlled by surface friction; a possible mechanism of super cloud cluster. Part I: Linear theory. *J. Meteor. Soc. Japan*, **75**, 497–511.

——, and ——, 2001: An MJO-like gravity wave and superclusters simulated in a two-dimensional cumulus-scale-resolving model. *J. Meteor. Soc. Japan*, **79**, 201–218, doi:10.2151/jmsj.79.201.

——, A. T. Noda, M. Satoh, H. Miura, H. Tomita, T. Nasuno, and S. Iga, 2009: A simulated preconditioning of typhoon genesis controlled by a boreal summer Madden-Julian Oscillation event in a global cloud-system-resolving model. *SOLA*, **5**, 65–68, doi:10.2151/sola.2009-017.

Rotunno, R., J. B. Klemp, and M. L. Weisman, 1988: A theory for strong, long-lived squall lines. *J. Atmos. Sci.*, **45**, 463–485, doi:10.1175/1520-0469(1988)045<0463:ATFSLL>2.0.CO;2.

Sato, T., H. Miura, M. Satoh, Y. N. Takayabu, and Y. Wang, 2009: Diurnal cycle of precipitation in the tropics simulated in a global cloud-resolving model. *J. Climate*, **22**, 4809–4826, doi:10.1175/2009JCLI2890.1.

Satoh, M., H. Tomita, H. Miura, S. Iga, and T. Nasuno, 2005: Development of a global cloud-resolving model—A multi-scale structure of tropical convections. *J. Earth Simul.*, **3**, 11–19.

——, T. Matsuno, H. Tomita, H. Miura, T. Nasuno, and S. Iga, 2008: Nonhydrostatic Icosahedral Atmospheric Model (NICAM) for global cloud-resolving simulations. *J. Comput. Phys.*, **227**, 3486–3514, doi:10.1016/j.jcp.2007.02.006.

Sui, C.-H., and K.-M. Lau, 1992: Multiscale phenomena in the tropical atmosphere over the western Pacific. *Mon. Wea. Rev.*, **120**, 407–430, doi:10.1175/1520-0493(1992)120<0407:MPITTA>2.0.CO;2.

Takahashi, M., 1987: A theory of the slow phase speed of the intraseasonal oscillation using the Wave-CISK. *J. Meteor. Soc. Japan*, **65**, 43–49.

Takayabu, Y. N., 1994: Large-scale cloud disturbances associated with equatorial waves. Part I: Spectral features of the cloud disturbances. *J. Meteor. Soc. Japan*, **72**, 433–448.

——, and M. Murakami, 1991: The structure of super cloud clusters observed in 1–20 June 1986 and their relationship to easterly waves. *J. Meteor. Soc. Japan*, **69**, 105–125.

——, K.-M. Lau, and C.-H. Sui, 1996: Observation of a quasi-2-day wave during TOGA COARE. *Mon. Wea. Rev.*, **124**, 1892–1913, doi:10.1175/1520-0493(1996)124<1892:OOAQDW>2.0.CO;2.

Takeda, T., 1971: Numerical simulation of a precipitating convective cloud: The formation of a "long-lasting" cloud. *J. Atmos. Sci.*, **28**, 350–376, doi:10.1175/1520-0469(1971)028<0350:NSOAPC>2.0.CO;2.

Taniguchi, H., W. Yanase, and M. Satoh, 2010: Ensemble simulation of Cyclone Nargis by a global cloud-system-resolving model—Modulation of cyclogenesis by the Madden-Julian oscillation. *J. Meteor. Soc. Japan*, **88**, 571–591, doi:10.2151/jmsj.2010-317.

Thorpe, A. J., M. J. Miller, and M. W. Moncrieff, 1982: Two-dimensional convection in non-constant shear: A model of midlatitude squall lines. *Quart. J. Roy. Meteor. Soc.*, **108**, 739–762, doi:10.1002/qj.49710845802.

Tomita, H., and M. Satoh, 2004: A new dynamical framework of nonhydrostatic global model using the icosahedral grid. *Fluid Dyn. Res.*, **34**, 357–400, doi:10.1016/j.fluiddyn.2004.03.003.

——, H. Miura, S. Iga, T. Nasuno, and M. Satoh, 2005: A global cloud-resolving Simulation: Preliminary results from an aquaplanet experiment. *Geophys. Res. Lett.*, **32**, L08805, doi:10.1029/2005GL022459.

Tulich, S. N., and B. E. Mapes, 2008: Multiscale convective wave disturbances in the tropics: Insights from a two-dimensional cloud-resolving model. *J. Atmos. Sci.*, **65**, 140–155, doi:10.1175/2007JAS2353.1.

Wang, B., and H. Rui, 1990: Synoptic climatology of transient tropical intraseasonal convective anomalies. *Meteor. Atmos. Phys.*, **44**, 43–61, doi:10.1007/BF01026810.

Wheeler, M. C., and G. N. Kiladis, 1999: Convectively coupled equatorial waves: Analysis of clouds and temperature in the wavenumber–frequency domain. *J. Atmos. Sci.*, **56**, 374–399, doi:10.1175/1520-0469(1999)056<0374:CCEWAO>2.0.CO;2.

——, and H. H. Hendon, 2004: An all-season real-time multivariate MJO index: Development of an index for monitoring and prediction. *Mon. Wea. Rev.*, **132**, 1917–1932, doi:10.1175/1520-0493(2004)132<1917:AARMMI>2.0.CO;2.

Yamasaki, M., 1984: Dynamics of convective clouds and "CISK" in vertical shear flow—With its application to easterly waves and squall-line systems. *J. Meteor. Soc. Japan*, **62**, 833–863.

——, 1988: Towards an understanding of the interaction between convection and the large-scale in the tropics. *Aust. Meteor. Mag.*, **36**, 171–182.

Yang, M.-J., and R. A. Houze, 1995: Multicell squall-line structure as a manifestation of vertically trapped gravity waves. *Mon. Wea. Rev.*, **123**, 641–661, doi:10.1175/1520-0493(1995)123<0641:MSLSAA>2.0.CO;2.

Yasunari, T., 1979: Cloudiness fluctuations associated with the northern hemisphere summer monsoon. *J. Meteor. Soc. Japan*, **57**, 227–242.

Yoneyama, K., C. Zhang, and C. N. Long, 2013: Tracking pulses of the Madden–Julian oscillation. *Bull. Amer. Meteor. Soc.*, **94**, 1871–1891, doi:10.1175/BAMS-D-12-00157.1.

Zipser, E. J., 1977: Mesoscale and convective-scale downdrafts as distinct components of squall-line structure. *Mon. Wea. Rev.*, **105**, 1568–1589, doi:10.1175/1520-0493(1977)105<1568:MACDAD>2.0.CO;2.

Chapter 15

Simulations of the Tropical General Circulation with a Multiscale Global Model

DAVID RANDALL,* CHARLOTTE DEMOTT,* CRISTIANA STAN,[+] MARAT KHAIROUTDINOV,[#]
JAMES BENEDICT,[@,&] RACHEL MCCRARY,** KATHERINE THAYER-CALDER,[++,##] AND MARK BRANSON*

*Colorado State University, Fort Collins, Colorado
[+] George Mason University, Fairfax, Virginia
[#] Stony Brook University, Stony Brook, New York
[@] Lawrence Berkeley National Laboratory, Berkeley, California
**NCAR, Boulder, Colorado
[++] University of Wisconsin–Milwaukee, Milwaukee, Wisconsin

ABSTRACT

Cloud processes play a central role in the dynamics of the tropical atmosphere, but for many years the shortcomings of cloud parameterizations have limited our ability to simulate and understand important tropical weather systems such as the Madden–Julian oscillation. Since about 2001, "superparameterization" has emerged as a new path forward, complementing but not replacing studies based on conventional parameterizations. This chapter provides an overview of work with superparameterization, including a discussion of the method itself and a summary of key results.

1. Introduction

The tropics strongly absorb solar radiation at all times of year. Much of that solar energy is used to evaporate water at Earth's surface. The water vapor thus added to the tropical atmosphere can be lofted to form beautiful convective cloud systems, which are organized on scales ranging from a few kilometers to many thousands of kilometers. The resulting weather systems include squall lines, tropical cyclones, and continent-spanning monsoons. The intense cloudy convective updrafts occupy only a tiny fraction of the tropics, but they carry prodigious amounts of energy upward through the depth of the troposphere. The outflows from the tops of the convective towers feed divergent upper-tropospheric winds, which carry energy poleward, and so contribute to Earth's global energy balance. This cloud-filled multiscale tropical circulation was the main subject of Michio Yanai's research.

In atmospheric global circulation models (GCMs), the small-scale processes at work in convective cloud systems must be included through simplified submodels called parameterizations (Fig. 15-1). Cloud parameterizations have been under development since the 1960s. A lot of progress has been achieved, but major difficulties remain (Randall et al. 2003a).

In 1999, National Center for Atmospheric Research (NCAR) scientists Wojciech Grabowski and Piotr Smolarkiewicz created a "multiscale" GCM in which the physical processes associated with clouds were represented by implementing a simple "cloud-resolving" model (CRM) within each grid column of a low-resolution global model (Grabowski and Smolarkiewicz 1999; Grabowski 2001, 2004). In this approach, parameterizations of radiation, cloud microphysics, and turbulence (including small clouds) are still needed (Fig. 15-2), but larger clouds and some mesoscale processes are explicitly (though crudely) simulated.

In highly idealized experiments, Grabowski and Smolarkiewicz found that their multiscale model

---

[&] Current affiliation: Rosenstiel School of Marine & Atmospheric Science, University of Miami
[##] Current affiliation: NCAR

Corresponding author address: David Randall, Atmospheric Science, Colorado State University, 200 W. Lake St., Fort Collins, CO 80523.
E-mail: randall@atmos.colostate.edu

DOI: 10.1175/AMSMONOGRAPHS-D-15-0016.1

Parameterized

FIG. 15-1. In conventional GCMs, the global circulation is explicitly simulated. Parameterized processes include cloud-scale and mesoscale dynamics, and also radiative transfer, microphysics, and turbulence.

produced promising simulations of organized tropical convection. In particular, the simulations showed propagating convective systems that resembled the Madden–Julian oscillation[1] (MJO; Madden and Julian 1971, 1972). The MJO is an eastward-propagating tropical disturbance that spans thousands of kilometers in the zonal direction, with an irregular period in the range 40–50 days. Despite its large spatial and temporal scales, and its powerful effects on tropical weather, the MJO has proven very difficult to simulate with GCMs (e.g., Lin et al. 2006; Kim et al. 2009).

Inspired by the results of Grabowski and Smolarkiewicz, Khairoutdinov and Randall (2001) created a multiscale version of the Community Atmosphere Model (CAM; Collins et al. 2006). They disabled the parameterizations of the CAM and replaced them with a simplified version of Khairoutdinov's CRM (Khairoutdinov and Randall 2003). One copy of the CRM runs in each grid column of the CAM. The CRM is two-dimensional (one horizontal dimension, plus the vertical) and uses periodic lateral boundary conditions. In the study of Khairoutdinov and Randall (2001), the CRM had a horizontal domain 64 grid columns wide, with a horizontal grid spacing of 4 km.

Khairoutdinov and Randall dubbed the embedded CRM a "superparameterization." The combination of a GCM with a superparameterization is now called a multiscale modeling framework (MMF), and the MMF based on the CAM is now called the SP-CAM. A second MMF was created by Tao et al. (2009), using a different GCM and a different CRM. As of spring 2014, three more MMFs are being tested, each based on a different GCM.

_____
[1] The senior author of this paper first learned about the MJO and entraining plumes in classes taught by Prof. Michio Yanai.

2. Coupling the CRM and the GCM

Figure 15-3 schematically illustrates the coupling of the embedded CRM with the GCM, as implemented in the SP-CAM. The GCM provides advective forcing to the CRM, much as observed advective forcing can be supplied to the "column physics" of a model that is being tested in simulations based on field data (e.g., Randall et al. 1996). The CRM provides heating and drying as feedback to the GCM, just as any conventional parameterization would do.

The GCM's time step is typically tens of minutes, whereas the CRM's time step is on the order of 10 seconds. It is assumed (and required) that the GCM's time step is an integer multiple of the CRM's time step. The CRM "subcycles" through a sequence of its short time steps, in order to determine the physical tendencies averaged over one of the longer GCM time steps. The CRM runs continuously throughout a simulation; it is not restarted between GCM time steps.

The coupling between the GCM and the CRM is very simple and has some nice properties. It was used by Khairoutdinov and Randall (2001) and in all subsequent work with the SP-CAM. Here is how it works: Let q be a generic scalar variable that is defined in both the GCM and CRM. We write

$$\widetilde{q_G^{n+1}} \equiv q_G^n + B_G \Delta t_G, \qquad (15\text{-}1)$$

so that

$$B_G = \frac{\widetilde{q_G^{n+1}} - q_G^n}{\Delta t_G}. \qquad (15\text{-}2)$$

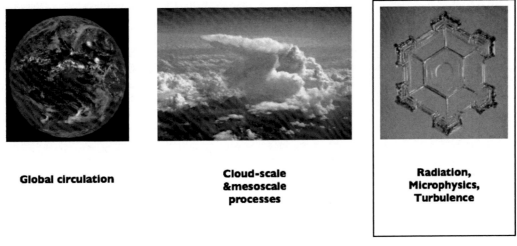

FIG. 15-2. When a CRM is used as a superparameterization in a GCM, the cloud and mesoscale dynamics are explicitly simulated, but radiation, microphysics, and turbulence must still be parameterized on the CRM's grid. Compare with Fig. 15-1.

Here B_G is the adiabatic tendency of q due to non-CRM effects, as determined by the GCM; the subscript G denotes a GCM value; the superscripts n and $n + 1$ denote successive GCM time steps; and Δt_G is the size of the GCM time step. The tilde on $\widetilde{q_G^{n+1}}$ denotes a provisional value. Equation (15-1) represents a "partial time step" before the diabatic tendency of q is accounted for by the CRM.

The CRM variables, denoted by q_C, are updated using CRM time steps of size Δt_C:

$$\frac{q_C^{m+1} - q_C^m}{\Delta t_C} = B_C + \left(\frac{\widetilde{q_G^{n+1}} - \langle q_C \rangle^n}{\Delta t_G} \right) + S_C. \quad (15\text{-}3)$$

Here the superscripts m and $m + 1$ denote successive CRM time steps. In (15-3), the term B_C represents the adiabatic tendency of q_C due to advection on the CRM's grid. Pointy brackets denote a horizontal average over the CRM's domain. The quantity $\langle q_C \rangle^n$ is the CRM-domain average of q_C at the beginning of the GCM time step. The term $[(\widetilde{q_G^{n+1}} - \langle q_C \rangle^n)/\Delta t_G]$ is the mechanism through which the GCM's advective tendencies are felt by the CRM; in other words, it represents the forcing of the CRM by the GCM. The forcing is held constant as the CRM subcycles through one GCM time step, and it is also independent of horizontal position on the CRM's grid. The source of q due to the CRM physics is denoted by S_C, which has no counterpart in (15-1) because the physics is computed only in the CRM, not in the GCM.

We take enough CRM time steps to span the longer GCM time step, finally arriving at GCM time level $n + 1$. Then we let the CRM feed back on the GCM. To implement the feedback, the GCM variables are updated using

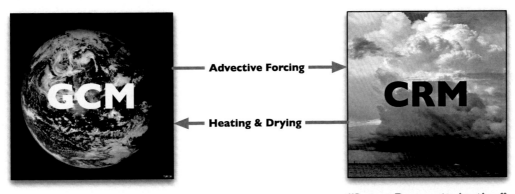

FIG. 15-3. Schematic illustrating the coupling of the GCM and the CRM, in the SP-CAM.

$$\frac{q_G^{n+1} - q_G^n}{\Delta t_G} = B_G + \frac{\langle q_C \rangle^{n+1} - \widetilde{q_G^{n+1}}}{\Delta t_G}$$

$$= B_G + \frac{\langle q_C \rangle^{n+1} - (q_G^n + B_G \Delta t_G)}{\Delta t_G}$$

$$= \frac{\langle q_C \rangle^{n+1} - q_G^n}{\Delta t_G} \qquad (15\text{-}4)$$

Here $\langle q_C \rangle^{n+1}$ represents the horizontal average of q_C at GCM time level $n + 1$, that is, at the end of the sequence of CRM time steps based on (15-2).

Comparing the left-hand side of (15-4) with the third line on the right-hand side, we see that

$$q_G^{n+1} = \langle q_C \rangle^{n+1}. \qquad (15\text{-}5)$$

Of course, we can also write

$$q_G^n = \langle q_C \rangle^n. \qquad (15\text{-}6)$$

The implication is that, at both the beginning and end of the GCM's time step, the GCM's value of q_G is guaranteed to agree with the horizontal average of the CRM values. The two models cannot "drift apart."

Use of (15-1) in (15-3) shows that the CRM variables evolve according to

$$\frac{q_C^{m+1} - q_C^m}{\Delta t_C} = B_C + B_G + S_C. \qquad (15\text{-}7)$$

This means that the CRM feels the GCM advection as a simple additive term. If we horizontally average (15-7) over the CRM's grid, we get

$$\frac{\langle q_C \rangle^{m+1} - \langle q_C \rangle^m}{\Delta t_C} = \langle B_C \rangle + B_G + \langle S_C \rangle. \qquad (15\text{-}8)$$

The CRM's periodic lateral boundary conditions imply that $\langle B_C \rangle$ involves only vertical transports. Finally, from (15-5), (15-6), and (15-8), we see that

$$\frac{q_G^{n+1} - q_G^n}{\Delta t_G} = B_G + \overline{\langle B_C + S_C \rangle}, \qquad (15\text{-}9)$$

where the overbar on the right-hand side represents an average over the GCM time step. The interpretation of (15-9) is simple and clear.

The key points to take away from the discussion above are that the coupling strategy represented by (15-1)–(15-3) leads to the result in (15-9), and also guarantees (15-5) and (15-6). The coupling strategy therefore guarantees consistency between the GCM fields and the CRM fields, as they coevolve during a simulation. See Grabowski (2004) for further discussion.

The CRM uses the anelastic system of equations with height as the vertical coordinate (Khairoutdinov and Randall 2003), while GCM uses the quasi-hydrostatic system with the terrain-following sigma-pressure coordinate (Collins et al. 2006). For each GCM grid column, the time-varying GCM sounding is used as the "reference sounding" in the anelastic system of the CRM. When the CRM is called by the GCM, it receives as input the provisional temperature and vapor profiles that have been updated by the GCM dynamics, as well as the height and pressure profiles. The CRM adjusts its vertical grid to agree with the GCM's, assuming that the reference-state pressures at midlevels and interface levels are the same as the GCM's. The CRM computes the reference-state air density from hydrostatic balance.

The heights of the GCM's grid levels are slightly different from those on the previous GCM time step, so there is a small spike at the beginning of each CRM subcycle, but because the grid height in the GCM evolves slowly, the spike does not significantly affect the CRM results.

At the beginning of a sequence of CRM time steps, the vertical integral of water vapor in the CRM is not the same as it was when the CRM finished the preceding sequence. This discrepancy is eliminated at the end of CRM call, however, because it is automatically corrected by the prescribed large-scale forcing. As a result, the vertical integral of total water in the CRM at the end of the GCM time step always equals the vertical integral of total water in the GCM after the provisional dynamics step, minus the precipitation that falls out during the CRM time steps. A similar result holds for the frozen static energy.

A complication is that water in the form of falling precipitation inside the grid columns of the CRM is not passed back to GCM at the end of the sequence of CRM time steps; some of the precipitation remains in the CRM grid, "hanging in the air." Although such hanging precipitation is not accounted for by the GCM, it is conserved by the CRM and can eventually reach the surface and be accounted for as rainfall, or else return to vapor through evaporation or sublimation.

Because the CRM is two-dimensional, we do not permit momentum feedback to the GCM winds. This means that the model cannot simulate the effects of vertical momentum transport by convection and gravity waves that are resolved on the CRM's grid (Khairoutdinov et al. 2005). We return to this point later.

Further discussion of the superparameterization and how it differs from conventional parameterizations is given near the end of this chapter. We now turn to an overview of results that have been produced through the use of the superparameterization.

3. Simulations of tropical variability

Many investigators have published the results of studies based on the SP-CAM, in more than 50 refereed journal articles. A recurring theme of these studies has been modes of variability in the climate system. In order of increasing time scale, the types of variability considered so far include the diurnal cycle (Khairoutdinov et al. 2005; Pritchard and Somerville 2009a,b; Pritchard et al. 2011; Kooperman et al. 2013), easterly waves and tropical cyclones (McCrary 2012; Stan 2012), fluctuations of the Asian summer monsoon (DeMott et al. 2011, 2013) and the Madden–Julian oscillation (Benedict and Randall 2009, 2011; Thayer-Calder and Randall 2009; Kim et al. 2009; Andersen and Kuang 2012; Arnold et al. 2013), El Niño and the Southern Oscillation (Stan et al. 2010), and anthropogenic climate change (Wyant et al. 2006, 2012; Stan and Xu 2014; Arnold et al. 2014). Here we provide a value-added overview of some of these results, with an emphasis on the effects of coupling the SP-CAM to an ocean model.

a. The diurnal cycle of precipitation

Solar heating and nocturnal cooling of the land surface are associated with a vigorous diurnal cycle of convective activity over the warm continents of the tropics and the summer hemisphere midlatitudes, with a maximum in late afternoon or early evening (e.g., Dai 2001). Many global models fail to simulate the observed afternoon maximum (Dai 2006; Dirmeyer et al. 2011). Khairoutdinov et al. (2005, their Fig. 13) pointed out that the SP-CAM produces fairly realistic diurnal cycles of precipitation over both the continents and the oceans. Pritchard and Somerville (2009a,b) Pritchard et al. (2011), and Kooperman et al. (2013) performed detailed analyses of the SP-CAM's simulations of the diurnal cycle of precipitation over central North America in summer. They showed that the model is able to simulate the observed propagation of diurnally forced convective systems from near the Rocky Mountains in the afternoon to near Omaha after midnight. An example is shown in Fig. 15-4, which depicts rapid eastward propagation of diurnally excited convective disturbances in both the observations and the SP-CAM, but not in the CAM.

The afternoon precipitation maximum over land is driven by solar heating of the ground, which leads to increasing sensible and latent heat fluxes from early morning until midafternoon. These energy fluxes directly modify the thermodynamic properties of the boundary layer. All models include these basic physical processes. To produce a realistic diurnal cycle of precipitation, however, deep convective clouds must respond realistically to the changing thermodynamic properties of the boundary layer. The coupling of the parameterized boundary layer with the parameterized deep convection is therefore key to the ability of a model to simulate the diurnal cycle of precipitation over heated

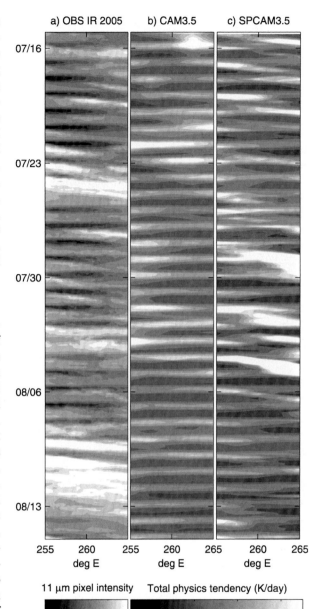

FIG. 15-4. Time–longitude structure of warm season diurnal convective activity in the lee of the Rocky Mountains between 35° and 45°N, and plotted for longitudes between 250° and 280°E, for (a) observations, (b) CAM3.5, and (c) SP-CAM3.5. Each panel shows the longwave cloud forcing, bandpass filtered to include periods between 12 and 48 h, plotted over a period of 30 days. Based on Kooperman et al. (2013).

land. This coupling includes the lofting of boundary layer air to form cumulus clouds in the free atmosphere, and the production of cold pools by precipitation-driven downdrafts that penetrate into the boundary layer. In many existing GCMs, the boundary layer and cumulus parameterizations have been developed independently, their coupling is perfunctory, and the processes mentioned above are either not explicitly formulated or missing altogether. Many models include rather arbitrary assumptions about "triggers" that enable the development of deep convection (e.g., the requirement of a minimum relative humidity in the boundary layer).

The situation is very different with SP-CAM, in which the "boundary layer parameterization" of the GCM is limited to a parameterization of the surface fluxes. Turbulent transport above the surface is computed on the CRM's relatively high-resolution grid, and includes both resolved and parameterized components. The coupling of the boundary layer with the cumulus layer is explicitly resolved, and does not rely on parameterizations, perfunctory or otherwise. Cumulus clouds form when warm, humid air near the surface is lofted by the vertical motion resolved on the CRM's grid. Downdrafts resolved on the CRM's grid can produce cold pools in the boundary layer. These processes are poorly resolved because of the CRM's 4-km horizontal grid spacing, but the results suggest that an explicit representation with poor resolution can be more successful than a current-generation parameterization.

b. The Great Red Spot

Early work with the SP-CAM was based on the use of prescribed seasonally varying sea surface temperatures (SSTs). As first pointed out by Khairoutdinov and Randall (2001), the SP-CAM produces an unrealistically active hydrologic cycle over the western North Pacific Ocean during the northern summer months (Fig. 15-5c). The precipitation rate is unrealistically high, the precipitable water is excessive, the surface pressure is lower than observed, and the cyclonic low-level winds are too strong. The problem appears as an unrealistic eastward extension of the Asian summer monsoon, which also produces excessively strong monsoon precipitation. It does not occur at other times of year. Because color plots of this phenomenon typically show strong red features over the western North Pacific Ocean, we call it the Great Red Spot (GRS). Interestingly, the MMF created by Tao et al. (2009) also produces a GRS, even though it is based on a different GCM and a different CRM.

A comparison of Figs. 15-5b and 15-5d shows that the GRS disappears when the SP-CAM is coupled with an ocean model. This was discovered when Stan et al. (2010) coupled the SP-CAM to a low-resolution version of POP (the Parallel Ocean Program); we call this coupled model the SP-CCSM.[2] As reported by Stan et al.

---

[2] The CCSM is the Community Climate System Model, which has recently been renamed as the Community Earth System Model (CESM). The SP-CCSM is a version of the CCSM that uses SP-CAM as its atmospheric component.

(2010), and further analyzed by DeMott et al. (2011, 2013, 2014), the SP-CCSM gives a more realistic simulation of the atmospheric circulation than the SP-CAM "right out of the box," without any tuning, a somewhat surprising result in view of earlier experiences of others (e.g., Sausen et al. 1988). Besides eliminating the GRS, the SP-CCSM also produces a more realistic simulation of the Asian summer monsoon. In addition, Stan et al. (2010) found that the coupled model gives a more realistic simulation of El Niño, La Niña, and the Southern Oscillation, relative to the same coupled model with conventional atmospheric parameterizations.

The price paid for these improved results is an error in the simulated SST distribution. The error takes the form of a cooling in many locations (not shown), with a root-mean-square (RMS) value of 2.08 K. This is actually slightly smaller than the RMS error of the SST in version 3 of the conventionally parameterized CCSM, which also tends to produce tropical SSTs that are cooler than observed.

In the uncoupled model, the prescribed SSTs represent an effectively infinite reservoir of sensible and latent heat, which is unaffected by what happens in the simulated atmosphere. The strong cyclonic winds of the GRS promote intense evaporation, which supplies energy to maintain the GRS, as discussed by Luo and Stephens (2006). In the coupled model, the surface winds cool the SSTs, and this negative feedback prevents the development of the GRS. We can interpret the GRS as an error that comes from failure to include air–sea interactions in the uncoupled model.

c. The MJO

The ability of the SP-CAM to simulate the MJO was first reported by Khairoutdinov and Randall (2001), although Grabowski (2001) found something similar in a simplified global model. An extensive and generally favorable comparison of the simulated MJO with observations was reported by Benedict and Randall (2009). Thayer-Calder and Randall (2009) analyzed the relationship between water vapor and precipitation rate in the SP-CAM, and compared with version 3 of the conventionally parameterized CAM. They showed that in the SP-CAM heavy rainfall is accompanied by a very humid troposphere, whereas in version 3 of the CAM the lower troposphere remains dry even during intense rainfall (Fig. 15-6). They argued that a prerequisite for a realistic simulation of the MJO is that water vapor must exhibit a realistically large dynamic range. As discussed by Derbyshire et al. (2004) and others, deep convection may be unable to form unless the middle troposphere becomes sufficiently moist. This moistening is created by the convection itself. There are thus two separate issues: On the one hand, the simulated middle troposphere

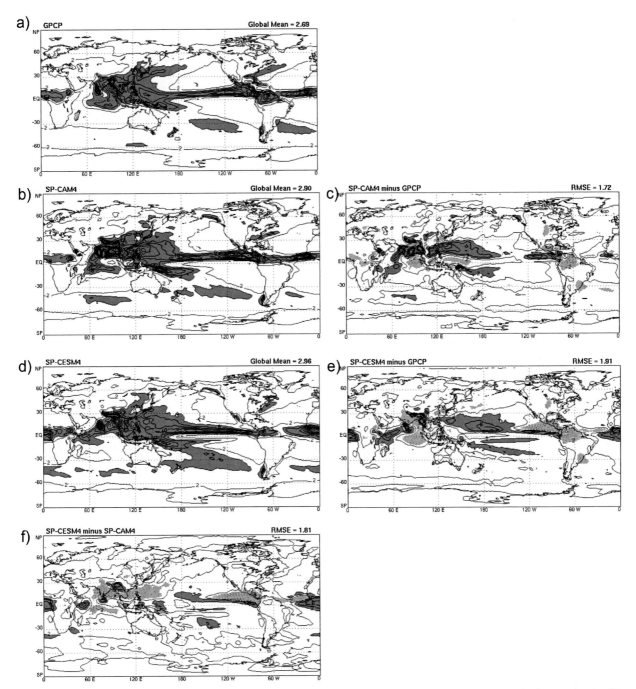

FIG. 15-5. (a) The observed climatological precipitation rate for July. (b) The simulated climatological precipitation rate for the SP-CAM. (c) The difference between the SP-CAM results and the observations. (d) The simulated climatological precipitation rate for the coupled ocean–atmosphere model, SP-CCSM. (e) The difference between the SP-CESM results and the observations. (f) The difference between the SP-CESM results and the SP-CAM results. Adapted from Stan et al. (2010). For precipitation rate maps, the contour interval is 2 mm day^{-1}, and values larger than 4 mm day^{-1} are shaded. For the difference maps, the contour interval is again 2 mm day^{-1}, values larger than 2 mm day^{-1} have dark shading, and values smaller than -2 mm day^{-1} have light shading.

must become humid enough to permit realistic deep convection. On the other hand, the parameterized convection must be capable of moistening the middle troposphere. We return to this point later.

The analysis of Benedict and Randall (2009) shows that it is possible to produce a reasonably successful simulation of the MJO without taking into account the effects of air–sea interactions. Nevertheless, it is

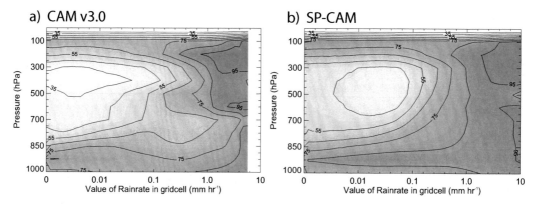

FIG. 15-6. Composite profiles of relative humidity binned by daily average rain rate in the region from 15°S to 15°N, and from 50°E to the date line, for (a) CAM3 and (b) SP-CAM. The contour interval is 10%, and the 70% contour is heavy. From Thayer-Calder and Randall (2009).

interesting to examine the effects of air–sea coupling on the MJO. Benedict and Randall (2011) used a slab ocean model (SOM) to study the effects of air–sea interactions on the MJO in a simplified framework. In the SOM, departures of the net surface energy flux from climatology can create SST anomalies, depending on a prescribed ocean mixed layer depth, but the SST anomalies are damped with a 50-day time scale to keep the SSTs of the SOM close to those of a control run. As shown in Fig. 15-7, the air–sea interactions promote coherent eastward propagation of the MJO. The air–sea interactions also strengthen the coupling between dynamics, as represented by the 850-hPa zonal wind, and moist processes, as represented by the precipitation rate at 90°E.

In an analysis of results from their fully coupled model, Stan et al. (2010) also found that the air–sea interactions improve the realism of the simulated MJO. Figure 15-8 shows the variance of precipitation in the range of frequencies and wavenumbers associated with the MJO, plotted as a function of longitude, from the western Indian Ocean to the central Pacific. In the uncoupled SP-CAM, the simulated variance takes large values not only west of the date line, as observed, but also considerably farther east. In the SP-CCSM, the simulated variance decreases realistically to the west of the date line. Further analysis of the results shows that the excessive variance in the SP-CAM results is found mainly in the region of the GRS. The elimination of the GRS in the coupled model leads to the more realistic variance distribution in SP-CCSM.

d. Yanai waves

Wheeler–Kiladis diagrams (Wheeler and Kiladis 1999) are ubiquitous now, but are most commonly shown only for modes that are symmetrical across the equator, such as the MJO and Kelvin waves. Figure 15-9 shows the antisymmetric power, which includes the westward-propagating

Yanai waves.[3] The SP-CCSM produces a realistic level of Yanai-wave activity, while the SP-CAM generates considerably less. This improvement in the coupled model is due to a more realistic "basic state," relative to the uncoupled model, as discussed by DeMott et al. (2011).

e. Monsoons

The Asian summer monsoon and its subseasonal variability attracted the attention of Prof. Yanai and his students (Luo and Yanai 1983, 1984; He et al. 1987; Yanai and Li 1994; Li and Yanai 1996; Hung et al. 2004). Recently, DeMott et al. (2011, 2013) have investigated the ability of the SP-CAM and the SP-CCSM to simulate these important tropical weather systems. Figure 15-10 shows the northward propagation (and also the weaker southward propagation) of precipitation anomalies that occur during the Asian summer monsoon. Both the SP-CAM and the SP-CCSM produce somewhat realistic northward propagation. The mechanisms of this variability are discussed in detail by DeMott et al. (2013).

Recently, McCrary (2012) has investigated the ability of the SP-CAM and the SP-CCSM to simulate the African summer monsoon and the associated African easterly waves, which are precursors to Atlantic tropical cyclones. Figure 15-11 shows the phase relationship between the vertically varying water vapor mixing ratio and the precipitation maximum associated with composite African easterly waves. The SP-CCSM correctly produces a very moist troposphere at the time of maximum rainfall, but the conventionally parameterized CAM3 produces a very unrealistic dry layer in the middle troposphere. This is strongly reminiscent of the earlier result of Thayer-Calder and Randall (2009), shown in Fig. 15-6, and suggests that the ability of models to produce realistic variations of water

---

[3] Also called mixed Rossby–gravity waves.

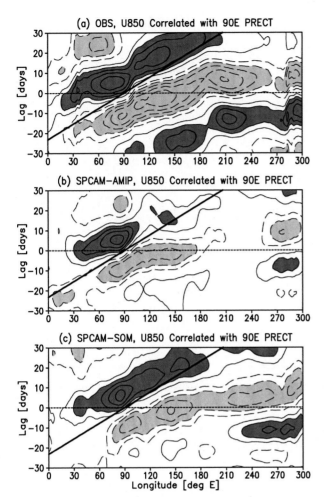

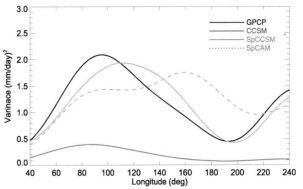

FIG. 15-8. May–October MJO precipitation variance as a function of longitude for GPCP, SP-CCSM3, SP-CAM3, and CCSM3. Precipitation is filtered to retain 20–100-day variability and eastward-propagating wavenumbers 1–6. Data are averaged over 5°–25°N. The "3" on the end of SP-CCSM3, SP-CAM3, and CCSM3 refers to version 3 of the CAM, which was used as the GCM in these simulations.

FIG. 15-7. Correlations of the lagged 850-hPa zonal wind with equatorial rainfall at 90°E for (a) observations (ERA-Interim and GPCP), (b) SP-CAM with prescribed seasonally varying sea surface temperatures, and (c) SP-CAM-SOM. All fields are filtered to pass only MJO time scales, and meridionally averaged near the equator. The horizontal axis is longitude relative to the Greenwich meridian, and the vertical axis is lag in days relative to the time of maximum precipitation at 90°E. The $5 \mathrm{~m~s}^{-1}$ phase speed is shown with a thick black line. The contour interval is 0.1, with positive contours solid, negative contours dashed, and the zero contour omitted. Dark (light) gray shading denotes correlations greater than 0.2 (less than −0.2). Adapted from Benedict and Randall (2011).

vapor is important not only for the MJO, but also for higher-frequency tropical disturbances. McCrary (2012) also shows that the SP-CCSM produces a somewhat realistic modulation of the African easterly waves by the MJO.

4. Concluding remarks

a. Parameterizations forever!

The results summarized in this paper demonstrate that superparameterization can simulate a wide variety of tropical variability more realistically than today's conventionally parameterized models. We are confident,[4] however, that future parameterizations can and will perform as well as or better than today's superparameterizations. Our hope is that work with superparameterizations can make that future happen sooner.

Apart from their practical use in simulations, parameterizations encapsulate our understanding of the way in which the parameterized physical processes interact with the large-scale circulation. Even with very successful simulations of tropical variability from global cloud-resolving models, we would need parameterizations, coupled with simplified models, to understand why the simulations succeeded.

b. Compared to what?

It is easy to criticize the superparameterization. The CRM is two-dimensional. The 4-km horizontal grid spacing that is typically used in the CRM is too coarse to accurately resolve even large cumulus clouds. The periodic boundary conditions are unrealistic. Given these weaknesses of the approach, why does super-parameterization work so well?

To answer this question, it is useful to review the key differences between an MMF and a conventionally parameterized GCM. First of all, the CRM used in an MMF simulates cloud dynamics by explicitly solving the (two-dimensional) equation of motion, thus tying back to the "basic physics" developed by Isaac Newton and others. In contrast, conventional GCMs parameterize all of the dynamical processes associated with cloud growth and decay. For example, most cumulus parameterizations

---

[4] Modelers have to be optimistic!

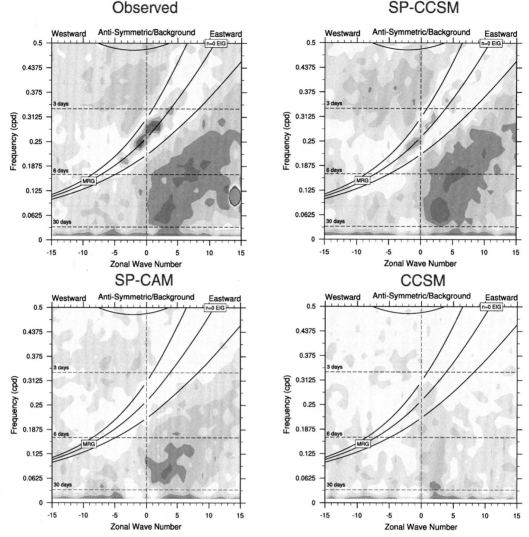

FIG. 15-9. Signal-to-noise ratio of the equatorially antisymmetric component of OLR in observations, SP-CCSM3, SP-CAM3, and CCSM3. The dispersion relations for typical equivalent depths are indicated by solid lines. The "observed" feature near zonal wavenumber 15 and frequency 0.12 day^{-1} is spurious, and arises from aliasing of the satellite data. Here MRG refers to mixed Rossby–gravity waves and EIG refers to eastward inertia–gravity waves.

represent cumulus clouds as one-dimensional entraining plumes (e.g., Arakawa and Schubert 1974). While it is true that the periodic lateral boundary conditions of the CRM are unrealistic, conventional parameterizations are so crude that the subject of lateral boundary conditions never even comes up.

In an MMF, the CRM simulates convective entrainment crudely but directly through its subgrid turbulence parameterization in combination with the simulated wind field resolved on the CRM grid. In contrast, entraining plume models often assume that the fractional entrainment rate is constant with height, which is in conflict with recent studies (e.g., Lin and Arakawa 1997a,b), or else they rely on entrainment

parameterizations (e.g., Gregory 2001; Chikira and Sugiyama 2010).

Because the CRM solves the equation of motion, an MMF has no need for closure assumptions (e.g., quasi-equilibrium), nor does it use assumptions about triggers that initiate cloud formation. In addition, the CRM explicitly simulates some aspects of mesoscale organization, such as aggregation or squall lines; conventional parameterizations omit these processes entirely.

The CRM relies on questionable parameterizations of microphysics, turbulence, and radiative transfer. Of course, conventional models must also parameterize these same processes. A key difference, however, is that the CRM provides relatively detailed information as input to

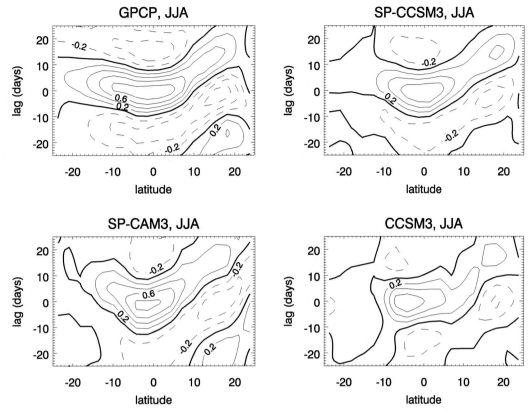

FIG. 15-10. June–August (JJA) lag correlation of 20–100-day filtered precipitation with itself in GPCP, SP-CCSM, SP-CAM, and CCSM. The base point time series is the filtered precipitation averaged over 10°S–5°N, 80°–100°E. The contour interval is 0.2, with positive (negative) values shown with solid (dashed) lines.

the parameterizations. For example, the CRM explicitly simulates (relatively) small-scale vertical velocities that can be used by microphysics parameterizations, whereas conventional parameterizations can only estimate cloud-scale vertical velocities using the entraining plume model.

The CRM simulates small-scale shear and buoyancy fluctuations that are key input to turbulence parameterizations. Last but not least, the CRM explicitly simulates fractional cloudiness and cloud overlap, which are critical inputs to a radiation parameterization.

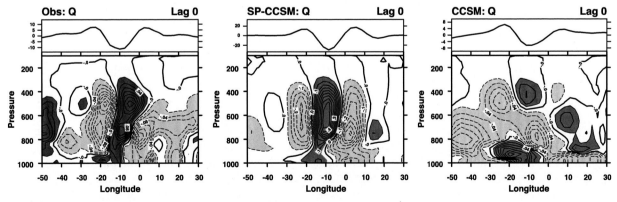

FIG. 15-11. Longitude–height cross section of the specific humidity anomaly (g kg^{-1}) in African easterly waves as (left) observed, (middle) simulated by the SP-CCSM, and (right) simulated by CCSM3. The latitude is 10°N. The heavy contour is zero, dashed contours are negative (lightly shaded), and solid contours are positive (darkly shaded). The contour interval for the SP-CCSM results is 0.05 g kg^{-1}, while the contour interval for the observations and for CCSM is 0.02 g kg^{-1}. The small plots at the top of each panel show anomalies of the outgoing longwave radiation (OLR; W m^{-2}). The anomalies of both water vapor and outgoing longwave radiation are relative to seasonal means. Based on the work of McCrary (2012).

In addition, the CRM is a nonlinear fluid-dynamical model that exhibits sensitive dependence on initial conditions; that is, it behaves chaotically, as does the real atmosphere (e.g., Hohenegger et al. 2006). As a result, the CRM-grid-averaged heating and drying rates that are passed to the GCM contain quasi-random components, which originate primarily on the smallest scales resolved by the CRM. In this sense, the super-parameterization is a stochastic parameterization. Efforts are under way to develop stochastic conventional parameterizations, but the superparameterization gener-ates stochastic heating and drying rates in a particularly natural way.

Finally, the SP-CAM is almost embarrassingly paral-lel, so that it can make efficient use of a very large number of processors. The reason is that the many copies of the CRM (one per CAM grid column) run independently, with no communication among them-selves. As a result, for a given GCM grid spacing the SP-CAM can use many more processors than a conven-tionally parameterized model. Although the SP-CAM does hundreds of times more arithmetic per simulated day than a conventionally parameterized GCM with the same resolution, the ability of the MMF to utilize more processors than a conventional GCM means that the wall-clock time required to complete a given simulation with the superparameterization is only moderately longer than that required with a conventional parameterization. An example is shown in Fig. 15-12. The SP-CAM is orders of magnitude less expensive than a global cloud-resolving model (GCRM; e.g., Tomita et al. 2005).

c. Process models and global models

Over the past two decades, cloud-parameterization testing has become organized on an international scale, beginning with the National Aeronautics and Space Administration (NASA) First International Satellite Cloud Climatology Project (ISCCP) Regional Experi-ment (FIRE) program in the 1980s (Cox et al. 1987), and continuing in the 1990s and beyond with the U.S. De-partment of Energy (DOE) Atmospheric Radiation Measurement Program (ARM; Stokes and Schwartz 1994) and the GCSS[5] activities (Randall et al. 2003b). One strategy for testing parameterizations is to drive both the parameterized column physics of a GCM and a high-resolution CRM with "forcing" databased on field observations, and then to intercompare the results of the two models with additional observations from the field

---

[5] GCSS was the GEWEX Cloud Systems Study; GEWEX was the Global Energy and Water Experiment. The work of GCSS is now continuing under the more general name GEWEX Atmo-spheric System Studies (GASS).

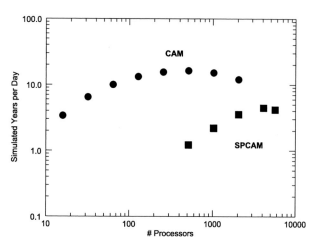

FIG. 15-12. Simulated years per wall-clock day vs the number of processors used, for the conventional CAM (dots) and the SP-CAM (squares). The GCM grid spacing is 2.5° longitude by 2° latitude for both models. Note that the axes are logarithmically scaled. The figure shows that, with this resolution, the SP-CAM can efficiently use thousands of processors, while the CAM is limited to hundreds. The timing tests were performed on Hopper, a Cray XE6 at the National Energy Research Scientific Computer Center (NERSC), by Mark Branson of Colorado State University.

(Randall et al. 1996). The column-physics is called a single-column model, and the high-resolution model is sometimes called a process model.

An important limitation of this strategy is that, be-cause single-column models and process models repre-sent only a small regional domain, they cannot "feed back" to the large-scale circulation; the effects of the large-scale circulation are simply prescribed and non-interactive. This limitation is now breaking down. Both MMFs and global cloud-resolving models are simulta-neously global models *and* process models (Fig. 15-13).

d. Improving the MMF

There are many ways to improve the MMF by en-hancing the CRM. Perhaps the highest priority is to in-corporate better parameterizations of shallow clouds and turbulence that cannot be resolved on the CRM's grid. Higher-order closure methods have been developed for the CRM and tested in the MMF (Cheng and Xu 2011, 2013; Xu and Cheng 2013a,b; Bogenschutz et al. 2012, 2013), with very promising results. These modern tur-bulence parameterizations predict (or diagnose) frac-tional cloudiness and the variability of temperature and water vapor at scales that are not resolved on the CRM's grid. With appropriate attention to coupling the parameterizations, this subgrid information can be used by improved parameterizations of microphysics (e.g., Morrison and Grabowski 2007, 2008) and radiation (e.g., Pincus and Stevens 2009). In addition, more advanced numerical methods have been implemented in the CRM

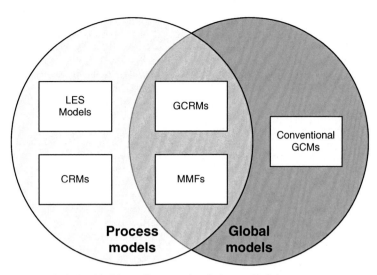

FIG. 15-13. In this Venn diagram, the circle on the left represents process models, and the circle on the right represents global atmospheric models. Until recently, these two classes of models did not overlap. Today, as shown in the figure, there is some intersection in the form of GCRMs and MMFs. This figure first appeared in Randall (2013).

(Yamaguchi et al. 2011). These various improvements are incremental, but very important.

More sweeping revisions of the MMF are also under development. The chapter of this book by Arakawa and colleagues describes a radically different "second-generation" MMF, called the quasi-three-dimensional (Q3D) MMF. The Q3D MMF partially removes the periodic boundary conditions of the CRM and includes the dynamical effects of three-dimensionality in a simplified way, while still consuming much less computer time than a GCRM. We hope that, in the years to come, the Q3D MMF will open the door to exciting new studies of global and tropical dynamics.

Acknowledgments. The lead author was lucky enough to take all of the classes that Professor Michio Yanai offered in the early and middle 1970s. Prof. Yanai was a great teacher, a superb scientist, and a very kind, humble man. Prof. Michael Pritchard of the University of California at Irvine provided Fig. 15-4. Initial development of the SP-CAM was supported by the U.S. Department of Energy through the ARM (Atmospheric Radiation Measurement) Program. More recently, the research has been supported by the National Science Foundation Science and Technology Center for Multi-Scale Modeling of Atmospheric Processes (CMMAP), managed by Colorado State University under Cooperative Agreement ATM-0425247.

REFERENCES

Andersen, J. A., and Z. Kuang, 2012: Moist static energy budget of MJO-like disturbances in the atmosphere of a zonally symmetric aquaplanet. *J. Climate*, **25**, 2782–2804, doi:10.1175/JCLI-D-11-00168.1.

Arakawa, A., and W. H. Schubert, 1974: Interaction of a cumulus cloud ensemble with the large-scale environment, Part I. *J. Atmos. Sci.*, **31**, 674–701, doi:10.1175/1520-0469(1974)031<0674:IOACCE>2.0.CO;2.

Arnold, N. P., Z. Kuang, and E. Tziperman, 2013: Enhanced MJO-like variability at high SST. *J. Climate*, **26**, 988–1001, doi:10.1175/JCLI-D-12-00272.1.

Benedict, J. J., and D. A. Randall, 2009: Structure of the Madden–Julian oscillation in the superparameterized CAM. *J. Atmos. Sci.*, **66**, 3277–3296, doi:10.1175/2009JAS3030.1.

——, and ——, 2011: Impacts of idealized air–sea coupling on Madden–Julian oscillation structure in the superparameterized CAM. *J. Atmos. Sci.*, **68**, 1990–2008, doi:10.1175/JAS-D-11-04.1.

Bogenschutz, P. A., A. Gettelman, H. Morrison, V. E. Larson, D. P. Schanen, N. R. Meyer, and C. Craig, 2012: Unified parameterization of the planetary boundary layer and shallow convection with a higher-order turbulence closure in the Community Atmosphere Model. *Geosci. Model Dev.*, **5**, 1407–1423, doi:10.5194/gmd-5-1407-2012.

——, ——, ——, ——, C. Craig, and D. P. Schanen, 2013: Higher-order turbulence closure and its impact on climate simulations in the Community Atmosphere Model. *J. Climate*, **26**, 9655–9676, doi:10.1175/JCLI-D-13-00075.1.

Cheng, A., and K.-M. Xu, 2011: Improved low-cloud simulation from a multiscale modeling framework with a third-order turbulence closure in its cloud-resolving model component. *J. Geophys. Res.*, **116**, D14101, doi:10.1029/2010JD015362.

——, and ——, 2013: Evaluating low cloud simulation from an upgraded multiscale modeling framework model. Part III: Tropical and subtropical cloud transitions over the northern Pacific. *J. Climate*, **26**, 5761–5781, doi:10.1175/JCLI-D-12-00650.1.

Chikira, M., and M. Sugiyama, 2010: A cumulus parameterization with state-dependent entrainment rate. Part I: Description and sensitivity to temperature and humidity profiles. *J. Atmos. Sci.*, **67**, 2171–2193, doi:10.1175/2010JAS3316.1.

Collins, W. D., and Coauthors, 2006: The Community Climate System Model version 3 (CCSM3). *J. Climate*, **19**, 2122–2143, doi:10.1175/JCLI3761.1.

Cox, S. K., D. McDougal, D. A. Randall, and R. A. Schiffer, 1987: FIRE—The First ISCCP Regional Experiment. *Bull. Amer. Meteor. Soc.*, **68**, 114–118, doi:10.1175/1520-0477(1987)068<0114:FFIRE>2.0.CO;2.

Dai, A., 2001: Global precipitation and thunderstorm frequencies. Part II: Diurnal variations. *J. Climate*, **14**, 1112–1128, doi:10.1175/1520-0442(2001)014<1112:GPATFP>2.0.CO;2.

——, 2006: Precipitation characteristics in eighteen coupled climate models. *J. Climate*, **19**, 4605–4630, doi:10.1175/JCLI3884.1.

DeMott, C. A., C. Stan, D. A. Randall, J. L. Kinter III, and M. Khairoutdinov, 2011: The Asian monsoon in the Super-Parameterized CCSM and its relation to tropical wave activity. *J. Climate*, **24**, 5134–5156, doi:10.1175/2011JCLI4202.1.

——, ——, and ——, 2013: Northward propagation mechanisms of the boreal summer intraseasonal oscillation in the ERA-Interim reanalysis and SP-CCSM. *J. Climate*, **26**, 1973–1992, doi:10.1175/JCLI-D-12-00191.1.

——, ——, ——, and M. Branson, 2014: Intraseasonal variability in coupled GCMs: The roles of ocean feedbacks and model physics. *J. Climate*, **27**, 4970–4994, doi:10.1175/JCLI-D-13-00760.1.

Derbyshire, S. H., I. Beau, P. Bechtold, J.-Y. Grandpeix, J.-M. Piriou, J. L. Redelsperger, and P. M. M. Soares, 2004: Sensitivity of moist convection to environmental humidity. *Quart. J. Roy. Meteor. Soc.*, **130**, 3055–3079, doi:10.1256/qj.03.130.

Dirmeyer, P. A., and Coauthors, 2011: Simulating the diurnal cycle of rainfall in global climate models: Resolution versus parameterization. *Climate Dyn.*, **39**, 399–418, doi:10.1007/s00382-011-1127-9.

Grabowski, W. W., 2001: Coupling cloud processes with the large-scale dynamics using the cloud-resolving convection parameterization (CRCP). *J. Atmos. Sci.*, **58**, 978–997, doi:10.1175/1520-0469(2001)058<0978:CCPWTL>2.0.CO;2.

——, 2004: An improved framework for superparameterization. *J. Atmos. Sci.*, **61**, 1940–1952, doi:10.1175/1520-0469(2004)061<1940:AIFFS>2.0.CO;2.

——, and P. K. Smolarkiewicz, 1999: CRCP: A cloud resolving convection parameterization for modeling the tropical convective atmosphere. *Physica D*, **133**, 171–178, doi:10.1016/S0167-2789(99)00104-9.

Gregory, D., 2001: Estimation of entrainment rate in simple models of convective clouds. *Quart. J. Roy. Meteor. Soc.*, **127**, 53–72, doi:10.1002/qj.49712757104.

He, H., J. W. McGinnis, Z. Song, and M. Yanai, 1987: Onset of the Asian summer monsoon in 1979 and the effect of the Tibetan Plateau. *Mon. Wea. Rev.*, **115**, 1966–1995, doi:10.1175/1520-0493(1987)115<1966:OOTASM>2.0.CO;2.

Hohenegger, C., D. Lüthi, and C. Schär, 2006: Predictability mysteries in cloud-resolving models. *Mon. Wea. Rev.*, **134**, 2095–2107, doi:10.1175/MWR3176.1.

Hung, C.-W., X. Liu, and M. Yanai, 2004: Symmetry and asymmetry of the Asian and Australian summer monsoons. *J. Climate*, **17**, 2413–2426, doi:10.1175/1520-0442(2004)017<2413:SAAOTA>2.0.CO;2.

Khairoutdinov, M. F., and D. A. Randall, 2001: A cloud resolving model as a cloud parameterization in the NCAR Community Climate System Model: Preliminary results. *Geophys. Res. Lett.*, **28**, 3617–3620, doi:10.1029/2001GL013552.

——, and ——, 2003: Cloud-resolving modeling of ARM summer 1997 IOP: Model formulation, results, uncertainties, and

sensitivities. *J. Atmos. Sci.*, **60**, 607–625, doi:10.1175/1520-0469(2003)060<0607:CRMOTA>2.0.CO;2.

——, ——, and C. A. DeMott, 2005: Simulation of the atmospheric general circulation using a cloud-resolving model as a super-parameterization of physical processes. *J. Atmos. Sci.*, **62**, 2136–2154, doi:10.1175/JAS3453.1.

Kim, D., and Coauthors, 2009: Application of MJO simulation diagnostics to climate models. *J. Climate*, **22**, 6413–6436, doi:10.1175/2009JCLI3063.1.

Kooperman, G. J., M. S. Pritchard, and R. C. J. Somerville, 2013: Robustness and sensitivities of central U.S. summer convection in the super-parameterized CAM: Multi-model intercomparison with a new regional EOF index. *Geophys. Res. Lett.*, **40**, 3287–3291, doi:10.1002/grl.50597.

Li, C., and M. Yanai, 1996: The onset and interannual variability of the Asian summer monsoon in relation to land–sea thermal contrast. *J. Climate*, **9**, 358–375, doi:10.1175/1520-0442(1996)009<0358:TOAIVO>2.0.CO;2.

Lin, C., and A. Arakawa, 1997a: The macroscopic entrainment processes of simulated cumulus ensemble. Part I: Entrainment sources. *J. Atmos. Sci.*, **54**, 1027–1043, doi:10.1175/1520-0469(1997)054<1027:TMEPOS>2.0.CO;2.

——, and ——, 1997b: The macroscopic entrainment processes of simulated cumulus ensemble. Part II: Testing the entraining-plume model. *J. Atmos. Sci.*, **54**, 1044–1053, doi:10.1175/1520-0469(1997)054<1044:TMEPOS>2.0.CO;2.

Lin, J., and Coauthors, 2006: Tropical intraseasonal variability in 14 IPCC AR4 climate models. Part I: Convective signals. *J. Climate*, **19**, 2665–2690, doi:10.1175/JCLI3735.1.

Luo, H., and M. Yanai, 1983: The large-scale circulation and heat sources over the Tibetan Plateau and surrounding areas during the early summer of 1979. Part I: Precipitation and kinematic analyses. *Mon. Wea. Rev.*, **111**, 922–944, doi:10.1175/1520-0493(1983)111<0922:TLSCAH>2.0.CO;2.

——, and ——, 1984: The large-scale circulation and heat sources over the Tibetan Plateau and surrounding areas during the early summer of 1979. Part II: Heat and moisture budgets. *Mon. Wea. Rev.*, **112**, 966–989, doi:10.1175/1520-0493(1984)112<0966:TLSCAH>2.0.CO;2.

Luo, J., and G. L. Stephens, 2006: An enhanced convection–wind–evaporation feedback in a superparameterization GCM (SP-GCM) depiction of the Asian summer monsoon. *Geophys. Res. Lett.*, **33**, L06707, doi:10.1029/2005GL025060.

Madden, R. A., and P. R. Julian, 1971: Detection of a 40–50 day oscillation in the zonal wind in the tropical Pacific. *J. Atmos. Sci.*, **28**, 702–708, doi:10.1175/1520-0469(1971)028<0702:DOADOI>2.0.CO;2.

——, and ——, 1972: Description of global-scale circulation cells in the tropics with a 40–50 day period. *J. Atmos. Sci.*, **29**, 1109–1123, doi:10.1175/1520-0469(1972)029<1109:DOGSCC>2.0.CO;2.

McCrary, R. R., 2012: Seasonal, synoptic and intraseasonal variability of the West African monsoon. Ph.D. dissertation, Colorado State University, 160 pp.

Morrison, H., and W. W. Grabowski, 2007: Comparison of bulk and bin warm-rain microphysics models using a kinematic framework. *J. Atmos. Sci.*, **64**, 2839–2861, doi:10.1175/JAS3980.

——, and ——, 2008: A novel approach for representing ice microphysics in models: Description and tests using a kinematic framework. *J. Atmos. Sci.*, **65**, 1528–1548, doi:10.1175/2007JAS2491.1.

Pincus, R., and B. Stevens, 2009: Monte Carlo spectral integration: A consistent approximation for radiative transfer in large eddy simulations. *J. Adv. Model. Earth Syst.*, **1**, doi:10.3894/JAMES.2009.1.1.

Pritchard, M. S., and R. C. J. Somerville, 2009a: Empirical orthogonal function analysis of the diurnal cycle of precipitation in a multi-scale climate model. *Geophys. Res. Lett.*, **36**, L05812, doi:10.1029/2008GL036964.

——, and ——, 2009b: Assessing the diurnal cycle of precipitation in a multi-scale climate model. *J. Adv. Model. Earth Syst.*, 1 (12), doi:10.3894/JAMES.2009.1.12.

——, M. W. Moncrieff, and R. C. J. Somerville, 2011: Orogenic propagating precipitation systems over the United States in a global climate model with embedded explicit convection. *J. Atmos. Sci.*, **68**, 1821–1840, doi:10.1175/2011JAS3699.1.

Randall, D. A., 2013: Beyond deadlock. *Geophys. Res. Lett.*, **40**, 5970–5976, doi:10.1002/2013GL057998.

——, K.-M. Xu, R. J. C. Somerville, and S. Iacobellis, 1996: Single-column models and cloud ensemble models as links between observations and climate models. *J. Climate*, **9**, 1683–1697, doi:10.1175/1520-0442(1996)009<1683:SCMACE>2.0.CO;2.

——, M. Khairoutdinov, A. Arakawa, and W. Grabowski, 2003a: Breaking the cloud-parameterization deadlock. *Bull. Amer. Meteor. Soc.*, **84**, 1547–1564, doi:10.1175/BAMS-84-11-1547.

——, and Coauthors, 2003b: Confronting models with data: The GEWEX Cloud Systems Study. *Bull. Amer. Meteor. Soc.*, **84**, 455–469, doi:10.1175/BAMS-84-4-455.

Sausen, R., K. Barthel, and K. Hasselmann, 1988: Coupled ocean–atmosphere models with flux correction. *Climate Dyn.*, **2**, 145–163, doi:10.1007/BF01053472.

Stan, C., 2012: Is cumulus convection the concertmaster of tropical cyclone activity in the Atlantic? *Geophys. Res. Lett.*, **39**, L19716, doi:10.1029/2012GL053449.

——, and L. Xu, 2014: Climate simulations and projections with a super-parameterized climate model. *Environ. Modell. Softw.*, **60**, 134–152, doi:10.1016/j.envsoft.2014.06.013.

——, M. Khairoutdinov, C. A. DeMott, V. Krishnamurthy, D. M. Straus, D. A. Randall, J. L. Kinter III, and J. Shukla, 2010: An ocean–atmosphere climate simulation with an embedded cloud resolving model. *Geophys. Res. Lett.*, **37**, L01702, doi:10.1029/2009GL040822.

Stokes, G. M., and S. E. Schwartz, 1994: The Atmospheric Radiation Measurement (ARM) Program: Programmatic background and design of the cloud and radiation test bed. *Bull. Amer. Meteor. Soc.*, **75**, 1201–1221, doi:10.1175/1520-0477(1994)075<1201:TARMPP>2.0.CO;2.

Tao, W.-K., and Coauthors, 2009: A multi-scale modeling system: Developments, applications, and critical issues. *Bull. Amer. Meteor. Soc.*, **90**, 515–534, doi:10.1175/2008BAMS2542.1.

Thayer-Calder, K., and D. A. Randall, 2009: The role of convective moistening in the Madden–Julian oscillation. *J. Atmos. Sci.*, **66**, 3297–3312, doi:10.1175/2009JAS3081.1.

Tomita, H., H. Miura, S. Iga, T. Nasuno, and M. Satoh, 2005: A global cloud-resolving simulation: Preliminary results from an aqua planet experiment. *Geophys. Res. Lett.*, **32**, L08805, doi:10.1029/2005GL022459.

Wheeler, M., and G. N. Kiladis, 1999: Convectively coupled equatorial waves: Analysis of clouds and temperature in the wavenumber–frequency domain. *J. Atmos. Sci.*, **56**, 374–399, doi:10.1175/1520-0469(1999)056<0374:CCEWAO>2.0.CO;2.

Wyant, M. C., M. Khairoutdinov, and C. S. Bretherton, 2006: Climate sensitivity and cloud response of a GCM with a super-parameterization. *Geophys. Res. Lett.*, **33**, L06714, doi:10.1029/2005GL025464.

——, C. S. Bretherton, P. N. Blossey, and M. Khairoutdinov, 2012: Fast cloud adjustment to increasing CO_2 in a super-parameterized climate model. *J. Adv. Model. Earth Syst.*, **4**, M05001, doi:10.1029/2011MS000092.

Xu, K.-M., and A. Cheng, 2013a: Evaluating low-cloud simulation with an upgraded multiscale modeling framework model. Part I: Sensitivity to spatial resolution and climatology. *J. Climate*, **26**, 5717–5740, doi:10.1175/JCLI-D-12-00200.1.

——, and ——, 2013b: Evaluating low-cloud simulation with an upgraded multiscale modeling framework model. Part II: Seasonal variations over the eastern Pacific. *J. Climate*, **26**, 5741–5760, doi:10.1175/JCLI-D-12-00276.1.

Yamaguchi, T., D. A. Randall, and M. Khairoutdinov, 2011: Cloud modeling tests of the ULTIMATE-MACHO scalar advection scheme. *Mon. Wea. Rev.*, **139**, 3248–3264, doi:10.1175/MWR-D-10-05044.1.

Yanai, M., and C. Li, 1994: Mechanism of heating and the boundary layer over the Tibetan Plateau. *Mon. Wea. Rev.*, **122**, 305–323, doi:10.1175/1520-0493(1994)122<0305:MOHATB>2.0.CO;2.

Chapter 16

Multiscale Modeling of the Moist-Convective Atmosphere

AKIO ARAKAWA

University of California, Los Angeles, Los Angeles, California

JOON-HEE JUNG

Colorado State University, Fort Collins, Colorado

CHIEN-MING WU

National Taiwan University, Taipei, Taiwan

ABSTRACT

One of the most important contributions of Michio Yanai to tropical meteorology is the introduction of the concepts of *apparent heat source* Q_1 and *apparent moisture sink* Q_2 in the large-scale heat and moisture budgets of the atmosphere. Through the inclusion of unresolved eddy effects, the vertical profiles of apparent sources (and sinks) are generally quite different from those of true sources taking place locally. In low-resolution models, such as the conventional general circulation models (GCMs), cumulus parameterization is supposed to determine the apparent sources for each grid cell from the explicitly predicted grid-scale processes. Because of the recent advancement of computer technology, however, increasingly higher horizontal resolutions are being used even for studying the global climate, and, therefore, the concept of apparent sources must be expanded rather drastically. Specifically, the simulated apparent sources should approach and eventually converge to the true sources as the horizontal resolution is refined. For this transition to take place, the conventional cumulus parameterization must be either generalized so that it is applicable to any horizontal resolutions or replaced with the mean effects of cloud-scale processes explicitly simulated by a cloud-resolving model (CRM). These two approaches are called ROUTE I and ROUTE II for unifying low- and high-resolution models, respectively. This chapter discusses the conceptual and technical problems in exploring these routes and reviews the authors' recent work on these subjects.

1. Introduction: Apparent sources and sinks

Michio Yanai worked on virtually all aspects of tropical meteorology, covering a broad range of the atmospheric spectrum from the cloud scale to the planetary scale. His contributions to tropical meteorology are unique in the following ways:

1) Opened new research areas in tropical meteorology
    Examples include the analysis of the formation stage of tropical cyclones, the discovery and extensive analysis of the equatorial waves, the quantitative analysis of heat source and moisture sink associated with cloud clusters, and the 3D heat and moisture budgets over the Tibetan plateau.
2) Developed new approaches in data analysis
    Examples include the power spectrum analysis combined with linearized equations and the analysis of the apparent heat source Q_1 and apparent moisture sink Q_2.
3) Deduced hidden reality from scarce data
    Examples include the transition of tropical disturbance from cold-core wave to warm-core vortex, the Yanai wave, and the heat, moisture, and momentum transports by cloud clusters.

Corresponding author address: Akio Arakawa, Dept. of Atmospheric and Oceanic Sciences, University of California, Los Angeles, 540 Euclid St., Santa Monica, CA 90402.
E-mail: aar@atmos.ucla.edu

DOI: 10.1175/AMSMONOGRAPHS-D-15-0014.1

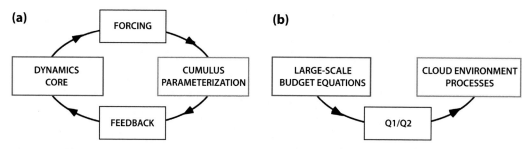

FIG. 16-1. (a) A schematic diagram showing the two-way interaction between the dynamics core and cumulus parameterization. (b) A schematic diagram showing the logical structure of the analysis performed by Yanai et al. (1973).

4) Made tropical meteorology more exciting to the general meteorological community

Examples include the Yanai wave and the role of moist convection in the large-scale dynamics.

5) Made unique contributions to international research programs

Examples include his contributions to GATE and the Monsoon Experiment (MONEX).

It is interesting to note that he never worked on forecast models. This does not mean that he was not interested in that subject. Although practically all of his papers are observationally oriented, they always went far beyond pure phenomenological descriptions and stimulated and inspired even those people whose primary interest was not in traditional tropical meteorology.

Except for the Yanai wave, his analyses of observed data almost always included moisture budgets, reflecting his recognition of the important role played by moist-convective processes in the tropical atmosphere. In Yanai (1961), which is based on his Ph.D. thesis, he presented a detailed analysis of typhoon formation and introduced the method of Q_1/Q_2 analysis to investigate the transition from a cold-core wave to a warm-core vortex. Here, Q_1 and Q_2 represent the source and sink terms in the large-scale heat and moisture budget equations, respectively. In this paper, he placed the emphasis on the similarity between the horizontal distributions of Q_1 and Q_2, rather than the difference in their vertical distributions, regarding them as independent estimates of the heat of condensation released in cloud clusters. In addition to the fascinating description of the transition the paper points out, "It is an important fact that the order of magnitude of local changes in potential temperature was very small compared with each of the heat source and the expected change from the dry adiabatic relation . . . this means that the stratification of the air was nearly neutral with respect to the moist-adiabatic process" (Yanai 1961, p. 210). This last point was later emphasized by Betts (1982) and Xu and Emanuel (1989) as a characteristic of the tropical

atmosphere and can be viewed as a prototype of the quasi-neutral hypothesis postulated by Arakawa (1969) and Arakawa and Schubert (1974).

During the late 1960s and early 1970s, Arakawa was struggling to formulate Q_1 and Q_2 for use in general circulation models (GCMs) in terms of the detrainment of cloud air into the environment and the cumulus-induced subsidence in the environment. This idea implicitly assumes that the gridpoint values of GCMs represent the environment of subgrid-scale clouds. The limit of this assumption will be discussed in detail in section 3. In any event, Arakawa's hope at that time was that the use of such a formulation in the cumulus parameterization could produce realistic profiles of Q_1 and Q_2, which are needed as "feedback," shown in Fig. 16-1a. For the loop in the figure to be closed, the model must formulate the effect of the dynamics core (and other modules, such as radiation and boundary layer processes) on the cumulus parameterization, which is called "forcing."

As pointed out by Arakawa (2004), the procedure followed by Yanai et al. (1973) reverses the lower half of the loop, as shown in Fig. 16-1b. First, Q_1 and Q_2 are obtained as the residuals in the large-scale budget

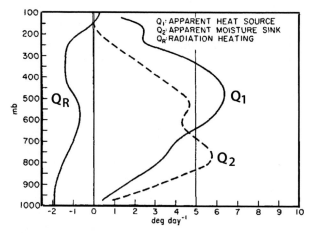

FIG. 16-2. Profiles of Q_1 and Q_2 obtained from large-scale budget analysis of the 1956 Marshall Islands data and an estimated profile of Q_R. From Yanai et al. (1973).

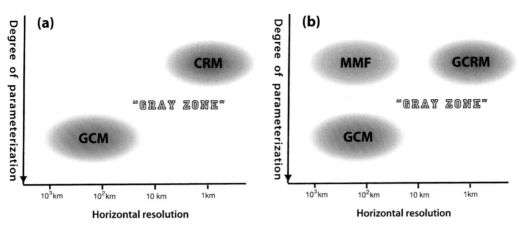

FIG. 16-3. Figures showing the (a) bi- and (b) tripolarizations of atmospheric models. See the text for further explanation. From Arakawa and Wu (2013).

equations. The terminologies "apparent heat source" and "apparent moisture sink" for these quantities are then introduced, recognizing that they are different from the true heat source and moisture sink taking place locally. These differences are due to the effects of eddies smaller than the size of the observation network. Their results obtained from the 1956 Marshall Islands data are shown in Fig. 16-2. Here, Q_R denotes the estimated radiation warming. Unlike in Yanai (1961), the difference between $Q_1 - Q_R$ and Q_2 is emphasized this time as a measure of convective activity. Note that $(Q_1 - Q_R) - Q_2$ is the apparent source of moist static energy, $c_p T + Lq + gz$, because of unresolved moist convection, where $c_p T$, Lq, and gz are sensible heat, latent heat, and geopotential energy, respectively. Subsequent to Yanai et al. (1973), the apparent heat source and apparent moisture sink became well-accepted concepts in tropical meteorology (see Johnson et al. 2016, chapter 1).

What made the paper Yanai et al. (1973) even more unique is that the authors went farther along the curve in Fig. 16-1b to quantitatively assess the cloud environment processes, such as the detrainment and cumulus-induced subsidence, through the use of a bulk cloud ensemble model. Yanai was at first a little hesitant doing this part of the analysis since the use of a model was against his philosophy on data analysis. Yet the approach worked beautifully, and the paper was highly applauded by the community. In spite of the reversing of the loop, or rather because of the reversing, the paper gave a rationale for the use of mass-flux formulation in cumulus parameterization. In this way, a closer tie was established between observed large-scale budget and cumulus parameterization studies almost for the first time.

The purpose of this chapter is to discuss the modeling aspects of apparent sources with an emphasis on their formulation applicable to any horizontal resolution between those typically used in GCMs and cloud-resolving models (CRMs). (Here and hereinafter, sources and sinks are simply called sources.) Section 2 presents a rationale for such formulations and points out that there are two routes to achieve the objective: "ROUTE I", following the conventional parameterization approach, and "ROUTE II", following the coupled GCM/CRM approach. Sections 3 and 4 review the authors' recent work on the ROUTE I and ROUTE II approaches, respectively, and section 5 presents concluding remarks.

2. Rationale for generalized formulations of apparent sources

In the conventional parameterizations of the apparent heat source Q_1 and apparent moisture sink Q_2, subgrid moist-convective processes are highly idealized inevitably. Not all of these idealizations are justifiable, and some can easily be criticized, especially from the observation side. But usually those criticisms point out only the existence of problems, not the solutions to the problems. Partly because of this, the rate of progress in representing cloud processes in climate models has been unacceptably slow, especially when it is compared to the rapid expansion of the scope of GCMs; so we may say, as Randall et al. (2003, p. 1548) do, that "the cloud parameterization problem is deadlocked."

As pointed out by Arakawa et al. (2011) and Arakawa and Wu (2013), one of the most important sides of the deadlock is that we currently have only two ways of representing deep moist-convective processes in numerical models of the atmosphere: one with highly parameterized apparent sources and the other with explicit simulation of true sources. The former is for low-horizontal-resolution models, such as the conventional GCMs, and the latter is for high-horizontal-resolution models, such as the CRMs. Correspondingly, besides those models that explicitly simulate turbulence, there have been two discrete families of atmospheric models, as shown in Fig. 16-3a. In this figure, the abscissa is the horizontal resolution, and the ordinate is a measure of the degree of parameterization, such as reduction in the degrees of freedom, increasing downward. These two families have been developed for applications with quite different ranges of horizontal resolution in mind. What we see here is a polarization of atmospheric models separated by a "gray zone." The difficulties in representing moist-convective processes in this zone have long been recognized among the mesoscale modeling community (e.g., Molinari and Dudek 1992; Molinari 1993; Frank 1993).

Recently, a new family of global models called the multiscale modeling framework (MMF) has been added, as shown in Fig. 16-3b. The MMF is a coupled GCM/CRM system that follows the superparameterization approach introduced and used by Grabowski and Smolarkiewicz (1999), Grabowski (2001), and Khairoutdinov and Randall (2001). In this approach, the cumulus parameterization in conventional GCMs is replaced with the mean effects of cloud-scale processes simulated by a 2D CRM embedded in each GCM grid cell. For more details of the MMF approach, including some results, see Randall et al. (2016, chapter 15).

Furthermore, because of the recent advancement of computer technology, straightforward applications of 3D CRMs to simulate true sources has begun to be feasible even for studying the global climate (e.g., Sato et al. 2009; Oouchi and Satoh 2016, chapter 14). In this way, global CRMs (GCRMs) joined the families of global models, as shown in Fig. 16-3b. This is an exciting development in the history of numerical modeling of the global atmosphere, in which the modular structure shown in Fig. 16-1a is completely abandoned as far as the scales comparable to or larger than deep moist convection are concerned.

It should be noted, however, that the gray zone still exists in the configuration shown in Fig. 16-3b. Thus, what we see now is a tripolarization of global models with differently formulated model physics for each of the three families. Consequently, both the conventional GCMs and the MMF do not converge to a GCRM as the horizontal grid size is reduced. This is not a scientifically healthy condition because the use of discretized equations is justified only when the error due to the discretization can be made arbitrarily small as the resolution is refined. This does not matter when the scales represented by the macroscopic and microscopic models are separated by many orders of magnitude, as in the case of continuum and molecular mechanics, because the refinement of the macroscopic resolution all the way down to the microscopic resolution will never be attempted. The situation in atmospheric modeling is quite different, because the spectrum is virtually

continuous. The physics of atmospheric models should then produce a smooth transition of the apparent sources from those required for the conventional GCMs to those required for the mesoscale models and, eventually, to the true sources that can be simulated by the CRMs. Thus, the traditional view of apparent sources must be expanded rather drastically to include their resolution dependence, even including the true sources in the limit.

Jung and Arakawa (2004) showed convincing evidence for the resolution dependence of apparent sources through budget analyses of data simulated by a CRM with different space/time resolutions. By comparing the results of low-resolution test runs without cloud microphysics over a selected time interval with those of a high-resolution run with cloud microphysics (control), they identified the apparent microphysical source required for the low-resolution solution to be equal to the space/time averages of the high-resolution solution. This procedure is repeated over many realizations selected from the control. Figure 16-4a shows examples of the domain- and ensemble-averaged profiles of the required source of moist static energy obtained in this way. The profiles shown in red and green are obtained using the horizontal grid size and the time interval of the test run given as (2 km, 2 min), approximately representing the true sources, and (32 km, 60 min), representing apparent sources for this resolution, respectively. The red profile shows a positive source due to freezing and a negative source due to melting immediately above and below the freezing level, respectively. There are practically no other sources and sinks, because moist static energy is conserved under moist-adiabatic processes. The green profile, on the other hand, shows marked negative values in the lower troposphere and positive values in the middle-to-upper troposphere, suggesting the dominant role played by the upward transport. Figure 16-4b is the same as Fig. 16-4a, but for the required source of total (airborne) water mixing ratio. The red profile shows dominant sinks in the middle troposphere due to the generation of precipitating particles and small peaks of source near the surface due to the evaporation from rain. The green profile again suggests the dominant role played by the upward transport.

Low-resolution models, such as the conventional GCMs, are supposed to produce profiles similar to the green profiles shown in Fig. 16-4, which we call the GCM type, while high-resolution models, such as the CRMs, are supposed to produce profiles similar to the red profiles, which we call the CRM type. As Arakawa (2004) emphasized, it is important to recognize that any space/time/ensemble average of the CRM-type profiles does not give a GCM-type profile. This means that the cumulus parameterization problem is more than a statistical theory of cloud microphysics. Also, it is not a

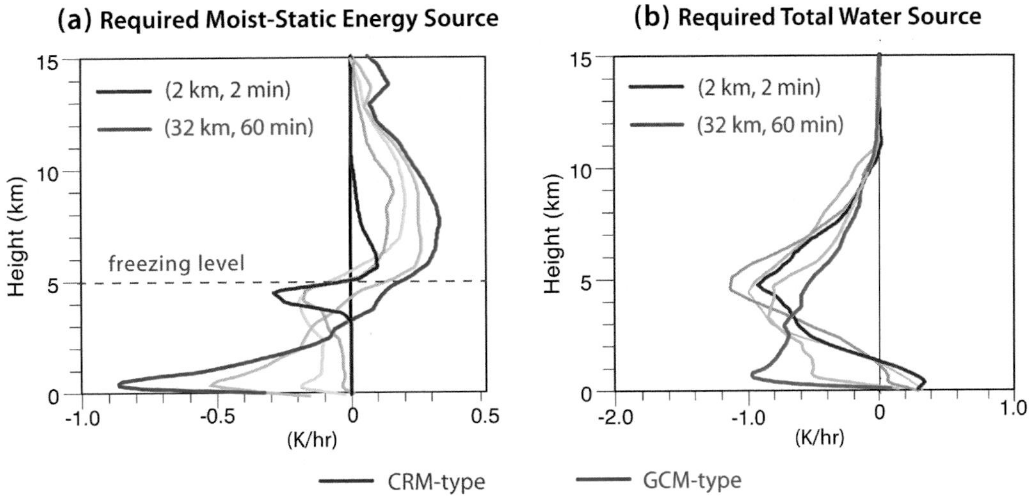

(a) Required Moist-Static Energy Source

- (2 km, 2 min)
- (32 km, 60 min)

freezing level

Height (km)

(K/hr)

(b) Required Total Water Source

- (2 km, 2 min)
- (32 km, 60 min)

Height (km)

(K/hr)

CRM-type   GCM-type

FIG. 16-4. Domain- and ensemble-averaged profiles of the required source for (a) moist static energy and (b) total (airborne) water as a result of cloud microphysics obtained with different horizontal grid sizes and time intervals. Redrawn from Jung and Arakawa (2004).

purely physical/dynamical problem, because it is needed as a consequence of mathematical truncation. Finally, it is not a purely mathematical problem because the use of a higher resolution or an improved numerical method while using the same formulation of model physics does not automatically improve the result. A complete theory for cumulus parameterization must address all of these aspects in a consistent manner.

As we have seen in Fig. 16-4, the required apparent sources highly depend on the horizontal resolution of the model, as well as the time interval for implementing model physics. The formulation of model physics should automatically produce this dependence as it is applied to different resolutions. Conventional cumulus parameterization schemes, however, cannot do this, because they assume either explicitly or implicitly that the horizontal grid size and the time interval for implementing physics are sufficiently larger and longer than the size and lifetime of individual moist-convective systems. If the model physics of GCMs is reformulated to produce such resolution dependence, GCMs and CRMs are unified to a single family of models that can be applicable to a wide range of horizontal resolutions, including the mesoscales with the same formulation of model physics.

Arakawa et al. (2011) and Arakawa and Jung (2011) discussed two routes to achieve the unification shown as ROUTE I and ROUTE II in Fig. 16-5. The departure points of these routes are the conventional GCMs and a new generation of MMF, respectively, but they share the same destination point, a GCRM. ROUTE I breaks through the gray zone by generalizing the conventional cumulus parameterization in such a way that it can be applied to any horizontal resolution. Consequently, the concept of

apparent sources is generalized to include their transition to the true sources. ROUTE II, on the other hand, bypasses the gray zone by replacing the conventional cumulus parameterization with explicit simulation of cloud-scale processes by a CRM. On this route, the transition of apparent sources to the true sources is numerically simulated. These two routes are discussed in the following two sections.

3. Route I: Unified parameterization

a. Identification of the problem

The first step to open ROUTE I is reexamination of the widely used assumption that convective updrafts

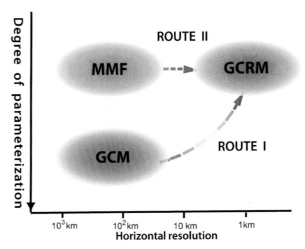

FIG. 16-5. Two routes for unifying the low-resolution and high-resolution models. See text for further explanations. From Arakawa and Wu (2013).

cover only a small fraction of the area represented by a GCM grid cell. Most conventional cumulus parameterizations assume this, at least implicitly, regarding the predicted thermodynamic fields as if they represent the environment of updrafts. Let σ be the "fractional convective cloudiness" or "fractional updraft area," which is the fractional area covered by convective updrafts in the grid cell. When the grid spacing is fixed, this is a measure of the population of updrafts. In terms of σ, the above assumption means $\sigma \ll 1$. In the limit as the grid spacing approaches zero, however, the grid cell is occupied either by an updraft or by its environment. Then σ becomes either 1 or 0, and thus the circulation associated with the updraft becomes the gird-scale circulation. More generally, the total effect of cumulus convection is the sum of its grid-scale and subgrid-scale effects, and it is important to remember that cumulus parameterization is supposed to formulate only the subgrid-scale effect of cumulus convection, not its total effect involving the grid-scale circulation. For vertical transport, then, what needs to be parameterized is the vertical eddy transport, not the total transport. If the parameterized transport exceeds the eddy transport, there may be either double counting of the same effect or spurious competition between the parameterized and explicitly simulated effects.

b. CRM simulations used for statistical analysis

To visualize the problem we are addressing, we have performed two numerical simulations using a CRM applied to an idealized horizontally periodic domain, one with and the other without background shear. The horizontal domain size and horizontal grid size are 512 km and 2 km, respectively. As expected, the two simulations represent quite different cloud regimes; but as far as the vertical transports of thermodynamic variables are concerned, we find only little qualitative difference between the two simulations. We therefore present only the shear case in this article.

To see the resolution dependence of the diagnosed statistics, we divide the original CRM domain into subdomains of identical size to represent a uniform GCM grid. We denote the side length of the subdomain by d, which ranges from the CRM grid size to the size of the entire domain.

c. Resolution dependence of ensemble average σ

In the following diagnosis of the simulated data, σ is determined for each subdomain by the fractional number of CRM grid points that satisfy $w \geq 0.5\,\mathrm{m\,s^{-1}}$. Let angle brackets $\langle \; \rangle$ denote the ensemble average over all updraft-containing subdomains (i.e., subdomains with $\sigma > 0$) during the analysis period (12 h). Figure 16-6

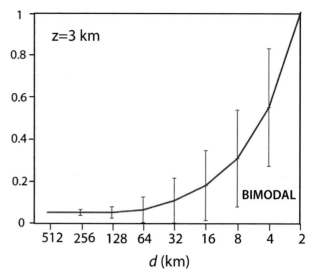

FIG. 16-6. The resolution dependence of $\langle\sigma\rangle$ and associated standard deviation at $z = 3$ km. Taken from Arakawa and Wu (2013).

shows the resolution dependence of $\langle\sigma\rangle$ and associated standard deviation at $z = 3$ km. In the figure, we see that $\langle\sigma\rangle$ drastically increases as the subdomain size d decreases. Clearly, the assumption $\sigma \ll 1$ can be justified only for low resolutions. For high resolutions, $\langle\sigma\rangle$ significantly deviates from 0 and becomes 1 for $d = 2$ km, which is the CRM grid spacing. Since there are a number of subdomains with $\sigma = 0$ that are not included in the ensemble average, the distribution of σ for high resolutions tends to be bimodal. Wu and Arakawa (2014) show that these features also appear practically at all levels in the vertical.

d. The ratio of the eddy to total vertical transports of moist static energy

We now look into the vertical transport of moist static energy diagnosed from this dataset. In view of the discussion presented in section 3a, our main interest here is the relative importance of the eddy transport $\langle \overline{w'h'} \rangle$ against the total transport $\langle \overline{wh} \rangle$. Here, h is the deviation of moist static energy from the average over the entire horizontal domain, the overbar denotes the average over all CRM grid points in the subdomain, $\langle \; \rangle$ is the ensemble average over all subdomains with $\sigma > 0$ as defined earlier, and $w' \equiv w - \overline{w}$ and $h' \equiv h - \overline{h}$ represent the eddy components of w and h. The large standard deviation shown in Fig. 16-6 indicates that there is a wide range of σ for each subdomain size. Figure 16-7 presents the ratio $\langle \overline{w'h'} \rangle / \langle \overline{wh} \rangle$ at $z = 1$, 3, and 6 km with the subdomain size d and the fractional updraft area σ. An empty box means that data are not sufficient for that combination of d and σ. For small values of σ, the total transport at $z = 3$ and 6 km is almost entirely due to the

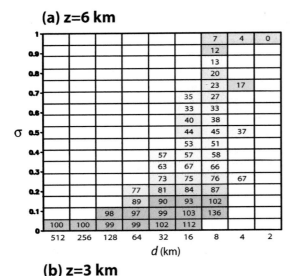

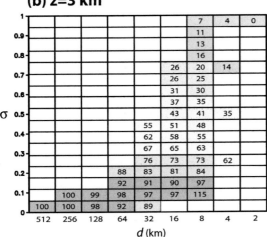

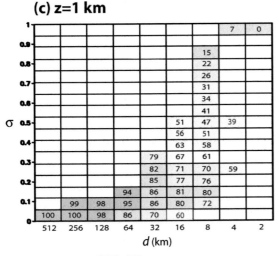

FIG. 16-7. The ratio $\langle \overline{w'h'} \rangle / \langle \overline{wh} \rangle$ (%) with various combinations of d and σ for (a) $z = 6$ km, (b) $z = 3$ km, and (c) $z = 1$ km. (The values exceeding 100% indicate that the ensemble-averaged grid-scale transport is negative for that box.) From Wu and Arakawa (2014).

eddy transport regardless of the resolution. For larger values of σ, however, the total transport is primarily due to explicitly simulated grid-scale vertical velocity. These features can also be seen at $z = 1$ km, although the resolution dependence is not negligible for small values of σ. Except for this height, Fig. 16-7 clearly shows that the ratio depends primarily on σ, not on d, so that what matters most in generalizing the conventional cumulus parameterization is to include the dependence on the fractional updraft area, not directly on the grid spacing. This is one of the most important results of the analysis presented in this article.

e. The σ dependence of the total, eddy, and modified eddy transports of moist static energy for $d = 8$ km

The last subsection points out that the magnitude of $\langle \overline{w'h'} \rangle$ relative to that of $\langle \overline{wh} \rangle$ crucially depends on the fractional convective area σ even when the subdomain size d is fixed. Taking the case of $d = 8$ km as an example, Figs. 16-8a and 16-8b show $\rho^* \langle \overline{wh} \rangle$ and $\rho^* \langle \overline{w'h'} \rangle$, respectively, as functions of height for each σ bin. Here, ρ^* is the density normalized by its value at $z = 3$ km. From these figures, we see that the total transport for small values of σ, say, for $\sigma \le 0.3$, is almost entirely due to the eddy transport at practically all height even for this relatively high resolution. For larger values of σ, however, a large part of the total transport is due to the grid-scale transport, which is mesoscale for this resolution. Correspondingly, the concept of apparent source must be generalized to include its σ dependence, converging to the true source in the high-resolution limit with $\sigma = 1$.

f. Parameterization of the σ dependence of vertical eddy transport by homogeneous updrafts/environment

Our main problem then becomes parameterization of the σ dependence of vertical eddy transports for use in a prognostic model. Most conventional parameterizations assume that the updrafts and the environment within each grid cell are individually horizontally homogeneous so that they can each be represented by a top-hat profile. Although this is a rather drastic idealization, we continue to use it in the first attempt of developing the unified parameterization. Then we can express properties of the updrafts and the environment by a single z-dependent variable for each. With this assumption, Arakawa and Wu (2013) show that the vertical eddy transport of an arbitrary variable ψ is given by

$$\overline{w'\psi'} = \sigma(1 - \sigma)\Delta w \Delta \psi, \qquad (16\text{-}1)$$

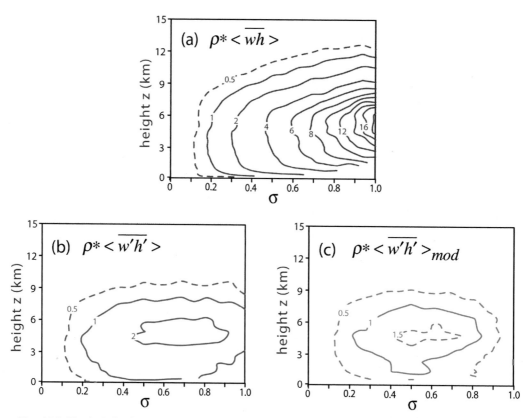

FIG. 16-8. Vertical distributions of the ensemble-average (a) total, (b) eddy, and (c) modified eddy vertical transports of moist static energy (m s^{-1} K) divided by c_p for $d = 8$ km with each bin of σ. See section 3f for an explanation of the modified eddy transport. From Wu and Arakawa (2014).

where Δ denotes the excess of the updraft value over the environment. Since Δw and $\Delta \psi$ depend only on the updraft properties relative to the environment, it is likely that they and their product $\Delta w \Delta \psi$ do not significantly depend on the population of updrafts and, therefore, not on σ. If this is the case, (16-1) shows that the σ dependence of the eddy transport is through the factor $\sigma(1 - \sigma)$. For small values of σ, the eddy transport increases with σ approximately linearly but reaches its maximum at $\sigma = 0.5$ and tends to vanish as σ approaches 1 because of the saturation of updrafts in the grid cell.

To see if the above reasoning with a top-hat profile is valid for the simulated dataset being used, we modify the data by replacing the prognostic variables at all updraft points with their averages over the subdomain and do the same for the environment points. Figure 16-8c shows the vertical distribution of the eddy transports of moist static energy diagnosed from the modified dataset. As anticipated, the shape of the σ dependence of the modified eddy transports is very close to the curve $\sigma(1 - \sigma)$ at all levels.

The difference between Figs. 16-8b and 16-8c represents the contribution from the inhomogeneous

structure of updrafts and the environment and/or from the existence of multiple cloud types. The difference is small for small values of σ, say, $\sigma < 0.3$. For larger values of σ, the difference is not small compared to the eddy transport based on the original dataset. However, since the eddy transport itself is small compared to the total transport for these values of σ, this difference may not be as important as it appears.

g. Determination of σ from given grid-scale processes

From the evidence presented above, it is clear that the fractional updraft area σ plays a key role in the unified parameterization. Prognostic models must be able to determine σ for each realization of grid-scale conditions as a closure.

Most cumulus parameterizations currently being used are adjustment schemes in which a measure of convective instability is at least partially adjusted to its equilibrium value [see Arakawa (2004) for a review]. Let the vertical eddy transport of moist static energy required for the full adjustment be $\overline{(w'h')}_E$. This can be considered as an external parameter since it represents convective forcing determined by the grid-scale destabilization.

The conventional parameterization schemes with full adjustment assume $\overline{w'h'} = (\overline{w'h'})_E$ in addition to $\sigma \ll 1$. Then (16-1) with $\psi = h$ implies

$$\sigma \simeq \frac{(\overline{w'h'})_E}{\Delta w \Delta h}. \qquad (16\text{-}2)$$

For (16-2) to be consistent with $\sigma \ll 1$, $(\overline{w'h'})_E \ll \Delta w \Delta h$ is necessary. Thus, such a scheme can be valid if either the destabilization rate $(\overline{w'h'})_E$ is small or the stratification is strongly unstable so that $\Delta w \Delta h$ is large.

The unified parameterization chooses

$$\sigma = \frac{(\overline{w'h'})_E}{\Delta w \Delta h + (\overline{w'h'})_E} \qquad (16\text{-}3)$$

instead of (16-2). This is the simplest choice to automatically satisfy the condition $0 \leq \sigma \leq 1$, as long as $(\overline{w'h'})_E$ and $\Delta w \Delta h$ have the same sign, while reducing to (16-2) when $(\overline{w'h'})_E \ll \Delta w \Delta h$. When $(\overline{w'h'})_E \geq \Delta w \Delta h$, (16-3) gives $\sigma \sim 1$ so that the grid cell is saturated with updrafts.

Eliminating $\Delta w \Delta h$ between (16-3) and (16-1), we obtain

$$\overline{w'h'} = (1 - \sigma)^2 (\overline{w'h'})_E. \qquad (16\text{-}4)$$

More generally, the unified parameterization uses

$$\overline{w'\psi'} = (1 - \sigma)^2 (\overline{w'\psi'})_E, \qquad (16\text{-}5)$$

where $(\overline{w'\psi'})_E$ is $\overline{w'\psi'}$ associated with the full adjustment. Since $\sigma > 0$, (16-5) shows $\overline{w'\psi'} < (\overline{w'\psi'})_E$. Thus, the practical application of the unified parameterization is a reduction of the eddy transports depending on the value of σ. Note that, unlike the commonly used relaxed adjustment schemes, the reduction is only for the transport effects, not for the sources due to the diabatic effects. In other words, the reduction is for the difference between the apparent and true sources. The diabatic effects have their own dependence on σ as discussed in section 3j.

In practical applications, however, it should be remembered that $\Delta w \Delta \psi$ in (16-3) is unknown, because what the model gives are the gridpoint values, not the environment values. Let δ denote the excess of the updraft value over the gridpoint values instead of the environment values. Since the relation between $\delta w \delta \psi$ and $\Delta w \Delta \psi$ involves σ, the problem becomes implicit; but as Arakawa and Wu (2013) show, it ends up solving the cubic equation given by

$$\sigma/(1 - \sigma)^3 = \lambda, \qquad (16\text{-}6)$$

where λ is a nondimensional parameter defined by

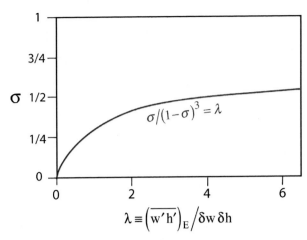

FIG. 16-9. Plot of σ given by (16-6) as a function of λ. From Arakawa and Wu (2013).

$$\lambda \equiv (\overline{w'h'})_E / \delta w \delta h. \qquad (16\text{-}7)$$

Recall that $(\overline{w'h'})_E$ in (16-7) is the destabilization rate due to the grid-scale processes, while $\delta w \delta h$ is a normalized eddy transport measuring the efficiency of the eddy transport. We see that λ is large either for a large destabilization rate, as is the case for mesoscale convective complex, or for small eddy transport efficiency, as is the case for stratocumulus clouds. Figure 16-9 shows the values of σ given by (16-6) as a function of the parameter λ. As (16-6) shows, $\sigma = 0$ when $\lambda = 0$, and $\sigma \to 1$ as $\lambda \to \infty$. The unified parameterization uses the value of σ determined in this way for all variables, assuming that variables other than h play only passive roles as far as the process of controlling σ is concerned.

h. A remark on stochastic parameterization

The unified parameterization can also provide a framework for including stochastic parameterization. Arakawa and Wu (2013) emphasized that stochastic formulation must be made under appropriate physical/dynamical/computational constraints that identify the source of uncertainty. In particular, the formulation must distinguish the uncertainty in formulating subgrid-scale processes from the irregular fluctuations of the grid-scale processes. Clearly, stochastic parameterization must deal with the former under a given grid-scale condition. Arakawa and Wu (2013) point out that different phases of cloud development are likely to be responsible for the main uncertainty.

i. Vertical transports of horizontal momentum

Up to this point we have discussed the transport of moist static energy. As shown by Arakawa and Wu

(2013), it is expected that the partitioning between the eddy- and grid-scale transports is similar for other thermodynamic variables. The situation can be essentially different, however, for the vertical transport of horizontal momentum for the following reasons:

(i) Horizontal momentum is not a conservative variable because of the pressure-gradient force.
(ii) Horizontal momentum is a vector with two independent components, and so is its vertical transport.

Most currently used parameterizations of the vertical transport of horizontal momentum recognize the importance of (i) but not necessarily that of (ii). These parameterizations are, therefore, effectively one-dimensional, in which the pressure gradient typically acts against the advection effect, maintaining the in-cloud horizontal velocity closer to the large-scale value (Gregory et al. 1997). Thus, the primary effect of the pressure gradient in those parameterizations is quantitative, rather than qualitative.

The point made in (ii) is well recognized in the parameterization presented by Wu and Yanai (1994). They considered a solution of the linearized diagnostic equation for pressure that can vary both in x and y with wavenumbers k and l, respectively. Here x and y are the components of arbitrarily chosen horizontal Cartesian coordinates. Using this solution, they showed that the net effect of the pressure gradient and advection depends on the ratio k^2/l^2. When $k^2 \sim l^2$, for example, the net effect is such that both the x and y components of the transport are downgradient. When $k^2 \ll l^2$, on the other hand, the x component of the transport representing the line-parallel component is downgradient while the y component representing the line-normal component can be upgradient. These results are generally consistent with the observational findings by LeMone (1983) and LeMone et al. (1984).

Wu and Arakawa (2014) used the shear case of the CRM-simulated dataset to analyze the vertical transports of horizontal momentum. The analysis shows that the momentum transport parameterized by analogy with the parameterizations of thermodynamic variables could work for the line-parallel component, but not for the line-normal component. For the latter, the transports due to the mesoscale organization of clouds are dominant, as has been consistently emphasized by Moncrieff (e.g., Moncrieff 1981, 1992). These results confirm the fact that the orientation of linearly organized convective systems must be known before any parameterization of the vertical transport of horizontal momentum is attempted. Unfortunately, this remains an extremely challenging task.

j. Implementation of cloud microphysical effects

To complete the unified representation of deep moist convection, this subsection discusses the implementation of cloud microphysical processes following Wu and Arakawa (2014). As in Gerard (2007), the unified parameterization uses a single formulation of cloud microphysics that can be applied to both convective updrafts and stratiform clouds. As is done for the transports of thermodynamic variables, we assume that the updraft air is horizontally homogeneous with a top-hat profile.

For the conversions taking place within updrafts, the cloud microphysical package is expected to determine the conversion rates for a given updraft vertical mass flux. The unified parameterization calculates the mass flux through $\rho \sigma w_c$, where w_c is the updraft vertical velocity predicted by the cloud model, and σ is, as defined earlier, the fractional convective area determined by the procedure presented in section 3g. Since w_c should not significantly depend on σ, the conversion rates should linearly increase as σ increases. When σ approaches 1, the mass flux approaches $\rho \overline{w}$, which is the grid-scale vertical mass flux.

The package is also expected to prognostically determine the conversions taking place outside of the updrafts, assuming $\sigma \sim 0$ temporarily. As in the existing parameterizations, the detrainment of hydrometeors from the updrafts can play a key role in this prediction. Then, since the fractional area outside of updrafts is given by $1 - \sigma$, the adjustment is made through multiplying the provisional conversion rates by $1 - \sigma$.

The above dependencies of the conversion rates are well supported by the analysis of the simulated data presented by Wu and Arakawa (2014), except that the melting of snow/graupel does not vanish even when $\sigma = 1$.

4. Route II: The quasi-3D multiscale modeling framework

a. Identification of the problem

As shown in Fig. 16-5, ROUTE II for unifying low- and high-resolution models follows the MMF approach using a coupled GCM/CRM system. In applied mathematics, a number of methods have been developed to couple macroscopic and microscopic models with objectives similar to that of the MMF. For example, E et al. (2007, p. 7), who proposed the heterogeneous multiscale method (HMM), state that their objective is "to design combined macroscopic–microscopic computational methods that are much more efficient than solving the full microscopic model and at the same time give the information we need." In the sense that the

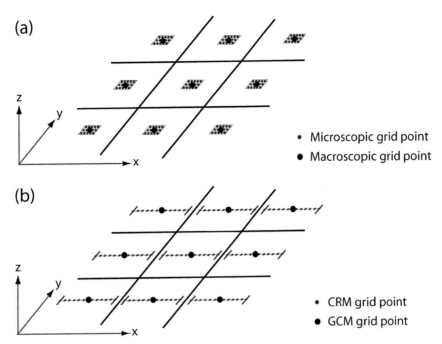

FIG. 16-10. Illustration of the grid systems used (a) in a typical coupled macroscopic/microscopic model and (b) in the prototype MMF.

parameterized equations are completely replaced with explicit simulations by the microscopic model, the MMF is similar to the equation-free approach (Kevrekidis et al. 2003; Li et al. 2007). The grid system used in these approaches may look like Fig. 16-10a. Typically, the embedded microscopic models are restarted and run until they reach equilibrium at each time step of the macroscopic model. In addition, the spatial domain of the microscopic model is very small for computing efficiency. Thus, the existence of large spectral gaps both in time and space is crucial in these methods.

In the MMF, on the other hand, the GCM and CRMs are run simultaneously. In this particular sense, the MMF is similar to the seamless multiscale methods (E et al. 2009). What makes the MMF unique is that the computing efficiency is gained through sacrificing three-dimensional representation of cloud-scale processes. This is achieved using 2D CRM grids such as those illustrated by Fig. 16-10b. The motivation for using 2D grids comes from the fact that 2D CRMs are reasonably successful in simulating the thermodynamic effects of deep moist convection (e.g., Tao et al. 1987; Grabowski et al. 1998; Xu et al. 2002).

The ROUTE II approach is an attempt to broaden the applicability of the prototype MMF without necessarily using a fully three-dimensional GCRM. As is the case for the prototype MMF, ROUTE II replaces the conventional cumulus parameterization with explicit

simulation of cloud-scale processes. Correspondingly, Fig. 16-1a is replaced with Fig. 16-11. Forcing and feedback are now the GCM's effect on the CRM and the CRM's effect on the GCM, respectively. Through forcing, the CRM recognizes

1) the vertical structure predicted by the GCM, including the GCM-scale convective instability, and
2) the horizontal inhomogeneity and anisotropy predicted by the GCM so that the CRM can respond to the GCM-scale three-dimensional processes.

Through feedback, the GCM recognizes

3) the mean "eddy" effects of cloud and associated processes simulated by CRM.

The precise definition of "eddy" is given later. Items 1 and 3 are parallel to the parameterization approach including the unified parameterization. What makes ROUTE II unique is item 2, which is difficult to achieve by the parameterization approach. Because of this feature, we call the ROUTE II approach the quasi-3D (Q3D) MMF.

Since the publication of Jung and Arakawa (2010), Arakawa et al. (2011), and Arakawa and Jung (2011), we have made considerable revisions to the Q3D algorithm, which is described in detail in Jung and Arakawa (2014). Here we present an outline of the revised algorithm.

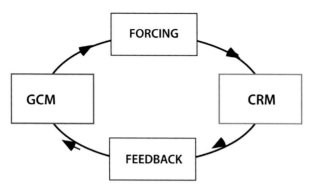

FIG. 16-11. As in Fig. 16-1a, but for ROUTE II.

b. Q3D grid system

Since CRM is used to simulate cloud-scale processes, the Q3D MMF is naturally more expensive than GCMs with conventional or unified parameterizations. Yet, when a standard resolution is used for the GCM component, the Q3D MMF is expected to be much less expensive than fully three-dimensional GCRMs. The Q3D MMF achieves this objective through the use of a grid system that has gaps in the horizontal space.

Figure 16-12 illustrates an example of the Q3D grid system. It consists of a GCM grid and two perpendicular sets of CRM gridpoint channels. Unlike the grid shown in Fig. 16-10b for the prototype MMF, each channel is extended beyond a GCM grid cell. In this way, the error due to the artificial confinement of cloud clusters within a single GCM grid cell is eliminated, as shown by Jung and Arakawa (2005). The essential difference between the two grids becomes apparent in the limit as the grid interval of GCM approaches that of the CRM. In this limit, each of the CRM grids shown in Fig. 16-10b shrinks to a single point. The gridpoint channels shown in Fig. 16-12, on the other hand, formally remain the same as the GCM resolution becomes higher, except that the interval between the parallel channels becomes

smaller. Thus, interaction of clouds with their environment is still simulated in this limit. The parallel channels are, however, not directly communicating with each other so that the convergence of the Q3D MMF to a 3D GCRM is not automatic, as discussed in section 4e.

The number of gridpoint arrays within each channel is optional. In practical applications, however, only a limited number of arrays are used for computing efficiency. In the example shown in Fig. 16-12, there are three gridpoint arrays in each channel so that there are three grid points in the lateral direction. The use of such a few grid points across the channel inevitably constrains simulation of cloud-scale three-dimensional processes. Interactions with larger-scale three-dimensionality, however, are maintained, as discussed in the following subsections.

c. The use of background fields

In the formulations of forcing and feedback shown in Fig. 16-11, the CRM component of the Q3D MMF uses background fields obtained through interpolations from the GCM grid points. Using these fields, an arbitrary CRM variable q is decomposed as

$$q = \bar{q} + q', \tag{16-8}$$

where $\bar{q}$ is the background value of q, and q' is the deviation of q from its background value. The background fields are used in the following three ways:

1) To specify lateral boundary conditions for the channel domains through which the CRM recognizes three-dimensional fields predicted by the GCM;
2) To provide reference fields to which CRM variables are relaxed to maintain compatibility of the two models;
3) To represent the subgrid components of CRM variables by the deviations from the background.

More details of these procedures are given in the next three subsections.

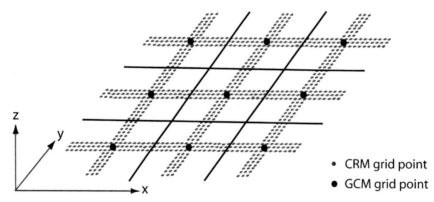

FIG. 16-12. Illustration of grid points used by the Q3D MMF.

d. Forcing: Lateral boundary condition

If the number of gridpoint arrays across the channel is not so few, our problem is basically that of limited-area modeling applied to the channel domain. Ideally, the lateral boundary condition should be nonreflecting for outflow and outward-propagating gravity waves, but designing a boundary condition that fully satisfies these requirements is a notoriously difficult problem. The problem we are facing is even more demanding because the two lateral boundaries are so close to each other. Under such a condition, Jung and Arakawa (2010) found that the cyclic boundary condition is the only one that is stable without excessive damping among many possibilities they examined. Based on this result, the CRM assumes that q' is cyclic across the channel, while its prediction recognizes the lateral as well as longitudinal gradients of $\overline{q}$. The use of the cyclic condition for q' means that the background field effectively determines the boundary condition. Thus, the parallel channels communicate only through the GCM component. We also let the perpendicular channels communicate with each other only through the GCM component so that they intersect only virtually with no singularity at their formal intersection points.

e. Forcing: Relaxation

Recall that $\overline{q}$ represents an interpolated GCM field, not a space-averaged CRM field. In the Q3D algorithm, however, these two are made sufficiently close through relaxation of q running averaged over a channel segment to $\overline{q}$ at the center of the segment. Here, the length of the channel segment is chosen to be the GCM grid spacing.

Jung and Arakawa (2014) discuss the problem of choosing the time scale for the relaxation. From a series of sensitivity experiments, they derive the following conclusions:

(i) If the relaxation time scale is very short, the CRM solution is too strongly constrained by the GCM solution, and, therefore, CRM loses its self-stabilization effect;

(ii) If the relaxation time scale is not very short, the self-stabilization effect can be maintained. But if it is shorter than a critical time scale τ_{crit}, the development of intermediate-scale cloud organization in CRM is still too strongly constrained by the GCM solution;

(iii) If the relaxation time scale is longer than τ_{crit}, the intermediate-scale cloud organization can take place in the CRM. But if the time scale is too long, compatibility between GCM and CRM solutions is not maintained.

Here, the intermediate scale refers to the scales between the GCM-resolvable scale and the cloud scale. If the GCM-resolvable scale is the synoptic scale, the intermediate scale is the mesoscale. Since the time scale associated with horizontal advection plays a crucial role in the development of mesoscale organization, the advective time scale given by d/V can be a good choice for τ_{crit}, where d and V denote the GCM horizontal grid spacing and the characteristic magnitude of horizontal velocity in the CRM solution, respectively. If $d = 96$ km, as is currently used, and $V \sim 15\,\mathrm{m\,s^{-1}}$, as is representative in these simulations, τ_{advec} is about 1.8 h, while the sensitivity experiments described by Jung and Arakawa (2014) suggest $\tau_{crit} \sim 2$ hr.

If the relaxation time scale is chosen near the advective time scale d/V, it becomes shorter as the GCM horizontal grid spacing decreases as long as the characteristic magnitude of V does not significantly change. Then the GCM and CRM components of the Q3D MMF tend to produce nearly identical solutions. This is crucially important for the convergence of the Q3D MMF to a GCRM, which is one of the most important objectives of the Q3D MMF. Although this assumption of nearly constant V may hold for most cases, it is likely that we need to adaptively determine the characteristic magnitude from the predicted fields in future applications.

f. Feedback

As stated in section 4a, feedback consists of the implementation of the mean effects of cloud and associated processes simulated by the CRM to the GCM component. For advective and dynamic processes, the CRM is supposed to implement only the eddy effects so that the CRM component does not overdo its job as a replacement for subgrid-scale parameterizations. This is in parallel to what the unified parameterization does for feedback. The Q3D MMF, however, uses its own definition of eddy: that is, the deviation from the background value rather than from a space-averaged value. Through the relaxation described in section 4e, however, this difference is made small at least for high resolutions. These eddy effects and the full diabatic effects calculated at each grid point are then averaged over all grid points in the channel segment centered at each GCM point. The length of the channel segment is again chosen to be equal to the GCM grid spacing. Finally, the effects of formally intersecting channel segments are averaged and then implemented into the GCM.

g. The base model

Although the strategy for the Q3D algorithm outlined above should be applicable to any base model, the model we have used for development of the Q3D MMF is based on the anelastic three-dimensional vorticity equation model developed by Jung and Arakawa (2008). In the model, the vertical component of vorticity is predicted at the top level, and the horizontal components of vorticity, potential temperature, and the mixing ratios of various phases of water are predicted at all levels. Vertical velocity is determined from the predicted horizontal components of vorticity by solving an elliptic equation. The horizontal components of velocity are then diagnostically determined from the known distributions of the horizontal components of vorticity and the vertical velocity. For more details, see Jung and Arakawa (2010).

h. Highlight of experimental results

Jung and Arakawa (2014) present a detailed evaluation of the Q3D MMF developed along the line described above. In this evaluation, an idealized horizontally periodic domain is used to simulate a wave-to-vortex transition in the tropics. In such a transition, the dynamics–convection interaction plays an essential role, as pointed out by Yanai (1961). A benchmark simulation (BM) is performed first with a fully three-dimensional CRM to provide a reference for Q3D MMF simulations and their initial conditions. The horizontal and vertical domain sizes are 3072 km × 3072 km and 30 km, respectively, and the horizontal grid spacing is 3 km. There are 34 layers in the vertical, using a stretched grid with the size ranging from about 0.1 km near the surface to about 2 km near the model top.

Using the horizontal grid network illustrated in Fig. 16-12 with the GCM and CRM grid spacing of 96 km and 3 km, respectively, several Q3D simulations are performed from selected dates of BM. Only one gridpoint array is used in each channel; therefore, only one grid point is independent across the channel. Because of the use of 96-km GCM grid spacing, Q3D solutions cannot be directly compared with BM solutions. Yet if the Q3D simulation produces the essential large-scale features of the BM, it should be considered as highly successful because the ratio of the number of CRM grid points used by the Q3D MMF and that by the BM is only 6% for this configuration of grid points. This ratio becomes even smaller if the GCM resolution is coarser, as is normally the case, or the CRM resolution is finer.

In the rest of this subsection, selected results of a 13-day Q3D simulation are shown. For the comparison with the Q3D simulation, the BM fields are averaged over the horizontal area represented by each GCM grid point. Figure 16-13a shows the time sequence of the vertical component of vorticity at $z = 3$ km taken from the BM. In this time sequence, we see that this period is

characterized by development and subsequent persistence of two intense vortices. Figure 16-13b shows the corresponding time sequence taken from the Q3D simulation. As in the BM, two intense vortices are developed and maintained, only with slight differences in the location and intensity of the vortices toward the end of the period.

The results from the Q3D simulation shown in Fig. 16-13b can be compared with those from a run shown in Fig. 16-13d in which the feedback from the CRM to the GCM is not included. In this simulation, the two vortices do not intensify and tend to merge into one vortex. These results confirm that the dynamics–convection interaction is crucial for the development and persistence of the vortices and the Q3D simulation is quite successful in representing the interaction.

To see the importance of using the background fields in the Q3D MMF, a test simulation is performed with a "2D MMF" in which 2D CRMs are used instead of 3D CRMs in the Q3D MMF. The CRM component of the 2D MMF still forms two perpendicular sets of gridpoint arrays virtually intersecting at the center of each GCM grid cell. Because of their two-dimensionality, however, the CRMs recognize the large-scale horizontal inhomogeneity only along the channel, not across the channel. Each CRM predicts only one horizontal vorticity component, the component normal to the gridpoint array. Otherwise, the prediction algorithm for this simulation is the same as that for the standard Q3D MMF. Figure 16-13c presents the evolution of the vertical component of vorticity simulated by the 2D MMF. (Note that this 2D MMF is not equivalent to the prototype MMF because of the use of different grid structures.) Comparing this sequence to that shown in Fig. 16-13b, we see that the recognition of large-scale inhomogeneity across the channel makes a drastic difference in the vortex development.

Figure 16-14 shows the time series of the precipitation and evaporation rates at the surface taken from the BM, Q3D MMF, and 2D MMF simulations averaged over their respective horizontal domains. Figure 16-14a shows that, after the initial adjustment period of a day or so, the surface precipitation rate of the Q3D simulation becomes very close to that of the BM, while that of the 2D MMF simulation is considerably underpredicted. Figure 16-14b presents similar results for the evaporation rate, showing that the 2D MMF predicts a significantly less-active hydrological balance.

The preliminary tests of the Q3D MMF presented above are quite encouraging, showing the big potential of the Q3D MMF as the basic framework for future NWP and climate models. More details on the Q3D MMF are presented by Jung and Arakawa (2014).

Vertical Component of Vorticity

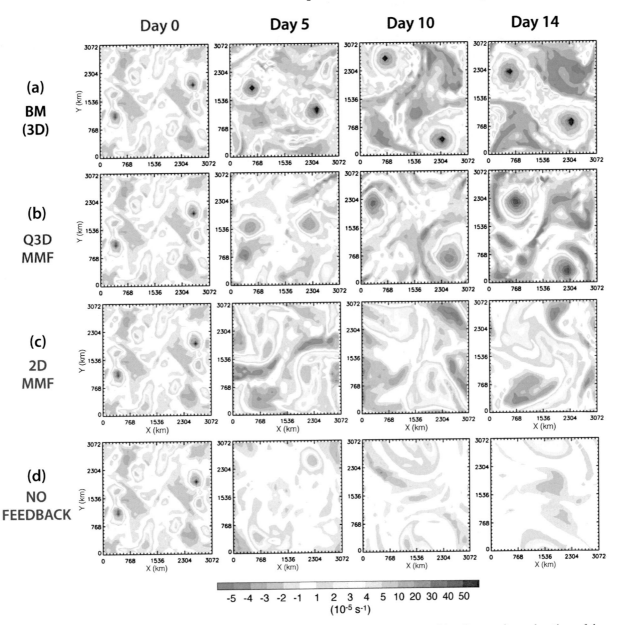

FIG. 16-13. Time sequences of the horizontal maps of the vertical component of vorticity at $z = 3$ km. See text for explanations of the sequences. Rearranged from Jung and Arakawa (2014).

5. Concluding remarks

Michio Yanai worked on virtually all aspects of tropical meteorology, covering a broad range of the spectrum from the cloud scale to the planetary scale. One of his most important contributions is the introduction of the concepts of apparent heat source Q_1 and apparent moisture sink Q_2 in the large-scale heat and moisture budgets of the atmosphere. Because of the inclusion of eddy effects, Q_1 and Q_2 are generally different from true heat sources and true moisture sinks taking place locally. In low-resolution models, such as the conventional general circulation models (GCMs), cumulus parameterization determines Q_1 and Q_2 from the gridpoint values of the model using a closure assumption. In the multiscale modeling framework (MMF), on the other hand, Q_1 and Q_2 are numerically simulated by a cloud-resolving model (CRM) coupled with the GCM.

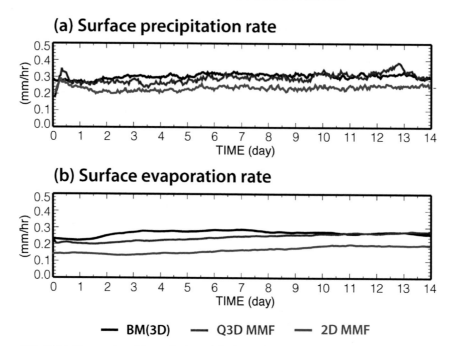

FIG. 16-14. Time series of the surface precipitation and evaporation rates. Rearranged from Jung and Arakawa (2014).

The main theme of this article is to build a bridge, or bridges, between apparent sources and true sources. Since the difference between them is the inclusion of eddy effects in the former, a generalized representation of those effects is crucial in building such a bridge. It is shown that there are two routes to do so: ROUTE I and ROUTE II. ROUTE I follows the parameterization approach, while ROUTE II follows the MMF approach.

We realize that more extensive and intensive work is essential to complete well-built bridges between various ways of representing moist-convective processes. Such work may well become a central issue in the coming era in the history of numerical modeling of the atmosphere. If Michio Yanai were still with us, he would certainly make valuable contributions to this era in his own way.

Acknowledgments. We wish to thank Professor David Randall for his interest and support of this work. The first and second authors are supported by the National Science Foundation Science and Technology Center for Multiscale Modeling of Atmospheric Processes, managed by Colorado State University under Cooperative Agreement ATM-0425247. This work utilized the CSU ISTeC Cray HPC System supported by NSF Grant CNS-0923386. The third author is supported by Taiwan's National Research Council through Grant 101-2111-M-002-006 to National Taiwan University.

REFERENCES

Arakawa, A., 1969: Parameterization of cumulus clouds. *Proc. WMO/IUGG Symp. on Numerical Weather Prediction*, Tokyo, Japan, Japan Meteorological Agency, IV-8-1–IV-8-6.

——, 2004: The cumulus parameterization problem: Past, present, and future. *J. Climate*, **17**, 2493–2525, doi:10.1175/1520-0442(2004)017<2493:RATCPP>2.0.CO;2.

——, and W. H. Schubert, 1974: Interaction of a cumulus cloud ensemble with the large-scale environment. Part I. *J. Atmos. Sci.*, **31**, 674–701, doi:10.1175/1520-0469(1974)031<0674:IOACCE>2.0.CO;2.

——, and J.-H. Jung, 2011: Multiscale modeling of the moist-convective atmosphere—A review. *Atmos. Res.*, **102**, 263–285, doi:10.1016/j.atmosres.2011.08.009.

——, and C.-M. Wu, 2013: A unified representation of deep moist convection in numerical modeling of the atmosphere. Part I. *J. Atmos. Sci.*, **70**, 1977–1992, doi:10.1175/JAS-D-12-0330.1.

——, J.-H. Jung, and C.-M. Wu, 2011: Toward unification of the multiscale modeling of the atmosphere. *Atmos. Chem. Phys.*, **11**, 3731–3742, doi:10.5194/acp-11-3731-2011.

Betts, A. K., 1982: Saturation point analysis of moist convective overturning. *J. Atmos. Sci.*, **39**, 1484–1505, doi:10.1175/1520-0469(1982)039<1484:SPAOMC>2.0.CO;2.

E, W., B. Engquist, X. Li, W. Ren, and E. Vanden-Eijnden, 2007: Heterogeneous multiscale methods: A review. *Commun. Comput. Phys.*, **2**, 367–450.

——, W. Ren, and E. Vanden-Eijnden, 2009: A general strategy for designing seamless multiscale methods. *J. Comput. Phys.*, **228**, 5437–5453, doi:10.1016/j.jcp.2009.04.030.

Frank, W. M., 1993: A hybrid parameterization with multiple closures. *The Representation of Cumulus Convection in Numerical Models*, Meteor. Monogr., No. 46, Amer. Meteor. Soc., 151–154.

Gerard, L., 2007: An integrated package for subgrid convection, clouds and precipitation compatible with the meso-gamma scales. *Quart. J. Roy. Meteor. Soc.*, **133**, 711–730, doi:10.1002/qj.58.

Grabowski, W. W., 2001: Coupling cloud processes with the large-scale dynamics using the cloud-resolving convective parameterization (CRCP). *J. Atmos. Sci.*, **58**, 978–997, doi:10.1175/1520-0469(2001)058<0978:CCPWTL>2.0.CO;2.

——, and P. K. Smolarkiewicz, 1999: CRCP: A cloud resolving convective parameterization for modeling the tropical convective atmosphere. *Physica D*, **133**, 171–178, doi:10.1016/S0167-2789(99)00104-9.

——, X. Wu, M. W. Moncrieff, and W. D. Hall, 1998: Cloud-resolving modeling of cloud systems during Phase III of GATE. Part II: Effects of resolution and the third spatial dimension. *J. Atmos. Sci.*, **55**, 3264–3282, doi:10.1175/1520-0469(1998)055<3264:CRMOCS>2.0.CO;2.

Gregory, D., R. Kershaw, and P. M. Inness, 1997: Parametrization of momentum transport by convection. II: Tests in single-column and general circulation models. *Quart. J. Roy. Meteor. Soc.*, **123**, 1153–1183, doi:10.1002/qj.49712354103.

Johnson, R. H., P. E. Ciesielski, and T. M. Rickenbach, 2016: A further look at Q_1 and Q_2 from TOGA COARE. *Multiscale Convection-Coupled Systems in the Tropics: A Tribute to Dr. Michio Yanai, Meteor. Monogr.*, No. 56, Amer. Meteor. Soc., doi:10.1175/AMSMONOGRAPHS-D-15-0002.1.

Jung, J.-H., and A. Arakawa, 2004: The resolution dependence of model physics: Illustrations from nonhydrostatic model experiments. *J. Atmos. Sci.*, **61**, 88–102, doi:10.1175/1520-0469(2004)061<0088:TRDOMP>2.0.CO;2.

——, and ——, 2005: A preliminary test of multiscale modeling with a two-dimensional framework: Sensitivity to coupling methods. *Mon. Wea. Rev.*, **133**, 649–662, doi:10.1175/MWR-2878.1.

——, and ——, 2008: A three-dimensional anelastic model based on the vorticity equation. *Mon. Wea. Rev.*, **136**, 276–294, doi:10.1175/2007MWR2095.1.

——, and ——, 2010: Development of a quasi-3D multiscale modeling framework: Motivation, basic algorithm and preliminary results. *J. Adv. Model. Earth Syst.*, **2** (4), doi:10.3894/JAMES.2010.2.11.

——, and ——, 2014: Modeling of the moist-convective atmosphere with a quasi-3-D multiscale modeling framework (Q3D MMF). *J. Adv. Model. Earth Syst.*, **6**, 185–205, doi:10.1002/2013MS000295.

Kevrekidis, I. G., C. W. Gear, J. M. Hyman, P. G. Kevrekidis, O. Runborg, and C. Theodoropoulos, 2003: Equation-free multiscale computation: Enabling microscopic simulators to perform system-level tasks. *Commun. Math. Sci.*, **1**, 715–762, doi:10.4310/CMS.2003.v1.n4.a5.

Khairoutdinov, M. F., and D. A. Randall, 2001: A cloud-resolving model as a cloud parameterization in the NCAR Community Climate System Model: Preliminary results. *Geophys. Res. Lett.*, **28**, 3617–3620, doi:10.1029/2001GL013552.

LeMone, M. A., 1983: Momentum transport by a line of cumulonimbus. *J. Atmos. Sci.*, **40**, 1815–1834, doi:10.1175/1520-0469(1983)040<1815:MTBALO>2.0.CO;2.

——, G. M. Barnes, and E. J. Zipser, 1984: Momentum flux by lines of cumulonimbus over the tropical oceans. *J. Atmos. Sci.*, **41**, 1914–1932, doi:10.1175/1520-0469(1984)041<1914:MFBLOC>2.0.CO;2.

Li, J., P. G. Kevrekidis, C. W. Gear, and I. G. Kevrekidis, 2007: Deciding the nature of the coarse equation through microscopic

simulations: The baby–bathwater scheme. *SIAM Rev.*, **49**, 469–487, doi:10.1137/070692303.

Molinari, J., 1993: An overview of cumulus parameterization in mesoscale models. *The Representation of Cumulus Convection in Numerical Models, Meteor. Monogr.*, No. 46, Amer. Meteor. Soc., 155–158.

——, and M. Dudek, 1992: Parameterization of convective precipitation in mesoscale numerical models: A critical review. *Mon. Wea. Rev.*, **120**, 326–344, doi:10.1175/1520-0493(1992)120<0326:POCPIM>2.0.CO;2.

Moncrieff, M. W., 1981: A theory of organized steady convection and its transport properties. *Quart. J. Roy. Meteor. Soc.*, **107**, 29–50, doi:10.1002/qj.49710745103.

——, 1992: Organized convective systems: Archetypal dynamical models, mass and momentum flux theory, and parametrization. *Quart. J. Roy. Meteor. Soc.*, **118**, 819–850, doi:10.1002/qj.49711850703.

Oouchi, K., and M. Satoh, 2016: A synoptic-scale cold-reservoir hypothesis on the origin of the mature-stage super cloud cluster: A case study with a global nonhydrostatic model. *Multiscale Convection-Coupled Systems in the Tropics: A Tribute to Dr. Michio Yanai, Meteor. Monogr.*, No. 56, Amer. Meteor. Soc., doi:10.1175/AMSMONOGRAPHS-D-15-0008.1.

Randall, D. A., M. Khairoutdinov, A. Arakawa, and W. Grabowski, 2003: Breaking the cloud parametrization deadlock. *Bull. Amer. Meteor. Soc.*, **84**, 1547–1564, doi:10.1175/BAMS-84-11-1547.

——, C. DeMott, C. Stan, M. Khairoutdinov, J. Benedict, R. McCrary, K. Thayer-Calder, and M. Branson, 2016: Simulations of the tropical general circulation with a multiscale global model. *Multiscale Convection-Coupled Systems in the Tropics: A Tribute to Dr. Michio Yanai, Meteor. Monogr.*, No. 56, Amer. Meteor. Soc., doi:10.1175/AMSM-D-15-0016.1.

Sato, T., H. Miura, M. Satoh, Y. N. Takayabu, and Y. Wang, 2009: Diurnal cycle of precipitation in the tropics simulated in a global cloud-resolving model. *J. Climate*, **22**, 4809–4826, doi:10.1175/2009JCLI2890.1.

Tao, W.-K., J. Simpson, and S.-T. Soong, 1987: Statistical properties of a cloud ensemble: A numerical study. *J. Atmos. Sci.*, **44**, 3175–3187, doi:10.1175/1520-0469(1987)044<3175:SPOACE>2.0.CO;2.

Wu, C.-M., and A. Arakawa, 2014: A unified representation of deep moist convection in numerical modeling of the atmosphere. Part II. *J. Atmos. Sci.*, **71**, 2089–2103, doi:10.1175/JAS-D-13-0382.1.

Wu, X., and M. Yanai, 1994: Effects of vertical wind shear on the cumulus transport of momentum: Observations and parameterization. *J. Atmos. Sci.*, **51**, 1640–1660, doi:10.1175/1520-0469(1994)051<1640:EOVWSO>2.0.CO;2.

Xu, K. M., and K. A. Emanuel, 1989: Is the tropical atmosphere conditionally unstable? *Mon. Wea. Rev.*, **117**, 1471–1479, doi:10.1175/1520-0493(1989)117<1471:ITTACU>2.0.CO;2.

——, and Coauthors, 2002: An intercomparison of cloud-resolving models with the atmospheric radiation measurement summer 1997 intensive observation period data. *Quart. J. Roy. Meteor. Soc.*, **128**, 593–624, doi:10.1256/003590002321042117.

Yanai, M., 1961: A detailed analysis of typhoon formation. *J. Meteor. Soc. Japan*, **39**, 187–214.

——, S. Esbensen, and J. Chu, 1973: Determination of bulk properties of tropical cloud clusters from large-scale heat and moisture budgets. *J. Atmos. Sci.*, **30**, 611–627, doi:10.1175/1520-0469(1973)030<0611:DOBPOT>2.0.CO;2.

Six Decades of Tropical Meteorology Research—A Retrospective on Michio Yanai's Life and Career

STEVEN K. ESBENSEN

College of Earth, Ocean and Atmospheric Sciences, Oregon State University, Corvallis, Oregon

JAN-HWA CHU

Defense Language Institute Foreign Language Center, Monterey, California

WEN-WEN TUNG

Earth, Atmospheric and Planetary Sciences, Purdue University, West Lafeyette, Indiana

ROBERT G. FOVELL

Department of Atmospheric and Oceanic Sciences, University of California, Los Angeles, Los Angeles, California

This monograph on convection-coupled systems in the tropics was inspired by the life and career of Professor Michio Yanai, whose major contributions to the subject spanned more than five decades. From a distant perspective, Professor Yanai's career can be understood in the context of Japanese scientists who immigrated to the United States in the decades of the 1950s and 1960s, enriching the meteorological research community in the United States as well as abroad (Lewis 1993). A closer look reminds us that the tapestry of scientific progress is created by the contributions of individual scientists with their unique backgrounds, motivations, and talents, and the serendipity of events that shape their lives.

---

1. Early encounters with typhoons

Michio Yanai was born 16 January 1934, in Tokyo, and grew up in Chigasaki, a relatively small town nearby. He was fascinated by weather—typhoons in particular—at an early age. Immediately after entering Shonan Middle School in 1946, he joined the meteorology club (otenkikai). The "meteorology boys" (Fig. 1) were most excited by the approach of typhoons that came every year, typically in September. The club issued storm warnings to the school principal. Teachers not only received the club's "unauthorized" warnings with enthusiasm, but also dismissed classes based on the meteorology club's judgment.

On one occasion, the meteorology club decided that they would determine the precise path of a typhoon that had just passed over Chigasaki (Fig. 2). Riding on a commuter train the next afternoon, a student got off at each station to investigate the direction of the fallen trees. In this way the club systematically determined the exact path of the typhoon's eye. Later in his career, Yanai was convinced that the meteorology boys invented this method before Professor Tetsuya (Ted) Fujita used a similar strategy to determine tornado tracks. The club reported its results to the Yokohama Weather Station (currently, the Yokohama District Observatory). They received thanks and praise for their investigation, and made the news the next morning.

2. Typhoon research at the University of Tokyo

Michio Yanai arrived as an undergraduate at the University of Tokyo in 1952. All students in their first and second years were required to attend the "Culture School" on the Komaba campus. Third-year studies focused on classical mechanics, fluid dynamics, thermodynamics and statistical mechanics, electromagnetism, and quantum mechanics. In the fourth and final year Yanai and his fellow geophysics majors studied

DOI: 10.1175/AMSMONOGRAPHS-D-15-0010.1

FIG. 1. Shonan meteorology boys (Michio Yanai in the back middle).

FIG. 2. Doing oil painting near Chigasaki (1954).

geodesy, seismology, geo-electromagnetism, oceanography, and meteorology.

In 1956, Yanai entered the University of Tokyo graduate school as a meteorology major. Professor Shigekata Syono suggested that he study typhoons, but gave no specific instructions. After publishing his master's thesis on a decaying typhoon (Yanai 1958), Yanai decided to study the formative process of typhoons. An intensive literature survey showed him that typhoons form either in high sea surface temperature regions in the ITCZ, frequently called the "equatorial front" (Palmén 1948), or from preexisting disturbances such as waves in the easterlies (e.g., Dunn 1940; Riehl 1948). But he found that there was not a single illustration of the actual formation process in the existing literature, even in Riehl's (1954) *Tropical Meteorology*.

The Japanese Meteorological Agency (JMA) loaned Yanai a set of teletyped station reports covering the period of Typhoon Doris (1958). Professor Syono assisted in the data collection by asking Lt. Col. Bundgaard (10th Weather Group, U.S. Air Force) to provide a more complete collection of teletyped reports. There were no copying machines in 1958 and the

loaned data forms had to be returned promptly, so Yanai copied the data by hand, day and night. The drawing of streamlines, isotherms, cross sections, and so forth, and the analysis of the formation process, occupied two full years of his graduate work. To study the heat source of this typhoon, Yanai defined the Q_1 (heat source) and Q_2 (moisture sink) terms (Fig. 3) to show their similarity. Later diagnostics developed at UCLA would emphasize the difference between Q_1 and Q_2 as a measure of convective activity and radiative effects.

In 1959, another typhoon would have a great impact on Yanai's life. High tides and the storm surge from the Ise-wan Taifu (Typhoon Vera) caused 5000 deaths along the coast of Ise-wan (bay) near Nagoya. The cabinet member in charge of Science and Technology (future Prime Minister Yasuhiro Nakasone) advocated the creation of the Typhoon Research Laboratory. In 1960, the Diet (legislature) instructed the government to

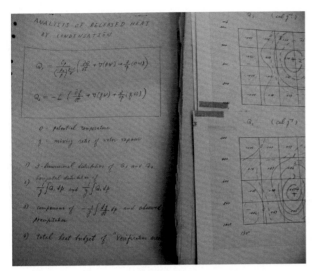

FIG. 3. (left) Notes on Q_1 and Q_2 and (right) hand-drawn contour plots for Typhoon Doris (1958).

form a typhoon research laboratory in the JMA's Meteorological Research Institute (MRI), and Yanai was invited to join after graduation and passing the civil service exam.

Meanwhile, Taroh Matsuno and Yanai (Fig. 4) formed a research team of eight graduate students to study the precipitation bands of Typhoon Vera (1959). They met once a week to discuss strategy and analyze

data. This activity was purely voluntary and not directly related to the students' thesis research. The group was so thorough in collecting data from non-JMA stations such as schools and railway stations that they received official protests from the JMA when the agency found nothing left at these stations when they attempted to collect the data. Results from the graduate students were published much later in two papers (Hamuro et al. 1969, 1970).

In 1960, Dr. Yanai presented his case study of Typhoon Doris at the First International Symposium on Numerical Weather Prediction in Tokyo. There he met Jule Charney, Akira Kasahara, Hsiao-Lan Kuo, and other scientists interested in numerical modeling of typhoons. Charney then visited Michio at the University of Tokyo to examine the original analyses of the typhoon. This occurred as the first attempts at numerically simulating tropical cyclone formation were being made. Simulations by Charney's group at the Massachusetts Institute of Technology (MIT), Kasahara's at the University of Chicago, and Syono's at the University of Tokyo were all failing to produce an organized tropical cyclone, producing model grid-size, Rayleigh-convection-like structures instead.

The main thrust of Yanai's graduate research to this point was to show the detailed transition process of a cold core easterly wave into an incipient typhoon with an organized surface circulation and a relatively warm

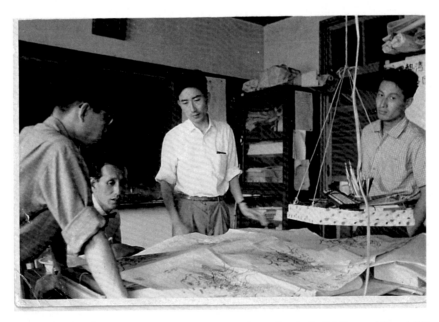

FIG. 4. Michio Yanai (center) and Taroh Matsuno (near, showing back) discussed the rainfall pattern brought about by Typhoon Vera (1959) with Associate Professor Kenji Isono (second left).

FIG. 5. At the University of Tokyo (1961).

FIG. 6. In Colorado during his visit to Fort Collins (1962–64).

core (Yanai 1961a). This work had been well received, but Professor Syono commented "your paper has no equations." Yanai's good-humored response was "OK, I will write another paper with 100 equations!" Soon he produced a second paper based on inertial instability (Yanai 1961b), extending the work of Kleinschmidt (1951), and received his D.Sc. degree from the University of Tokyo (Fig. 5).

3. A visit to Fort Collins and the National Hurricane Research Project

Shortly after taking up his first position at the MRI's Typhoon Research Laboratory, Yanai received the 1962 Meteorological Society of Japan Award for his study of typhoon formation, and was invited by Professor Herbert Riehl to visit Colorado State University (CSU) in Fort Collins for two years (Fig. 6) as a visiting assistant meteorologist. Although this period did not result in many publications, his 1964 review paper on tropical cyclone formation (Yanai 1964) served as the most comprehensive review of the field for over a decade. This work was the result of a direct meeting in 1963 with Professor Joanne Simpson. She had recently been appointed as a UCLA professor and was visiting her former advisor, Professor Riehl. She encouraged Riehl's young postdoc to write a review paper on the formation of tropical cyclones. Without hesitation Yanai accepted her invitation.

Professor Riehl arranged for his postdoc to visit the National Hurricane Research Project (NHRP) in Miami. By an agreement between CSU and NHRP, Michio was stationed in Miami during the 1962 hurricane season. There, he met Ed Zipser, then a student at Florida State University (FSU), who was also visiting NHRP and helped Yanai find an affordable apartment. He met NHRP scientists Craig Gentry, Stan Rosenthal, and

others, and took in as many experiences as possible, including a hurricane reconnaissance flight over the Gulf of Mexico. In August 1962, Hurricane Alma nearly hit Miami. Yanai studied this hurricane in depth and subsequently published the work as Yanai and Nitta (1967) and Yanai (1968).

Dr. Yanai would meet many scientists in the United States who would become valued colleagues throughout his career. These included of course his hosts Herb Riehl, Bill Gray (Fig. 7), Elmar Reiter, and Ferdinand Baer while he was at CSU. Yanai also attended meetings in New York (1963) and Mexico City (1963) where he met Professors Jordan and LaSeur (FSU), Krishnamurti [University of California, Los Angeles (UCLA)], Ooyama (New York University), and Miyakoda, Platzman, and Fultz (Chicago). On his way home to Tokyo in 1964, he visited Professors Yale Mintz at UCLA and James Sadler at the University of Hawaii.

4. Tropical wave and cloud cluster studies in Tokyo

In 1965, Yanai returned to the University of Tokyo as an assistant professor at the request of Professor Syono. This was to fill the position vacated by Professor Kikuro Miyakoda, who had moved to the Geophysical Fluid Dynamics Laboratory in Princeton. Yanai's tenure at the University of Tokyo was very productive. Together with outstanding graduate students Taketo Maruyama, Tsuyoshi Nitta, Yoshikazu Hayashi, Tatsushi Tokioka, Masato Murakami, and Masanori Yamasaki, the group (Fig. 8) published 25 papers.

The discovery of the Yanai–Maruyama wave in the equatorial stratosphere (Yanai and Maruyama 1966) was motivated by Professor Richard Reed's review paper on the quasi-biennial oscillation (QBO). Reed had suggested relating the time change of the stratospheric zonal wind to the convergence of the eddy

FIG. 7. With Dr. Bill Gray in Fort Collins (1972).

transport of momentum. This led Yanai and Maruyama to examine the vertical structure of the wave to examine its potential for vertical momentum transport. Soon the Yanai–Maruyama wave was identified as the mixed Rossby–gravity wave in the theory of equatorially trapped waves (Matsuno 1966; Lindzen and Matsuno 1968).

Professor Yanai's group extended their earlier work on tropical easterly waves by quantitative analysis of the structure and evolution of Caribbean wave disturbances, and examined the potential for initial growth of easterly waves by barotropic instability processes. They also applied spectral analysis to time series of atmospheric sounding data throughout the tropical Pacific to study the kinetic and potential energy of energetic wave disturbances as a function of their frequency, from the surface to the lower stratosphere. They used increasingly sophisticated methods of cross-spectral analysis to sort out the vertical and horizontal structure of the energetic wave modes and to obtain clues about their excitation mechanisms. As stated in the introduction of Madden and Julian (1971), the studies by the Tokyo group motivated the work that led to the discovery of the Madden–Julian oscillation (MJO).

In 1967, the Global Atmospheric Research Program (GARP) was initiated and Professors Yanai, G. Yamamoto, and Yoshi Ogura were invited to its first international study conference held in Stockholm. Michio was appointed as a member of the Study Group on Tropical Disturbances in 1968, together with Professors Pisharoty (India) and Fujita (University of Chicago). The group met over the period of a month at the University of Wisconsin to conduct a census of cloud systems in the global tropics using geostationary satellite pictures, and to formulate the scientific requirements for the tropical GARP subprogram. The group classified cloud systems into "cloud clusters," "monsoon clusters," and "popcorn cumulonimbi." They recommended the western North Pacific as the site for a tropical GARP field program, in agreement with recommendations of several other groups. In December 1969, Michio attended the GARP Joint Organizing Committee (JOC) planning meeting held in Miami for the First Tropical Experiment.

At this critical moment in Professor Yanai's research and international activity, student riots broke out at the University of Tokyo and lasted for a year. Radical students influenced by China's Great Cultural Revolution and Berkeley's radical student movement occupied many key buildings and soon the teaching and research functions of the university ceased. Offers of "rescue" from his American friends and colleagues began to arrive.

In 1968, Professor Akio Arakawa attended the WMO–International Union of Geodesy and Geophysics (IUGG) Numerical Weather Prediction Symposium at JMA and communicated an offer of a full professorship at UCLA to Professor Yanai that he quickly accepted. Yanai requested a one-year leave of absence from UCLA to take care of his students and finish the remaining work in Tokyo. He departed the University of Tokyo for UCLA in September 1970, with his wife Yoko and their two young sons, Takashi and Satoshi (Fig. 9).

5. Cloud cluster, tropical waves, and monsoon studies in Los Angeles

On his arrival at UCLA in the fall of 1970, Professor Yanai became immersed in the tasks of building a new research group and participating in the teaching activities of the department, then known as the Department of Meteorology. The new courses that he would develop in tropical meteorology, tropical waves, and tropical cyclones were encyclopedic in their scope. And it was time to consider the lines of research that would define his UCLA research group.

FIG. 8. Tokyo, June 1966. Group members of assistant professor M. Yanai and Dr. Matsuno
(back and front center). Graduate student Tsuyoshi Nitta is on the far left.

a. The Q_1 and Q_2 studies

Soon after his arrival in Los Angeles, Professor Arakawa persuaded Yanai to begin a new research effort based on the diagnosis of the large-scale heat source and moisture sink, Q_1 and Q_2. Arakawa was developing the Arakawa–Schubert cumulus parameterization based on the concept that cumulus clouds modify the large-scale temperature and moisture fields through detrainment and cumulus-induced subsidence in their environment. He felt that many other cumulus parameterization schemes were using rather arbitrary assumptions and he encouraged Yanai to develop an analysis scheme that could diagnose the cumulus mass flux and detrainment from observed Q_1 and Q_2, based on fundamental physical laws. They became aware that the difference between Q_1 and Q_2 is a measure of eddy vertical transport of total heat, and thus it contains information about cumulus activity.

Yanai started work on the diagnostic scheme immediately. On 18 November 1970, he prepared a table of all upper-air stations in the Marshall Islands that participated in a special observation program in 1956 in support of nuclear tests (Operation Redwing, Joint Task Force Seven). From the data books for all 11 stations, sounding data were punched on cards by an undergraduate student, and then written on magnetic tapes.

Meanwhile, Yanai designed line-integral (horizontal) and finite-difference (vertical) schemes to calculate mass, heat, and moisture budgets, and began to construct

a bulk cumulus ensemble model to derive the cumulus mass flux, detrainment, cumulus-induced heating, and moistening. Graduate students Steve Esbensen and Jan-Hwa Chu were interested in what he was doing and joined the project. By September 1971, Professor Yanai was writing up the early results using the Marshall Islands data.

The most important conclusion from this initial study was that the cloud mass flux exceeds the mean vertical mass flux required by the large-scale convergence, thereby causing a compensating sinking motion between the active clouds. The detrainment of large amounts of cloud water acts to cool the environment, counteracting the effect of warming due to adiabatic compression in the sinking environmental air. The detrainment of cloud water and water

FIG. 9. Sons and father in 1970, prior to leaving Tokyo.

FIG. 10. AMS Tropical Meteorology Conference in Miami, at Steve Lord's home. From left: T. Nitta, M.-M. Lu, B. Esbensen, M. Yanai, S. Esbensen, S. Lord, J. Lord, and L. Shapiro (1983).

vapor counteracts the drying due to the environmental sinking motion. The calculation further demonstrated the feasibility of using cumulus ensemble models and observed Q_1 and Q_2 to quantify the various effects of clouds on the large-scale thermodynamic fields. Some of these conclusions were also reached by Gray (1973), using an independent approach.

The research on the Marshall Island data by Professor Yanai's group was taking place at the same time as his participation in a flurry of scientific planning and organizational activities that would become the GARP Atlantic Tropical Experiment (GATE). The interim Scientific and Management Group for GATE (ISMG) was established with Joachim Kuettner as the United States member. Yanai became a consultant to the ISMG and a member of the U.S. GATE panel, attending the second meeting of the JOC Study Group in Geneva in January 1971. By request of ISMG, Michio wrote "A review of recent studies of tropical meteorology relevant to the planning of GATE" in September 1971. This manuscript included the early results from the UCLA project's Q_1 and Q_2 calculations and was widely distributed within the GATE community (Annex I, Experimental Design Proposal for the GARP Atlantic Tropical Experiment, GATE Report 1, June 1972). In December 1971, early results from the Q_1 and Q_2 study were presented at the AMS tropical meteorology meeting in Barbados.

Dr. Kuettner nominated Professor Yanai to become the Chief Scientist of GATE. He was, however, unable to accept. Foreshadowing health issues that would periodically limit his activities for the remainder of his career, Yanai became ill in 1972 and was hospitalized due to exhaustion from overwork.

A year was needed to polish the Marshall Islands diagnostic work, refining computational schemes as well as the English writing and organizational structure of the paper with the help of Professor Yale Mintz. The Yanai, Esbensen, and Chu paper was finally submitted to the *Journal of Atmospheric Sciences* in December 1972. Between the editor's receipt of the manuscript and the time the paper was sent to the technical editor, the authors heard nothing from the editor. Concerned, they inquired about its status, only to learn that the paper was "unconditionally" accepted; that is, no changes were required.

b. Beyond Q_1 and Q_2

The Yanai et al. (1973) paper inspired Yanai's group to push the method further by diagnosing detailed properties of cumulus ensembles, not only in large-scale heat and moisture budgets (Yanai et al. 1976), but in large scale momentum, vorticity, and potential vorticity budgets as well. Yanai and his graduate students wrote a series of papers diagnosing cloud properties of the

FIG. 11. With UCLA colleagues and spouses. From left: J.D. Neelin, M. Ghil, K. Arakawa, A. Arakawa, V. Magaña, M. Yanai, and Y. Yanai.

vorticity and momentum fields (Chu et al. 1981; Yanai et al. 1982; Sui and Yanai 1986; Sui et al. 1989; Tung and Yanai 2002a,b) and introduced refinements in the cumulus ensemble model that allowed for diagnosis of the roles of cumulus downdrafts and mesoscale effects (Cheng and Yanai 1989) and vertical wind shear (Wu and Yanai 1994). In addition, there was an ongoing effort to diagnose the vertical structure and spatial distribution of Q_1 and Q_2 variability over the tropical oceans (Johnson et al. 1987; Tung et al. 1999; Yanai and Tomita 1998).

Professor Yanai viewed opportunities to interact with visiting postdocs and scholars, attending professional meetings and research proposal writing as part of integrated graduate student training (Fig. 10). His former graduate student Tsuyoshi Nitta was invited to make two short but, as it turned out, extremely productive, visits to UCLA, shortly after Nitta had obtained his degree from the University of Tokyo. Professor Yanai suggested that Dr. Nitta think about converting the bulk cloud model in Yanai et al. (1973) to a spectral model that allows the diagnosis of cloud types categorized by the cloud-top heights. Within days, the problem was solved, and Yanai's group used this formalism for all of its cloud diagnostic work over the following decade. Dr. Nitta also worked closely with Steve Esbensen on the analysis of the structure and variability of the trade wind boundary layer during BOMEX and its cloud ensemble properties (Nitta

FIG. 12. After B. Chen's Ph.D. defense, 1998, at UCLA.

FIG. 13. The second U.S.–People's Republic of China Workshop on Cooperation in Monsoon Research, held at the Naval Postgraduate School in June 1985.

and Esbensen 1974a,b; Nitta 1975). Dr. Nitta later became a much respected professor at the University of Tokyo. Sadly, Professor Nitta would die at a young age, as would former University of Tokyo students Dr. Masato Murakami (MRI) and Dr. Yoshikazu Hayashi (GFDL).

c. Tropical waves and the Madden–Julian oscillation

Tropical atmospheric waves and the MJO continued to fascinate Professor Yanai after his arrival at UCLA, and became a major focus of his newly formed research group (Fig. 11). His first student to receive a Ph.D. at UCLA was Abraham Zangvil. Together they studied the dynamics of upper tropospheric waves and their association with clouds in the wavenumber–frequency domain (Zangvil 1975; Zangvil and Yanai 1980, 1981), extending the analysis beyond individual station data (e.g., Yanai et al. 1968; Wallace and Kousky

1968) to the entire tropical domain and applying Dr. Hayashi's new space–time analysis techniques (Hayashi 1977, then at GFDL) to create wavenumber–frequency spectra of the latest 200-mb wind and satellite brightness fields. The same techniques were made even better known by Professor Nitta's Ph.D. student Yukari Takayabu (Takayabu 1994a,b) in the study of convection-coupled equatorial waves.

A particular emphasis of the UCLA group (Fig. 12) was quantifying the importance of tropical–midlatitude interactions over the central and eastern Pacific. Professor Yanai and graduate students Mong-Ming Lu and Victor Magaña calculated lateral energy fluxes by large-scale disturbances in the upper troposphere and found evidence for selective forcing of equatorially trapped planetary waves in the upper troposphere depending on season and the characteristics of waves propagating energy into the tropics (Yanai and Lu

FIG. 14. In Michio's office, with Chinese visitors and his Ph.D. students. From left: M.-M. Lu,
C.-H. Sui, Dr. B. Chen, M. Yanai, and Dr. H. He (1984).

1983; Magaña and Yanai 1995). Magaña and Yanai (1991) also presented evidence for a summertime modulation of tropical–midlatitude interaction in the mid-Pacific by the MJO. And although analyses over the warm pool region of the western Pacific showed that heating perturbations were responsible for generating available potential energy in disturbances with 30–60-day periods, they found strong horizontal convergence of wave energy flux into the upper tropospheric energy duct over the central-eastern Pacific

from extratropical latitudes, suggesting a link between the MJO and the midlatitudes (Yanai et al. 2000; Chen and Yanai 2000).

d. The Tibetan Plateau and Asian monsoon

In the 1980s, a decade after arriving at UCLA, Professor Yanai and his Chinese colleagues began a series of diagnostic studies that would quantify the heating effects of the Tibetan Plateau in the Asian monsoon. One can view this work as a logical extension

FIG. 15. With four former Ph.D. students in Taiwan. From left: M.-M. Lu, C.-W. Hung, M.-D. Cheng, M. Yanai, and
C.-H. Sui. (2003).

FIG. 16. At his retirement party, receiving a gift from his sons (shown is Satoshi Yanai, 1999).

FIG. 17. In Lhasa, Tibet, with Yoko Yanai (2004).

of Yanai's computations of Q_1 and Q_2, the large-scale heat source and moisture sink, but from a phenomenological point of view this was a new line of scientific research.

From the beginning, all of Professor Yanai's monsoon research activities were conducted as cooperative research projects with Chinese scientists (Fig. 13), and he frequently went to China on official visits together with his American colleagues C.-P. Chang, T.N. Krishnamurti, Takio Murakami, Peter Webster, Pam Stephens, Gabriel Lau, and Bill Lau. Many of Yanai's papers on this subject were written with invited Chinese scientists (Luo, He, Song, Liu) (Fig. 14) and Professor Guo-Xiong Wu.

The first set of monsoon papers (Luo and Yanai 1983, 1984; He et al. 1987; Yanai and Li 1994) quantified heat and moisture budgets, circulation, and precipitation patterns over the Tibetan Plateau during the period surrounding the 1979 summer monsoon onset, using gridded data from the First GARP Global Experiment. The results show intense heating of the air over the elevated plateau before the arrival of the monsoon rain, with a deep diurnally varying mixed layer extending almost to the tropopause, formed every day by the late afternoon. This suggests that the dry thermal convection originating near the heated surface of the plateau is responsible for atmospheric heating found there during the preonset phase. After the rains begin over the eastern plateau, the Luo and Yanai analysis shows that the heating due to the dry convection in that region is replaced by condensation heating associated with cumulus convection.

A second group of papers addresses the roles of heating over the Tibetan Plateau and land–sea contrast

in the evolution and interannual variability of Asian summer monsoon (Yanai et al. 1992; Li and Yanai 1996). Other investigations included the relationship between Indian monsoon rainfall and tropospheric temperature over the Eurasian continent (Liu and Yanai 2001), and between Eurasian spring snow cover and Asian summer rainfall (Liu and Yanai 2002). Professor Yanai, graduate student Chih-Wen Hung (Fig. 15), and their Chinese colleague Xiaodong Liu identified factors contributing to the Australian summer monsoon and examined its symmetries and asymmetries with the Asian summer monsoon (Hung and Yanai 2004; Hung et al. 2004). The effects of the Tibetan Plateau on the Asian monsoon were summarized in a review article by Yanai and Wu (2006).

6. Postretirement activities

Professor Yanai's seminal research contributions were recognized with two major awards, the American Meteorological Society's (AMS's) Jule Charney award

FIG. 18. Group picture at the AMS Yanai Symposium in Seattle, Washington, on 27 January 2011 (http://91photos.ametsoc.net/albums/
thursday/yanai-symposium/ams1126.jpg28496.jpg).

in 1986 and the Fujiwara Award from the Meteorological Society of Japan in 1993. A few years later, in 1999, Professor Yanai formally retired from UCLA (Fig. 16) but remained an active and valued member of his department and the global tropical meteorological community. He was particularly fond of communicating by e-mail with his many friends and colleagues not only on scientific matters but also on the arts, especially opera (Fig. 17).

A great example of Yanai's postretirement service to the tropical research community was his UCLA Tropical Meteorology and Climate Newsletter, an e-mail digest of announcements, paper abstracts, and short articles that he started in 1996. Originally named the UCLA Tropical Meteorology Newsletter, it was created as a means of communication among the past, present, and future members of his research group. However, the topics covered by the newsletter and the number of contributors and recipients expanded very rapidly, becoming a valuable forum for more than 800 scientists worldwide by the time of his passing. The last newsletter was issued on 8 October 2010.

A symposium in Professor Yanai's honor was held at the American Meteorological Society's annual meeting

in January 2011 (Fig. 18). The occasion of the symposium led him to reminisce not only about his life and career but also about the histories and contributions of his colleagues, especially his fellow meteorologists who emerged from postwar Japan. Indeed, much of this epilogue is based on notes he wrote in the last year of his life, allowing us to tell and preserve his story. Professor Yanai was in the midst of collecting oral histories of the UCLA department from past and present members of the UCLA family when he passed away on 13 October 2010.

Those who had the pleasure to know him as a colleague, teacher, and friend will remember Professor Yanai's passion and good humor, and his concern for all those around him. He was not only a superb scientist but also a renaissance man with an infectious interest in politics and the arts. This volume is dedicated to Professor Yanai and to the better understanding of tropical multiscale cloud systems that motivated so much of his research.

Acknowledgments. This epilogue is based on the narrative in the program booklet for the AMS Yanai

Symposium held in Seattle, Washington, on 27 January 2011. Many friends offered their help to enhance the narrative, including C. P. Chang (Fig. 13), Baode Chen (Fig. 12), Bill Gray (Fig. 7), Chih-Wen Hung, Mong-Ming Lu (Figs. 10, 14, and 15), and Taroh Matsuno (Figs. 4 and 8). Most importantly, we thank the Yanai family, Yoko, Takashi, and Satoshi, who shared the most wonderful pictures of Michio (Figs. 1–6, 8, 9, 11, 12, 16, and 17) and helped get the factual details straight. Finally, we thank the amazing and supportive AMS organizers and participants at the symposium (Fig. 18), who completed the narrative.

REFERENCES

Chen, B., and M. Yanai, 2000: Comparison of the Madden–Julian oscillation (MJO) during the TOGA COARE IOP with a 15-year climatology. *J. Geophys. Res.*, **105**, 2139–2149, doi:10.1029/1999JD901045.

Cheng, M.-D., and M. Yanai, 1989: Effects of downdrafts and mesoscale convective organization on the heat and moisture budgets of tropical cloud clusters. Part III: Effects of mesoscale convective organization. *J. Atmos. Sci.*, **46**, 1566–1588, doi:10.1175/1520-0469(1989)046<1566:EODAMC>2.0.CO;2.

Chu, J.-H., M. Yanai, and C.-H. Sui, 1981: Effects of cumulus convection on the vorticity field in the tropics. Part I: The large-scale budget. *J. Meteor. Soc. Japan*, **59**, 535–546.

Dunn, G. E., 1940: Cyclogenesis in the tropical Atlantic. *Bull. Amer. Meteor. Soc.*, **21**, 215–229.

Gray, W. M., 1973: Cumulus convection and larger scale circulations. I. Broadscale and mesoscale considerations. *Mon. Wea. Rev.*, **101**, 839–855, doi:10.1175/1520-0493(1973)101<0839:CCALSC>2.3.CO;2.

Hamuro, M., and Coauthors, 1969: Precipitation bands of typhoon Vera in 1959 (Part I). *J. Meteor. Soc. Japan*, **47**, 298–309.

——, and Coauthors, 1970: Precipitation bands of typhoon Vera in 1959 (Part II). *J. Meteor. Soc. Japan*, **48**, 103–117.

Hayashi, Y., 1977: Space-time spectral analysis using the maximum entropy method. *J. Meteor. Soc. Japan*, **55**, 415–420.

He, H., J. W. McGinnis, Z. Song, and M. Yanai, 1987: Onset of the Asian monsoon in 1979 and the effect of the Tibetan Plateau. *Mon. Wea. Rev.*, **115**, 1966–1995, doi:10.1175/1520-0493(1987)115<1966:OOTASM>2.0.CO;2.

Hung, C.-W., and M. Yanai, 2004: Factors contributing to the onset of the Australian summer monsoon. *Quart. J. Roy. Meteor. Soc.*, **130**, 739–761, doi:10.1256/qj.02.191.

——, X. Liu, and M. Yanai, 2004: Symmetry and asymmetry of the Asian and Australian summer monsoons. *J. Climate*, **17**, 2413–2426, doi:10.1175/1520-0442(2004)017<2413:SAAOTA>2.0.CO;2.

Johnson, D. R., M. Yanai, and T. K. Schaack, 1987: Global and regional distributions of atmospheric heat sources and sinks during the GWE. *Monsoon Meteorology*, C.-P. Chang and T. N. Krishnamurti, Eds., Oxford University Press, 271–297.

Kleinschmidt, E., Jr., 1951: Grundlagen einer Theorie der tropischen Zyklonen. *Arch. Meteor. Geophys. Bioklimatol.*, **4A**, 53–72, doi:10.1007/BF02246793.

Lewis, J. M., 1993: Meteorologists from the University of Tokyo: Their exodus to the United States following World War II.

Bull. Amer. Meteor. Soc., **74**, 1351–1360, doi:10.1175/1520-0477(1993)074<1351:MFTUOT>2.0.CO;2.

Li, C., and M. Yanai, 1996: The onset and interannual variability of the Asian summer monsoon in relation to land–sea thermal contrast. *J. Climate*, **9**, 358–375, doi:10.1175/1520-0442(1996)009<0358:TOAIVO>2.0.CO;2.

Lindzen, R. S., and T. Matsuno, 1968: On the nature of large scale wave disturbances in the equatorial lower stratosphere. *J. Meteor. Soc. Japan*, **46**, 215–221.

Liu, X., and M. Yanai, 2001: Relationship between the Indian monsoon rainfall and the tropospheric temperature over the Eurasian continent. *Quart. J. Roy. Meteor. Soc.*, **127**, 909–937, doi:10.1002/qj.49712757311.

——, and ——, 2002: Influence of Eurasian spring snow cover on Asian summer rainfall. *Int. J. Climatol.*, **22**, 1075–1089, doi:10.1002/joc.784.

Luo, H., and M. Yanai, 1983: The large-scale circulation and heat sources over the Tibetan Plateau and surrounding areas during the early summer of 1979. Part I: Precipitation and kinematic analyses. *Mon. Wea. Rev.*, **111**, 922–944, doi:10.1175/1520-0493(1983)111<0922:TLSCAH>2.0.CO;2.

——, and ——, 1984: The large-scale circulation and heat sources over the Tibetan Plateau and surrounding areas during the early summer of 1979. Part II: Heat and moisture budgets. *Mon. Wea. Rev.*, **112**, 966–989, doi:10.1175/1520-0493(1984)112<0966:TLSCAH>2.0.CO;2.

Madden, R. A., and P. R. Julian, 1971: Detection of a 40–50 day oscillation in the zonal wind in the tropical Pacific. *J. Atmos. Sci.*, **28**, 702–708, doi:10.1175/1520-0469(1971)028<0702:DOADOI>2.0.CO;2.

Magaña, V., and M. Yanai, 1991: Tropical–midlatitude interaction on the time scale of 30 to 60 days during the northern summer of 1979. *J. Climate*, **4**, 180–201, doi:10.1175/1520-0442(1991)004<0180:TMIOTT>2.0.CO;2.

——, and ——, 1995: Mixed Rossby–gravity waves triggered by lateral forcing. *J. Atmos. Sci.*, **52**, 1473–1486, doi:10.1175/1520-0469(1995)052<1473:MRWTBL>2.0.CO;2.

Matsuno, T., 1966: Quasi-geostrophic motions in the equatorial area. *J. Meteor. Soc. Japan*, **44**, 25–43.

Nitta, T., 1975: Observational determination of cloud mass flux distributions. *J. Atmos. Sci.*, **32**, 73–91, doi:10.1175/1520-0469(1975)032<0073:ODOCMF>2.0.CO;2.

——, and S. Esbensen, 1974a: Diurnal variations in the western Atlantic trades during the BOMEX. *J. Meteor. Soc. Japan*, **52**, 254–257.

——, and ——, 1974b: Heat and moisture budget analyses using BOMEX data. *Mon. Wea. Rev.*, **102**, 17–28, doi:10.1175/1520-0493(1974)102<0017:HAMBAU>2.0.CO;2.

Palmén, E., 1948: On the formation and structure of the tropical hurricane. *Geophysica*, **3**, 26–38.

Riehl, H., 1948: On the formation of typhoons. *J. Meteor.*, **5**, 247–265, doi:10.1175/1520-0469(1948)005<0247:OTFOT>2.0.CO;2.

——, 1954: *Tropical Meteorology*. McGraw-Hill, 392 pp.

Sui, C.-H., and M. Yanai, 1986: Cumulus ensemble effects on the large-scale vorticity and momentum fields of GATE. Part I: Observational evidence. *J. Atmos. Sci.*, **43**, 1618–1642, doi:10.1175/1520-0469(1986)043<1618:CEEOTL>2.0.CO;2.

——, M.-D. Cheng, X. Wu, and M. Yanai, 1989: Cumulus ensemble effects on the large-scale vorticity and momentum fields of GATE. Part II: Parameterization. *J. Atmos. Sci.*, **46**, 1609–1629, doi:10.1175/1520-0469(1989)046<1609:CEEOTL>2.0.CO;2.

Takayabu, Y. N., 1994a: Large-scale cloud disturbances associated with equatorial waves. Part I: Spectral features of the cloud disturbances. *J. Meteor. Soc. Japan*, **72**, 433–449.

——, 1994b: Large-scale cloud disturbances associated with equatorial waves. Part II: Westward-propagating inertia–gravity waves. *J. Meteor. Soc. Japan,* **72,** 451–465.

Tung, W.-W., and M. Yanai, 2002a: Convective momentum transport observed during the TOGA COARE IOP. Part I: General features. *J. Atmos. Sci.,* **59,** 1857–1871, doi:10.1175/1520-0469(2002)059<1857:CMTODT>2.0.CO;2.

——, and ——, 2002b: Convective momentum transport observed during the TOGA COARE IOP. Part II: Case studies. *J. Atmos. Sci.,* **59,** 2535–2549, doi:10.1175/1520-0469(2002)059<2535:CMTODT>2.0.CO;2.

——, C. Lin, B. Chen, M. Yanai, and A. Arakawa, 1999: Basic modes of cumulus heating and drying observed during TOGA-COARE IOP. *Geophys. Res. Lett.,* **26,** 3117–3120, doi:10.1029/1999GL900607.

Wallace, J. M., and V. E. Kousky, 1968: Observational evidence of Kelvin waves in the tropical stratosphere. *J. Atmos. Sci.,* **25,** 900–907, doi:10.1175/1520-0469(1968)025<0900:OEOKWI>2.0.CO;2.

Wu, X., and M. Yanai, 1994: Effects of vertical wind shear on the cumulus transport of momentum: Observations and parameterization. *J. Atmos. Sci.,* **51,** 1640–1660, doi:10.1175/1520-0469(1994)051<1640:EOVWSO>2.0.CO;2.

Yanai, M., 1958: On the changes in thermal and wind structure in a decaying typhoon. *J. Meteor. Soc. Japan,* **36,** 141–155.

——, 1961a: A detailed analysis of typhoon formation. *J. Meteor. Soc. Japan,* **39,** 187–214.

——, 1961b: Dynamical aspects of typhoon formation. *J. Meteor. Soc. Japan,* **39,** 282–309.

——, 1964: Formation of tropical cyclones. *Rev. Geophys.,* **2,** 367–414.

——, 1968: Evolution of a tropical disturbance in the Caribbean Sea region. *J. Meteor. Soc. Japan,* **46,** 86–109.

——, and T. Maruyama, 1966: Stratospheric wave disturbances propagating over the equatorial Pacific. *J. Meteor. Soc. Japan,* **44,** 291–294.

——, and T. Nitta, 1967: Computation of vertical motion and vorticity budget in a Caribbean easterly wave. *J. Meteor. Soc. Japan,* **45,** 444–466.

——, and M.-M. Lu, 1983: Equatorially trapped waves at the 200 mb level and their association with meridional convergence of wave energy flux. *J. Atmos. Sci.,* **40,** 2785–2803, doi:10.1175/1520-0469(1983)040<2785:ETWATM>2.0.CO;2.

——, and C. Li, 1994: Mechanism of heating and the boundary layer over the Tibetan Plateau. *Mon. Wea. Rev.,* **122,** 305–323, doi:10.1175/1520-0493(1994)122<0305:MOHATB>2.0.CO;2.

——, and T. Tomita, 1998: Seasonal and interannual variability of atmospheric heat sources and moisture sinks as determined from NCEP–NCAR reanalysis. *J. Climate,* **11,** 463–482, doi:10.1175/1520-0442(1998)011<0463:SAIVOA>2.0.CO;2.

——, and G.-X. Wu, 2006: Effects of the Tibetan Plateau. *The Asian Monsoon,* B. Wang, Ed., Springer, 513–549.

——, T. Maruyama, T. Nitta, and Y. Hayashi, 1968: Power spectra of large-scale disturbances over the tropical Pacific. *J. Meteor. Soc. Japan,* **46,** 308–323.

——, S. Esbensen, and J.-H. Chu, 1973: Determination of bulk properties of tropical cloud clusters from large-scale heat and moisture budgets. *J. Atmos. Sci.,* **30,** 611–627, doi:10.1175/1520-0469(1973)030<0611:DOBPOT>2.0.CO;2.

——, J.-H. Chu, T. E. Stark, and T. Nitta, 1976: Response of deep and shallow tropical maritime cumuli to large-scale processes. *J. Atmos. Sci.,* **33,** 976–991, doi:10.1175/1520-0469(1976)033<0976:RODAST>2.0.CO;2.

——, C.-H. Sui, and J.-H. Chu, 1982: Effects of cumulus convection on the vorticity field in the tropics. Part II: Interpretation. *J. Meteor. Soc. Japan,* **60,** 411–424.

——, C. Li, and Z. Song, 1992: Seasonal heating of the Tibetan Plateau and its effects on the evolution of the Asian summer monsoon. *J. Meteor. Soc. Japan,* **70,** 319–351.

——, B. Chen, and W.-W. Tung, 2000: The Madden–Julian oscillation observed during the TOGA-COARE IOP: Global view. *J. Atmos. Sci.,* **57,** 2374–2396, doi:10.1175/1520-0469(2000)057<2374:TMJOOD>2.0.CO;2.

Zangvil, A., 1975: Temporal and spatial behavior of large-scale disturbances in tropical cloudiness deduced from satellite brightness data. *Mon. Wea. Rev.,* **103,** 904–920, doi:10.1175/1520-0493(1975)103<0904:TASBOL>2.0.CO;2.

——, and M. Yanai, 1980: Upper tropospheric waves in the tropics. Part I: Dynamical analysis in the wavenumber–frequency domain. *J. Atmos. Sci.,* **37,** 283–298, doi:10.1175/1520-0469(1980)037<0283:UTWITT>2.0.CO;2.

——, and ——, 1981: Upper tropospheric waves in the tropics. Part II: Association with clouds in the wavenumber–frequency domain. *J. Atmos. Sci.,* **38,** 939–953, doi:10.1175/1520-0469(1981)038<0939:UTWITT>2.0.CO;2.